Chromatin Signaling and Neurological Disorders

Translational Epigenetics Series

Trygve O. Tollefsbol, Series Editor

Transgenerational Epigenetics
Edited by Trygve O. Tollefsbol, 2014

Personalized Epigenetics
Edited by Trygve O. Tollefsbol, 2015

Epigenetic Technological Applications
Edited by Y. George Zheng, 2015

Epigenetic Cancer Therapy
Edited by Steven G. Gray, 2015

DNA Methylation and Complex Human Disease
By Michel Neidhart, 2015

Epigenomics in Health and Disease
Edited by Mario F. Fraga and Agustin F.F. Fernández, 2015

Epigenetic Gene Expression and Regulation
Edited by Suming Huang, Michael Litt and C. Ann Blakey, 2015

Epigenetic Biomarkers and Diagnostics
Edited by Jose Luis García-Giménez, 2015

Drug Discovery in Cancer Epigenetics
Edited by Gerda Egger and Paola Barbara Arimondo, 2015

Medical Epigenetics
Edited by Trygve O. Tollefsbol, 2016

Chromatin Signaling and Diseases
Edited by Olivier Binda and Martin Fernandez-Zapico, 2016

Genome Stability
Edited by Igor Kovalchuk and Olga Kovalchuk, 2016

Chromatin Regulation and Dynamics
Edited by Anita Göndör, 2016

Neuropsychiatric Disorders and Epigenetics
Edited by Dag H. Yasui, Jacob Peedicayil and Dennis R. Grayson, 2016

Polycomb Group Proteins
Edited by Vincenzo Pirrotta, 2016

Epigenetics and Systems Biology
Edited by Leonie Ringrose, 2017

Cancer and Noncoding RNAs
Edited by Jayprokas Chakrabarti and Sanga Mitra, 2017

Nuclear Architecture and Dynamics
Edited by Christophe Lavelle and Jean-Marc Victor, 2017

Epigenetic Mechanisms in Cancer
Edited by Sabita Saldanha, 2017

Epigenetics of Aging and Longevity
Edited by Alexey Moskalev and Alexander M. Vaiserman, 2017

The Epigenetics of Autoimmunity
Edited by Rongxin Zhang, 2018

Epigenetics in Human Disease, Second Edition
Edited by Trygve O. Tollefsbol, 2018

Epigenetics of Chronic Pain
Edited by Guang Bai and Ke Ren, 2018

Epigenetics of Cancer Prevention
Edited by Anupam Bishayee and Deepak Bhatia, 2018

Computational Epigenetics and Diseases
Edited by Loo Keat Wei, 2019

Translational Epigenetics

Volume 12

Chromatin Signaling and Neurological Disorders

Edited by

Olivier Binda

Institut NeuroMyoGène
Université Claude Bernard Lyon 1
Lyon
France

ACADEMIC PRESS

An imprint of Elsevier

Academic Press is an imprint of Elsevier
125 London Wall, London EC2Y 5AS, United Kingdom
525 B Street, Suite 1650, San Diego, CA 92101, United States
50 Hampshire Street, 5th Floor, Cambridge, MA 02139, United States
The Boulevard, Langford Lane, Kidlington, Oxford OX5 1GB, United Kingdom

Library of Congress Cataloging-in-Publication Data
A catalog record for this book is available from the Library of Congress

British Library Cataloguing-in-Publication Data
A catalogue record for this book is available from the British Library

ISBN: 978-0-12-813796-3

For information on all Academic Press publications visit our website at https://www.elsevier.com/books-and-journals

Publisher: John Fedor
Acquisition Editor: Peter B. Linsley
Editorial Project Manager: Jennifer Horigan
Production Project Manager: Poulouse Joseph
Cover designer: Mark Rogers

Typeset by TNQ Technologies

Contents

Contributors xiii
Preface xvii

CHAPTER 1 Chromatin and epigenetic signaling pathways 1

Catherine A. Musselman

1.1 Chromatin signaling and epigenetics 1
1.2 Chromatin organization 2
1.3 Histone posttranslational modifications and the histone code 4
1.4 Functions of histone posttranslational modifications 5
1.5 DNA methylation 6
1.6 Writers, erasers, and readers 7
1.6.1 Histone writers 7
1.6.2 DNA writers 7
1.6.3 Histone erasers 8
1.6.4 DNA erasers 8
1.6.5 Histone readers 8
1.6.6 DNA readers 9
1.7 Modification cross talk 10
1.8 Effects of metabolism on histone and DNA modifications 11
1.9 Epigenetic inheritance 13
1.10 Summary 15
References 16

SECTION 1 Neurodegenerative disorders

CHAPTER 2 Into the unknown: chromatin signaling in spinal muscular atrophy 27

Marc-Olivier Deguise and Rashmi Kothary

2.1 Spinal muscular atrophy: prevalence, genetic basis, clinical features, and pathogenesis 27
2.2 The survival motor neuron protein: localization, structure, and function 30
2.3 Epigenetic landscape in spinal muscular atrophy pathogenesis 31
2.4 Targeting epigenetic factors as potential therapeutics in spinal muscular atrophy 32
2.4.1 Histone deacetylase inhibitors as regulators of the survival motor neuron gene 32
2.4.2 The nonspecific effect of histone deacetylase inhibitors 33

2.4.3 The potential protective effect of histone deacetylase inhibitors in the pathogenesis of spinal muscular atrophy 41
2.5 Conclusion 42
Acronyms and abbreviations 43
Acknowledgments 44
References 44

CHAPTER 3 Charcot-Marie-Tooth disease 53

Phu Duong and John Svaren

3.1 Introduction 53
3.2 Epigenetic regulation of Schwann cell development 55
3.3 Epigenetic regulation of dosage-sensitive genes 57
3.4 Epigenetic regulators targeted by CMT mutations 60
3.4.1 DNMT1 60
3.4.2 LMNA 61
3.4.3 SYNE1 62
3.4.4 MED25 62
3.4.5 SETX 62
3.4.6 MORC2 63
3.4.7 PRDM12 64
3.5 Novel mechanisms for CMT mutations 64
3.6 Summary 64
Acknowledgments 65
References 65

CHAPTER 4 Epigenetic mechanisms in Huntington's disease 73

Elizabeth A. Thomas

4.1 Introduction 74
4.2 Huntington's disease 74
4.2.1 Neuropathology of HD 75
4.3 Transcriptional dysregulation in HD 75
4.4 Altered epigenetic marks in HD 76
4.4.1 Histone modifications 76
4.4.2 DNA methylation 83
4.4.3 DNA methylation changes in HD 84
4.5 Epigenetic-based therapies 86
4.5.1 HDAC inhibitors as a treatment for HD 86
4.5.2 Methylation-inhibiting drugs 87
4.6 Concluding remarks 87
4.7 Abbreviations 87
References 88

CHAPTER 5 The epigenetics of multiple sclerosis **97**
Tove Christensen and Christian Muchardt
5.1 Multiple sclerosis, the knowns and the unknowns 98
5.1.1 A chronic progressive disease of the central nervous system 98
5.1.2 The genetics of MS 98
5.1.3 Gender bias and parent-of-origin effect 98
5.1.4 A role for environmental factors 99
5.1.5 A viral component in MS 99
5.2 MS as an epigenetic disorder 100
5.2.1 Nongenetic causes of MS and their link to chromatin and transcription 100
5.2.2 DNA and histone modifications are footprints of transcriptional regulation 101
5.3 DNA and histone modifications linked to MS 101
5.3.1 DNA methylation 102
5.3.2 A possibly reduced efficiency of the H3K9me/HP1 axis of transcriptional repression in MS 103
5.4 Epigenetics beyond transcription 108
5.4.1 Exosomal miRNA silencing 109
5.4.2 Microbiota 109
5.4.3 Environment: pollutants that may interfere with silencing machineries 110
5.5 Conclusions 110
Acknowledgments 111
References 111

CHAPTER 6 Alterations in epigenetic regulation contribute to neurodegeneration of ataxia-telangiectasia **119**
Jiali Li and Karl Herrup
6.1 Decreased level of histone acetylation induced by nuclear accumulation of HDAC4 drives A-T neurodegeneration 120
6.2 Dysfunction of polycomb repressive complex 2 involved in A-T neurodegeneration 123
6.3 Selective loss of 5-hmC is associated with purkinje cell vulnerability in A-T brain 124
6.4 TETs-mediated DNA oxidation regulates ATM/ATR-dependent DDR 126
6.5 Conclusion 128
6.6 Future perspective 128
Acknowledgments 129
References 129

CHAPTER 7 Cockayne syndrome **135**

Luca Proietti-De-Santis, Vincent Laugel and Giorgio Prantera

7.1 Clinical phenotypes 135
7.1.1 Classical (moderate) type I cockayne syndrome 136
7.1.2 Early-onset (severe) subtypes 138
7.1.3 Late-onset subtypes 139
7.2 Genetics 140
7.3 CSA and CSB proteins 140
7.4 Cellular and molecular aspects 140
7.5 The molecular basis of neurodegeneration 141
7.6 Concluding remarks 146
References 147

CHAPTER 8 Epigenetic processes in Alzheimer's disease **153**

Isabel Castanho and Katie Lunnon

8.1 Alzheimer's disease: a need for new drug targets 153
8.2 Alzheimer's disease: the genomic era 155
8.3 An additional layer of information: Alzheimer's disease from an epigenetic perspective 156
8.3.1 DNA modifications 157
8.3.2 Histone modifications 160
8.3.3 Regulatory RNA—based mechanisms 162
8.3.4 Epigenetic signatures as blood biomarkers 164
8.4 Modeling Alzheimer's disease: mouse models as powerful tools 166
8.5 Current challenges and future directions 169
8.6 Final considerations 171
References 172

SECTION 2 Neurodevelopmental disorders

CHAPTER 9 Genetic and epigenetic influences on the phenotype of Rett syndrome **183**

Jenny Downs, Janine M. LaSalle and Helen Leonard

9.1 Introduction 184
9.2 The genetic cause of Rett syndrome 185
9.3 The biology of MeCP2 186
9.3.1 Neurobiology of MeCP2 186
9.3.2 Molecular functioning of MeCP2 187
9.4 The phenotype of Rett syndrome 189
9.4.1 Early development and regression 189

9.4.2 Diagnostic criteria ... 190
9.4.3 Functional impairments ... 190
9.4.4 Stereotypies ... 193
9.4.5 Comorbidities ... 193
9.4.6 Epidemiology ... 197
9.5 Evidence for epigenetic mechanisms affecting *MECP2* function and expression ... 198
9.5.1 DNA methylation ... 199
9.5.2 Histone modifications and nucleosome and higher-order chromatin remodeling ... 200
9.5.3 Noncoding RNAs ... 201
9.5.4 RNA splicing ... 201
9.6 Epigenetic regulation of MeCP2 expression or phenotypes ... 202
9.6.1 X chromosome inactivation ... 202
9.6.2 IGF1/mTOR pathway ... 203
9.6.3 Enriched environments ... 204
9.7 Inclusion of epigenetic data collection in epidemiological studies ... 206
9.8 Summary ... 207
References ... 207

CHAPTER 10 Sotos syndrome ... 219

Chloe Lane and Megan Freeth

10.1 Introduction ... 219
10.2 The genetic basis of Sotos syndrome ... 220
10.3 Comparing Sotos syndrome with other single-gene overgrowth syndromes ... 221
10.4 Neurological profile of Sotos syndrome ... 221
10.5 The cognitive and behavioral profile of Sotos syndrome ... 222
10.5.1 Cognition ... 222
10.5.2 Behavior ... 225
10.5.3 Aggression and tantrums ... 226
10.6 Sotos syndrome and autism spectrum disorder ... 226
10.7 Nuclear receptor–binding SET domain methyltransferases modify histones and affect epigenetics ... 228
10.8 Limitations and future research directions ... 229
10.9 Summary and conclusions ... 230
References ... 230

CHAPTER 11 ATRX tames repetitive DNA within heterochromatin to promote normal brain development and regulate oncogenesis **235**

Valerie Turcotte-Cardin, Kevin G. Young and David J. Picketts

11.1 Introduction 235
11.2 Biochemical and molecular functions of ATRX 236
11.2.1 ATRX protein structure 236
11.2.2 ATRX is a heterochromatin interacting protein 238
11.2.3 Other critical interactions and functions of ATRX 239
11.2.4 ATRX interactions with RNA 241
11.3 Neurologic deficits and phenotypic variability in *ATRX*-associated syndromes 242
11.4 Delineating a role for ATRX in cancer 244
11.4.1 Cancer profiling identifies ATRX as a common mutation target 244
11.4.2 Cancers with ATRX mutations are alternative lengthening of telomere positive 245
11.4.3 Understanding the alternative lengthening of telomeres pathway 247
11.4.4 ATRX is a suppressor of the alternative lengthening of telomeres pathway 249
11.5 Conclusion 251
List of abbreviations 251
References 252

SECTION 3 Neuropsychiatric disorders

CHAPTER 12 Epigenetic dysregulation in the fragile X-related disorders **261**

Karen Usdin and Daman Kumari

12.1 Introduction 261
12.2 Clinical features of the FXDs 262
12.2.1 Fragile X syndrome 262
12.2.2 Fragile X-associated tremor/ataxia syndrome 263
12.2.3 Fragile X-associated primary ovarian insufficiency 263
12.3 Genetics of the FXDs 264
12.4 The pathological basis of FXTAS 265
12.5 The pathological basis of FXPOI 267
12.6 The pathological basis of FXS 268
12.7 Epigenetic abnormalities associated with the FXDs 268
12.8 Resolving the repeat paradox 270
12.9 Prospects and challenges for epigenetic therapies for the FXDs 271
12.10 Concluding remarks 273
References 273

CHAPTER 13 The epigenetics of autism **285**

Aicha Massrali, Varun Warrier, Arkoprovo Paul, Dwaipayan Adhya, Deepak P. Srivastava, Mark Kotter and Simon Baron-Cohen

13.1 Autism 285
13.1.1 Heritability and genetics 286
13.2 Epigenetics of autism 287
13.2.1 DNA methylation and hydroxymethylation in autism 288
13.2.2 Histone modifications 292
13.2.3 Risk factors affecting the epigenetics of autism 294
13.3 Discussion 296
Acknowledgments 297
References 297

CHAPTER 14 Chromatin modification and remodeling in schizophrenia **303**

Jubao Duan

14.1 Introduction 304
14.2 SZ GWAS implicate gene expression and chromatin regulation as a possible causal molecular mechanism 305
14.3 SZ and DNA methylation 306
14.3.1 Aberrant DNA methylation in SZ 307
14.3.2 Genetic control of DNA methylation and its relevance to SZ 308
14.4 SZ and histone modifications 309
14.4.1 Histone acetylation 310
14.4.2 Histone methylation 311
14.5 SZ and 2D chromatin structure 312
14.6 SZ and higher-order chromatin structure 313
14.7 SZ genetic risk variants affect chromatin remodeling gene pathway 315
14.7.1 Analysis of common SZ variants implicates the dysregulated chromatin-signaling pathway 315
14.7.2 Rare SZ coding risk variants and chromatin remodeling 316
14.7.3 SZ-associated CNVs and abnormal chromatin organization 317
14.8 hiPSC model combined with CRISPR editing for studying SZ-relevant chromatin function 318
14.8.1 hiPSC-derived neurons as a cellular model for neurodevelopmental disorder 318
14.8.2 CRISPR-based approaches for genome/epigenome perturbation 319
14.8.3 CRISPR-based 2D and 3D chromatin perturbation relevant to SZ in hiPSC models 320

14.9 Therapeutic drugs that target chromatin structure and activity in SZ 321
14.10 Conclusion and perspectives 322
Acknowledgments 323
References 323

CHAPTER 15 Gilles de la Tourette syndrome 331

Claudia Selvini, Stefano Cavanna and Andrea E. Cavanna

15.1 Introduction: Gilles de la Tourette syndrome and other tic disorders 332
15.1.1 Definition and diagnostic criteria of Gilles de la Tourette syndrome and other tic disorders 332
15.1.2 Epidemiology 333
15.2 Clinical presentation of tics 333
15.2.1 Shared characteristics of tics 333
15.2.2 Characteristics of motor tics 334
15.2.3 Characteristics of vocal/phonic tics 334
15.2.4 Characteristics of cognitive tics 334
15.3 Tic-related behavioral symptoms and health-related quality of life 334
15.3.1 Behavioral spectrum of Gilles de la Tourette syndrome 334
15.3.2 Obsessive—compulsive disorder 335
15.3.3 Attention-deficit and hyperactivity disorder 335
15.3.4 Health-related quality of life 335
15.4 Etiology and pathophysiology 335
15.4.1 Genetic factors 335
15.4.2 Environmental factors 336
15.4.3 Role of dopamine and cortico-striato-thalamo-cortical pathways 336
15.4.4 Possible role of chromatin regulation 337
15.5 Treatment strategies 339
15.5.1 Psychoeducation 339
15.5.2 Behavioral therapy 339
15.5.3 Pharmacotherapy 340
15.5.4 Other approaches 340
15.6 Conclusions: open questions and suggestions for future research 340
Acknowledgments 341
References 341

Index 347

Contributors

Dwaipayan Adhya
Autism Research Centre, Department of Psychiatry, University of Cambridge, Cambridge, United Kingdom

Simon Baron-Cohen
Autism Research Centre, Department of Psychiatry, University of Cambridge, Cambridge, United Kingdom

Isabel Castanho
University of Exeter Medical School, University of Exeter, Devon, United Kingdom

Stefano Cavanna
Department of Radiology, University of Turin, Turin, Italy

Andrea E. Cavanna
Department of Neuropsychiatry, BSMHFT and University of Birmingham, Birmingham, United Kingdom; School of Life and Health Sciences, Aston University, Birmingham, United Kingdom; Sobell Department of Motor Neuroscience and Movement Disorders, UCL and Institute of Neurology, London, United Kingdom

Tove Christensen
Department of Biomedicine, Aarhus University, Bartholins Allé 6, DK-8000 Aarhus C, Denmark

Marc-Olivier Deguise
Regenerative Medicine Program, Ottawa Hospital Research Institute, Ottawa, ON, Canada; Department of Cellular and Molecular Medicine, University of Ottawa, Ottawa, ON, Canada; Centre for Neuromuscular Disease, University of Ottawa, Ottawa, ON, Canada

Jenny Downs
Telethon Kids Institute, The University of Western Australia, Perth, WA, Australia; School of Physiotherapy and Exercise Science, Curtin University, Perth, WA, Australia

Jubao Duan
Center for Psychiatric Genetics, NorthShore University HealthSystem, Evanston, IL, United States; Department of Psychiatry and Behavioral Neurosciences, The University of Chicago, Chicago, IL, United States

Phu Duong
Waisman Center, University of Wisconsin–Madison, Madison, WI, United States; Cellular and Molecular Pathology Graduate Program, University of Wisconsin–Madison, Madison, WI, United States

Megan Freeth
Psychology Department, University of Sheffield, Sheffield, United Kingdom

Karl Herrup
Division of Life Science and the State Key Laboratory of Molecular Neuroscience, The Hong Kong University of Science and Technology, Kowloon, Hong Kong; Department of Neurobiology, University of Pittsburgh School of Medicine, Pittsburgh, Pennsylvania, United States

Rashmi Kothary
Regenerative Medicine Program, Ottawa Hospital Research Institute, Ottawa, ON, Canada; Department of Cellular and Molecular Medicine, University of Ottawa, Ottawa, ON, Canada; Centre for Neuromuscular Disease, University of Ottawa, Ottawa, ON, Canada; Department of Medicine, and Department of Biochemistry, Microbiology, and Immunology, University of Ottawa, Ottawa, ON, Canada

Mark Kotter
Department of Clinical Neurosciences, Ann McLaren Laboratory of Regenerative Medicine, University of Cambridge, Cambridge, United Kingdom

Daman Kumari
Section on Gene Structure and Disease, Laboratory of Cell and Molecular Biology, National Institute of Diabetes, Digestive and Kidney Diseases, National Institutes of Health, Bethesda, MD, United States

Chloe Lane
Psychology Department, University of Sheffield, Sheffield, United Kingdom

Janine M. LaSalle
Medical Microbiology and Immunology, Genome Center, MIND Institute, University of California, Davis, CA, United States

Vincent Laugel
Laboratoire de Génétique Médicale - INSERM U1112, Institut de Génétique Médicale d'Alsace (IGMA), Faculté de médecine de Strasbourg, Strasbourg, France

Helen Leonard
Telethon Kids Institute, The University of Western Australia, Perth, WA, Australia

Jiali Li
National Institute on Drug Dependence, Peking University, Beijing, China; PKU/McGovern Institute for Brain Research, Peking University, Beijing, China; Key Laboratory of Animal Models and Human Disease Mechanisms of the Chinese Academy of Sciences & Yunnan Province, Kunming Institute of Zoology, Chinese Academy of Sciences, Kunming, Yunnan, China; CAS Center for Excellence in Animal Evolution and Genetics, Chinese Academy of Sciences, Kunming, Yunnan, China

Katie Lunnon
University of Exeter Medical School, University of Exeter, Devon, United Kingdom

Aicha Massrali
Autism Research Centre, Department of Psychiatry, University of Cambridge, Cambridge, United Kingdom

Christian Muchardt
Institut Pasteur, Département de Biologie du Développement et Cellules Souches, Unité de Régulation Epigénétique, Paris, France; CNRS UMR 3738, Paris, France

Catherine A. Musselman
University of Iowa Carver College of Medicine, Iowa City, IA United States

Arkoprovo Paul
Autism Research Centre, Department of Psychiatry, University of Cambridge, Cambridge, United Kingdom

David J. Picketts
Regenerative Medicine Program, Ottawa Hospital Research Institute, Ottawa, ON, Canada; Departments of Cellular and Molecular Medicine, University of Ottawa, Ottawa, ON, Canada; Departments of Biochemistry, Microbiology & Immunology, University of Ottawa, Ottawa, ON, Canada

Giorgio Prantera
Laboratorio di Epigenetica, Dipartimento di Scienze Ecologiche e Biologiche, Università della Tuscia, Viterbo, Italy

Luca Proietti-De-Santis
Laboratorio di Genetica Molecolare dell'Invecchiamento, Dipartimento di Scienze Ecologiche e Biologiche, Università della Tuscia, Viterbo, Italy

Claudia Selvini
Child Neuropsychiatry Unit, Department of Experimental Medicine, University of Insubria, Varese, Italy

Deepak P. Srivastava
Department of Basic and Clinical Neuroscience, Maurice Wohl Clinical Neuroscience Institute, Institute of Psychiatry, Psychology and Neuroscience, King's College London, London, United Kingdom; MRC Centre for Neurodevelopmental Disorders, King's College London, London, United Kingdom

John Svaren
Waisman Center, University of Wisconsin—Madison, Madison, WI, United States; Department of Comparative Biosciences, University of Wisconsin—Madison, Madison, WI, United States

Elizabeth A. Thomas
Department of Neuroscience, The Scripps Research Institute, La Jolla, CA, United States

Valerie Turcotte-Cardin
Regenerative Medicine Program, Ottawa Hospital Research Institute, Ottawa, ON, Canada; Departments of Cellular and Molecular Medicine, University of Ottawa, Ottawa, ON, Canada

Karen Usdin
Section on Gene Structure and Disease, Laboratory of Cell and Molecular Biology, National Institute of Diabetes, Digestive and Kidney Diseases, National Institutes of Health, Bethesda, MD, United States

Varun Warrier
Autism Research Centre, Department of Psychiatry, University of Cambridge, Cambridge, United Kingdom

Kevin G. Young
Regenerative Medicine Program, Ottawa Hospital Research Institute, Ottawa, ON, Canada

Preface

Eukaryotic genomes are generally embedded and compacted in proteinaceous macromolecules scaffolds. As such, the genetic information is not readily accessible. To circumvent this, several molecular mechanisms are involved in regulating access to genetic information. These processes need to be tightly regulated in specific cell types, during development, and throughout lifespan. Dysregulation of the mechanisms of access to genetic information can lead to severe consequences. One notorious and well established example involves the silencing of tumor suppressors, which lead to cancer development. However, disorganized access to genetic information is starting to be recognized as a major contributor to other pathologies including neurodegenerative (e.g., spinal muscle atrophy, Huntington's disease, multiple sclerosis), neurodevelopmental (e.g., Rett syndrome), and neuropsychiatric (e.g., Schizophrenia) disorders. Within the chapters of this textbook, you will discover, through leading researchers, how chromatin and epigenetics are involved in neurological pathologies.

Chromatin and epigenetics

The genomic DNA of eukaryotic organisms is twisted around a core of eight histone proteins composed of two copies of each H2A, H2B, H3, and H4 at a rate of 146 base pairs of DNA per histone octamer, forming the basic repetitive unit of chromatin called nucleosome. Each histone is characterized by an amino terminal tail and a globular domain. The amino terminal portion protrudes outside of the nucleosome and is thus amenable to posttranslational modifications. Indeed, histones are modified by methylation on arginine (R^{me}) and lysine (K^{me}) residues, acetylation on lysine residues (K^{ac}), and phosphorylation on serine residues (S^{phos}), among others. These modifications, or **marks**, are laid down by enzymes colloquially called **writers**, which allow through posttranslational modifications protein–protein interactions between histone mark **readers** and chromatin. Reader–chromatin

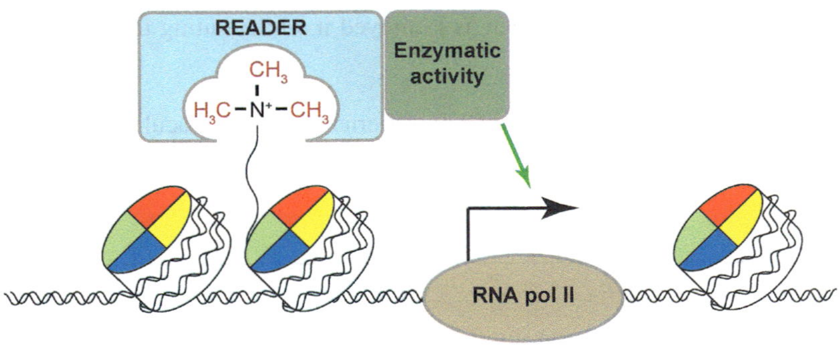

FIGURE 1

Reader–chromatin interactions are critical to regulate access to genetic information. Histone octamers are illustrated as red–yellow–green–blue drums and surrounded by dsDNA strands. An example of reader is depicted to interact with a trimethylated-lysine residue, allowing the recruitment of an enzymatic activity that activates transcription [3]. However, many other outcomes are possible, such as direct anchoring of transcriptional machinery to chromatin [4], recruitment of enzymatic activities that either silence gene expression [5], stimulate DNA repair [6–8], or facilitate V(D)J recombination [9].

interactions facilitate the recruitment of enzymatic activities (Figure 1) to regulate DNA-templated transactions (e.g., transcription, replication, repair) (see Chapter 1 or [1] for more details).

In addition to histone marks, access to genetic information is regulated by other molecular mechanisms, including ones that can be inherited and referred to as *epigenetic* mechanisms. By definition, epigenetic refers to **inheritable** changes that do not alter the genome itself. These inheritable changes include chemical modifications such as DNA methylation at CpG dinucleotides and some modifications to histone proteins, but very few histone modifications have been rigorously defined as inheritable.

Chromatin signaling and neurological diseases

As chromatin is so intimately woven into the regulation of genetic information, any defects can and do lead to detrimental consequences such as aging, cancer, neurological diseases, and many more human pathologies. Normal human developmental processes require near absolute timing. During normal neurological development, stem cells give rise to more specialized cells, including neurons, which form the central nervous system and the peripheral nervous system. As depicted at the end of 2018 [2], the involvement of chromatin and epigenetic defects are more and more recognized behind neurological diseases.

Herein, I have recruited illustrious contributors to discuss our current knowledge of the implication of chromatin in neurological diseases. I have decided to divide the present textbook into three distinct sections roughly based on NIH classification of neurological disorders. An introductory BACKGROUND section provides a broad outline on chromatin and how access to genetic information is regulated in cells to enable scholars from all background to understand the chromatin and epigenetic aspects of the subsequent chapters. This is followed by a section on NEURODEGENERATIVE DISORDERS (e.g., Alzheimer's disease, multiple sclerosis, Parkinson's disease), then NEURODEVELOPMENTAL DISORDERS (e.g., Rett syndrome, Sotos syndrome), and finally NEUROPSYCHIATRIC DISORDERS (e.g., Autism, Schizophrenia, Gilles de la Tourette Syndrome).

I hope you will enjoy your read as much as I enjoyed it while editing this very special textbook.

Olivier Binda
Université Claude Bernard Lyon 1, Faculté de Médecine Lyon Est,
Institut NeuroMyoGène (INMG), Lyon, France

References

[1] Binda O, Fernandez-Zapico ME. Chromatin signaling and diseases. Academic Press; 2016. https://doi.org/10.1016/C2014-0-02211-3.

[2] PsychENCODE Consortium. Revealing the brain's molecular architecture. Science 2018;362(6420):1262–3. https://doi.org/10.1126/science.362.6420.1262.

[3] Hung T, et al. ING4 mediates crosstalk between histone H3 K4 trimethylation and H3 acetylation to attenuate cellular transformation. Mol Cell 2009;33(2):248–56. https://doi.org/10.1016/j.molcel.2008.12.016.

[4] Vermeulen M, et al. Selective anchoring of TFIID to nucleosomes by trimethylation of histone H3 lysine 4. Cell 2007;131(1):58–69. https://doi.org/10.1016/j.cell.2007.08.016.

[5] Shi X, et al. ING2 PHD domain links histone H3 lysine 4 methylation to active gene repression. Nature 2006; 442(7098):96–9. https://doi.org/10.1038/nature04835.
[6] Botuyan MV, et al. Structural basis for the methylation state-specific recognition of histone H4-K20 by 53BP1 and Crb2 in DNA repair. Cell 2006;127(7):1361–73. https://doi.org/10.1016/j.cell.2006.10.043.
[7] Mallette FA, et al. RNF8- and RNF168-dependent degradation of KDM4A/JMJD2A triggers 53BP1 recruitment to DNA damage sites. EMBO J 2012;31:1865–78. https://doi.org/10.1038/emboj.2012.47.
[8] Lu R, Wang GG. Tudor: a versatile family of histone methylation readers. Trends Biochem Sci 2013;38(11): 546–55. https://doi.org/10.1016/j.tibs.2013.08.002.
[9] Matthews AG, et al. RAG2 PHD finger couples histone H3 lysine 4 trimethylation with V(D)J recombination. Nature 2007;450(7172):1106–10. https://doi.org/10.1038/nature06431.

CHAPTER

Chromatin and epigenetic signaling pathways

1

Catherine A. Musselman
University of Iowa Carver College of Medicine, Iowa City, IA United States

Chapter Outline

1.1 **Chromatin signaling and epigenetics** ... 1
1.2 **Chromatin organization** ... 2
1.3 **Histone posttranslational modifications and the histone code** ... 4
1.4 **Functions of histone posttranslational modifications** ... 5
1.5 **DNA methylation** ... 6
1.6 **Writers, erasers, and readers** ... 7
  1.6.1 Histone writers ... 7
  1.6.2 DNA writers ... 7
  1.6.3 Histone erasers ... 8
  1.6.4 DNA erasers ... 8
  1.6.5 Histone readers ... 8
  1.6.6 DNA readers ... 9
1.7 **Modification cross talk** ... 10
1.8 **Effects of metabolism on histone and DNA modifications** ... 11
1.9 **Epigenetic inheritance** ... 13
1.10 **Summary** ... 15
References ... 16

1.1 Chromatin signaling and epigenetics

The term chromatin was coined around the year 1880 by Walther Flemming. Flemming noted that "The word chromatin may stand until its chemical nature is known, and meanwhile stands for that substance in the cell nucleus which is readily stained" [1,2]. Today we know that chromatin is composed of genomic DNA in complex with histone proteins, but the original name still holds. Beyond packaging the genome into the nucleus, chromatin provides an elegant mechanism for regulating all DNA-templated processes. There are a large number of factors that go into defining and regulating chromatin structure, including the extensive modification of the histones and DNA.

The modification of histones and the potential for effect on genome regulation was first recognized in 1964. In particular, Allfrey and Mirsky [3] noted that the posttranslational acetylation of histones led

Chromatin Signaling and Neurological Disorders. https://doi.org/10.1016/B978-0-12-813796-3.00001-8

to changes in transcription. Some years later, it was proposed that DNA methylation may also affect gene regulation [4,5]. However, it was not until several years later that the study of chromatin modifications entered the spotlight. This coincided with three key discoveries. It was found that the activities of histone acetyltransferase and deacetylase (enzymes responsible for catalyzing the placement and removal of acetyl groups, respectively) were directly associated with changes in transcription, followed by the discovery that an acetyltransferase subdomain, known as the bromodomain, could specifically recognize this histone modification [6,7]. These discoveries definitively linked histone modification to gene regulation, confirmed the reversible nature of histone modifications, and demonstrated that these modifications could be functionally recognized. The parallel to classical signal transduction was evident and the defined field of chromatin signaling was born. In the 20 years since these landmark reports, a wealth of discoveries has been made regarding chromatin signaling pathways. In addition, it has been determined that dysregulation of chromatin modifications contributes to a wide range of diseases and disorders, including neurodegenerative, neurodevelopmental, and neuropsychiatric disorders.

The term epigenetics is often associated with chromatin signaling pathways. This term is usually credited to Waddington, who described it in 1942 as the complex process between genotype and phenotype, especially as it relates to development [8]. As it became apparent that DNA and histone modifications could alter gene expression and change phenotype, these pathways begun to be referred to as epigenetic pathways. Epigenetic processes have historically been discussed in the context of heritability and in terms of environmental influences on gene expression. However, there has been much debate over whether or not these are requirements for something to be considered truly epigenetic [9]. The plasticity of this term in the field has led to some consternation among researchers, and the debate over what makes something truly epigenetic continues to evolve. As the mechanisms of chromatin signaling, the influence of environment on these pathways, and the potential for heritability of chromatin states are uncovered, this term may be defined more concretely. In this chapter the basics of chromatin structure and signaling will be presented, as well as the most recent findings on the influence of environment and metabolism, and the potential for heritability, with an overall emphasis on mechanism.

1.2 Chromatin organization

The fundamental subunit of chromatin is the nucleosome (Fig. 1.1). The nucleosome was first identified through enzymatic digestion of chromatin and was characterized through microscopy and X-ray diffraction [10,11]. These studies were later followed by a high-resolution crystal structure of the nucleosome core particle (NCP) [12]. The NCP consists of an octamer of the histone proteins H2A, H2B, H3, and H4 that is wrapped by ~147 base pairs of DNA, with the DNA that bridges adjacent nucleosomes referred to as the linker DNA. The N-terminus of each of the histone proteins and the C-terminus of H2A protrude from the core and are referred to as the histone tails. These tails do not resolve in the majority of crystal structures of the NCP and are sensitive to protease digestion [13,14], leading to the common model that they are largely disordered and flexible, although some reports suggest transient interaction of the tails with the linker DNA [15–17].

Although the whole of chromatin consists of repeats of nucleosomes, the local chromatin structure is actually quite diverse. This diversity arises from a variety of factors and includes DNA sequence,

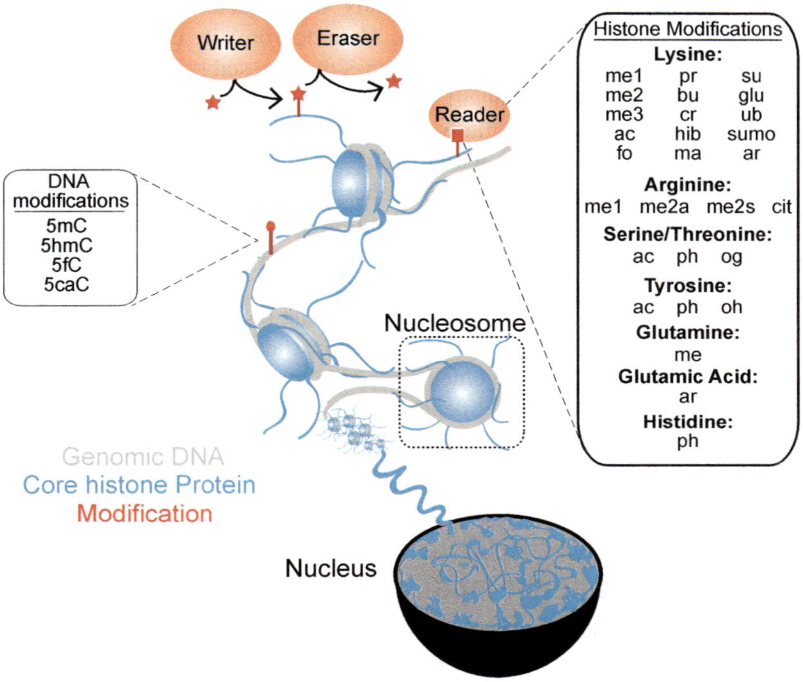

FIGURE 1.1 Chromatin and chromatin signaling.

The most basic description of chromatin as arrays of nucleosomes (outlined by dashed box), which are composed of histone proteins (blue) wrapped by segments of DNA (gray). Nucleosomes compact into the nucleus as either euchromatin (less dense) or heterochromatin (more dense). The histone tails protrude from the nucleosome core and are thought to be largely unstructured. Studies show evidence of both high flexibility and solvent exposure, as well as interactions with DNA. Both the histones and DNA can be chemically modified (modifications shown in red). Modifications are placed by enzymes known as writers, removed by enzymes known as writers, and read by subdomains known as readers (shown as mauve ovals). Boxes describe modifications that include (1) for DNA, 5-methylcytosine (5mA), 5-hydroxymethylcytosine (5hmC), 5-formylcytosine (5fC), and 5-carboxylcytosine (5caC) and (2) for histones, monomethylation (me1), dimethylation (me2), trimethylation (me3), acetylation (ac), formylation (fo), propionylation (pr), butyrylation (bu), crotonylation (cr), 2-hydroxylisobutyrylation (hib), malonylation (ma), succinylation (su), glutarylation (glu), ubiquitylation (ub), sumoylation (sumo), ADP ribosylation (ar), symmetric dimethylation (me2s), asymmetric dimethylation (me2a), citrullination (cit), phosphorylation (ph) and O-GlcNAcylation (og), and hydroxylation (oh).

nucleosome density and positioning, the presence or absence of histone H1 (which binds to the linker DNA), incorporation of histone variants, and chemical modification of the histones and DNA. Chromatin can be classified into two general categories: euchromatin and heterochromatin. Euchromatin has a lower density of nucleosomes, adopts a more open structure, and is associated with transcriptionally active genes. In contrast, heterochromatin contains a higher density of nucleosomes, is substantially more compact, and is transcriptionally silent. Heterochromatin is also enriched in linker histone, which contributes to compaction. Heterochromatin can either be constitutive or facultative. Constitutive heterochromatin contains genes that are stably repressed, whereas facultative heterochromatin retains the ability to convert between states. In interphase and postmitotic nuclei, euchromatin and heterochromatin are spatially segregated, where the former resides away from the nuclear periphery and the latter associates with the nuclear lamina. It is thought that this organization contributes to gene regulation [18]. In fact the higher order organization of chromatin within the nucleus into territories, domains, and subdomains and the role of this organization in genome regulation are now being explored in detail through advanced microscopic, imaging, and chromatin capture methods [19–21].

Chromatin structure is overall very dynamic [22]. It must be locally remodeled during all DNA-templated processes as well as during DNA repair. It must also be dramatically remodeled during cell cycle. These local and global rearrangements are mediated by a number of chromatin regulators that can remodel nucleosomes in an ATP-dependent manner (chromatin remodelers), incorporate various histone variants (histone chaperones), structurally organize chromatin, or modify the histones or DNA (chromatin modifiers). These regulators work in a cooperative manner to define the local and global chromatin architecture both spatially and temporally in order to properly regulate the genome.

1.3 Histone posttranslational modifications and the histone code

The histone proteins can be heavily posttranslationally modified (Fig. 1.1). To date the posttranslational modifications (PTMs) that have been identified include methylation (mono, di, and tri), short-chain acylation (acetylation, formylation, propionylation, butyrylation, crotonylation, 2-hydroxylisobutyrylation, malonylation, succinylation, glutarylation), ubiquitylation, SUMOylation (where SUMO stands for small ubiquitin-like modifier), and ADP ribosylation of lysine; methylation (monomethylation and asymmetric or symmetric dimethylation) and citrullination of arginine; acetylation, phosphorylation, and O-GlcNAcylation of serine and threonine; phosphorylation, acetylation, and hydroxylation of tyrosine; methylation of glutamine; phosphorylation of histidine; and ADP ribosylation of glutamic acid [23,24]. Biotinylation of lysine has also been detected, but its biological importance is still debated [25,26]. Genome wide, many of these modifications are correlated with specific genomic states and elements [27–32]. For instance, acetylation of the H3 and H4 tails is generally associated with open chromatin and transcriptional activation. Trimethylation of histone H3 at lysine 27 (H3K27me3) is a marker of facultative heterochromatin, whereas H3K9me3 and H4K20me3 are markers of constitutive heterochromatin. The promoters of active genes contain H3K4me3, whereas enhancers are enriched for H3K4me1 and H3K27ac.

It has been proposed that histone modifications may act as a code, the "histone code hypothesis" [33,34]. This hypothesis, formulated by Strahl and Allis, states that "multiple histone modifications, acting in a combinatorial or sequential fashion on one or multiple histone tails, specify unique downstream functions". Although not contrary to this, it became clear that the relationship between any single modification and function is complex and that it is critical to take into account the context both within chromatin and the regulatory process being carried out [35]. However, the idea of a histone code has been somewhat controversial, with some arguing that histone modifications may simply be a consequence of transcriptional processes, in turn altering nucleosome stability and DNA accessibility, and thus acting more like a "cog" in the function of the transcriptional machinery [36]. Indeed, to date it has proven difficult to ascertain the function of most of these modifications, especially to determine if they are causative (i.e., play a role in determining the gene regulatory process) or simply correlative.

An elegant study early in this debate provided evidence of both specific and nonspecific effects of histone PTMs on transcription. In this work, all four modifiable lysines on H4 (K5, K8, K12, and K16) in budding yeasts were mutated to arginine, retaining the positive charge, but making it so that they could not be acetylated. Only mutation of H4K16 had specific transcriptional consequences on a unique subset of genes, whereas mutation of the other lysines led to nonspecific changes in transcription scaling with the number of lysines mutated [37]. Application of newly developed CRISPR/dCas9 methods for targeted epigenome editing have great potential to provide insight into the role of histone modifications in gene regulation [38]. Indeed, studies with dCas9 fused to enzymes that catalyze certain histone modifications, namely, p300 acetyltransferase, HDAC3 deacetylase, and PRDM9 methyltransferase, suggest a causal role for H3K27ac and H3K4me3 in activating target genes [39–42]. However, much work remains to be done in order to elucidate the existence, extent, or rules of a histone code. Although the exact role of histone modifications has been a matter of debate, there is clear consensus on the fact that histone modifications are critical in the regulation of the genome and that dysregulation of the so-called epigenome can have diverse and devastating consequences [43–45].

1.4 Functions of histone posttranslational modifications

The functional consequence of histone modifications can be categorized into two major mechanisms: (1) a direct effect on chromatin structure and (2) an indirect effect by contributing to the function of chromatin regulatory proteins and protein complexes. Several histone modifications alter the electrostatic property of residues. These include all the acyl modifications on lysine, namely, phosphorylation, citrullination, and ADP ribosylation. Acetylation of H3K122, H3K64, and H3K56, which reside within the nucleosome core (i.e., the folded regions of the histones within the wrapped DNA), can decrease the stability of the nucleosome, promoting a more open chromatin state, and in some cases nucleosome disassembly [46–50]. Modification of residues within the tail domains can also have a direct effect on chromatin structure. Acetylation of the H2A and H4 tails has been shown to alter nucleosome array compaction, while acetylation of the H3 tail has been shown to decrease the stability of the nucleosome [51]. In addition, SUMOylation of H4K12 can also disrupt chromatin compaction [52]. These effects are consistent with the general correlation of acetylation with open chromatin and gene activation. In contrast, H4K20me3 has been found to increase the stability of nucleosome arrays [53], consistent with this modification being found at pericentric heterochromatin.

The best characterized example of a histone tail modification having a direct effect on chromatin structure is H4K16ac. In the first high-resolution crystal structure of the nucleosome, residues K16–N25 of the H4 tail make an interaction with the acidic patch on the H2A/H2B dimer of an adjacent nucleosome in the crystal (i.e., crystallographic symmetry mate) [12]. The authors hypothesized that this interaction may be important in higher order chromatin compaction, and indeed, it has been shown that acetylation of H4K16 disrupts the compaction of nucleosome arrays [12,54–57]. In fact, it was found that acetylation of H4K16 in a single nucleosome in the middle of an array was sufficient to cause localized changes in DNA accessibility [58]. Although the tails are often viewed as fully solvent exposed, accumulating evidence strongly suggests that the histone tails interact with DNA within the chromatin context and stabilize the nucleosome itself [16,17,59–63]. Thus, modifications in the tail domains may act by altering local nucleosome dynamics as well as the observed internucleosomal effects.

The indirect functional effects of histone modifications, at least within the tail domains, are currently better understood than the direct effects. Specific recognition of histone modifications by chromatin regulatory proteins or protein complexes has now been shown to have a number of functional outcomes. The most straightforward outcome is in targeting these regulators to regions of chromatin enriched in particular histone PTMs. Alternatively, PTMs may act to retain the regulator at chromatin after initial targeting. PTMs have been shown to play a role in regulating (through allosteric mechanisms or otherwise) the activity of chromatin regulators. These functions will be discussed further in the following sections.

1.5 DNA methylation

DNA can also be modified (Fig. 1.1) through methylation of carbon 5 on cytosine (5-methylcytosine, 5mC). Though methylation of carbon 6 on adenine (6mA) is a putative additional modification, identification of 6mA is new in higher eukaryotes, occurs at low levels, and remains to be fully elucidated [64–68]. The most common form of 5mC is in the context of CpG dinucleotides (mCpG). In fact, it is estimated that ~70%–80% of the CGs in mammalian cells are methylated, and most of these sites are symmetrically methylated (i.e., methylated on both DNA strands) [69]. The exception to this is regions strongly enriched in CpG content known as CpG islands (CPIs), which are resistant to methylation. Many genes contain CPIs at their transcription start sites (TSSs), the large majority of which remain demethylated throughout all stages of development and in all tissue types, independent of whether the gene is active or repressed. However, a fraction of CPIs at TSSs are methylated and in these cases are correlated with long-term stable repression, for instance, in imprinting, X-chromosome inactivation, or repression of transposable elements [70,71]. In these cases, CpG methylation appears to act as a lock, besides other repressive mechanisms including the presence of H3K9me3. Notably, there are also CPIs found in the body of genes that are resistant to methylation and they appear to be positively correlated with transcription. Evidence suggests that CpG methylation in the gene body may play a role in splicing [72].

Non-CpG methylation has also been identified. The so-called CpH methylation (where H = A/T/C) is found to be enriched in pluripotent cells and is found in the mouse germline, but is missing in most somatic tissues [73–75]. The exception to this is mouse and human neurons, in which CpH methylation makes up almost 25% of methylated cytosine, with mCpA being the most abundant [76–78]. In these contexts, CpH methylation appears to be correlated with gene repression. Oxidized versions of 5mC may also soon be added to the list of stable DNA modifications as will be discussed in the following.

Similar to histone modifications, it appears that DNA methylation can also have direct and indirect effects on chromatin structure. It has been shown that DNA methylation increases nucleosome stability [79,80]. It can also negatively modulate the binding of the transcription machinery and inhibit transcription elongation [81]. Specific recognition of methylated DNA can also lead to recruitment or regulation of a variety of chromatin regulators as discussed in the following sections.

1.6 Writers, erasers, and readers

With the histone code hypothesis came a nomenclature that is now commonly used to describe the players in chromatin signaling. Specifically, enzymes that place modifications are referred to as "writers", those that remove modifications are referred to as "erasers", and the subdomains that specifically recognize modifications are referred to as "readers" (Fig. 1.1). It should be noted that writers and erasers themselves can, and often do, contain reader domains.

1.6.1 Histone writers

Writers have been identified for many, but not all, histone modifications. These include histone methyltransferases (HMTs), histone acyltransferases (responsible for acetylation, butyrylation, crotonylation, 2-hydroxylisobutyrylation, β-hydroxybutyrylation, succinylation, glutarylation), kinases, poly-ADP-ribose-polymerases, E1/E2/E3 ubiquitin and SUMO ligases, and O-GlcNAc transferase. The catalytic writer proteins often exist in large multiprotein complexes. The composition of these complexes can vary by cell type and developmental stage, and their differential composition is thought to be functionally significant [82].

For some histone residues a particular modification is generated by a unique writer, whereas for other residues the same modification can be generated by several different writers. For example, trimethylation on H3K27 is only generated by EZH1/2, which resides in the polycomb repressive complex 2 (PRC2) [83]. In contrast, H3K9 can be trimethylated by eight different enzymes [84]. Similarly, while some writers have a unique histone substrate, others are far more promiscuous [84—89]. Successive methylation of arginine and lysine either can be carried out by a single enzyme or may require multiple enzymes depending on the specific arginine or lysine residue. For instance, PRC2 can generate mono-, di-, and trimethylated H3K27, whereas H3K36 is mono- and dimethylated by the NSD (nuclear receptor-binding SET domain) family of methyltransferases but trimethylated by SETD2. Notably, these enzymes have many nonhistone targets [90]. As such, there is a push to move away from the histone-specific nomenclature to one that is more general. For instance, referring to classical HMTs as KMTs (for lysine) or RMTs (for arginine), or even more generic as PMTs (for protein).

1.6.2 DNA writers

DNA methylation is established *de novo* by the DNA methyltransferase 3 (DNMT3) enzymes DNMT3A and DNMT3B [91,92]. These enzymes do not discriminate between unmethylated and hemimethylated DNA, allowing them to fully methylate unmodified DNA. A related, but catalytically inactive, protein, DNMT3L, associates with DNMT3A and DNMT3B and upregulates their catalytic activity. DNMT3L is primarily expressed in undifferentiated cells and is important for establishing DNA methylation patterns in early development [93]. Methylation is maintained through cell division by DNMT1, which preferentially methylates hemimethylated DNA generated after replication. DNMT1 requires the histone E3 ubiquitin ligase, UHRF1, for proper maintenance of DNA methylation [94,95]. As mCpH is inherently asymmetric, it must be established *de novo* after cell division and has been shown to be dependent on DNMT3A [73,77].

1.6.3 Histone erasers

Although there were early debates on the reversibility of histone modifications, most PTMs now have identified erasers. These include histone lysine demethylases; histone deacylases, including Sirtuins, which are responsible for removing acetylation, butyrylation, crotonylation, 2-hydroxylisobutyrylation, succinylation, glutarylation; phosphatases; deubiquitinases (DUBs); sentrin-specific proteases, which remove SUMO; and O-GlcNAcases (OGAs). Enzymes that degrade ADP-ribose polymer chains have also been identified. In addition, some enzymes can remove the mono(ADP-ribosyl) group from a modified residue. Notably, the ability for methylated arginine to be enzymatically reverted to unmodified arginine has been somewhat controversial. Although a subset of JMJC demethylases have been shown to demethylate arginine *in vitro* and potentially *in vivo*, the validity of these results remains a matter of debate [96–100]. However, conversion of a methylated arginine to citrulline is well established and is catalyzed by peptidylarginine deiminases. The resultant citrulline has no net charge, giving it properties distinct from those of arginine. Histone modifications can also be removed *en masse* through histone tail clipping; however, the functional significance of this activity is still under study [101].

1.6.4 DNA erasers

There are currently no known enzymes that can directly remove the methyl group from DNA. DNA demethylation occurs through one of two mechanisms, either active or passive [102]. Passive DNA demethylation occurs through dilution of the methyl marks by way of replication in the absence of maintenance methylation pathways (via DNMT1). Active demethylation is a more elaborate pathway that involves chemical conversion of 5mC and removal of the entire base. Conversion of 5mC to thymine can occur spontaneously or through the action of cytosine deaminase. Alternatively, 5mC can be oxidized by the ten-eleven translocation (TET) methylcytosine dioxygenase enzymes [103]. These enzymes lead to the iterative oxidation of 5mC to 5-hydroxymethylcytosine (5hmC), 5-formylcytosine (5fC), and 5-carboxylcytosine (5caC). As DNMT1 is substantially less efficient on oxidized 5mC, these oxidation mechanisms can lead to passive demethylation during replication. As an alternative mechanism of removal, thymine DNA glycosylase (TDG) can excise 5fC and 5caC, producing abasic sites, which is followed by base excision repair leading to restoration of unmethylated cytosine [104]. TDG can also repair the T:G mismatch created by the cytosine to thymine conversion.

Notably, 5hmC is enriched in the central nervous system and in murine cerebellar Purkinje neurons [105–107], and aberrant levels of 5hmC have been shown to be associated with early- and late-onset mental illnesses [108]. This has led to the proposal that 5hmC is, in fact, a stable modification with functions distinct from those of 5mC. Interestingly, oxidization of methylcytosine has been shown to have variable effects on nucleosome stability, depending on the oxidation state [109,110], but the role of these modifications in chromatin structure regulation is still unclear.

1.6.5 Histone readers

The readout of histone modifications is accomplished through specific recognition by protein subdomains, known as histone mark reader domains (or simply readers) [111,112]. The first identified reader domain was the bromodomain, when the P/CAF bromodomain was shown to recognize acetylated lysines [6]. A couple of years later the chromodomain of HP1 was shown to bind methylated lysines, with specificity for H3K9me2 and H3K9me3 [113]. Since then, a large number and diversity of reader domains have been identified. These domains can be categorized into families based on structural fold. Although the fold is highly conserved, the sequence conservation among members of each family is often quite low, except for critical residues in the histone-binding pocket.

Specificity of reader domains occurs on two levels: (1) specificity for a particular PTM such as methylated lysine as opposed to acetylated lysine and (2) specificity for a particular modified amino acid residue such as methylated H3K9 as opposed to methylated H3K36. Generalizations can be made regarding the specificity of families of domains for a particular PTM; for example, bromodomains generally recognize acetylated lysines, whereas chromodomains generally recognize methylated lysines. However, there are numerous variations to this and the specificity of each domain should be considered individually [24]. Specificity for a particular modified amino acid residue is achieved through unique recognition of the surrounding residues. While some domains demonstrate strict specificity, for example, chromodomains can generally differentiate even between methylated H3K9 and H3K27 despite similar surrounding sequences, others are far more promiscuous as in the case of bromodomains, which can often recognize several acetylated lysines with nearly equal affinity. Notably, many of these domains have now been identified to have DNA- or RNA-binding capability in addition to their histone-binding activity, which can lead to multivalent chromatin interaction and increased selectivity [114–123].

The functional consequence of these recognition events varies and includes targeting, retention, and regulation of the host protein or complex. For instance, the BPTF (Bromodomain PHD [plant homeodomain] Finger Transcription Factor) C-terminal PHD finger recognizes H3K4me3 and is important in targeting the NURF (nucleosome remodeling factor) remodeling complex to the promoter of target genes [124,125]. The PHD finger of BHC80, on the other hand, associates with unmodified H3K4. BHC80 is a component of the LSD1 H3K4 demethylase complex, which catalyzes the removal of H3K4me3 for transcriptional repression. Recognition of the newly demethylated histone tail by the BHC80 PHD finger is important for retaining the complex at the target promoter [126]. Furthermore, several instances have been reported where binding to a specific modification leads to allosteric regulation of the host complex. This is the case for the H3K27 PRC2 methyltransferase complex. Here, recognition of the PRC2 product, H3K27me3, by the PRC2 subunit EED leads to allosteric upregulation of methyltransferase activity [127,128]. This regulation is thought to be important for spreading the H3K27me3 modification along chromatin.

1.6.6 DNA readers

Reader domains of DNA methylation have also been identified. To date, these include two families of well-characterized domains: the methyl-CpG-binding domain (MBD) and the SET- and RING finger–associated (SRA) domain [129]. MBDs bind specifically to symmetrically methylated CpGs. Studies have indicated that nucleotides adjacent to the 5mC nucleotide can alter binding affinity, suggesting moderate sequence preference. A higher density of mCpG often correlates with increased association. It has been found that the methyl-CpG-binding protein 2 (MECP2) MBD can recognize mCpA and hmCpA in addition to mCpG [78,130]. Notably, the MECP2 MBD is expressed at very high levels in the brain, coinciding with the high levels of 5hmC found in neurons. Mutations in MECP2 are associated with the X-linked neurologic disorder Rett syndrome, a progressive autism spectrum disorder that is one of the most common causes of neurologic impairment in girls [131] (see also Chapter 9). Together, this strongly suggests that 5hmC may indeed be a functional intermediate along the DNA demethylation pathway. In contrast to the MBD, the SRA domain specifically binds to hemimethylated DNA [132]. This domain is found in UHRF1, which is important in the DNMT1 maintenance of DNA methylation.

DNA methylation has also been shown to positively alter, or alternatively regulate, the action of some transcription factors [129,133]. Several transcription factors, including many pioneer transcription factors, have been identified that specifically recognize methylated DNA. Notably, the presence of a methyl group can alter the sequence specificity of the transcription factor. In other words a sequence that would not be preferred when it is unmethylated may be preferred once methylated. For a few of these transcription factors, methyl DNA binding has been found to be mediated by C2H2-type zinc finger domains. Other putative methyl DNA–binding domains include the homeobox, bHLH (basic helix-loop-helix), forkhead, bZIP, and HMG (high-mobility group) box domains, but these remain to be confirmed.

Recognition of methylated DNA by MBDs is generally associated with targeting chromatin regulators to regions enriched in methylation [129]. The SRA domain of UHRF1 plays a more complex role. Although it helps UHRF1 to target DNA, it is only modestly sensitive to the DNA methylation state. However, recognition of hemimethylated DNA allosterically upregulates the ubiquitin ligase activity of UHRF1, which subsequently recruits DNMT1 to the sites of hemimethylated DNA [134–136].

1.7 Modification cross talk

Genome-wide analysis of chromatin modifications reveals clear positive and negative correlations. Mechanistically, much of this is mediated through the effect of modifications on the action of the writers, erasers, and readers. For instance, DNA methylation is known to be negatively correlated with H3K4me3. Indeed, the ADD domain of DNMT3A is a reader for unmodified H3K4 [137]. Within DNMT3A the ADD domain forms an intramolecular interaction with the catalytic domain blocking the DNA-binding pocket [138]. Upon association with the unmodified H3 tail, DNMT3A undergoes a conformational change releasing the catalytic domain and leading to upregulation of DNA methylation. If H3K4 is methylated, however, the ADD domain will not bind, and DNMT3A will remain inhibited [138,139]. HMTs can also be regulated by the surrounding histone modification state [140]. For at least one of the H3K9 methyltransferases, G9a, activity is inhibited by H3K4me3 [141]. Structural studies revealed that the catalytic domain associates with the H3 tail in part through contacts with unmodified H3K4 and that these contacts are inhibited by methylation of H3K4 [142], mediating a negative correlation between H3K4me3 and H3K9me3.

A positive correlation is seen between H3K9me3 and H3S10ph during mitosis, which negatively regulates HP1 temporally. HP1 proteins (-α, -β, and -γ) associate with H3K9me3 through their chromodomain and are important in heterochromatin organization. However, during mitosis, HP1 proteins largely dissociated from chromatin, which is thought to be important for allowing the large-scale chromatin remodeling necessary for chromatin condensation and chromosome segregation. It was found that phosphorylation of H3S10 during mitosis inhibits the association of the HP1 chromodomain with H3K9me3, causing the protein to dissociate [143]. These types of cross talk are thought to be critical in defining chromatin states and in the regulation of the genome. In fact, it is a particular pattern of histone and DNA modifications that is thought to be important in regulating the local and global chromatin structure. As mentioned already, many reader domains (as well as the catalytic pocket of writers and erasers) recognize a substantial stretch of histone tails and are thus sensitive to the modification state of several residues. In addition, many chromatin regulators contain multiple reader domains providing the ability to respond to a complex and dynamic chromatin environment [111,144,145].

1.8 Effects of metabolism on histone and DNA modifications

The metabolic state of a cell can impact the chromatin landscape through a number of pathways, including by affecting chromatin regulator activity either through its PTM or at the level of its transcription or cellular localization. However, it can also have a direct effect on the chromatin landscape, as several metabolites are substrates and cofactors of the chromatin regulators themselves [146,147]. Notably, some metabolites can also inhibit the activity of chromatin regulators. Nutrient imbalances caused by dietary deficiencies or lifestyle choices, such as a high-fat diet, lead to improper DNA and histone modifications and susceptibility to disease. If these occur during pregnancy, they can also lead to severe developmental defects in the child.

One-carbon metabolism is the pathway that ultimately provides methyl groups for methylation of histones and DNA (Fig. 1.2). S-adenosylmethionine (SAM) is the major methyl donor in a cell. Cellular levels of SAM have an effect on both histone and DNA methylation [148,149]. Interestingly, different histone residues are differentially sensitive to the levels of SAM, with H3K4me2/3 showing the highest sensitivity [150–152]. Several vitamins are involved in the pathway that produce SAM. In particular, vitamins B_2, B_6, and B_{12}, as well as folate (vitamin B_9), and methionine, threonine, and choline have been shown to be important in chromatin regulation. Deficiencies in these vitamins, amino acids, and nutrients can alter DNA and histone methylation patterns and lead to increased risk of diseases. For instance, folate deficiency is well known to cause neural tube defects, due in part to aberrant DNA methylation. In addition, choline is critical for fetal neurogenesis, affecting both histone and DNA methylation [149].

Histone acylation is established by acyltransferases, with the specific acyl-CoA substrate determining the acylation state (Fig. 1.2). The best characterized acylated state, acetylation, is established using acetyl-CoA as the cofactor, the major source of which is citrate. Citrate is derived through glycolysis and the tricarboxylic acid (TCA) cycle and is thus influenced by glucose intake. Nonacetyl acylations have been shown to be derived, at least in part, from short-chain fatty acids through the action of acyl-CoA synthetase short-chain family member 2 (ACSS2). Notably, acetyl-CoA can also be derived in this manner from acetate, which mechanism is upregulated under hypoxic conditions. In addition, in the liver the short-chain fatty acid β-hydroxybutyrate (bhb) can be converted to bhb-CoA through ketogenesis, which is activated under low-glucose conditions [153]. Other pathways of nonacetyl acyl-CoA production may have yet to be identified. It has been proposed that the relative levels of acylation depend on the relative availability of the different acyl-CoAs and thus would be influenced strongly by glucose availability [89]. Several studies have demonstrated that the different proportion of these acyl states has important functional consequences. For instance, in mice that were

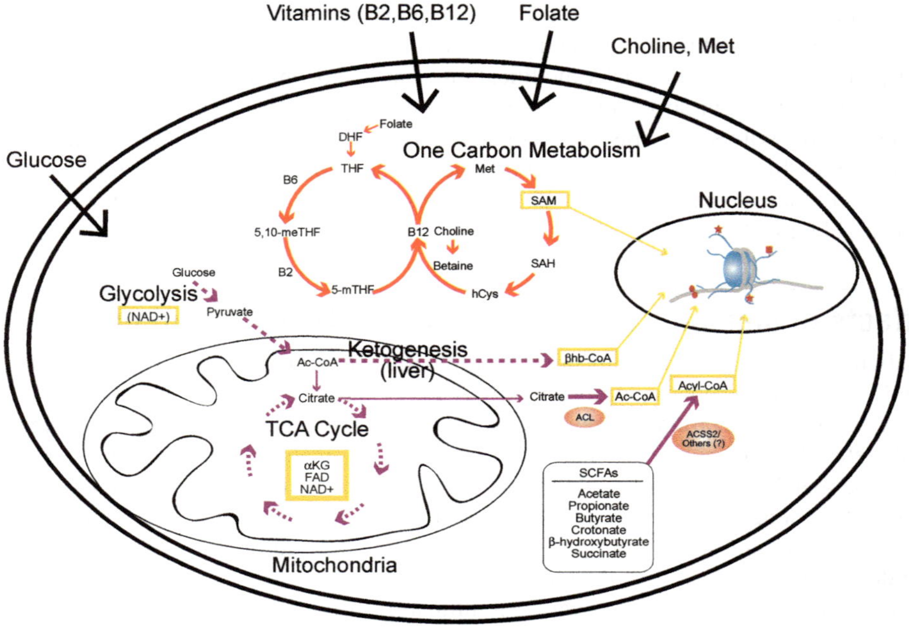

FIGURE 1.2 Metabolism and chromatin signaling.

The metabolic inputs and pathways that are known to influence the levels of cofactors (boxed in yellow) used by chromatin writers and erasers in the nucleus. Vitamins B_2, B_6, and B_{12}; folate; choline; and methionine (Met) feed into the one-carbon metabolism (red), which is the major source of S-adenosylmethionine (SAM). Glucose feeds into the tricarboxylic acid (TCA) cycle, producing citrate, the major source of acetyl-CoA (Ac-CoA). Ac-CoA and other acyl-CoAs can also be synthesized by acyl-CoA synthetase short-chain family member 2 (ACSS2) and potentially other synthetases from short-chain fatty acids (SCFAs). Specific to the liver, ketogenesis leads to the production of β-hydroxybutyrate (bhb), which can be converted to bhb-CoA under low-glucose conditions. Dashed arrows indicate that not all factors are shown for a given pathway.

subjected to prolonged fasting or induced diabetic ketoacidosis, there was found to be an upregulation in β-hydroxybutyrylation of lysine at some gene promoters. Notably, those genes with increased levels of H3K9bhb were found to be associated with starvation-induced genes that were upregulated during the ketogenic response [153].

Histone acetylation as well as histone and DNA methylation levels can also be affected by the levels of cofactors available for the eraser enzymes (Fig. 1.2). Most histone deacetylases depend on Zn^{2+}, however, the Sirtuin deacetylases use NAD^+ as a cofactor. Thus an increase in NAD^+, either through calorie restriction or external administration, can increase Sirtuin activity and lead to histone deacetylation. JMJC demethylases and TET enzymes involved in DNA demethylation are dependent on α-ketoglutarate, whereas the LSD1 histone demethylases rely on flavin adenosine dinucleotide (FAD), both of which are produced via the TCA cycle and are thus influenced by glucose availability [154].

1.9 Epigenetic inheritance

Epigenetic inheritance is thought to occur on many levels, from maintenance of cell identity through somatic cell division to transgenerational inheritance of chromatin states (Fig. 1.3). However, uncovering the extent to which histone and DNA modifications are inherited, and the underlying mechanisms, has been challenging.

Inheritance of DNA methylation through cell division is perhaps the best understood mechanism of epigenetic inheritance. In this process, DNA methylation is semiconservatively split during DNA replication, and DNMT1 restores symmetric DNA methylation, using the postreplicative hemimethylated DNA as a template. Inheritance of histone modifications through mitosis is less well understood. It has been shown that parental histones are at least in part retained through replication, and stable inheritance of histone methylation has been shown in yeast and mammalian cells [155,156]. The mechanism behind maintenance of histone modifications, on the other hand, has remained elusive. It has been established that about half of the histones on a newly replicated DNA are parental and retain the parental histone modification signature. The deposition of these parental histones is thought to follow a conservative mechanism in which most of the time the full H3-H4 tetramer is deposited onto the nascent DNA rather than splitting into two dimers [157,158]. However, splitting of the tetramer is sometimes observed, and it has been proposed that euchromatic chromatin may undergo more extensive H3-H4 tetramer splitting [157,159], suggesting that the mechanism of inheritance might not be identical for every histone modification signature. Restoration of the fully modified chromatin domain in either case could be restored through spreading of the mark by the appropriate writer, either to adjacent nucleosomes or to nucleosomes across the replication fork [158].

An unfortunate effect of epigenetic inheritance through mitosis is that nutritional deficiencies or metabolic disorders can be remembered. This is thought to be the case in type-2 diabetes, where even after controlling blood glucose levels, patients suffer from inflammation and vascular complications. Studies in diabetic mice and cells from diabetic human patients demonstrate aberrant histone methylation and acetylation as compared to healthy cells. Importantly, overexpression of writers or small molecule activators of writers largely reversed this effect, suggesting that these epigenetic states may be reset therapeutically [154,160,161].

Intergenerational and transgenerational inheritance of histone and DNA modifications could provide a memory of the current environmental conditions to future generations (Fig. 1.3). Although this could be a positive avenue for fast adaptation, it could also make future generations susceptible to the negative effects of various environmental insults, including increased susceptibility to diseases. In fact, in mammals, there is significant reprogramming of both histone and DNA modifications after fertilization and even more extensive reprogramming in the germline that would appear to reset almost all these modifications [162–164]. This includes passive and active DNA demethylation and changes in histone modifications. In sperm, histones are extensively replaced with protamines, effectively removing all histone modifications. These mechanisms have long appeared to completely erase any epigenetic memory of the environment from future generations. However, exceptions are becoming apparent and may allow for some transmission of chromatin states. One known exception to reprogramming is found in parental imprinting. Imprinted genes are resistant to postfertilization reprogramming, retaining their DNA and histone methylation status, in order to enable sex-specific expression. Although these imprints are largely removed during reprogramming in the germline, a

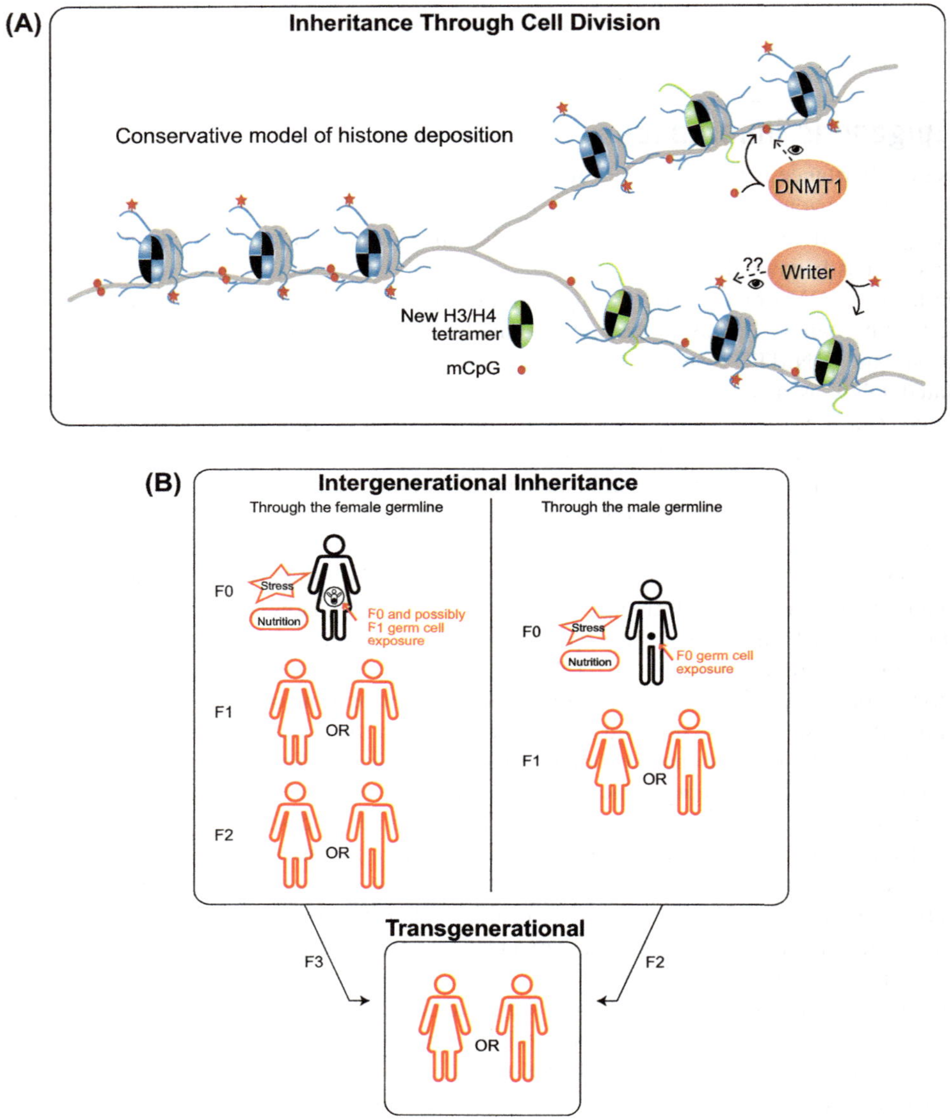

FIGURE 1.3 Epigenetic inheritance.

(A) Inheritance through cell division. Histone and DNA modifications (red) are known to be inherited through replication. Shown is a replication fork. DNA methylation is semiconservatively split. CpG methylation is maintained through the action of DNMT1, which senses and methylates the hemimethylated DNA. The current model of histone retention is that the H3-H4 tetramers are conservatively split. It is thought that histone writers may recognize modifications and modify adjacent nucleosomes accordingly, but this is not fully understood. (B) Intergenerational inheritance. Epigenetic modifications may provide a memory of current conditions, including nutrition and stress, to future generations. Through the maternal line, these inputs will affect the mother (F0) and her germline (F1), or if pregnant the germline of F1 (F2). Through the paternal line, these inputs will affect the father (F0) and his germline (F1). If these epigenetic marks are transgenerational, heritable effects would be seen in offspring never exposed to the original input, which is the F3 on the maternal side or the F2 on the paternal side.

subset of genes have been found to be resistant to this reprogramming as well [162,165]. In addition, it has been found that a small subset of genes retain histone proteins in sperm at some developmentally important loci, along with extensive histone modifications [166]. These exceptions suggest that there may indeed be mechanisms for transmission of histone and DNA modifications that escape reprogramming.

Several studies have revealed transgenerational inheritance of epigenetic states in *Caenorhabditis elegans* and *Drosophila*, which include changes in direct response to the environment [167–169]. The extent of transgenerational inheritance in mammals and the mechanisms by which this might occur are not yet fully understood. However, several studies in mice and rats suggest intergenerational and potentially transgenerational effects of nutrition and stress on the offspring through epigenetic mechanisms [167]. For instance, in mice fed with a low-protein diet, there were changes in chromatin modifications in the sperm. In addition, there was an increase in the expression of genes involved in lipid and cholesterol biosynthesis in the liver of the offspring of these males. Stress was also seen to have heritable epigenetic effects in mice [170]. Mice exposed to unpredictable maternal separation combined with unpredictable maternal stress (MSUS) demonstrated depression-like behavior into adulthood. These behaviors were retained in the offspring of the studied males, despite the next generation being raised normally. Notably, changes in DNA methylation were observed in the sperm of the MSUS-exposed males [171]. Specifically, an increase in DNA methylation was observed at the promoters of the cannabinoid receptor 1, which is associated with emotionality, and *MECP2*, which is expressed at high levels in the brain. In contrast the stress hormone receptor, corticotrophin-releasing factor receptor 2, showed a decrease in methylation. Future generations would need to be studied to establish the presence of *bona fide* transgenerational inheritance.

1.10 Summary

Over the past two decades, there have been dramatic advances in our understanding of chromatin and epigenetic signaling pathways. Genome-wide studies have revealed that specific modifications are correlated with specific genomic states and elements, and great strides are now being made to understand the function of these modifications in regulating the genome. A large number of histone and DNA writers, erasers, and readers have been identified, with more likely to come, and their mechanism of function is beginning to be determined. In addition, important patterns of modifications are now being defined and the mechanisms of cross talk between them is being revealed. Finally, there are great gains being made in our understanding of how the environment and metabolism alter chromatin states, leading to changes in genome regulation, and the extent to which these states can be inherited. It is becoming clear that the so-called epigenome contributes a great deal to disease susceptibility. Importantly, the dynamic nature of histone and DNA modifications makes these pathways attractive therapeutic targets. Indeed, there are several advantages in the development of drugs to treat epigenetic disorders [172,173], providing hope for new therapeutic avenues in the treatment of a wide range of diseases.

References

[1] Georg thieme verlag, stuttgart. *Walther flemming, zellsubstanz, Kern und zelltheilung*. Dtsch Med Wochenschr 1883;9:342. 342 (Georg Thieme Verlag, Stuttgart).

[2] Olins DE, Olins AL. Chromatin history: our view from the bridge. Nat Rev Mol Cell Biol 2003;4:809–14.

[3] Allfrey VG, Faulkner R, Mirsky AE. Acetylation and Methylation of histones and their possible role in the regulation of rna synthesis. Proc Natl Acad Sci USA 1964;51:786–94.

[4] Riggs AD. X inactivation, differentiation, and DNA methylation. Cytogenet Cell Genet 1975;14:9–25.

[5] Holliday R, Pugh JE. DNA modification mechanisms and gene activity during development. Science 1975; 187:226–32.

[6] Dhalluin C, Carlson JE, Zeng L, He C, Aggarwal AK. Structure and ligand of a histone acetyltransferase bromodomain. Nature 1999;399(6735):491–6.

[7] Winston F, Allis CD. The bromodomain: a chromatin-targeting module? Nat Struct Biol 1999;6:601–4.

[8] Waddington CH. The epigenotype. Endeavour 1942;1:18–20.

[9] Haig D. Commentary: the epidemiology of epigenetics. Int J Epidemiol 2012;41:13–6.

[10] Olins AL, Olins DE. Spheroid chromatin units (v bodies). Science 1973;183(4122):330–2.

[11] Kornberg RD. Chromatin structure: a repeating unit of histones and DNA. Science 1974;184:868–71.

[12] Luger K, Mäder AW, Richmond RK, Sargent DF. Crystal structure of the nucleosome core particle at 2.8 Å resolution. Nature 1997;389(6648):251–60.

[13] Böhm L, Crane-Robinson C. Proteases as structural probes for chromatin: the domain structure of histones. Biosci Rep 1984;4:365–86.

[14] Rosenberg NL, Smith RM, Rill RL. The action of chymotrypsin on nucleosome cores. Histone products and conformational effects of limited digestion. J Biol Chem 1986;261:12375–83.

[15] Biswas M, Langowski J, Bishop TC. Atomistic simulations of nucleosomes. WIREs Comput Mol Sci 2013; 3:378–92.

[16] Cutter AR, Hayes JJ. A brief review of nucleosome structure. FEBS Lett 2015. https://doi.org/10.1016/j.febslet.2015.05.016.

[17] Stützer A, et al. Modulations of DNA contacts by linker histones and post-translational modifications determine the mobility and modifiability of nucleosomal H3 tails. Mol Cell 2016;61:247–59.

[18] Cabianca DS, Gasser SM. Spatial segregation of heterochromatin: uncovering functionality in a multicellular organism. Nucleus 2016;7:301–7.

[19] Sewitz SA, Fahmi Z, Lipkow K. Higher order assembly: folding the chromosome. Curr Opin Struct Biol 2017;42:162–8.

[20] Nozaki T, et al. Dynamic organization of chromatin domains revealed by super-resolution live-cell imaging. Mol Cell 2017;67:282–93. e7.

[21] Ou HD, et al. ChromEMT: visualizing 3D chromatin structure and compaction in interphase and mitotic cells. Science 2017;357. eaag0025.

[22] Henikoff S. Mechanisms of nucleosome dynamics in vivo. Cold Spring Harb Perspect Med 2016;6:a026666.

[23] Huang H, Lin S, Garcia BA, Zhao Y. Quantitative proteomic analysis of histone modifications. Chem Rev 2015;115:2376–418.

[24] Andrews FH, Strahl BD, Kutateladze TG. Insights into newly discovered marks and readers of epigenetic information. Nat Chem Biol 2016;12:662–8.

[25] Kuroishi T, Rios-Avila L, Pestinger V, Wijeratne SSK, Zempleni J. Biotinylation is a natural, albeit rare, modification of human histones. Mol Genet Metabol 2011;104:537–45.

[26] León-Del-Río A, Valadez-Graham V, Gravel RA. Holocarboxylase synthetase: a moonlighting transcriptional coregulator of gene expression and a cytosolic regulator of biotin utilization. Annu Rev Nutr 2017;37: 207–23.

[27] Barski A, et al. High-resolution profiling of histone methylations in the human genome. Cell 2007;129: 823–37.
[28] Mikkelsen TS, et al. Genome-wide maps of chromatin state in pluripotent and lineage-committed cells. Nature 2007;448:553–60.
[29] Zhou VW, Goren A, Bernstein BE. Charting histone modifications and the functional organization of mammalian genomes. Nat Rev Genet 2010;12:7–18.
[30] Ernst J, et al. Mapping and analysis of chromatin state dynamics in nine human cell types. Nature 2011;473: 43–9.
[31] Creyghton MP, et al. Histone H3K27ac separates active from poised enhancers and predicts developmental state. Proc Natl Acad Sci Unit States Am 2010;107:21931–6.
[32] Rivera CM, Ren B. Mapping human epigenomes. Cell 2013;155(1).
[33] Strahl BD, Allis CD. The language of covalent histone modifications. Nature 2000;403:41–5.
[34] Gardner KE, Allis CD, Strahl BD. OPERating ON chromatin, a colorful language where context matters. J Mol Biol 2011;409(1):36–46.
[35] Berger SL. The complex language of chromatin regulation during transcription. Nature 2007;447:407–12.
[36] Henikoff S, Shilatifard A. Histone modification: cause or cog? Trends Genet 2011;27(10):389–96.
[37] Dion MF, Altschuler SJ, Wu LF, Rando OJ. Genomic characterization reveals a simple histone H4 acetylation code. Proc Natl Acad Sci USA 2005;102:5501–6.
[38] Pulecio J, Verma N, Mejía-Ramírez E, Huangfu D, Raya A. CRISPR/Cas9-Based engineering of the epigenome. Cell Stem Cell 2017;21:431–47.
[39] Hilton IB, et al. Epigenome editing by a CRISPR-Cas9-based acetyltransferase activates genes from promoters and enhancers. Nat Biotechnol 2015;33:510–7.
[40] Cano-Rodriguez D, et al. Writing of H3K4Me3 overcomes epigenetic silencing in a sustained but context-dependent manner. Nat Commun 2016;7:12284.
[41] Kwon DY, Zhao Y-T, Lamonica JM, Zhou Z. Locus-specific histone deacetylation using a synthetic CRISPR-Cas9-based HDAC. Nat Commun 2017;8:15315.
[42] Chen T, et al. Chemically controlled epigenome editing through an inducible dCas9 system. J Am Chem Soc 2017;139:11337–40.
[43] Chi P, Allis CD, Wang GG. Covalent histone modifications — miswritten, misinterpreted and mis-erased in human cancers. Nat Rev Canc 2010;10:457–69.
[44] Portela A, Esteller M. Epigenetic modifications and human disease. Nat Biotechnol 2010;28:1057–68.
[45] Brookes E, Shi Y. Diverse epigenetic mechanisms of human disease. Annu Rev Genet 2014;48:237–68.
[46] Manohar M, et al. Acetylation of histone H3 at the nucleosome dyad alters DNA-histone binding. J Biol Chem 2009;284:23312–21.
[47] Simon M, et al. Histone fold modifications control nucleosome unwrapping and disassembly. Proc Natl Acad Sci USA 2011;108:12711–6.
[48] Tropberger P, et al. Regulation of transcription through acetylation of H3K122 on the lateral surface of the histone octamer. Cell 2013;152:859–72.
[49] Di Cerbo V, et al. Acetylation of histone H3 at lysine 64 regulates nucleosome dynamics and facilitates transcription. Elife 2014;3:e01632.
[50] Bernier M, et al. Linker histone H1 and H3K56 acetylation are antagonistic regulators of nucleosome dynamics. Nat Commun 2015;6:10152.
[51] Wang X, Hayes JJ. Acetylation mimics within individual core histone tail domains indicate distinct roles in regulating the stability of higher-order chromatin structure. Mol Cell Biol 2007;28:227–36.
[52] Dhall A, et al. Sumoylated human histone H4 prevents chromatin compaction by inhibiting long-range internucleosomal interactions. J Biol Chem 2014. https://doi.org/10.1074/jbc.M114.591644.

[53] Lu X, et al. The effect of H3K79 dimethylation and H4K20 trimethylation on nucleosome and chromatin structure. Nat Struct Mol Biol 2008;15:1122–4.
[54] Dorigo B, Schalch T, Bystricky K. Chromatin fiber folding: requirement for the histone H4 N-terminal tail. J Mol 2003;327(1):85–96.
[55] Shogren-Knaak M, et al. Histone H4-K16 acetylation controls chromatin structure and protein interactions. Science 2006;311:844–7.
[56] Kan P-Y, Caterino TL, Hayes JJ. The H4 tail domain participates in intra- and internucleosome interactions with protein and DNA during folding and oligomerization of nucleosome arrays. Mol Cell Biol 2009;29: 538–46.
[57] Zhou B-R, et al. Histone H4 K16Q mutation, an acetylation mimic, causes structural disorder of its N-terminal basic patch in the nucleosome. J Mol Biol 2012;421:30–7.
[58] Mishra LN, Pepenella S, Rogge R, Hansen JC, Hayes JJ. Acetylation mimics within a single nucleosome alter local DNA accessibility in compacted nucleosome arrays. Sci Rep 2016;6:34808.
[59] Ausio J, Dong F, van Holde KE. Use of selectively trypsinized nucleosome core particles to analyze the role of the histone 'tails' in the stabilization of the nucleosome. J Mol Biol 1989;206:451–63.
[60] Polach KJ, Lowary PT, Widom J. Effects of core histone tail domains on the equilibrium constants for dynamic DNA site accessibility in nucleosomes. J Mol Biol 2000;298(2):211–23.
[61] Yang Z, Zheng C, Thiriet C, Hayes JJ. The core histone N-terminal tail domains negatively regulate binding of transcription factor IIIA to a nucleosome containing a 5S RNA gene via a novel mechanism. Mol Cell Biol 2004;25:241–9.
[62] Andresen K, Jimenez-Useche I, Howell SC, Yuan C, Qiu X. Solution scattering and FRET studies on nucleosomes reveal DNA unwrapping effects of H3 and H4 tail removal. PLoS One 2013;8:e78587.
[63] Rhee HS, Bataille AR, Zhang L, Pugh BF. Subnucleosomal structures and nucleosome asymmetry across a genome. Cell 2014;159:1377–88.
[64] O'Brown ZK, Greer EL. N6-Methyladenine: a conserved and dynamic DNA mark. Adv Exp Med Biol 2016;945:213–46.
[65] Koziol MJ, et al. Identification of methylated deoxyadenosines in vertebrates reveals diversity in DNA modifications. Nat Struct Mol Biol 2016;23:24–30.
[66] Wu TP, et al. DNA methylation on N(6)-adenine in mammalian embryonic stem cells. Nature 2016;532: 329–33.
[67] Parashar NC, Parashar G, Nayyar H, Sandhir R. N(6)-adenine DNA methylation demystified in eukaryotic genome: from biology to pathology. Biochimie 2017. https://doi.org/10.1016/j.biochi.2017.10.014.
[68] Luo G-Z, He C. DNA N(6)-methyladenine in metazoans: functional epigenetic mark or bystander? Nat Struct Mol Biol 2017;24:503–6.
[69] Ehrlich M, et al. Amount and distribution of 5-methylcytosine in human DNA from different types of tissues of cells. Nucleic Acids Res 1982;10:2709–21.
[70] Jones PA. Functions of DNA methylation: islands, start sites, gene bodies and beyond. Nat Rev Genet 2012; 13:484–92.
[71] Patil V, Ward RL, Hesson LB. The evidence for functional non-CpG methylation in mammalian cells. Epigenetics 2014;9:823–8.
[72] Shukla S, et al. CTCF-promoted RNA polymerase II pausing links DNA methylation to splicing. Nature 2011;479:74–9.
[73] Ramsahoye BH, et al. Non-CpG methylation is prevalent in embryonic stem cells and may be mediated by DNA methyltransferase 3a. Proc Natl Acad Sci USA 2000;97:5237–42.
[74] Ziller MJ, et al. Genomic distribution and inter-sample variation of non-CpG methylation across human cell types. PLoS Genet 2011;7:e1002389.

[75] Smith ZD, et al. A unique regulatory phase of DNA methylation in the early mammalian embryo. Nature 2012;484:339–44.
[76] Lister R, et al. Global epigenomic reconfiguration during mammalian brain development. Science 2013; 341:1237905. 1237905.
[77] Guo JU, et al. Distribution, recognition and regulation of non-CpG methylation in the adult mammalian brain. Nat Neurosci 2014;17:215–22.
[78] Kinde B, Gabel HW, Gilbert CS, Griffith EC, Greenberg ME. Reading the unique DNA methylation landscape of the brain: non-CpG methylation, hydroxymethylation, and MeCP2. Proc Natl Acad Sci Unit States Am 2015;112:6800–6.
[79] Choy JS, et al. DNA methylation increases nucleosome compaction and rigidity. J Am Chem Soc 2010;132: 1782–3.
[80] Collings CK, Waddell PJ, Anderson JN. Effects of DNA methylation on nucleosome stability. Nucleic Acids Res 2013;41:2918–31.
[81] Tate PH, Bird AP. Effects of DNA methylation on DNA-binding proteins and gene expression. Curr Opin Genet Dev 1993;3(2):226–31.
[82] Wu JI, Lessard J, Crabtree GR. Understanding the words of chromatin regulation. Cell 2009;136:200–6.
[83] Margueron R, Reinberg D. The Polycomb complex PRC2 and its mark in life. Nature 2011;469:343–9.
[84] Boriack-Sjodin PA, Swinger KK. Protein methyltransferases: a distinct, diverse, and dynamic family of enzymes. Biochemistry 2016;55:1557–69.
[85] Rossetto D, Avvakumov N, Côté J. Histone phosphorylation: a chromatin modification involved in diverse nuclear events. Epigenetics 2012;7:1098–108.
[86] Marmorstein R, Zhou M-M. Writers and readers of histone acetylation: structure, mechanism, and inhibition. Cold Spring Harb Perspect Biol 2014;6:a018762. a018762.
[87] Fuhrmann J, Thompson PR. Protein arginine methylation and citrullination in epigenetic regulation. ACS Chem Biol 2016;11:654–68.
[88] DesJarlais R, Tummino PJ. The role of histone-modifying enzymes and their complexes in regulation of chromatin biology. Biochemistry 2016;55:1584–99.
[89] Sabari BR, Zhang D, Allis CD, Zhao Y. Metabolic regulation of gene expression through histone acylations. Nat Rev Mol Cell Biol 2017;18:90–101.
[90] Wu Z, Connolly J, Biggar KK. Beyond histones - the expanding roles of protein lysine methylation. FEBS J 2017;284:2732–44.
[91] Law JA, Jacobsen SE. Establishing, maintaining and modifying DNA methylation patterns in plants and animals. Nat Rev Genet 2010;11:204–20.
[92] Lyko F. The DNA methyltransferase family: a versatile toolkit for epigenetic regulation. Nat Rev Genet 2017. https://doi.org/10.1038/nrg.2017.80.
[93] Jeltsch A, Jurkowska RZ. Allosteric control of mammalian DNA methyltransferases - a new regulatory paradigm. Nucleic Acids Res 2016;44:8556–75.
[94] Bostick M, et al. UHRF1 plays a role in maintaining DNA methylation in mammalian cells. Science 2007; 317:1760–4.
[95] Sharif J, et al. The SRA protein Np95 mediates epigenetic inheritance by recruiting Dnmt1 to methylated DNA. Nature 2007;450:908–12.
[96] Chang B, Chen Y, Zhao Y, Bruick RK. JMJD6 is a histone arginine demethylase. Science 2007;318:444–7.
[97] Webby CJ, et al. Jmjd6 catalyses lysyl-hydroxylation of U2AF65, a protein associated with RNA splicing. Science 2009;325:90–3.
[98] Han G, et al. The hydroxylation activity of Jmjd6 is required for its homo-oligomerization. J Cell Biochem 2012;113:1663–70.

[99] Böttger A, Islam MS, Chowdhury R, Schofield CJ, Wolf A. The oxygenase Jmjd6–a case study in conflicting assignments. Biochem J 2015;468:191−202.
[100] Walport LJ, et al. Arginine demethylation is catalysed by a subset of JmjC histone lysine demethylases. Nat Commun 2016;7:11974.
[101] Azad GK, Tomar RS. Proteolytic clipping of histone tails: the emerging role of histone proteases in regulation of various biological processes. Mol Biol Rep 2014;41:2717−30.
[102] Wu X, Zhang Y. TET-mediated active DNA demethylation: mechanism, function and beyond. Nat Rev Genet 2017;18:517−34.
[103] Rasmussen KD, Helin K. Role of TET enzymes in DNA methylation, development, and cancer. Genes Dev 2016;30:733−50.
[104] Drohat AC, Coey CT. Role of base excision 'repair' enzymes in erasing epigenetic marks from DNA. Chem Rev 2016;116:12711−29.
[105] Kriaucionis S, Heintz N. The nuclear DNA base 5-hydroxymethylcytosine is present in Purkinje neurons and the brain. Science 2009;324:929−30.
[106] Globisch D, et al. Tissue distribution of 5-hydroxymethylcytosine and search for active demethylation intermediates. PLoS One 2010;5:e15367.
[107] Münzel M, et al. Quantification of the sixth DNA base hydroxymethylcytosine in the brain. Angew Chem, Int Ed Engl 2010;49:5375−7.
[108] Madrid A, Papale LA, Alisch RS. New hope: the emerging role of 5-hydroxymethylcytosine in mental health and disease. Epigenomics 2016;8:981−91.
[109] Mendonca A, Chang EH, Liu W, Yuan C. Hydroxymethylation of DNA influences nucleosomal conformation and stability in vitro. Biochim Biophys Acta 2014;1839:1323−9.
[110] Ngo TTM, et al. Effects of cytosine modifications on DNA flexibility and nucleosome mechanical stability. Nat Commun 2016;7:10813.
[111] Musselman CA, Lalonde M-E, Côté J, Kutateladze TG. Perceiving the epigenetic landscape through histone readers. Nat Struct Mol Biol 2012;19:1218−27.
[112] Patel DJ, Wang Z. Readout of epigenetic modifications. Annu Rev Biochem 2013;82:81−118. https://doi.org/10.1146/annurev-biochem-072711-165700.
[113] Bannister AJ, et al. Selective recognition of methylated lysine 9 on histone H3 by the HP1 chromo domain. Nature 2001;410:120−4.
[114] Charier G, et al. The Tudor tandem of 53BP1: a new structural motif involved in DNA and RG-rich peptide binding. Structure 2004;12:1551−62.
[115] Kim D, et al. Corecognition of DNA and a methylated histone tail by the MSL3 chromodomain. Nat Struct Mol Biol 2010;17:1027−9.
[116] Yap KL, et al. Molecular interplay of the noncoding RNA ANRIL and methylated histone H3 lysine 27 by polycomb CBX7 in transcriptional silencing of INK4a. Mol Cell 2010;38:662−74.
[117] Qiu Y, et al. Solution structure of the Pdp1 PWWP domain reveals its unique binding sites for methylated H4K20 and DNA. Biochem J 2012;442:527−38.
[118] van Nuland R, et al. Nucleosomal DNA binding drives the recognition of H3K36-methylated nucleosomes by the PSIP1-PWWP domain. Epigenet Chromatin 2013;6:12.
[119] Musselman CA, et al. Binding of PHF1 Tudor to H3K36me3 enhances nucleosome accessibility. Nat Commun 2013;4:2969.
[120] Eidahl JO, et al. Structural basis for high-affinity binding of LEDGF PWWP to mononucleosomes. Nucleic acids 2013;41(6):3924−36.
[121] Miller TCR, et al. A bromodomain-DNA interaction facilitates acetylation-dependent bivalent nucleosome recognition by the BET protein BRDT. Nat Commun 2016;7:13855.

[122] Filippakopoulos P. Multivalent histone and DNA engagement by a PHD/BRD/PWWP triple reader cassette recruits ZMYND8 to K14ac-rich chromatin. Cell Rep 2016;17(10):2724−37.

[123] Morrison EA, et al. DNA binding drives the association of BRG1/hBRM bromodomains with nucleosomes. Nat Commun 2017;8:16080.

[124] Li H, et al. Molecular basis for site-specific read-out of histone H3K4me3 by the BPTF PHD finger of NURF. Nature 2006;442:91−5.

[125] Wysocka J, et al. A PHD finger of NURF couples histone H3 lysine 4 trimethylation with chromatin remodelling. Nature 2006. https://doi.org/10.1038/nature04815.

[126] Lan F, et al. Recognition of unmethylated histone H3 lysine 4 links BHC80 to LSD1-mediated gene repression. Nature 2007;448:718−22.

[127] Margueron R, et al. Role of the polycomb protein EED in the propagation of repressive histone marks. Nature 2009;461:762−7.

[128] Xu C, et al. Binding of different histone marks differentially regulates the activity and specificity of polycomb repressive complex 2 (PRC2). Proc Natl Acad Sci U S A 2010;107(45):19266−71.

[129] Shimbo T, Wade PA. Proteins that read DNA methylation. Adv Exp Med Biol 2016;945:303−20.

[130] Sperlazza MJ, Bilinovich SM, Sinanan LM, Javier FR, Williams DC. Structural basis of MeCP2 distribution on non-CpG methylated and hydroxymethylated DNA. J Mol Biol 2017;429:1581−94.

[131] Amir RE, et al. Rett syndrome is caused by mutations in X-linked MECP2, encoding methyl-CpG-binding protein 2. Nat Genet 1999;23:185−8.

[132] Avvakumov GV, et al. Structural basis for recognition of hemi-methylated DNA by the SRA domain of human UHRF1. Nature 2008;455:822−5.

[133] Zhu H, Wang G, Qian J. Transcription factors as readers and effectors of DNA methylation. Nat Rev Genet 2016;17:551−65.

[134] Nishiyama A, et al. Uhrf1-dependent H3K23 ubiquitylation couples maintenance DNA methylation and replication. Nature 2013;502:249−53.

[135] Qin W, et al. DNA methylation requires a DNMT1 ubiquitin interacting motif (UIM) and histone ubiquitination. Cell Res 2015;25:911−29.

[136] Harrison JS, et al. Hemi-methylated DNA regulates DNA methylation inheritance through allosteric activation of H3 ubiquitylation by UHRF1. Elife 2016;5.

[137] Otani J, et al. Structural basis for recognition of H3K4 methylation status by the DNA methyltransferase 3A ATRX-DNMT3-DNMT3L domain. EMBO Rep 2009;10:1235−41.

[138] Guo X, et al. Structural insight into autoinhibition and histone H3-induced activation of DNMT3A. Nature 2014:1−20. https://doi.org/10.1038/nature13899.

[139] Zhang Y, et al. Chromatin methylation activity of Dnmt3a and Dnmt3a/3L is guided by interaction of the ADD domain with the histone H3 tail. Nucleic Acids Res 2010;38:4246−53.

[140] Binda O. On your histone mark, SET, methylate! Epigenetics 2013;8:457−63.

[141] Binda O, et al. Trimethylation of histone H3 lysine 4 impairs methylation of histone H3 lysine 9: regulation of lysine methyltransferases by physical interaction with their substrates. Epigenetics 2010;5: 767−75.

[142] Wu H, et al. Structural biology of human H3K9 methyltransferases. PLoS One 2010;5:e8570.

[143] Fischle W, et al. Regulation of HP1−chromatin binding by histone H3 methylation and phosphorylation. Nat Cell Biol 2005;438:1116−22.

[144] Rothbart SB, Strahl BD. Interpreting the language of histone and DNA modifications. Biochim Biophys Acta 2014;1839:627−43.

[145] Su Z, Denu JM. Reading the combinatorial histone language. ACS Chem Biol 2015;11:564−74.

[146] Fan J, Krautkramer KA, Feldman JL, Denu JM. Metabolic regulation of histone post-translational modifications. ACS Chem Biol 2015;10:95−108.

[147] Sharma U, Rando OJ. Metabolic inputs into the epigenome. Cell Metabol 2017;25:544–58.
[148] Su X, Wellen KE, Rabinowitz JD. Metabolic control of methylation and acetylation. Curr Opin Chem Biol 2016;30:52–60.
[149] Friso S, Udali S, De Santis D, Choi S-W. One-carbon metabolism and epigenetics. Mol Aspect Med 2017; 54:28–36.
[150] Shyh-Chang N, et al. Influence of threonine metabolism on S-adenosylmethionine and histone methylation. Science 2013;339:222–6.
[151] Shiraki N, et al. Methionine metabolism regulates maintenance and differentiation of human pluripotent stem cells. Cell Metabol 2014;19:780–94.
[152] Mentch SJ, et al. Histone methylation dynamics and gene regulation occur through the sensing of one-carbon metabolism. Cell Metabol 2015;22:861–73.
[153] Xie Z, et al. Metabolic regulation of gene expression by histone lysine β-hydroxybutyrylation. Mol Cell 2016;62:194–206.
[154] Xu W, Wang F, Yu Z, Xin F. Epigenetics and cellular metabolism. Genet Epigenet 2016;8. GEG.S32160.
[155] Ragunathan K, Jih G, Moazed D. Epigenetics. Epigenetic inheritance uncoupled from sequence-specific recruitment. Science 2015;348:1258699.
[156] Hathaway NA, et al. Dynamics and memory of heterochromatin in living cells. Cell 2012;149:1447–60.
[157] Xu M, et al. Partitioning of histone H3-H4 tetramers during DNA replication-dependent chromatin assembly. Science 2010;328:94–8.
[158] Campos EI, Stafford JM, Reinberg D. Epigenetic inheritance: histone bookmarks across generations. Trends Cell Biol 2014;24:664–74.
[159] Huang C, et al. H3.3-H4 tetramer splitting events feature cell-type specific enhancers. PLoS Genet 2013;9: e1003558.
[160] Villeneuve LM, et al. Epigenetic histone H3 lysine 9 methylation in metabolic memory and inflammatory phenotype of vascular smooth muscle cells in diabetes. Proc Natl Acad Sci USA 2008;105:9047–52.
[161] Vecellio M, et al. The histone acetylase activator pentadecylidenemalonate 1b rescues proliferation and differentiation in the human cardiac mesenchymal cells of type 2 diabetic patients. Diabetes 2014;63: 2132–47.
[162] Hackett JA, et al. Germline DNA demethylation dynamics and imprint erasure through 5-hydroxymethylcytosine. Science 2013;339:448–52.
[163] Heard E, Martienssen RA. Transgenerational epigenetic inheritance: myths and mechanisms. Cell 2014; 157:95–109.
[164] Hackett JA, Surani MA. DNA methylation dynamics during the mammalian life cycle. Philos Trans R Soc Lond, B, Biol Sci 2013;368. 20110328.
[165] Seisenberger S, et al. The dynamics of genome-wide DNA methylation reprogramming in mouse primordial germ cells. Mol Cell 2012;48:849–62.
[166] Hammoud SS, et al. Distinctive chromatin in human sperm packages genes for embryo development. Nature 2009;460:473–8.
[167] Lim JP, Brunet A. Bridging the transgenerational gap with epigenetic memory. Trends Genet 2013;29: 176–86.
[168] Coleman RT, Struhl G. Causal role for inheritance of H3K27me3 in maintaining the OFF state of a Drosophila HOX gene. Science 2017. https://doi.org/10.1126/science.aai8236.
[169] Zenk F, et al. Germ line-inherited H3K27me3 restricts enhancer function during maternal-to-zygotic transition. Science 2017;357:212–6.
[170] Carone BR, et al. Paternally induced transgenerational environmental reprogramming of metabolic gene expression in mammals. Cell 2010;143:1084–96.

[171] Franklin TB, Mansuy IM. Epigenetic inheritance in mammals: evidence for the impact of adverse environmental effects. Neurobiol Dis 2010;39:61–5.
[172] Arrowsmith CH, Bountra C, Fish PV, Lee K, Schapira M. Epigenetic protein families: a new frontier for drug discovery. Nat Rev Drug Discov 2012;11:384–400.
[173] Dawson MA, Kouzarides T. Cancer epigenetics: from mechanism to therapy. Cell 2012;150:12–27.

SECTION

1

Neurodegenerative disorders

CHAPTER

Into the unknown: chromatin signaling in spinal muscular atrophy

2

Marc-Olivier Deguise[1,2,3], Rashmi Kothary[1,2,3,4]

Regenerative Medicine Program, Ottawa Hospital Research Institute, Ottawa, ON, Canada[1]; Department of Cellular and Molecular Medicine, University of Ottawa, Ottawa, ON, Canada[2]; Centre for Neuromuscular Disease, University of Ottawa, Ottawa, ON, Canada[3]; Department of Medicine, and Department of Biochemistry, Microbiology, and Immunology, University of Ottawa, Ottawa, ON, Canada[4]

Chapter outline

2.1 Spinal muscular atrophy: prevalence, genetic basis, clinical features, and pathogenesis 27
2.2 The survival motor neuron protein: localization, structure, and function 30
2.3 Epigenetic landscape in spinal muscular atrophy pathogenesis 31
2.4 Targeting epigenetic factors as potential therapeutics in spinal muscular atrophy 32
2.4.1 Histone deacetylase inhibitors as regulators of the survival motor neuron gene 32
2.4.2 The nonspecific effect of histone deacetylase inhibitors 33
2.4.3 The potential protective effect of histone deacetylase inhibitors in the pathogenesis of spinal muscular atrophy 41
2.5 Conclusion 42
Acronyms and abbreviations 43
Acknowledgments 44
References 44

2.1 Spinal muscular atrophy: prevalence, genetic basis, clinical features, and pathogenesis

Spinal muscular atrophy (SMA) is a devastating neurological disease that can affect individuals of any age, but most commonly diagnosed in babies and young infants. In the latter, it is considered as the most common genetic disorder leading to death after cystic fibrosis. In fact, 1 in 11,000 live births will be diagnosed with SMA, the consequent result of a homozygous mutation or deletion in the *Survival Motor Neuron 1* (*SMN1*) gene [1,2]. The carrier frequency for deleterious *SMN* mutations is relatively high, albeit variable between ethnicities, ranging from 1 in 40 to 1 in 125 [2,3]. *SMN1* is found on chromosome 5q13, a region deemed unstable [4]. Over the course of evolution, two distinct genetic events had particular importance for the disease. First, a duplication event in the chimpanzee led to a 500-kb inverted repeat element containing a second copy of the *SMN* gene, termed *SMN2* [5]. Second, a

Chromatin Signaling and Neurological Disorders. https://doi.org/10.1016/B978-0-12-813796-3.00002-X

single nucleotide (c.C680T) substitution arose in the duplicated *SMN2* gene, a characteristic unique to *Homo sapiens* [5]. Although this nucleotide substitution is translationally silent, it is not without consequences. Indeed the substitution is found in a splicing enhancer region of exon 7, which favors exon 7 exclusion in 90% of the transcripts from the *SMN2* gene [6]. Consequently, the protein product from this gene is truncated, unstable, and rapidly degraded, leading to the production of low amounts of functional SMN protein [6,7]. Therefore, SMA is a disease of low levels of SMN rather than complete absence, due to the loss of *SMN1* but presence of the intact *SMN2* gene. In fact, complete depletion of SMN is incompatible with life [8]. In the same manner, the *SMN2* copy number modulates disease severity, where a high copy number leads to more SMN protein and consequently a milder disease outcome [9,10]. Therefore, *SMN2* acts as natural disease modifier in the context of SMA.

Traditionally, SMA is characterized as a motor neuron disorder because of the preferential susceptibility of alpha motor neurons, which ultimately results in paralysis and muscle weakness to a varying degree. This weakness is more prominent in the proximal muscle, more so in the legs than in the arms. The clinical spectrum of SMA disease is mainly divided into four types [4]. Patients with SMA type I (also known as Werdnig—Hoffman disease) have generally two *SMN2* copies [9] and are diagnosed before 6 months of age. They show the most severe phenotype, with babies not reaching most motor milestones, including sitting [4]. Without supportive therapy, or the new FDA-approved drug Spinraza, the patients usually succumb to disease before their second year of life because of respiratory distress. However, supportive therapy offers significant lifespan extension in patients with severe SMA [11]. Although patients with SMA type I represent about 50% of the diagnosed cases, they make up for a small proportion of the SMA population (12%), given the fatality of their symptoms [4]. Patients with SMA type II (also known as Dubowitz disease) have three copies of *SMN2* and are diagnosed between 6 and 18 months of age [9]. These children usually gain the ability to sit independently but most will not walk [4]. Unlike patients with SMA type I, they are more prone to develop scoliosis, which may result in respiratory compromise. Patients afflicted with SMA type II are the most prevalent (52%), given their less severe clinical presentation [4]. Patients with SMA type III (also known as Kugelberg—Welander disease) have three to four copies of *SMN2* and are diagnosed at the age of 18 months or older [4,9]. They account for 36% of patients with SMA [4,9]. These infants will be able to walk independently but may lose this ability later in life [4]. Patients with type IV SMA have the least severe symptoms. The disease manifests after 21 years of age and progresses very slowly [4]. These patients will have a normal life expectancy because of the presence of more than four copies of *SMN2* but will eventually develop muscle weakness [4].

Even though SMN is ubiquitously expressed and has unequivocal housekeeping functions (discussed more extensively later), SMA has long been considered a motor neuron disorder. However, SMA is more recently being accepted as a multiorgan disorder [12,13] (Fig. 2.1). Indeed, initial studies in SMA research debated the contribution of neurons and muscles as the initiating event in disease pathogenesis. While interesting findings highlight potential defects in cells on both sides of the neuromuscular junction (NMJ), conditional repletion of Smn in either cell type highlighted the particular susceptibility of motor neurons to low Smn levels and the necessity for therapeutic intervention [14]. For this reason, most of the SMA research has focused on the preferential susceptibility of motor neurons and potential therapies that could abrogate their loss. The SMA research community has yet to identify the molecular pathways at the core of this phenomenon. Despite all the attention given to the motor neuron, muscle cell defects are present [15—18] and multiple groups thereafter nicely highlighted intrinsic impairment, more particularly at the level of myogenesis [13,19—22].

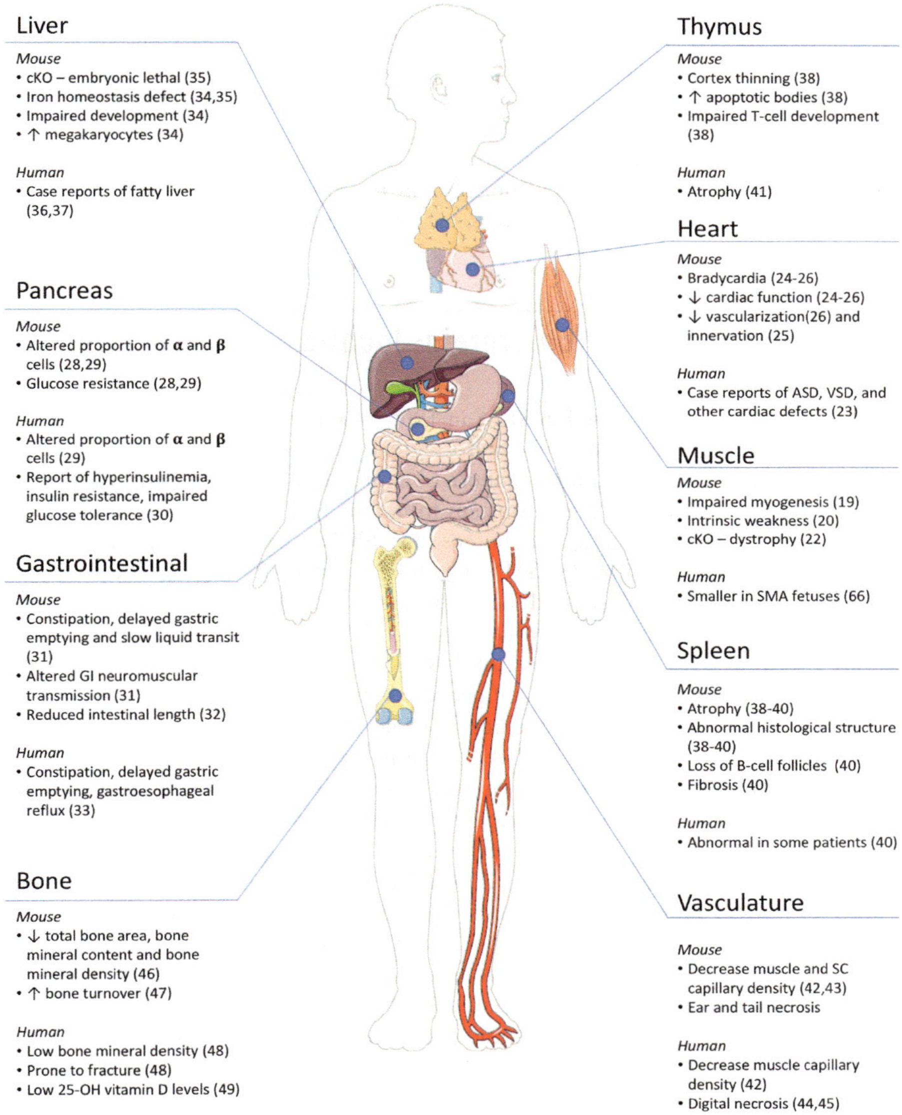

FIGURE 2.1

Contributions of non-neuronal organs in spinal muscular atrophy disease. This schematic highlights the major findings to date of non-neuronal organs in both human patients and mouse models. This figure is by no means inclusive of all the findings in the different systems. *ASD*, atrial septal defect; *cKO*, conditional knockout; *GI*, gastrointestinal; *SC*, spinal cord; *VSD*, ventricular septal defect.

Heart problems first surfaced in multiple case studies of patients with SMA [23] and were later more thoroughly studied in mouse models [24–27]. Major findings included bradycardia, reduced cardiac function, and reduced vascularization and innervation [24–27]. Many other studies highlight defects in other organ systems such as the pancreas [28–30], gastrointestinal system [31–33], liver [34–37], spleen [38–40], thymus [38,41], vasculature [42–45], and bones [46–49]. In favor of the multiorgan involvement in SMA, it was repeatedly shown that systemic delivery of therapeutic compounds results in better outcome than neuronal-restricted delivery in preclinical models [50,51]. Altogether, this highlights the complex contribution of multiple organ systems in SMA and their importance for complete recovery.

2.2 The survival motor neuron protein: localization, structure, and function

Since the discovery of the gene mutated in SMA in 1995, research may not have yielded a conclusive mechanism for the disease, but it has informed us greatly about SMN localization, structure, and function. The SMN protein has a ubiquitous pattern of expression and is temporally regulated [52,53]. Within the cell, SMN shuttles from the cytoplasm to the nucleus, and vice versa, and aggregates in nuclear structures called gems [54] and Cajal bodies [55]. SMN is naturally found as part of a complex called the SMN complex, which includes SMN, Gemins2–8, and Unrip [56]. The first, and the most widely, studied function of SMN described the protein as essential for the assembly of small nuclear ribonucleoproteins (snRNPs), which are involved in pre-mRNA splicing [56]. Ever since, studies have implicated SMN in a growing array of mechanisms mediating RNA metabolism that extends from telomere synthesis to mRNA transport [57]. Of particular interest to chromatin, SMN is also involved in histone mRNA 3′ processing [58]. Canonical histone (H2A, H2B, H3, H4, and H1) genes are replication dependent [59]. Unlike other mRNA, their transcripts do not contain any introns and do not harbor a poly-A tail but instead a stem loop on their 3′ end [59]. U7 snRNPs are responsible for the cleavage of the 3′ end of the histone transcript. As with other snRNPs, a modified SMN complex, which includes Lsm10 and Lsm11 instead of SmD1 and SmD2 [60,61], assist in the U7 snRNP assembly [58]. SMN depletion leads to loss of U7 snRNP formation and consequently impairs histone mRNA 3′ processing [58]. This has a tremendous effect on histone expression and protein levels [58]. It is possible that a minimal change in the pool of histones may lead to a significant change in the proportion of canonical histone to noncanonical histone variants, a type of histone that is not replication dependent and processed as regular mRNA transcripts [59]. As some histone variants regulate transcription [62], they could have fundamental effects on the expression of the rest of the genome. Furthermore, low levels of U7 snRNPs or abnormal levels of replication-dependent histones required to pack DNA following replication may have major consequences in cellular proliferation. Indeed, reducing the levels of U7 snRNP proteins or U7 snRNA is enough to slow the progression during cell cycle [63,64], while low proliferative rates have been reported in SMN-depleted cells [65,66]. Interestingly, SMN is required during early development, a time marked by increased proliferation. Moreover, this mechanism could also explain pathologic findings in fetuses with SMA [67]. Given its ubiquitous trait, it also fits the current view that SMA is a multiorgan disorder (Fig. 2.1). However, even though this histone dysregulation defect is appealing as a potential explanation for the pathogenesis of SMA, the leading pathogenic mechanism involving mRNA mis-splicing is also consistent with the observations made earlier. It is therefore possible that both histone dysregulation and mRNA mis-splicing work in tandem to cause SMA disease.

The SMN protein is relatively small and is composed of four main domains: a Gemin2-binding domain, a Tudor domain, a proline-rich domain, and a YG domain [56]. Many studies have established that the primary function of these domains mediates the interaction between SMN and its numerous binding partners [56]. In the context of chromatin signaling, the Tudor domain of SMN may hold significant importance. Indeed, the Tudor domains of multiple proteins act as histone mark readers, more specifically in the recognition of methyl-lysine and methyl-arginine [68]. While it is well recognized that the Tudor domain of SMN binds to methylated Sm proteins [69,70], very little is known about its potential ability to bind to methylated histones. Interestingly, the preferential binding of the SMN protein to dimethylated arginine is limited not only to Sm proteins but also to other proteins that have this posttranslational modification [69]. A report highlighted the capability of SMN to bind mono- and dimethylated histone H3 at lysine 79 (H3K79me1/2) in response to interphase centromere damaged by the viral E3 ubiquitin ligase protein ICP0 [71]. This was mediated by the Tudor domain of SMN and the knockdown of disruptor of telomeric silencing 1-like (DOT1L), a methyltransferase targeting this site, abrogated appropriate SMN localization [71]. It is important to note that the methylated H3K79 epigenetic mark has been associated with a wide array of processes including increased transcription, cell cycle regulation, and DNA repair [72]. Intriguingly, SMN depletion leads to defects in all these pathways [65,66,73]. Of particular interest, chromatin regulator and chromatin methylation pathways were identified in a screen aimed at identifying transcriptional differences in SMA vulnerable motor neurons [73]. The Tudor domain of SMN was also found to bind to the R1810 dimethylated arginine residue of the RNA polymerase II (RNAP II) subunit POLR2A, which might impart a new function [74]. Indeed, SMN associates with β-actin and glyceraldehyde-3-phosphate dehydrogenase (GAPDH) genes by recruiting POLR2A to the RNAP II elongation complexes [74]. Of note, actin dynamic defects and reduced GAPDH expression levels have been reported at multiple occasions [75–79]. In combination with POLR2A and senataxin, SMN appears to play a role in the termination and release of RNAP II from the DNA by resolving R-loop formation [74]. Knockdown of SMN leads to RNAP II accumulation in termination regions across the genome, a feature that was also identified in SMA patient fibroblasts and B cells [74]. The authors further elegantly showed that unresolved R-loops may contribute to genome instability by increasing γH2AX levels at termination regions [74]. However, another report found that γH2AX levels were reduced in SMA motor neurons [73]. Nonetheless, such novel and significant interactions will require some attention. Indeed, it places SMN as having a role in other essential cellular functions and as a potential mediator of yet additional levels of gene expression regulation. Research in this field could yield significant results in deciphering potential pathogenic events that have been overlooked so far.

2.3 Epigenetic landscape in spinal muscular atrophy pathogenesis

As a monogenic disorder of known origin, the SMA research field has focused on genetic characterization of the various regulatory elements present in the *SMN1 and SMN2* genes in order to enhance the full-length production of SMN by *SMN2* and increase SMN protein levels in patients with SMA [80]. Despite our increasing understanding of the genetic regulation of *SMN1 and SMN2*, very little is known about the changes in epigenetic landscape during SMA disease progression. Most of what we know about epigenetic signaling in SMA comes from studies trying to identify the mechanism underlying the effect of histone deacetylase inhibitor (HDACi) on *SMN2* gene expression (discussed in more detail later). For example, Kernochan et al. [81] identified that histones H3 and H4 were highly acetylated in two regions found on either side of the *SMN2* transcription start sites (HuSP3 and HuSP4) and hypoacetylated in two regions upstream (HuSP1 and HuSP2). HDACi suberoylanilide hydroxamine (SAHA) and valproic acid (VPA) increased acetylation of HuSP1 and HuSP2 and were consequently believed to enhance the promoter activity of the *SMN2* gene [81]. The authors suggest that deacetylation is likely caused by histone deacetylases (HDACs) 1 and 2 and this may be responsible for the decrease in SMN protein levels observed during development [81]. This ultimately infers that HDAC1 and HDAC2 are prime therapeutic targets. Nonetheless, VPA or SAHA did not appear to be working on the same sites as HDAC1 and 2, leaving the possibility that other mechanisms of regulation at the *SMN2* locus may be involved [81]. Another study identified that the *SMN2* gene has four CpG islands [82]. Interestingly, these methylation marks reduced *SMN2* expression, a process mediated by methyl-CpG-binding-protein 2 (MeCP2) [82]. Of note, some HDACi could overcome this inhibitory signal without affecting the methylation status, potentially because MeCP2 requires HDAC activity for its proper function as a repressor [82,83]. Interestingly, methylation marks localized near the transcription start site of *SMN2* correlated with the severity of disease in patients, implying that not all *SMN2* genes are equivalent [82]. This partially explains why patients with SMA of the same genotype (for example, two copies of *SMN2*) can present with type I, II, or III disease features [9,82]. The first long noncoding RNA (lncRNA) involved in SMN regulation was recently described [84]. This lncRNA, termed SMN-AS1, is located in the antisense strand of intron 1 of the *SMN* gene and is preferentially expressed in neurons [84]. More specifically, SMN-AS1 recruits the polycomb repressive 2 complex to the *SMN* promoter, which leads to a trimethylation event at lysine 27 of the histone 3 (H3K27me3) [84]. Ultimately, this results in inhibition of transcription at the *SMN* promoter [84]. It was suggested that this mechanism may also be responsible for the developmental regulation of SMN [84]. Strikingly, the inhibitory effect of SMN-AS1 could be reversed upon administration of an antisense oligonucleotide (ASO) against SMN-AS1 [84]. Furthermore, combinatorial therapy using this ASO and the new FDA-approved therapeutic drug Spinraza showed promising results in a mouse model of SMA [84].

While very little evidence is present, it is likely that these initial observations are simply the beginning of a much more complex chromatin signaling network that could contribute to SMN regulation and/or SMA pathogenesis. Chromatin immunoprecipitation-sequencing studies will be an important tool in identifying whether the SMA epigenetic landscape is perturbed at different stages of the disease and in different organs. Such studies could provide us with a new angle in our understanding of the etiology of SMA.

2.4 Targeting epigenetic factors as potential therapeutics in spinal muscular atrophy

2.4.1 Histone deacetylase inhibitors as regulators of the survival motor neuron gene

The primary objective in the search for treatment of SMA is to restore the SMN imbalance in patients. The *SMN2* gene provides an excellent target for therapeutic agents. Multiple mechanisms could be used to increase full-length SMN production from *SMN2*. For example, a compound that could increase transcription and/or modify the aberrant splicing of exon 7 of *SMN2* could lead to significant increase in SMN levels and therefore would be of clinical benefit for patients. For this reason, HDAC inhibitors were solicited as potential therapeutic candidates in treating SMA because of their ability to relax the chromatin, thereby leading to an increase in the overall transcription of *SMN2*. Indeed the potential benefit of HDAC inhibitors in modulating SMA was highlighted for the first time in SMA-like animals that were treated with sodium butyrate (SB) [85] (see Table 2.1). Interest in these compounds peaked very quickly and multiple studies reported the potential of HDAC inhibitors in modulating SMN protein expression (all preclinical studies [21,25,82,85–112] using HDAC inhibitors in the context of SMA have been summarized in Tables 2.1–2.4). Valproate (valproic acid, VPA) was by far the most extensively studied in the context of SMA (Table 2.2). Evaluation of individual class I and II HDAC proteins showed that only HDAC5 inhibition did not increase the production of SMN full-length protein [106]. More specifically, inhibition of HDAC1, 2, 3, and 8 appeared to preferentially increase transcriptional activity, while class II HDAC 5 and 6 inhibition might have the potential to modulate exon 7 inclusion [106]. It was reasoned that an increase in SMN production would subsequently lead to improvements in SMA disease phenotype in mouse models, and indeed this was the case. Most studies on SMA mouse models were performed using HDAC inhibitors classified as hydroxamic acids (Table 2.3). Unfortunately, clinical trials with VPA and sodium phenylbutyrate (4-phenylbutyric acid [4-PBA]) yielded mixed results [113–123] (see Table 2.5). In fact, it appears that only one-third of patients were "responders" [110,115]. It was suggested that CD36, a fatty acid translocase, might be responsible for the "nonresponsiveness" in some patients with SMA who were treated with short-chain fatty acid HDAC inhibitors [90]. Interestingly, HDACi LBH589, a hydroxamic acid, was able to increase SMN production in "nonresponder" fibroblasts [90], leaving the possibility that failure in trials may not have been for HDAC inhibitors in general, but rather uniquely for short-chain fatty acid HDAC inhibitors. However, it is important to note that 4-PBA, an aromatic acid HDACi, leads to modest improvements in patients with SMA [113,114]. Screening for new HDAC inhibitors and their potential in treating SMA continues [124]. Of note, some natural compounds commonly found in food may lead to increased SMN expression through potential HDAC inhibition [125–129] (see Table 2.6). Nonetheless, ever since the results of the clinical trials, the excitement for HDAC inhibitors progressively faded in the shadow of new and more effective SMN-inducing therapies such as ASOs and gene therapy.

2.4.2 The nonspecific effect of histone deacetylase inhibitors

As the HDACi mechanism of action is not specific to *SMN2*, but rather relaxes the chromatin throughout the genome, unintended effects are likely to occur (Fig. 2.2). This is particularly relevant in the context of SMA. Since SMN is a splicing modulator, there are likely numerous aberrant splice products in many cell types as a result of SMN depletion in SMA. As such, the widespread effect of HDAC inhibitors on genome regulation may actually help alleviate some of the defects caused by the mis-splicing. In addition, HDACs can deacetylate many other proteins that are not histones, including p53, GATA transcription factors, MyoD, α-tubulin, actin, SOD1/2, and profilin 1, thereby modulating their function [130]. Interestingly, many of these proteins have been implicated in SMA (p53 [73], MyoD [19], actin [77], profilin 1 [131,132]). Not surprisingly, it appears that HDAC inhibitors may also mediate a therapeutic benefit through these off-target effects. Using the $Smn^{2B/-}$ mouse model, which does not harbor the *SMN2* gene, we were able to demonstrate that trichostatin A (TSA)-treated

Table 2.1 Overview of preclinical studies using short-chain fatty acid–based histone deacetylase inhibitors.

Compound	Experimental setting	Observations	FL-SMN transcript	SMN protein	References
Sodium butyrate	Lymphoid cells Mouse (SMA-like mice)	Modest ↑ in SMN levels in cells and in mice, apparent reduction in severity of SMA-like mice from treated pregnant mother	↑	↑ 1.5- to 2-fold	[85]
	NSC-34 cells	Two fold ↑ in reporter gene assay of SMN2	N.A.	N.A.	[86]
	Mouse (*SMNΔ7*)	No or very modest change in survival, weight, motor function, or MN count	N.A.	N.A.	[87]
Sodium phenylbutyrate (4-PBA)	Fibroblasts of patients with SMA	↑ SMN transcription and ↑ gems	↑ 1.8– to 5–fold	↑ 1.5- to 1.8-fold	[88]
	Lymphoblast cells from SMA type III	N.C. in SMN levels	N.C.	N.C.	[89]
	Fibroblasts of patients with SMA	N.C. in FL-SMN2 transcripts	N.C.	N.A.	[82]
	Fibroblasts of patients with SMA	↑ SMN in fibroblasts of "nonresponders"	N.A.	↑ 1.2- to 1.4-fold	[90]
	Mouse (*SMNΔ7*)	Modest change in survival, weight and motor function, ↑ MN count and pAkt and pGSK3β protein levels	N.C.	N.C.	[87]
Glyceryl tributyrate (BA3G)	Mouse (*SMNΔ7*)	Modest change in survival, N.C. in weight	N.A.	N.A.	[87]
VX563	Mouse (*SMNΔ7*)	↑ Survival, weight, and motor function, MN count, and pAkt and pGSK3β protein levels. No improvement if drug was given after phenotype manifestation.	N.C.	N.C.	[87]

↑, *Increase;* 4-PBA, *4-phenylbutyric acid;* FL, *full length;* MN, *motor neuron;* N.A., *not assessed;* N.C., *no change;* SMA, *spinal muscular atrophy;* SMN, *survival motor neuron.*

Table 2.2 Overview of preclinical studies using the short-chain fatty acid HDACi VPA.

Compound	Experimental setting	Observations	FL-SMN transcript	SMN protein	References
VPA	Fibroblasts of patients with SMA Rat OHSC	Possibly ↑ SMN transcription by SP1 and AP1 and correct splicing by ↑ level of htra2-β1	↑ 1.8- to 5.2-fold	↑ 2- to 4-fold	[91]
	Fibroblasts of patients with SMA	↑ SMN transcription and exon 7 inclusion, ↑ gem number	↑ 1.75-fold	↑ 2- to 3-fold	[92]
	F98 rat glioma cells Rat OHSC MN/SC coculture Human OHSC	↑ SMN transcription	↑ 1.4- to 1.8-fold	↑ 1.5- to 3-fold	[93]
	Mouse (SMA-like type III)	↑ Motor function, electrophysiology, MN count	N.A.	↑ 2-fold	[94]
	PC12 cells	↑ Neurite outgrowth without ↑ in SMN protein levels	N.A.	↑ 1- to 1.5-fold[a]	[95]
	Mouse (SMA-like type III)	↑ Motor function, electrophysiology, MN count, muscle pathology, NMJ pathology, ↑ antiapoptotic factors, and possible astrocytic proliferation in SC	↑ 1.5-fold	↑ 1.3-fold	[96]
	Fibroblasts of patients with SMA Human OHSC	N.C. in SMN2 FL transcript	↑ 1- to 2-fold[a]	N.A.	[82]
	NSC Spinal motor neuron (*Smn*$^{-/-}$;*SMN2*)	↑ SMN protein levels, worsening of axon growth, reduced growth cone size, and reduced excitability of MN	N.A.	↑ 1.2- to 2-fold	[97]
	Fibroblasts of patients with SMA	↑ SMN transcription	↑	↑ 1-1.5- fold[a]	[98]
	Human (types I and II) Fibroblasts of patients with SMA iPSC derived GABAergic neurons	One-third patients were responsive to VPA. Responsiveness was mediated by CD36 and may apply to all short-chain fatty acid HDACis.	↑ 1- to 3.98- fold[a]	N.A.	[90]

↑, *Increase,* FL, *full length;* HDACi, *histone deacetylase inhibitor;* iPSC, *induced pluripotent stem cell;* MN, *motor neuron;* MN/SC, *motor neuron/Schwann cell;* N.A., *not assessed;* N.C., *no change;* NMJ, *neuromuscular junction;* NSCs, *neuronal precursor cells;* OHSC, *organotypic hippocampal slice culture;* SC, *spinal cord;* SMA, *spinal muscular atrophy;* SMN, *survival motor neuron* VPA, *valproic acid.*

[a] *One fold denotes no change.*

Table 2.3 Overview of pre-clinical studies using hydroxamic acid–based histone deacetylase inhibitors.

Compound	Experimental setting	Observations	FL-SMN transcript	SMN protein	References
TSA	NSC-34 cells	Twofold ↑ in SMN2 reporter gene assay	N.A.	N.A.	[86]
	Mouse (*SMNΔ7*) treatment TSA + nutrition	↑ Survival, weight, and motor function and improved MN size but not numbers, myofiber size, and snRNP assembly	↑ 1.5- to 5-fold	↑ 1.5- to 3-fold	[99]
	Mouse (*SMNΔ7*)	↑ Survival, weight, and motor function and improved NMJ	N.A.	↑ 1.2- to 1.5-fold	[100]
	Mouse (*SMNΔ7*)	↑ Survival, weight, and motor function and partially restored heart size and cardiac defects	N.A.	↑ 1.2- to 2.7-fold	[25]
	Mouse (*SMNΔ7*) TSA + bortezomib	↑ Survival, weight, motor function, MN number, NMJ pathology	N.A.	↑ 2- to 6-fold	[101]
	Mouse (*SMNΔ7*)	improved NMJ pathology	N.A.	N.A.	[102]
	Mouse (*Smn2B/−*)	↑ Survival, motor function, MN number, NMJ pathology despite no increase in Smn levels	N.C.	N.C.	[103]
	Mouse (*SMNΔ7*)	Reverse MyoG-dependent muscle atrophy	N.A.	N.A.	[21]
	Mouse (*Smn2B/−*)	Reversal of proteasomal and autophagosomal protein degradation mediated by forkhead box O	N.C.	N.C.	[104]
SAHA (vorinostat)	Mouse (*Smn−/−*; *SMN2* & Taiwanese)	Modest ↑ in survival, weight, motor function, NMJ and muscle pathology	↑ 1- to 2.5-fold[a]	↑ 2.7- to 17.6-fold	[105]
	F98 rat glioma cells Rat OHSC MN/SC coculture Fibroblasts of patients with SMA Human OHSC	↑ SMN transcription	↑ 1.9- to 2.55-fold	↑ 1- to 3-fold[a]	[93]
	Fibroblasts of patients with SMA Human OHSC	↑ SMN transcription	↑ 2- to 5-fold	N.A.	[82]
	HEK293 cells	↑ SMN protein levels, possible splicing modulation	↑ 100-fold	↑ 10-fold	[106]

SAHA (vorinostat)	Mouse (Taiwanese)	improvement in weight and motor function and restoration of vascular density in muscles	N.A.	↑ 2-fold	[107]
	Mouse (Taiwanese)	Amelioration of muscle molecular defects (VDAC2, parvalbumin, and H2AX)	N.A.	↑ 2-fold	[108]
	Fibroblasts of patients with SMA	↑ SMN transcription and exon 7 inclusion, ↓ methylation at the *SMN2* promoter	↑ 2- to 6-fold	↑ 2-fold	[109]
LBH589 (panobinostat)	Fibroblasts of patients with SMA Human NSC Mouse MEFs	↑ SMN transcription, exon 7 inclusion, gem number, ↓ SMN ubiquitinylation	↑ 2- to 3-fold	↑	[110]
	Fibroblasts of patients with SMA	LBH589 ↑ SMN in fibroblasts of "nonresponders"	N.A.	↑ 2.6- to 5.9-fold	[90]
JNJ-26481585 (quisinostat)	Mouse (Taiwanese)	Slight improvement in weight, motor function, and NMJ pathology without change in survival, vascular density in muscle, and pathology of other organs	↑ 1.2- to 2.25-fold	↑ 2- to 3-fold	[111]
LAQ824 (dacinostat)	Fibroblasts of patients with SMA	↑ SMN transcription and exon 7 inclusion, ↓ methylation at the *SMN2* promoter	↑ 3- to 4-fold	↑ 1- to 2.5-fold[a]	[109]

↑, *increase;* ↓, *decrease;* FL, *full length;* iPSC, *induced pluripotent stem cell;* MEFs, *murine embryonic fibroblasts;* MN, *motor neuron;* MN/SC, *motor neuron/Schwann cell;* MyoG, *myogenin;* N.A., *not assessed;* N.C., *no change;* NMJ, *neuromuscular junction;* NSCs, *neuronal precursor cells;* OHSC, *organotypic hippocampal slice culture;* SAHA, *suberoylanilide hydroxamine;* SC, *spinal cord;* SMA, *spinal muscular atrophy;* SMN, *survival motor neuron;* snRNP, *small nuclear ribonucleoprotein;* TSA, *trichostatin A.*

[a] *One fold denotes no change.*

Table 2.4 Overview of preclinical studies using benzamide-based histone deacetylase inhibitor.

Compound	Experimental setting	Observations	FL-SMN transcript	SMN protein	References
FK-228 (romidepsin)	Fibroblasts of patients with SMA	↑ SMN transcription	↑ 3.5- to 4-fold	N.A.	[82]
M344	Fibroblasts of patients with SMA	↑ SMN transcription, exon 7 inclusion, ↑ gems	↑ 1.6- to 2-fold	↑ 3- to 7-fold	[112]
	F98 rat glioma cells Rat OHSC MN/SC coculture Fibroblasts of patients with SMA	↑ SMN transcription	↑ 1.5- to 2.3-fold	↑ 1- to 2.7-fold[a]	[93]
	Fibroblasts of patients with SMA Human OHSC	↑ SMN transcription	↑ 2- to 5.5-fold	N.A.	[82]
	HEK293 cells	↑ SMN protein levels, possible splicing modulation	↑ 30-fold	↑ 10-fold	[106]
MS-275 (entinostat)	F98 rat glioma cells Rat OHSC Fibroblasts of patients with SMA	N.C. in SMN transcription	N.C.	N.C.	[93]
	Fibroblasts of patients with SMA	N.C. in SMN transcription	N.C.	N.A.	[82]
	HEK293 cells	↑ SMN protein levels by transcription	↑ 10-fold	↑ 4-fold	[106]

↑, *increase;* FL, *full length;* iPSC, *induced pluripotent stem cell;* MEFs, *murine embryonic fibroblasts;* MN, *motor neuron;* MN/SC, *motor neuron/Schwann cell;* N.A., *not assessed;* N.C., *no change;* NSCs, *neuronal precursor cells;* OHSC, *organotypic hippocampal slice culture;* SC, *spinal cord;* SMA, *spinal muscular atrophy;* SMN, *survival motor neuron.*
[a]*One fold denotes no change.*

SMA mice showed extended survival, increase in weight and myofiber size, and more mature NMJs [103]. It was noted that motor neurons were also modestly protected in treated animals [103]. Interestingly, these benefits were observed despite no increase in SMN protein levels [103]. A subsequent report showed that TSA may mitigate the myogenic defects observed in SMA by modulating myogenic regulatory factors in muscles [19]. In a similar manner, TSA-treated *Smn*$^{2B/-}$ mice showed complete reversal of forkhead box O (FoxO)-dependent atrophy pathways otherwise observed in non-treated mice [104]. It was proposed that TSA could change FoxO acetylation status, which can restrict its function in the nucleus [104]. However, it is possible that amelioration of the NMJ phenotype and a

Table 2.5 Overview of histone deacetylase inhibitor clinical trials.

Compound	Experimental setting	Observations	FL-SMN transcript	SMN protein	References
Sodium phenylbutyrate (4-PBA)	Human (type II)	Modest ↑ in motor function	N.A.	N.A.	[114]
	Human (types II and III)	↑ SMN transcription and slight improvement in motor function	↑ 1- to 2.4-fold[a]	N.A.	[113]
Valproic acid (VPA)	Human (types I and III)	↑ SMN transcription (↑ in 7 of 20 patients with SMA)	↑ 1- to 3.4-fold in carriers[a]	↑ 1- to 13.7-fold in carriers[a]	[115]
	Human (types II and III)	Some motor improvement in some patients	N.A.	N.A.	[116]
	Human (types I and III)	Some toxicity, better motor function in type II, ↑ bone density	N.C.	N.A.	[119]
	Human (types II and III)	↑ SMN protein levels in some patients	N.A.	↑ 1.2- to 2.9-fold	[117]
	Human (types II and III)	N.C. in motor function, electrophysiologic studies, bone density, and QOL	N.C.	N.A.	[118]
	Human (type III)	No or very modest change in motor function, electrophysiologic studies, and QOL	N.C.	N.A.	[120]
	Human (types II and III)	Some motor benefits in type II	N.A.	N.A.	[121]
	Human (adult patients with SMA)	N.C. in motor function or electrophysiologic studies	N.C.	N.C.	[122]
	Human (adult patients with SMA)	↑ H4 acetylation without ↑ SMN expression	N.C.	N.C.	[123]

↑, *increase;* 4-PBA, *4-phenylbutyric acid;* N.A., *not assessed;* N.C., *no change;* QOL, *quality of life;* SMA, *spinal muscular atrophy;* SMN, *survival motor neuron.*
[a]*One fold denotes no change.*

modest increase in motor neuron numbers in these mice lead to ablation of FoxO-dependent atrophy [103]. Furthermore, in a second study on atrophy in SMA, it was proposed that HDAC4 is responsible for myogenin (MyoG)-dependent neurogenic atrophy [133]. TSA administration reduced the presence of MyoG at the muscle-specific RING finger protein 1 (MuRF1) and atrogin-1 promoter, two

Table 2.6 Natural compounds that may provide a beneficial effect by HDAC inhibition in SMA.

Compound	Experimental setting	Observations	FL-SMN transcript	SMN protein	References
Resveratrol	Fibroblasts of patients with SMA	↑ Exon 7 inclusion, SMN protein levels, and gems	N.A.	↑ 2.5- to 3.5-fold	[125]
	Fibroblasts of patients with SMA	Modest ↑ in SMN transcript and protein levels	↑ 1- to 1.3-fold[a]	↑ 1- to 1.2-fold[a]	[126]
	PC12 cells	N.C. in neurite length	N.A.	N.A.	[127]
Curcumin	Fibroblasts of patients with SMA	↑ Exon 7 inclusion, SMN protein levels, and gems	N.A.	↑ 2-fold	[125]
	Fibroblasts of patients with SMA	Modest ↑ in SMN transcription and exon 7 inclusion	↑ 1.7-fold	N.A.	[128]
	PC12 cells	N.C. in neurite length	N.A.	N.C.	[127]
Caffeic acid	Fibroblasts of patients with SMA	Modest ↑ in SMN transcription and exon 7 inclusion	↑ 1.5-fold	N.A.	[128]
Chlorogenic acid	Fibroblasts of patients with SMA	Modest ↑ in SMN transcription and exon 7 inclusion	↑ 1.3-fold	N.A.	[128]
Quercetin	Fibroblasts of patients with SMA	↑ in SMN2 transcription but not in SMN protein levels	↑ 3- to 4-fold	N.C.	[129]
Epigallocatechin gallate (EGCG)	Fibroblasts of patients with SMA	↑ Exon 7 inclusion, SMN protein levels, and gems	N.A.	↑ 1.4-fold	[125]

↑, *increase;* N.A., *not assessed;* N.C., *no change;* SMA, *spinal muscular atrophy;* SMN, *survival motor neuron.*
[a]*One fold denotes no change.*

important proteins involved in atrophy [133]. Indeed, this mechanism may explain why the pathologic condition of muscles was improved despite the finding that motor neuron numbers were not significantly increased in TSA-treated *SmnΔ7* mice [99,133]. Tsai et al. [96] also observed restored protein levels of B-cell CLL/lymphoma 2 (Bcl-2) and bcl-2-like protein 1 (Bcl-x_L) to be higher than those in wild-type upon VPA treatment, which may provide an antiapoptotic protection to the motor neurons. Butchbach et al. [87] also showed modest improvements in *SmnΔ7* mouse model treated with 4-PBA or VX563, despite no increase in SMN levels.

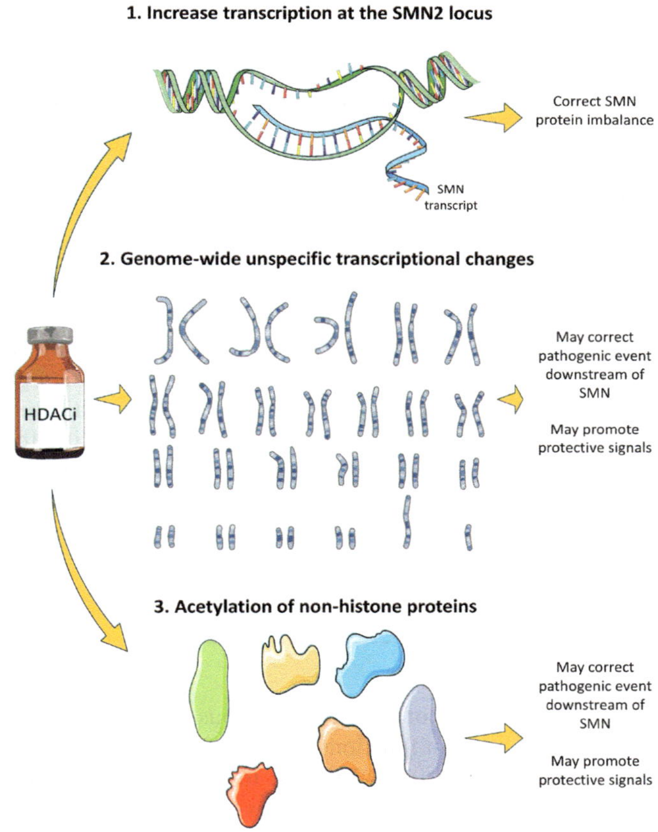

FIGURE 2.2

Mechanisms by which histone deacetylase inhibitors (HDACis) can exert a beneficial effect in the context of spinal muscular atrophy. (A) The initial mechanism sought after was the increased transcription at the *SMN2* locus. (B) However, it is likely that HDAC inhibitors interfere with the transcription of other genes and (C) deregulate the acetylation of non-histone proteins.

2.4.3 The potential protective effect of histone deacetylase inhibitors in the pathogenesis of spinal muscular atrophy

Studies in other diseases gave us some hints into possible HDACi therapeutic mechanisms that could be applied to SMA. For example, VPA, 4-PBA, and TSA increased α-synuclein protein production and protected against glutamate excitotoxicity [134]. Interestingly, virus-mediated therapy with α-synuclein was able to improve lifespan, weight, and NMJ pathology in the $Smn^{2B/-}$ mouse model [135]. Moreover, class I/II HDAC inhibitors have the capability to interfere with apoptosis through the regulation of p53 activity [136]. Another study showed that VPA, SB, and TSA treatment could reduce p53 induction in a permanent middle cerebral artery occlusion model [137]. Intriguingly, this pathway was identified as misregulated in microdissected SMA motor neurons [73]. Multiple HDAC inhibitors are also able to increase expression of neurotrophic factors such as brain-derived neurotrophic factor (BDNF) and glial cell–derived neurotrophic factor (GDNF) [138–140]. In the context of SMA, GDNF production and secretion is reduced in SMA induced pluripotent stem cell (iPSC)-derived astrocytes in comparison to controls [141]. However, restoring GDNF levels does not seem to give significant benefits to SMA iPSC-derived astrocytes or motor neurons *in vitro* [142], but its potential has not been thoroughly studied *in vivo*. TSA-treated *SmnΔ7* mice have increased BDNF transcripts but not proteins in the spinal cord and muscle [99].

Since 2017, large body of evidence has implicated immune organs in SMA [38–40]. This led to the speculation that neuroinflammation may be a factor contributing to SMA, a mechanism currently understudied in the field [143]. This is supported by astrocytic and microglial activation [141,143–145] and proinflammatory cytokine profile in the spinal cord [38,146]. Interestingly, it appears that HDAC inhibition could lead to an anti-inflammatory effect through reduction of microglial activation, cyclooxygenase 2, nitric oxide synthase 2, and tumor necrosis factor α, likely mitigating neuroinflammation [137,147,148]. Although it is clear that HDAC inhibitors lead to nonspecific therapeutic effects, the underlying molecular pathways remain largely unidentified in SMA. Importantly, getting a better insight into the mechanism of HDAC inhibitors will require a holistic approach by including many of the tissues involved in SMA pathogenesis, as they may not lead to similar molecular changes in different cell types. Ultimately, such an endeavor could help us identify disease modifiers previously unknown that could be easily targeted for combinatorial therapies.

2.5 Conclusion

Since the discovery of the genetic basis of SMA, tremendous advances have been made in understanding the *SMN* locus, the SMN protein functions, the contribution of SMN in various tissues, and how to increase SMN protein levels *in vivo*. This led to the approval of the first SMA-targeted therapeutic agent, called Spinraza, in 2016, which uses ASOs to increase SMN protein levels. Although increasing the levels of the depleted protein alleviates symptoms, it does not completely restore function [149]. The pathogenic events leading to SMA remain mysterious. While SMN is widely accepted as an important splicing regulator, it is not yet clear how this function is implicated in the pathogenic mechanism. Indeed, other essential functions have emerged for SMN. Its role in 3′ end processing of canonical histones and its ability to bind methylated histones are only beginning to be appreciated, and how these contribute to SMA pathogenesis is unknown. Undoubtedly, both these functions can have a tremendous effect on transcription regulation and cause widespread effects on multiple cell types. Research targeting these pathways will require a holistic approach to ensure a broad understanding of the consequence of SMN depletion. To this end, much of the epigenetic landscape in SMA has focused on the *SMN2* locus. Obtaining a better comprehension of the epigenetic changes in different cell types during SMA disease progression will lead to new research avenues. In the same manner, HDAC inhibitors have long been used in preclinical trials for their potential to increase *SMN2* transcription and ultimately SMN protein production. However, it is slowly becoming recognized that HDAC inhibition may result in beneficial effects through an SMN-independent manner, such as widespread transcriptional changes and acetylation of non-histone proteins. Indeed, HDAC inhibitors could be used as a proxy to identify pathogenic events or protective pathways in the context of SMA. In the future, it will be important to gain better appreciation of the chromatin signaling events that SMN might be directly involved in, the epigenetic landscape over SMA progression, and how HDAC inhibitors may provide beneficial effects.

Acronyms and abbreviations

4-PBA	sodium phenylbutyrate or 4-phenylbutyric acid
ASD	atrial septal defect
ASO	antisense oligonucleotide
Bcl-2	B-cell CLL/lymphoma 2
Bcl-x_L	bcl-2-like protein 1
BDNF	brain-derived neurotrophic factor
ChIP	chromatin immunoprecipitation
cKO	conditional knockout
DOT1L	disruptor of telomeric silencing 1-like
FoxO	forkhead box O
FL	full length
GAPDH	glyceraldehyde-3-phosphate dehydrogenase
GDNF	glial cell–derived neurotrophic factor
GI	gastrointestinal
H3K27m3	trimethylated histone 3 at lysine 27
H3K79	histone H3 at lysine 79
H3K79me1	monomethylated histone H3 at lysine 79
H3K79me2	dimethylated histone H3 at lysine 79
HDAC	histone deacetylase
HDACi	histone deacetylase inhibitor
iPSC	induced pluripotent stem cell
lncRNA	long noncoding RNA
MeCP2	methyl-CpG-binding protein 2
MEFs	murine embryonic fibroblasts
MN	motor neuron
MN/SC	motor neuron/Schwann cell
MuRF1	muscle-specific RING finger protein 1
MyoG	myogenin
N.A.	not assessed
N.C.	no change
NMJ	neuromuscular junction
NSC	neuronal precursor cells
OHSC	organotypic hippocampal slice
QOL	quality of life
RNAP II	RNA polymerase II
SAHA	suberoylanilide hydroxamine
SMA	spinal muscular atrophy
SMN1	survival motor neuron 1
SMN2	survival motor neuron 2
SmnΔ7	$Smn^{-/-};SMN2^{+/+};Smn\Delta7^{+/+}$
snRNP	small nuclear ribonucleoprotein
SB	sodium butyrate
SC	spinal cord
TSA	trichostatin A
VPA	valproic acid or valproate
VSD	ventricular septal defect
↑	increase
↓	decrease

Acknowledgments

We would like to thank Servier for the artwork used in the figures of this manuscript (Servier Medical Art https://smart.servier.com). Servier Medical Art by Servier is licensed under a Creative Commons Attribution 3.0 Unported License. This work was supported by Cure SMA/Families of SMA Canada, the Muscular Dystrophy Association (USA) (grant number 294568), the Canadian Institutes of Health Research (CIHR) (grant number MOP-130279), and the E-Rare-2 program from the CIHR (grant number ERL-138414). M-O.D. was supported by the Frederick Banting and Charles Best CIHR Doctoral Research Award.

References

[1] Lefebvre S, Bürglen L, Reboullet S, Clermont O, Burlet P, Viollet L, et al. Identification and characterization of a spinal muscular atrophy-determining gene. Cell 1995;80(1):155–65.

[2] Sugarman EA, Nagan N, Zhu H, Akmaev VR, Zhou Z, Rohlfs EM, et al. Pan-ethnic carrier screening and prenatal diagnosis for spinal muscular atrophy: clinical laboratory analysis of >72,400 specimens. Eur J Hum Genet 2012;20(1):27–32.

[3] Hendrickson BC, Donohoe C, Akmaev VR, Sugarman EA, Labrousse P, Boguslavskiy L, et al. Differences in SMN1 allele frequencies among ethnic groups within North America. J Med Genet 2009;46(9):641–4.

[4] Oskoui M, Darras B, De Vivo D. Spinal muscular atrophy: 125 Years later and on the verge of a cure. In: Sumner CJ, Paushkin S, Ko CP, editors. Spinal muscular atrophy: disease mechanisms and therapy. 1st ed. ed. Elsevier; 2017. p. 3–19.

[5] Rochette CF, Gilbert N, Simard LR. SMN gene duplication and the emergence of the SMN2 gene occurred in distinct hominids: SMN2 is unique to *Homo sapiens*. Hum Genet 2001;108(3):255–66.

[6] Lorson CL, Hahnen E, Androphy EJ, Wirth B. A single nucleotide in the SMN gene regulates splicing and is responsible for spinal muscular atrophy. Proc Natl Acad Sci USA 1999;96(11):6307–11.

[7] Lorson CL, Strasswimmer J, Yao JM, Baleja JD, Hahnen E, Wirth B, et al. SMN oligomerization defect correlates with spinal muscular atrophy severity. Nat Genet 1998;19(1):63–6.

[8] Schrank B, Götz R, Gunnersen JM, Ure JM, Toyka KV, Smith AG, et al. Inactivation of the survival motor neuron gene, a candidate gene for human spinal muscular atrophy, leads to massive cell death in early mouse embryos. Proc Natl Acad Sci USA 1997;94(18):9920–5.

[9] Feldkotter M, Schwarzer V, Wirth R, Wienker TF, Wirth B. Quantitative analyses of SMN1 and SMN2 based on real-time lightCycler PCR: fast and highly reliable carrier testing and prediction of severity of spinal muscular atrophy. Am J Hum Genet 2002;70(2):358–68.

[10] Wirth B, Brichta L, Schrank B, Lochmuller H, Blick S, Baasner A, et al. Mildly affected patients with spinal muscular atrophy are partially protected by an increased SMN2 copy number. Hum Genet 2006;119(4):422–8.

[11] Oskoui M, Levy G, Garland CJ, Gray JM, O'Hagen J, Vivo DDC, et al. The changing natural history of spinal muscular atrophy type 1. Neurology 2007;69(20):1931–6.

[12] Shababi M, Lorson CL, Rudnik-Schöneborn SS. Spinal muscular atrophy: a motor neuron disorder or a multi-organ disease? J Anat 2014;224(1):15–28.

[13] Deguise MO, Patitucci TN, Ebert AD, Lorson C, Kothary R. Contributions of different cell types to spinal muscular atrophy pathogenesis. In: Sumner CJ, Paushkin S, Ko CP, editors. Spinal muscular atrophy: disease mechanisms and therapy. 1st ed. London: Academic Press; 2017. p. 167–90.

[14] Gavrilina TO, McGovern VL, Workman E, Crawford TO, Gogliotti RG, DiDonato CJ, et al. Neuronal SMN expression corrects spinal muscular atrophy in severe SMA mice while muscle-specific SMN expression has no phenotypic effect. Hum Mol Genet 2008;17(8):1063–75.

[15] Braun S, Croizat B, Lagrange MC, Warter JM, Poindron P. Constitutive muscular abnormalities in culture in spinal muscular atrophy. Lancet 1995;345(8951):694–5.

[16] Guettier-Sigrist S, Coupin G, Braun S, Rogovitz D, Courdier I, Warter JM, et al. On the possible role of muscle in the pathogenesis of spinal muscular atrophy. Fundam Clin Pharmacol 2001;15(1):31–40.

[17] Guettier-Sigrist S, Coupin G, Braun S, Warter JM, Poindron P. Muscle could be the therapeutic target in SMA treatment. J Neurosci Res 1998;53(6):663–9.

[18] Shafey D, Côté PD, Kothary R. Hypomorphic Smn knockdown C2C12 myoblasts reveal intrinsic defects in myoblast fusion and myotube morphology. Exp Cell Res 2005;311(1):49–61.

[19] Boyer JG, Deguise MO, Murray LM, Yazdani A, De Repentigny Y, Boudreau-Lariviere C, et al. Myogenic program dysregulation is contributory to disease pathogenesis in spinal muscular atrophy. Hum Mol Genet 2014;23(16):4249–59.

[20] Boyer JG, Murray LM, Scott K, De Repentigny Y, Renaud J-MM, Kothary R. Early onset muscle weakness and disruption of muscle proteins in mouse models of spinal muscular atrophy. Skeletal Muscle 2013;3(1):24.

[21] Bricceno KV, Martinez T, Leikina E, Duguez S, Partridge TA, Chernomordik LV, et al. Survival motor neuron protein deficiency impairs myotube formation by altering myogenic gene expression and focal adhesion dynamics. Hum Mol Genet 2014;23(18):4745–57.

[22] Cifuentes-Diaz C, Frugier T, Tiziano FD, Lacène E, Roblot N, Joshi V, et al. Deletion of murine SMN exon 7 directed to skeletal muscle leads to severe muscular dystrophy. J Cell Biol 2001;152(5):1107–14.

[23] Rudnik-Schöneborn S, Heller R, Berg C, Betzler C, Grimm T, Eggermann T, et al. Congenital heart disease is a feature of severe infantile spinal muscular atrophy. J Med Genet 2008;45(10).

[24] Bevan AK, Hutchinson KR, Foust KD, Braun L, McGovern VL, Schmelzer L, et al. Early heart failure in the SMNDelta7 model of spinal muscular atrophy and correction by postnatal scAAV9-SMN delivery. Hum Mol Genet 2010;19(20):3895–905.

[25] Heier CR, Satta R, Lutz C, DiDonato CJ. Arrhythmia and cardiac defects are a feature of spinal muscular atrophy model mice. Hum Mol Genet 2010;19(20):3906–18.

[26] Shababi M, Habibi J, Ma L, Glascock JJ, Sowers JR, Lorson CL. Partial restoration of cardio-vascular defects in a rescued severe model of spinal muscular atrophy. J Mol Cell Cardiol 2012;52(5):1074–82.

[27] Shababi M, Habibi J, Yang HT, Vale SM, Sewell WA, Lorson CL. Cardiac defects contribute to the pathology of spinal muscular atrophy models. Hum Mol Genet 2010;19(20):4059–71.

[28] Bowerman M, Michalski J-PP, Beauvais A, Murray LM, DeRepentigny Y, Kothary R. Defects in pancreatic development and glucose metabolism in SMN-depleted mice independent of canonical spinal muscular atrophy neuromuscular pathology. Hum Mol Genet 2014;23(13):3432–44.

[29] Bowerman M, Swoboda KJ, Michalski J-PP, Wang G-SS, Reeks C, Beauvais A, et al. Glucose metabolism and pancreatic defects in spinal muscular atrophy. Ann Neurol 2012;72(2):256–68.

[30] Davis R, Miller EA, Zhang R, Swoboda KJ. Responses to fasting and glucose loading in a cohort of well children with spinal muscular atrophy type II. J Pediatr 2015;167(6):1362–1368.e1.

[31] Gombash SE, Cowley CJ, Fitzgerald JA, Iyer CC, Fried D, McGovern VL, et al. SMN deficiency disrupts gastrointestinal and enteric nervous system function in mice. Hum Mol Genet 2015;24(13):3847–60.

[32] Sintusek P, Catapano F, Angkathunkayul N, Marrosu E, Parson SH, Morgan JE, et al. Histopathological defects in intestine in severe spinal muscular atrophy mice are improved by systemic antisense oligonucleotide treatment. PLoS One 2016;11(5).

[33] Wang CH, Finkel RS, Bertini ES, Schroth M. Consensus statement for standard of care in spinal muscular atrophy. J Child Neurol 2007;22(8):1027–49.

[34] Szunyogova E, Zhou H, Maxwell GK, Powis RA, Francesco M, Gillingwater TH, et al. Survival Motor Neuron (SMN) protein is required for normal mouse liver development. Sci Rep 2016;6:34635.

[35] Vitte JMM, Davoult B, Roblot N, Mayer M, Joshi V, Courageot S, et al. Deletion of murine Smn exon 7 directed to liver leads to severe defect of liver development associated with iron overload. Am J Pathol 2004;165(5):1731–41.
[36] Kelley RI, Sladky JT. Dicarboxylic aciduria in an infant with spinal muscular atrophy. Ann Neurol 1986; 20(6):734–6.
[37] Zolkipli Z, Sherlock M, Biggar WD, Taylor G, Hutchison JS, Peliowski A, et al. Abnormal fatty acid metabolism in spinal muscular atrophy may predispose to perioperative risks. Eur J Paediatr Neurol 2012; 16(5):549–53.
[38] Deguise MO, De Repentigny Y, McFall E, Auclair N, Sad S, Kothary R. Immune dysregulation may contribute to disease pathogenesis in spinal muscular atrophy mice. Hum Mol Genet 2017;26(4):801–19.
[39] Khairallah MT, Astroski J, Custer SK, Androphy EJ, Franklin CL, Lorson CL. SMN deficiency negatively impacts red pulp macrophages and spleen development in mouse models of Spinal Muscular Atrophy. Hum Mol Genet 2017;26(5):932–41.
[40] Thomson AK, Somers E, Powis RA, Shorrock HK, Murphy K, Swoboda KJ, et al. Survival of motor neurone protein is required for normal postnatal development of the spleen. J Anat 2017;230(2):337–46.
[41] Ryniewicz B, Pawinska M. Preliminary immunological studies in spinal muscular atrophy. Eur J Pediatr 1978;128(1):57–60.
[42] Somers E, Lees RD, Hoban K, Sleigh JN, Zhou H, Muntoni F, et al. Vascular defects and spinal cord hypoxia in spinal muscular atrophy. Ann Neurol 2016;79(2):217–30.
[43] Somers E, Stencel Z, Wishart TM, Gillingwater TH, Parson SH. Density, calibre and ramification of muscle capillaries are altered in a mouse model of severe spinal muscular atrophy. Neuromuscul Disord 2012; 22(5):435–42.
[44] Araujo A, Araujo M, Swoboda KJ. Vascular perfusion abnormalities in infants with spinal muscular atrophy. J Pediatr 2009;155(2):292–4.
[45] Rudnik-Schöneborn S, Vogelgesang S. Digital necroses and vascular thrombosis in severe spinal muscular atrophy. Muscle Nerve 2010;42(1):144–7.
[46] Shanmugarajan S, Swoboda KJ, Iannaccone ST, Ries WL, Maria BL, Reddy SV. Congenital bone fractures in spinal muscular atrophy: functional role for SMN protein in bone remodeling. J Child Neurol 2007;22(8): 967–73.
[47] Shanmugarajan S, Tsuruga E, Swoboda KJ, Maria BL, Ries WL, Reddy SV. Bone loss in survival motor neuron (Smn(-/-) SMN2) genetic mouse model of spinal muscular atrophy. J Pathol 2009;219(1):52–60.
[48] Wasserman HM, Hornung LN, Stenger PJ, Rutter MM, Wong BL, Rybalsky I, et al. Low bone mineral density and fractures are highly prevalent in pediatric patients with spinal muscular atrophy regardless of disease severity. Neuromuscul Disord 2017;27(4):331–7.
[49] Vai S, Bianchi ML, Moroni I, Mastella C, Broggi F, Morandi L, et al. Bone and spinal muscular atrophy. Bone 2015;79:116–20.
[50] Hua Y, Liu YH, Sahashi K, Rigo F. Motor neuron cell-nonautonomous rescue of spinal muscular atrophy phenotypes in mild and severe transgenic mouse models. Genes 2015;29(3):288–97.
[51] Hua Y, Sahashi K, Rigo F, Hung G, Horev G, Bennett CF, et al. Peripheral SMN restoration is essential for long-term rescue of a severe spinal muscular atrophy mouse model. Nature 2011;478(7367):123–6.
[52] Burlet P, Huber C, Bertrandy S, Ludosky MA, Zwaenepoel I, Clermont O, et al. The distribution of SMN protein complex in human fetal tissues and its alteration in spinal muscular atrophy. Hum Mol Genet 1998; 7(12):1927–33.
[53] La Bella V, Cisterni C, Salaün D, Pettmann B. Survival motor neuron (SMN) protein in rat is expressed as different molecular forms and is developmentally regulated. Eur J Neurosci 1998;10(9):2913–23.
[54] Liu Q, Fischer U, Wang F, Dreyfuss G. The spinal muscular atrophy disease gene product, SMN, and its associated protein SIP1 are in a complex with spliceosomal snRNP proteins. Cell 1997;90(6):1013–21.

[55] Hebert MD, Szymczyk PW, Shpargel KB, Matera AG. Coilin forms the bridge between Cajal bodies and SMN, the spinal muscular atrophy protein. Genes Dev 2001;15(20):2720–9.

[56] So BR, Zhang Z, Dreyfuss G. The function of survival motor neuron complex and its role in spinal muscular atrophy pathogenesis. In: Sumner CJ, Paushkin S, Ko CP, editors. Spinal muscular atrophy: disease mechanisms and therapy. 1st ed. ed. Elsevier; 2016. p. 99–111.

[57] Tisdale S, Pellizzoni L. RNA-processing dysfunction in spinal muscular atrophy. In: Sumner CJ, Paushkin S, Ko CP, editors. Spinal muscular atrophy: disease mechanisms and therapy. 1st ed. ed. Elsevier; 2016. p. 75–97.

[58] Tisdale S, Lotti F, Saieva L, Van Meerbeke JP, Crawford TO, Sumner CJ, et al. SMN is essential for the biogenesis of U7 small nuclear ribonucleoprotein and 3'-end formation of histone mRNAs. Cell Rep 2013; 5(5):1187–95.

[59] Marzluff WF, Wagner EJ, Duronio RJ. Metabolism and regulation of canonical histone mRNAs: life without a poly(A) tail. Nat Rev Genet 2008;9(11):843–54.

[60] Pillai RS, Grimmler M, Meister G, Will CL, Luhrmann R, Fischer U, et al. Unique Sm core structure of U7 snRNPs: assembly by a specialized SMN complex and the role of a new component, Lsm11, in histone RNA processing. Genes Dev 2003;17(18):2321–33.

[61] Pillai RS, Will CL, Luhrmann R, Schumperli D, Muller B. Purified U7 snRNPs lack the Sm proteins D1 and D2 but contain Lsm10, a new 14 kDa Sm D1-like protein. EMBO J 2001;20(19):5470–9.

[62] Weber CM, Henikoff S. Histone variants: dynamic punctuation in transcription. Genes Dev 2014;28(7): 672–82.

[63] Wagner EJ, Marzluff WF. ZFP100, a component of the active U7 snRNP limiting for histone pre-mRNA processing, is required for entry into S phase. Mol Cell Biol 2006;26(17):6702–12.

[64] Ideue T, Hino K, Kitao S, Yokoi T, Hirose T. Efficient oligonucleotide-mediated degradation of nuclear noncoding RNAs in mammalian cultured cells. RNA 2009;15(8):1578–87.

[65] Lotti F, Imlach WL, Saieva L, Beck ES, Hao le T, Li DK, et al. An SMN-dependent U12 splicing event essential for motor circuit function. Cell 2012;151(2):440–54.

[66] Grice SJ, Liu JL. Survival motor neuron protein regulates stem cell division, proliferation, and differentiation in Drosophila. PLoS Genet 2011;7(4):e1002030.

[67] Martínez-Hernández R, Soler-Botija C, Also E, Alias L, Caselles L, Gich I, et al. The developmental pattern of myotubes in spinal muscular atrophy indicates prenatal delay of muscle maturation. J Neuropathol Exp Neurol. 2009;68(5):474-481.

[68] Botuyan MV, Mer G. Tudor domains as methyl-lysine and methyl-arginine readers. In: Binda O, Fernandez-Zapico ME, editors. Chromatin signaling and diseases. 1st ed. ed. Elsevier; 2016. p. 149–65.

[69] Friesen WJ, Massenet S, Paushkin S, Wyce A, Dreyfuss G. SMN, the product of the spinal muscular atrophy gene, binds preferentially to dimethylarginine-containing protein targets. Mol Cell 2001;7(5):1111–7.

[70] Brahms H, Meheus L, de Brabandere V, Fischer U, Luhrmann R. Symmetrical dimethylation of arginine residues in spliceosomal Sm protein B/B' and the Sm-like protein LSm4, and their interaction with the SMN protein. RNA 2001;7(11):1531–42.

[71] Sabra M, Texier P, El Maalouf J, Lomonte P. The Tudor protein survival motor neuron (SMN) is a chromatin-binding protein that interacts with methylated lysine 79 of histone H3. J Cell Sci 2013; 126(Pt 16):3664–77.

[72] Nguyen AT, Zhang Y. The diverse functions of Dot1 and H3K79 methylation. Genes Dev 2011;25(13): 1345–58.

[73] Murray LM, Beauvais A, Gibeault S, Courtney NL, Kothary R. Transcriptional profiling of differentially vulnerable motor neurons at pre-symptomatic stage in the Smn 2b/- mouse model of spinal muscular atrophy. Acta Neuropathologica Communications 2015;3(1).

[74] Zhao DY, Gish G, Braunschweig U, Li Y, Ni Z, Schmitges FW, et al. SMN and symmetric arginine dimethylation of RNA polymerase II C-terminal domain control termination. Nature 2016;529(7584): 48–53.
[75] Bowerman M, Anderson CL, Beauvais A, Boyl PP, Witke W, Kothary R. SMN, profilin IIa and plastin 3: a link between the deregulation of actin dynamics and SMA pathogenesis. Molecular and cellular neurosciences 2009;42(1):66–74.
[76] Bowerman M, Beauvais A, Anderson CL, Kothary R. Rho-kinase inactivation prolongs survival of an intermediate SMA mouse model. Hum Mol Genet 2010;19(8):1468–78.
[77] Bowerman M, Murray LM, Boyer JG, Anderson CL, Kothary R. Fasudil improves survival and promotes skeletal muscle development in a mouse model of spinal muscular atrophy. BMC Medicine 2012;10:24.
[78] Bowerman M, Shafey D, Kothary R. Smn depletion alters profilin II expression and leads to upregulation of the RhoA/ROCK pathway and defects in neuronal integrity. J Mol Neurosci 2007;32(2):120–31.
[79] Fuller HR, Gillingwater TH, Wishart TM. Commonality amid diversity: multi-study proteomic identification of conserved disease mechanisms in spinal muscular atrophy. Neuromuscul Disord 2016;26(9):560–9.
[80] Singh NN, Howell MD, Singh RN. Transcriptional and splicing regulation of spinal muscular atrophy genes. In: Sumner CJ, Paushkin S, Ko CP, editors. Spinal muscular atrophy: disease mechanisms and therapy. 1st ed. ed. Elsevier; 2016.
[81] Kernochan LE, Russo ML, Woodling NS, Huynh TN, Avila AM, Fischbeck KH, et al. The role of histone acetylation in SMN gene expression. Hum Mol Genet 2005;14(9):1171–82.
[82] Hauke J, Riessland M, Lunke S, Eyupoglu IY, Blumcke I, El-Osta A, et al. Survival motor neuron gene 2 silencing by DNA methylation correlates with spinal muscular atrophy disease severity and can be bypassed by histone deacetylase inhibition. Hum Mol Genet 2009;18(2):304–17.
[83] Nan X, Ng HH, Johnson CA, Laherty CD, Turner BM, Eisenman RN, et al. Transcriptional repression by the methyl-CpG-binding protein MeCP2 involves a histone deacetylase complex. Nature 1998;393(6683): 386–9.
[84] d'Ydewalle C, Ramos DM, Pyles NJ, Ng SY, Gorz M, Pilato CM, et al. The antisense transcript SMN-AS1 regulates SMN expression and is a novel therapeutic target for spinal muscular atrophy. Neuron 2017;93(1): 66–79.
[85] Chang JG, Hsieh-Li HM, Jong YJ, Wang NM, Tsai CH, Li H. Treatment of spinal muscular atrophy by sodium butyrate. Proc Natl Acad Sci USA 2001;98(17):9808–13.
[86] Jarecki J, Chen X, Bernardino A, Coovert DD, Whitney M, Burghes A, et al. Diverse small-molecule modulators of SMN expression found by high-throughput compound screening: early leads towards a therapeutic for spinal muscular atrophy. Hum Mol Genet 2005;14(14):2003–18.
[87] Butchbach MER, Lumpkin CJ, Harris AW, Saieva L, Edwards JD, Workman E, et al. Protective effects of butyrate-based compounds on a mouse model for spinal muscular atrophy. Exp Neurol 2016;279:13–26.
[88] Andreassi C, Angelozzi C, Tiziano FD, Vitali T, De Vincenzi E, Boninsegna A, et al. Phenylbutyrate increases SMN expression in vitro: relevance for treatment of spinal muscular atrophy. Eur J Hum Genet 2004;12(1):59–65.
[89] Dayangac-Erden D, Topaloglu H, Erdem-Yurter H. A preliminary report on spinal muscular atrophy lymphoblastoid cell lines: are they an appropriate tool for drug screening? Adv Ther 2008;25(3):274–9.
[90] Garbes L, Heesen L, Holker I, Bauer T, Schreml J, Zimmermann K, et al. VPA response in SMA is suppressed by the fatty acid translocase CD36. Hum Mol Genet 2013;22(2):398–407.
[91] Brichta L, Hofmann Y, Hahnen E, Siebzehnrubl FA, Raschke H, Blumcke I, et al. Valproic acid increases the SMN2 protein level: a well-known drug as a potential therapy for spinal muscular atrophy. Hum Mol Genet 2003;12(19):2481–9.
[92] Sumner CJ, Huynh TN, Markowitz JA, Perhac JS, Hill B, Coovert DD, et al. Valproic acid increases SMN levels in spinal muscular atrophy patient cells. Ann Neurol 2003;54(5):647–54.

[93] Hahnen E, Eyupoglu IY, Brichta L, Haastert K, Trankle C, Siebzehnrubl FA, et al. In vitro and ex vivo evaluation of second-generation histone deacetylase inhibitors for the treatment of spinal muscular atrophy. J Neurochem 2006;98(1):193–202.
[94] Tsai L-KK, Tsai M-SS, Lin T-BB, Hwu W-LL, Li H. Establishing a standardized therapeutic testing protocol for spinal muscular atrophy. Neurobiol Dis 2006;24(2):286–95.
[95] van Bergeijk J, Haastert K, Grothe C, Claus P. Valproic acid promotes neurite outgrowth in PC12 cells independent from regulation of the survival of motoneuron protein. Chem Biol Drug Des 2006;67(3): 244–7.
[96] Tsai LK, Tsai MS, Ting CH, Li H. Multiple therapeutic effects of valproic acid in spinal muscular atrophy model mice. J Mol Med 2008;86(11):1243–54.
[97] Rak K, Lechner BD, Schneider C, Drexl H, Sendtner M, Jablonka S. Valproic acid blocks excitability in SMA type I mouse motor neurons. Neurobiol Dis 2009;36(3):477–87.
[98] Harahap IS, Saito T, San LP, Sasaki N, Gunadi, Nurputra DK, et al. Valproic acid increases SMN2 expression and modulates SF2/ASF and hnRNPA1 expression in SMA fibroblast cell lines. Brain Dev 2012; 34(3):213–22.
[99] Avila AM, Burnett BG, Taye AA, Gabanella F, Knight MA, Hartenstein P, et al. Trichostatin A increases SMN expression and survival in a mouse model of spinal muscular atrophy. J Clin Invest 2007;117(3): 659–71.
[100] Narver HL, Kong L, Burnett BG, Choe DW, Bosch-Marcé M, Taye AA, et al. Sustained improvement of spinal muscular atrophy mice treated with trichostatin A plus nutrition. Ann Neurol 2008;64(4):465–70.
[101] Kwon DY, Motley WW, Fischbeck KH, Burnett BG. Increasing expression and decreasing degradation of SMN ameliorate the spinal muscular atrophy phenotype in mice. Hum Mol Genet 2011;20(18):3667–77.
[102] Ling KK, Gibbs RM, Feng Z, Ko C-PP. Severe neuromuscular denervation of clinically relevant muscles in a mouse model of spinal muscular atrophy. Hum Mol Genet 2012;21(1):185–95.
[103] Liu H, Yazdani A, Murray LM, Beauvais A, Kothary R. The Smn-independent beneficial effects of trichostatin A on an intermediate mouse model of spinal muscular atrophy. PLoS One 2014;9(7).
[104] Deguise M-OO, Boyer JG, McFall ER, Yazdani A, De Repentigny Y, Kothary R. Differential induction of muscle atrophy pathways in two mouse models of spinal muscular atrophy. Sci Rep 2016;6:28846.
[105] Riessland M, Ackermann B, Förster A, Jakubik M, Hauke J, Garbes L, et al. SAHA ameliorates the SMA phenotype in two mouse models for spinal muscular atrophy. Hum Mol Genet 2010;19(8):1492–506.
[106] Evans MC, Cherry JJ, Androphy EJ. Differential regulation of the SMN2 gene by individual HDAC proteins. Biochem Biophys Res Commun 2011;414(1):25–30.
[107] Somers E, Riessland M, Schreml J, Wirth B, Gillingwater TH, Parson SH. Increasing SMN levels using the histone deacetylase inhibitor SAHA ameliorates defects in skeletal muscle microvasculature in a mouse model of severe spinal muscular atrophy. Neurosci Lett 2013;544:100–4.
[108] Mutsaers CA, Wishart TM, Lamont DJ, Riessland M, Schreml J, Comley LH, et al. Reversible molecular pathology of skeletal muscle in spinal muscular atrophy. Hum Mol Genet 2011;20(22):4334–44.
[109] Mohseni J, Al-Najjar BO, Wahab HA, Zabidi-Hussin ZA, Sasongko TH. Transcript, methylation and molecular docking analyses of the effects of HDAC inhibitors, SAHA and Dacinostat, on SMN2 expression in fibroblasts of SMA patients. J Hum Genet 2016;61(9):823–30.
[110] Garbes L, Riessland M, Holker I, Heller R, Hauke J, Trankle C, et al. LBH589 induces up to 10-fold SMN protein levels by several independent mechanisms and is effective even in cells from SMA patients non-responsive to valproate. Hum Mol Genet 2009;18(19):3645–58.
[111] Schreml J, Riessland M, Paterno M, Garbes L, Roßbach K, Ackermann B, et al. Severe SMA mice show organ impairment that cannot be rescued by therapy with the HDACi JNJ-26481585. Eur J Hum Genet. 2012. 21(6):643-652.

[112] Riessland M, Brichta L, Hahnen E, Wirth B. The benzamide M344, a novel histone deacetylase inhibitor, significantly increases SMN2 RNA/protein levels in spinal muscular atrophy cells. Hum Genet 2006; 120(1):101–10.

[113] Brahe C, Vitali T, Tiziano FD, Angelozzi C, Pinto AM, Borgo F, et al. Phenylbutyrate increases SMN gene expression in spinal muscular atrophy patients. Eur J Hum Genet 2005;13(2):256–9.

[114] Mercuri E, Bertini E, Messina S, Pelliccioni M, D'Amico A, Colitto F, et al. Pilot trial of phenylbutyrate in spinal muscular atrophy. Neuromuscul Disord 2004;14(2):130–5.

[115] Brichta L, Holker I, Haug K, Klockgether T, Wirth B. In vivo activation of SMN in spinal muscular atrophy carriers and patients treated with valproate. Ann Neurol 2006;59(6):970–5.

[116] Tsai LK, Yang CC, Hwu WL, Li H. Valproic acid treatment in six patients with spinal muscular atrophy. Eur J Neurol 2007;14(12):e8–9.

[117] Piepers S, Cobben JM, Sodaar P, Jansen MD, Wadman RI, Meester-Delver A, et al. Quantification of SMN protein in leucocytes from spinal muscular atrophy patients: effects of treatment with valproic acid. J Neurol Neurosurg Psychiatry 2011;82(8):850–2.

[118] Swoboda KJ, Scott CB, Crawford TO, Simard LR, Reyna SP, Krosschell KJ, et al. SMA CARNI-VAL trial part I: double-blind, randomized, placebo-controlled trial of L-carnitine and valproic acid in spinal muscular atrophy. PLoS One 2010;5(8):e12140.

[119] Swoboda KJ, Scott CB, Reyna SP, Prior TW, LaSalle B, Sorenson SL, et al. Phase II open label study of valproic acid in spinal muscular atrophy. PLoS One 2009;4(5):e5268.

[120] Kissel JT, Scott CB, Reyna SP, Crawford TO, Simard LR, Krosschell KJ, et al. SMA CARNIVAL TRIAL PART II: a prospective, single-armed trial of L-carnitine and valproic acid in ambulatory children with spinal muscular atrophy. PLoS One 2011;6(7):e21296.

[121] Darbar IA, Plaggert PG, Resende MB, Zanoteli E, Reed UC. Evaluation of muscle strength and motor abilities in children with type II and III spinal muscle atrophy treated with valproic acid. BMC Neurol 2011; 11:36.

[122] Kissel JT, Elsheikh B, King WM, Freimer M, Scott CB, Kolb SJ, et al. SMA valiant trial: a prospective, double-blind, placebo-controlled trial of valproic acid in ambulatory adults with spinal muscular atrophy. Muscle Nerve 2014;49(2):187–92.

[123] Renusch SR, Harshman S, Pi H, Workman E, Wehr A, Li X, et al. Spinal muscular atrophy biomarker measurements from blood samples in a clinical trial of valproic acid in ambulatory adults. J Neuromuscul Dis 2015;2(2):119–30.

[124] Lai JI, Leman LJ, Ku S, Vickers CJ, Olsen CA, Montero A, et al. Cyclic tetrapeptide HDAC inhibitors as potential therapeutics for spinal muscular atrophy: screening with iPSC-derived neuronal cells. Bioorg Med Chem Lett 2017;27(15):3289–93.

[125] Sakla MS, Lorson CL. Induction of full-length survival motor neuron by polyphenol botanical compounds. Hum Genet 2008;122(6):635–43.

[126] Dayangac-Erden D, Bora G, Ayhan P, Kocaefe C, Dalkara S, Yelekci K, et al. Histone deacetylase inhibition activity and molecular docking of (e)-resveratrol: its therapeutic potential in spinal muscular atrophy. Chem Biol Drug Des 2009;73(3):355–64.

[127] Bora-Tatar G, Erdem-Yurter H. Investigations of curcumin and resveratrol on neurite outgrowth: perspectives on spinal muscular atrophy. BioMed Res Int 2014;2014. 709108.

[128] Dayangac-Erden D, Bora-Tatar G, Dalkara S, Demir AS, Erdem-Yurter H. Carboxylic acid derivatives of histone deacetylase inhibitors induce full length SMN2 transcripts: a promising target for spinal muscular atrophy therapeutics. Arch Med Sci 2011;7(2):230–4.

[129] Uzunallı G, Bora-Tatar G, Dayangaç-Erden D, Erdem-Yurter H. Effects of flavonoid quercetin on survival of motor neuron gene expression. Cell Biology International 2015;39(3):350–4.

[130] Spange S, Wagner T, Heinzel T, Kramer OH. Acetylation of non-histone proteins modulates cellular signalling at multiple levels. Int J Biochem Cell Biol 2009;41(1):185–98.
[131] Giesemann T, Rathke-Hartlieb S, Rothkegel M, Bartsch JW, Buchmeier S, Jockusch BM, et al. A role for polyproline motifs in the spinal muscular atrophy protein SMN. Profilins bind to and colocalize with smn in nuclear gems. J Biol Chem 1999;274(53):37908–14.
[132] Shafey D, Boyer JG, Bhanot K, Kothary R. Identification of novel interacting protein partners of SMN using tandem affinity purification. J Proteome Res 2010;9(4):1659–69.
[133] Bricceno KV, Sampognaro PJ, Van Meerbeke JP, Sumner CJ, Fischbeck KH, Burnett BG. Histone deacetylase inhibition suppresses myogenin-dependent atrogene activation in spinal muscular atrophy mice. Hum Mol Genet 2012;21(20):4448–59.
[134] Leng Y, Chuang DM. Endogenous alpha-synuclein is induced by valproic acid through histone deacetylase inhibition and participates in neuroprotection against glutamate-induced excitotoxicity. J Neurosci 2006; 26(28):7502–12.
[135] Kline RA, Kaifer KA, Osman EY, Carella F, Tiberi A, Ross J, et al. Comparison of independent screens on differentially vulnerable motor neurons reveals alpha-synuclein as a common modifier in motor neuron diseases. PLoS Genet 2017;13(3):e1006680.
[136] Uo T, Veenstra TD, Morrison RS. Histone deacetylase inhibitors prevent p53-dependent and p53-independent Bax-mediated neuronal apoptosis through two distinct mechanisms. J Neurosci 2009;29(9): 2824–32.
[137] Kim HJ, Rowe M, Ren M, Hong JS, Chen PS, Chuang DM. Histone deacetylase inhibitors exhibit anti-inflammatory and neuroprotective effects in a rat permanent ischemic model of stroke: multiple mechanisms of action. J Pharmacol Exp Therapeut 2007;321(3):892–901.
[138] Yasuda S, Liang MH, Marinova Z, Yahyavi A, Chuang DM. The mood stabilizers lithium and valproate selectively activate the promoter IV of brain-derived neurotrophic factor in neurons. Mol Psychiatr 2009; 14(1):51–9.
[139] Chen PS, Peng GS, Li G, Yang S, Wu X, Wang CC, et al. Valproate protects dopaminergic neurons in midbrain neuron/glia cultures by stimulating the release of neurotrophic factors from astrocytes. Mol Psychiatr 2006;11(12):1116–25.
[140] Wu X, Chen PS, Dallas S, Wilson B, Block ML, Wang CC, et al. Histone deacetylase inhibitors up-regulate astrocyte GDNF and BDNF gene transcription and protect dopaminergic neurons. Int J Neuropsychopharmacol 2008;11(8):1123–34.
[141] McGivern JV, Patitucci TN, Nord JA, Barabas M-EAE, Stucky CL, Ebert AD. Spinal muscular atrophy astrocytes exhibit abnormal calcium regulation and reduced growth factor production. Glia 2013;61(9): 1418–28.
[142] Sison SL, Patitucci TN, Seminary ER, Villalon E, Lorson CL, Ebert AD. Astrocyte-produced miR-146a as a mediator of motor neuron loss in spinal muscular atrophy. Hum Mol Genet 2017;26(17):3409–20.
[143] Deguise MO, Kothary R. New insights into SMA pathogenesis: immune dysfunction and neuroinflammation. Ann Clin Transl Neurol 2017;4(7):522–30.
[144] Ling KK, Lin MY, Zingg B, Feng Z, Ko CP. Synaptic defects in the spinal and neuromuscular circuitry in a mouse model of spinal muscular atrophy. PLoS One 2010;5(11):e15457.
[145] Tarabal O, Caraballo-Miralles V, Cardona-Rossinyol A, Correa FJ, Olmos G, Llado J, et al. Mechanisms involved in spinal cord central synapse loss in a mouse model of spinal muscular atrophy. J Neuropathol Exp Neurol 2014;73(6):519–35.
[146] Rindt H, Feng Z, Mazzasette C, Glascock JJ, Valdivia D, Pyles N, et al. Astrocytes influence the severity of spinal muscular atrophy. Hum Mol Genet 2015;24(14):4094–102.

[147] Peng GS, Li G, Tzeng NS, Chen PS, Chuang DM, Hsu YD, et al. Valproate pretreatment protects dopaminergic neurons from LPS-induced neurotoxicity in rat primary midbrain cultures: role of microglia. Brain Res Mol Brain Res 2005;134(1):162–9.
[148] Huuskonen J, Suuronen T, Nuutinen T, Kyrylenko S, Salminen A. Regulation of microglial inflammatory response by sodium butyrate and short-chain fatty acids. Br J Pharmacol 2004;141(5):874–80.
[149] Finkel RS, Chiriboga CA, Vajsar J, Day JW, Montes J, De Vivo DC, et al. Treatment of infantile-onset spinal muscular atrophy with nusinersen: a phase 2, open-label, dose-escalation study. Lancet 2016; 388(10063):3017–26.

CHAPTER

Charcot-Marie-Tooth disease

3

Phu Duong[1,2], John Svaren[1,3]

Waisman Center, University of Wisconsin—Madison, Madison, WI, United States[1]; Cellular and Molecular Pathology Graduate Program, University of Wisconsin—Madison, Madison, WI, United States[2]; Department of Comparative Biosciences, University of Wisconsin—Madison, Madison, WI, United States[3]

Chapter Outline

3.1 Introduction 53
3.2 Epigenetic regulation of Schwann cell development 55
3.3 Epigenetic regulation of dosage-sensitive genes 57
3.4 Epigenetic regulators targeted by CMT mutations 60
3.4.1 DNMT1 60
3.4.2 LMNA 61
3.4.3 SYNE1 62
3.4.4 MED25 62
3.4.5 SETX 62
3.4.6 MORC2 63
3.4.7 PRDM12 64
3.5 Novel mechanisms for CMT mutations 64
3.6 Summary 64
Acknowledgments 65
References 65

3.1 Introduction

The axons of peripheral nerves connect the central nervous system to the muscles and sensory organs of the body. Axons larger than ~1 μm in diameter are myelinated by Schwann cells (Fig. 3.1). In addition to their well-known function of increasing conduction velocity by enabling saltatory conduction of action potentials, myelinating Schwann cells also provide metabolic support to axons. Smaller diameter axons (of sensory and autonomic neurons) are ensheathed but not myelinated by Schwann cells, forming Remak bundles [reviewed in Refs. [1,2].

A broad spectrum of genetic, inflammatory, mechanical, metabolic, and toxic insults results in peripheral neuropathy. The hereditary peripheral neuropathies, usually called Charcot-Marie-Tooth disease (CMT; also known as hereditary motor sensory neuropathies), are among the most common

Chromatin Signaling and Neurological Disorders. https://doi.org/10.1016/B978-0-12-813796-3.00003-1

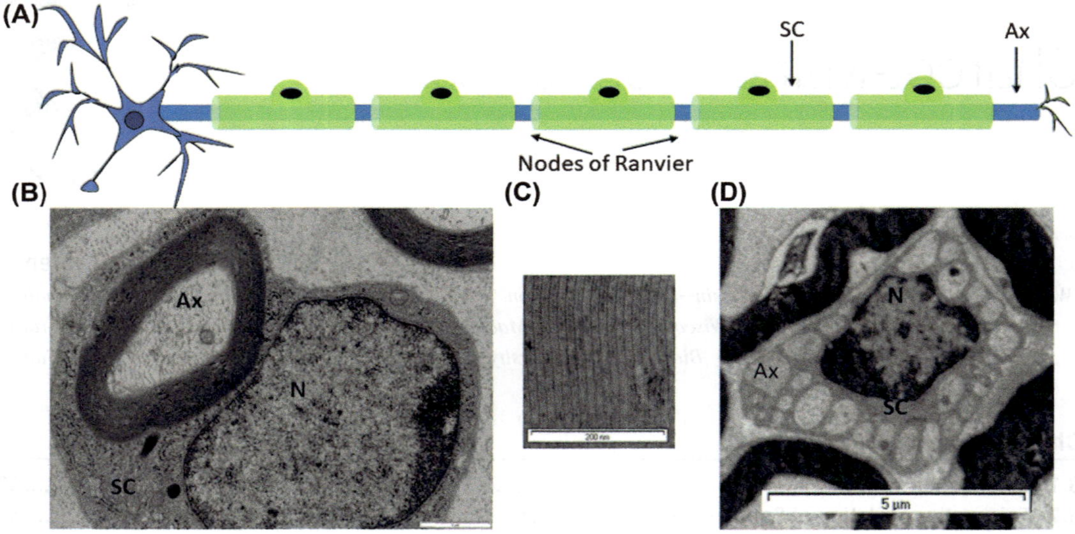

FIGURE 3.1 Structure of axons and Schwann cells in the peripheral nervous system.

(A) The diagram shows a motor neuron, where the cell body lies within the spinal cord, and the peripheral axon projecting to a muscle is myelinated by Schwann cells. The gaps are known as nodes of Ranvier. (B) Electron microscopy image shows a cross-section of a myelinating Schwann cell (SC), where the axon (Ax) is encased in multiple layers of myelin membrane. The image also shows the nucleus (N) of the Schwann cell. (C) Higher resolution image of the layers of myelin membrane is shown. (D) The EM image shows a Remak (nonmyelinating) Schwann cell that is associated with several small diameter axons, which are typically sensory neurons.

genetic diseases that principally affect the nervous system, with an estimated prevalence of 1 in 2500 individuals [3,4]. Although it varies by the genetic cause, typical onset is in childhood/adolescence, and the progression of the disease continues throughout adulthood.

CMT is generally divided into two types: (1) demyelinating forms, which reduce the speed of nerve conduction, and are caused by mutations of genes important for myelinating Schwann cells, and (2) axonal forms, manifested by reduced amplitude of electrophysiological signals owing to axonal loss, and are caused by mutations of genes affecting motor and/or sensory axons. The dominantly inherited demyelinating and axonal forms are called CMT1 and CMT2, respectively, while recessively inherited demyelinating and axonal forms are designated as CMT4 and AR-CMT, respectively. This is a somewhat artificial separation, because myelinated axons are a functional interaction between axons and Schwann cells. In particular, defective myelination results in axonal degeneration so that the clinical progression of both axonal and demyelinating forms is ultimately caused by axonal loss.

Approximately 40% of all cases of CMT are caused by a 1.4 Mb duplication on chromosome 17 [5,6], resulting in trisomy of a critical myelin gene, peripheral myelin protein 22 (*PMP22*) (4-6); this is classified as CMT1A. Conversely, the loss of one *PMP22* allele results in another, milder neuropathy known as hereditary neuropathy with liability to pressure palsies. A variety of studies have demonstrated that the increased level of PMP22 expression is sufficient to cause a demyelinating phenotype in rodent models [7–11]. Importantly, lowering expression of PMP22 improves myelination in rodent models of CMT1A [12–14].

Aside from the copy number variation (CNV) of *PMP22*, mutations in >80 genes cause other forms of CMT [15,16]. Some of these mutations can be grouped into functional classes of molecules that are the targets of such mutations, including myelin membrane proteins, endosomal proteins, aminoacyl transferases, axonal transport, and mitochondrial proteins [6,16–18]. The most prevalent causes of CMT aside from PMP22 duplication are CMT1X caused by mutations in *GJB1* (gap junction beta 1, which encodes a gap junction protein known as connexin 32), CMT1B caused by *MPZ* mutations (myelin protein zero, a major myelin membrane protein involved in myelin compaction), and CMT2A caused by *MFN2* mutations (encoding the mitofusin 2, which is involved in mitochondrial fusion). This chapter will focus on epigenetic mechanisms that are affected in CMT and their influence on gene regulatory processes that are critical to proper formation and maintenance of peripheral nerve structure and function.

3.2 Epigenetic regulation of Schwann cell development

A number of studies in the last several years have identified the epigenetic mechanisms that shape Schwann cell development [19,20]. Schwann cells are derived from the neural crest, and their development is shaped by axonal signals especially neuregulin 1, which is required for different aspects of Schwann cell development [21,22]. The trajectory of Schwann cell development is also directed by critical transcription factors, including EGR2/Krox20 and SOX10 [23–25]. SOX10 is a high mobility group (HMG) box transcription factor that is required for Schwann cell specification, but also continues to be expressed in adult Schwann cells. SOX10 belongs to a subclass of the SOX family, containing SOX8, SOX9, and SOX10, which have a dimerization domain enabling them to bind to dimeric sites. SOX10 is important for many neural crest-derived cell types, including not only Schwann cells but also melanocytes and enteric glia. In addition, SOX10 is also required for the development of oligodendrocytes, the myelinating glia of the central nervous system [50]. The *SOX10* gene is also mutated in a syndromic condition (peripheral demyelinating neuropathy, central dysmyelinating leukodystrophy, Waardenburg syndrome, and Hirschsprung disease [PCWH]) that affects virtually all these cell types, including a peripheral neuropathy [51]. Interestingly, some of the mutations are considered to act in a dominant fashion, while others cause nonsense-mediated decay. The development of melanocytes and enteric glia is most sensitive to SOX10 dosage, as mice with heterozygous mutations display defects in these cell types. In Schwann cells, SOX10 interacts with many transcription factors and the combined action of SOX10 along with POU3F1/YY1/NFATC3/TEAD factors result in activation of a downstream regulatory element of *Egr2* [26–30], a required transcription factor for the terminal differentiation of myelinating Schwann cells. Both SOX10 and EGR2 are required for the maintenance of myelinating Schwann cells, as induced deletion in adult mice leads to demyelination [31,32]. *EGR2* is mutated in a rare form of dominantly inherited demyelinating neuropathy (CMT1D), or more severe forms, Dejerine Sottas neuropathy or congenital hypomyelinating neuropathy [33–35]. One recessive mutation has also been identified (described later).

The activity of EGR2 and SOX10 in coordinating Schwann cell development is dependent on epigenetic mechanisms, and these factors also play a role in targeting epigenetic regulators to various genes involved in myelination. One of the first investigations of epigenetic regulation was the Schwann cell−specific deletion of histone deacetylases (HDAC) 1 and 2 [36,37], which generally remove acetylation from lysine residues in histone proteins and thereby reduce transcriptional activation. In mice, deletion of *Hdac1* and *Hdac2* in early Schwann cell development causes a dramatic arrest in further differentiation, including the axonal sorting required for promyelinating Schwann cells to isolate large diameter (1 μm) axons for myelination. In vitro studies suggested that HDAC1 and HDAC2 play nonredundant roles in SC development, although either single knockout does not appear to have any phenotype. Part of the developmental arrest is due to loss of SOX10 expression, which impairs induction of promyelinating transcription factors, EGR2 and POU3F1. A subsequent study using deletion in neural crest with Wnt1-cre showed that HDAC activity is required for Pax3 induction [38], which, in turn, is required for SOX10 expression in developing Schwann cells. Although histones may be the target of HDAC, they also regulate transcription factors such as nuclear factor κB (NF-kB) that may play a role in myelination [36].

Histone deacetylases are often components of larger chromatin modification complexes, such as the mSin3 and NuRD complexes. Investigation of the NuRD complex was initiated by the characterization of NGFI-A binding (NAB) corepressors (NAB1/NAB2), which interact with EGR2 [39−41]. Interestingly, a recessive mutation of the NAB-binding domain of EGR2 was found in a family with congenital hypomyelinating neuropathy [33]. Further studies in mice found that a knockin *Egr2* mutation in the NAB-binding domain and double deletion of *Nab1/Nab2* resulted in similarly profound demyelinating neuropathies that were generally similar to the phenotype of *Egr2* null mice [42−44]. To identify the mechanism of NAB repression, yeast two-hybrid studies identified an interaction of NAB proteins with the CHD4 ATPase (chromodomain helicase DNA-binding protein) subunit of the NuRD (nucleosome remodeling and deacetylase) complex [39]. The NuRD chromatin-remodeling complex also contains HDAC1 and HDAC2, which along with SWI/SNF-like ATPase subunit (CHD3 or CHD4) can reposition nucleosomes. Because CHD4 is highly expressed in Schwann cells, a Schwann cell−specific deletion of *Chd4* was created and analyzed. Although this knockout was not as severe as the *Egr2* knockout or the combined *Nab1/Nab2* knockout, myelination was transiently delayed during development and demyelination was found in mature nerves [45]. The relatively mild phenotype may be due to residual CHD3, which is also expressed in Schwann cells. Therefore, the existence of the *EGR2* mutation ultimately highlighted a link through NAB corepressors to the NuRD chromatin-remodeling complex, which is required for normal myelination.

Some of the effect of HDAC may be through the NuRD complex, although the *Nab1/2* and *Chd4* knockouts did not show the same profound loss of SOX10 and EGR2 transcription factors as the *Hdac1/2* double knockout. Although the NuRD complex was originally considered to be primarily repressive, either CHD4 or the associated NuRD complex also activates transcription. Accordingly, the CHD4 conditional knockout nerves showed deficient repression of EGR2 target genes that are normally repressed during the course of Schwann cell development, but also reduced activation of some major myelin genes [45]. Chromatin immunoprecipitation (ChIP) assays showed colocalization of CHD4 and NuRD components at EGR2-binding sites of both repressed and active genes. As EGR2 and NAB proteins appear to have roles in both gene activation and gene repression [43,46−48], it is plausible that CHD4 plays a cooperating role in both contexts. However, it is important to note that ChIP assays do not necessarily connote function, and it is also possible that inappropriate induction of inhibitory genes could have indirect effects on myelin gene activation. Nonetheless, ChIP-seq studies have shown widespread binding of HDAC to most active promoters [49]. More recent studies have examined the role of HDAC1/2 in regulation of POU3F1 and EGR2 expression. These studies were conducted in the context of nerve injury, when Schwann cells initiate a program of demyelination and myelin debris clearance followed by stimulation of axon regeneration and finally remyelination of regenerated axons. These studies identified a functional linkage between HDAC activity and two demethylases that remove repressive histone H3K9 methylation, JMJD2C and KDM3A, which may constitute a complex required for activation of genes such as *Egr2* and *Pou3f1* [50].

NuRD likely acts with other transcription factors, and recently, the ZEB2/Sip1/Zfhx1b transcription factor has been studied in Schwann cell development, and found to be required for myelination by Schwann cells after conditional deletion [51,52]. Dominant (mostly loss of function) *ZEB2* mutations cause Mowat—Wilson syndrome, which has a number of nervous system defects, including ones in neural crest-derived cell types. Some point mutations in *ZEB2* interrupt an interaction with the NuRD chromatin-remodeling complex, and some of these effects may be attributed to its ability to silence genes by recruiting NuRD-associated HDAC.

As shown for EGR2 earlier, the function of SOX10 is also highly related to epigenetic regulators, and recent studies have probed the role of these in the development of Schwann cells [20]. SOX10 has been shown to be associated with HDAC1 in regulation of its target genes, and this appears to be required for maintenance of *Mpz* expression even in mature Schwann cells (described later) [37,53]. SOX10 also interacts with multisubunit BRG1-associated factors (BAF) complexes, which have tissue-specific compositions and play important roles in development. Core subunits of BAF complexes, BRG1 and BRM, are ATPase-dependent chromatin-remodeling enzymes [54—56], which are members of the SWI/SNF family of remodelers that were characterized in yeast and Drosophila studies. A Schwann cell—specific deletion of *Brg1* resulted in reduced expression of both POU3F1 and EGR2, as well as impaired axonal sorting and an almost complete absence of myelin. ChIP analysis revealed colocalization of BRG1 with SOX10-binding sites in the *Egr2* and *Pou3f1* genes. Although loss of EGR2 expression may largely account for the failure to myelinate, BRG1 interacts with a variety of other transcription factors, and neuregulin-regulated interaction of BRG1 with NF-kB in Schwann cells may also be important [56].

Finally, SOX10 was also found to interact with the mediator complex [57], which is involved in the interface between transcriptional activators and the transcription initiation complex and formation of looping interactions between enhancers [58]. Yeast two-hybrid studies identified an interaction with the MED12 and MED12L, subunits of mediator complex. MED12 is a mediator complex subunit, but is also associated with CDK8, which can form complexes that are independent from mediator. However, SOX10 does interact with other mediator subunits as well. Schwann cell—specific elimination of MED12 did not affect SOX10 or POU3F1 expression, but induction of EGR2 was largely absent, showing that the mediator complex is required for the onset of myelination, but not for the SOX10-dependent activation of *Pou3f1*.

3.3 Epigenetic regulation of dosage-sensitive genes

Although transcription factors and chromatin regulators have been defined in Schwann cell development, many of the profound effects of epigenetic regulation are manifested at the single gene level where such factors become bound during development. One important aspect of myelinating Schwan cells is not only that lipid production is dramatically upregulated, but there are also very highly expressed myelin genes that are constituents of the myelin membrane. Some of the most highly expressed genes are *Mpz* and *Pmp22*, but there are also high levels of other myelin genes, such as *Mag, Periaxin, Mbp, Ndrg1*, and many of these are genes mutated in human CMT. Moreover, some of these genes are dosage sensitive. In addition to *PMP22,* where monosomy or trisomy both cause neuropathy [5,6], the dosage of the *MPZ* gene is also sensitive. Heterozygous loss of *Mpz* has been shown to cause a mild neuropathy in mice [59], and in humans, some *MPZ* mutations likely cause a loss-of-function disease resulting in a milder neuropathy than other *MPZ* mutations [60,61]. In addition, *Mpz* overexpression in mice causes a dramatic neuropathy [62], and there has been a recent report of an *MPZ* duplication associated with neuropathy [63].

Interestingly, epigenetic manipulation has been shown to cause neuropathy through deregulation of *Mpz*. As discussed earlier, deletion of *Hdac1/2* in developing Schwann cells causes profound defects in Schwann cell development, but the same deletion in adult Schwann cells caused more restricted effects on integrity of the protein complexes that form the nodes of Ranvier [53], which are required for saltatory conduction by virtue of clustering of ion channels in the nodal regions. Investigation of this model revealed a 50% reduction in MPZ level, and it was subsequently found that MPZ interacts with neurofascin components of the nodal protein complex.

Investigation of these critical dosage-sensitive genes using epigenomic analysis has provided a number of insights. For example, initial studies of *Pmp22* regulation had focused on *Pmp22* promoters [64], and then subsequent transgenic analysis identified an enhancer region upstream of the gene [65], known as the late myelinating Schwann cell enhancer (LMSE). However, the LMSE did not recapitulate the developmental expression pattern of *Pmp22* in transgenic assays, suggesting that there must be other critical regulatory elements. To perform a more comprehensive analysis of *Pmp22* regulation, ChIP techniques were used to map binding sites of EGR2 and SOX10 [48]. Peripheral nerve is an excellent substrate for ChIP analysis: the sciatic nerve do not contain neuronal nuclei, and is highly enriched in Schwann cells, which selectively express *Pmp22*. Moreover, most Schwann cells are myelinating, whereas the remaining nonmyelinating Remak Schwann cells do not express high levels of myelin genes.

Our analysis identified binding sites for EGR2 and SOX10 within an intronic regulatory element in *Pmp22*. The intronic site responds to EGR2 and SOX10 activity in transient transfection assays, and also drives tissue-specific expression to peripheral nerve in mouse transgenic assays [66]. The initial identification of this novel enhancer relied on ChIP-chip analysis, which could only cover selected genomic windows, but the extension of this approach with ChIP-seq identified a further cluster of enhancers upstream of *Pmp22* [29,48,67]. The three upstream elements (labeled A, B, C in Fig. 3.2) have several characteristics of typical enhancer elements. First, all three elements show binding of EGR2 and/or SOX10 and have conserved consensus binding sites for these factors. Second, the three genomic segments show EGR2-and/or SOX10-dependent activity in reporter assays. Third, chromatin structure analysis of the *Pmp22* locus revealed that regions A, B, and C bear chromatin modifications typically found in regulatory regions [29,68], such as histone H3K27 acetylation (H3K27ac, Fig. 3.2), a marker of actively engaged enhancers. This cluster of enhancers actually meets the criteria of super enhancers, which are extended genomic regions with high levels of transcription factor binding and histone modifications associated with active transcription.

Interestingly, the human sequences of these enhancers reside within two independent duplications (of <200 kb) identified by CNV analysis of patients with a mild form of CMT [69,70]. In these patients, only these enhancers are duplicated, not the *PMP22* gene itself. The two duplicated regions

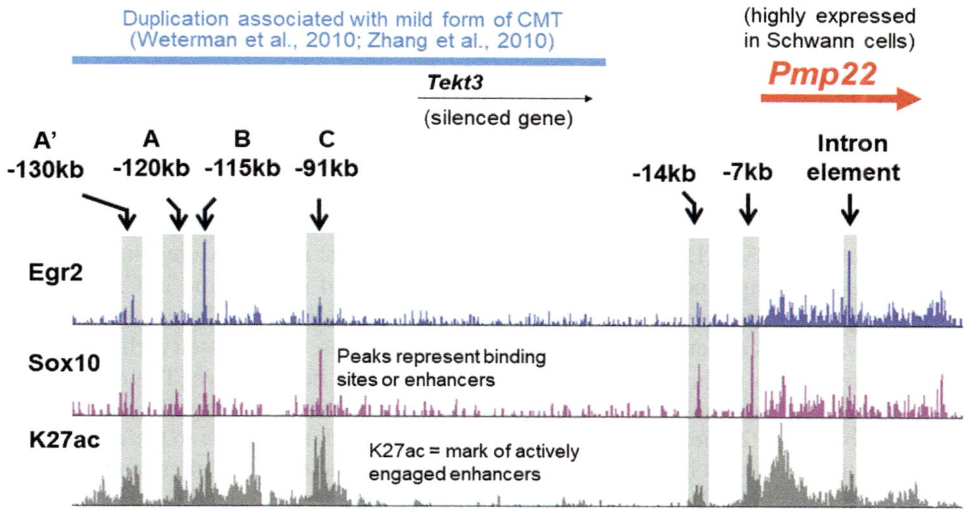

FIGURE 3.2 Chromatin structure of a dosage-sensitive myelin gene.

The diagram shows ChIP-seq data for the *Pmp22* gene, which is duplicated in the most common form of CMT, known as CMT1A. The rat *Pmp22* gene is shown by the red arrow and the tracks show ChIP-seq data for the Egr2 and Sox10 transcription factors, along with a marker of actively engaged enhancers, histone H3K27 acetylation. Enhancers are shaded by gray, and a cluster of upstream enhancers (A′, A, B, and C) constitute a domain that qualifies as a super enhancer. The position of the *Tekt3* gene is indicated, which is silent in Schwann cells. Although the normal duplication in CMT1A is 1.4 Mb surrounding the entire *PMP22* gene, there have been smaller duplications associated with mild forms of CMT, indicated by a blue bar, which would duplicate the super enhancer.

overlap by 168 kb and encompass not only the three enhancers, but also contain the intervening *TekT3* gene, which is primarily expressed in the testes [71]. The *Tekt3* gene is inactive in Schwann cells, and therefore is not likely the target of these enhancers. Our data are consistent with the possibility that the upstream duplications alter *PMP22* regulation by inserting an extra set of enhancers upstream of the gene. The upstream cluster of enhancers is not absolutely required for *Pmp22* expression, because genomic clones of *Pmp22* lacking the upstream cluster have been used to make transgenic models of CMT1A [72]. However, the lack of the upstream cluster in these genomic clones may explain why several copies (>3) of the locus must be inserted to significantly increase *PMP22* mRNA levels.

As *Pmp22* regulation appears to involve several enhancers, several of which are remote, we initiated a project to use genome editing to embed reporters in the endogenous *Pmp22* locus, so that reporters would reflect the long-range enhancers and also epigenetic regulation (e.g., chromatin structure, microRNA regulation) of the *Pmp22* gene. We used TALEN-mediated recombination to insert reporters in the S16 rat Schwann cell line as a fusion within the endogenous *Pmp22* gene with an intervening 2a self-cleaving sequence [73], and this approach was able to identify candidate compounds that were not identified using single enhancer constructs. These studies identified HDAC inhibitors as regulators of *Pmp22* expression; romidepsin was a particularly potent inhibitor of expression. Follow-up studies showed that other HDAC inhibitors could reduce *Pmp22* expression, although several others seemed ineffective, suggesting that specific HDAC subtypes are involved (unpublished data).

Genome-wide data of histone H3K27 acetylation in peripheral nerve [68] provide enhancer maps of all the regulatory elements associated with myelin genes. Their analysis indicates binding sites of well-established regulators of Schwann cell development, and provides clues regarding novel regulators of Schwann cell differentiation. Although tremendous progress has been made in the identification of mutations in genes associated with neuropathy, mutations have not been identified for some cases of demyelinating CMT. However, it is important to note that mutation screening has principally relied on exome screening, and analysis of the "dark matter" of the genome in CMT is still in its infancy. Although the most prevalent cause of CMT is a gene duplication, there may be a number of other structural variations affecting noncoding regulatory elements, as suggested earlier for *PMP22*. In addition, point mutations within regulatory elements could also be causative. The only known such causes for CMT are mutations associated with EGR2 and SOX10-binding sites in the Schwann cell–specific promoter (P2) the *GJB1* gene [74,75]. These were identified due to their proximity to the transcriptional start site, and it is also important to note that characterization of enhancers in the *GJB1* gene suggests that there is only one major enhancer. Nonetheless, it is hoped that epigenomic annotation of the histone modifications will help identify mutations within enhancer regions that diminish expression of Schwann cell–specific genes.

3.4 Epigenetic regulators targeted by CMT mutations

Although epigenetic signaling is disrupted in different mutations causing demyelinating forms of CMT, there is a greater diversity of genetic mutations associated with axonal CMT2 neuropathies [15,16], and mutations of epigenetic regulators form a significant subset in this type of inherited neuropathy (see Table 3.1). The following summarizes the genes mutated in axonal neuropathies and what is known about pathomechanisms of such mutations.

3.4.1 DNMT1

DNMT1 is a DNA methyltransferase that functions during DNA replication to methylate hemimethylated sites, that is, DNMT1 methylates cytosines in CpG dinucleotides on newly replicated strands in the wake of the replication fork through methylated DNA regions. It is therefore quite important in maintaining the epigenetic aspect of DNA methylation as it preserves methylation that would otherwise be lost as a function of DNA replication. DNA methylation is generally associated with gene silencing, and methylation plays a profound role in development of the nervous system. Because of the ubiquitous nature of DNMT1 activity, and because neurons do not divide, it was somewhat surprising that dominant mutations in *DNMT1* cause peripheral and central neurodegeneration in a syndrome also involving dementia and hearing loss, classified as Hereditary sensory and autonomic neuropathy (HSAN I) [76–79], or an alternate clinical presentation, cerebellar ataxia along with deafness and narcolepsy, which share a triad of core symptoms including hearing loss, pan sensory neuropathy, and cognitive decline. While Tyr495 is a hot spot for HSANIE, the mutations appear to cause unstable protein and therefore may cause loss-of-function, although haploinsufficiency

Table 3.1 CMT mutations affecting epigenetic regulation.

Gene	Clinical classification	Heritability
LMNA, lamin A	CMT2B1	Recessive
DNMT1, DNA methyltransferase 1	HSAN1E	Dominant
MED25, mediator 25	AR-CMT2B/CMT2B2	Recessive
MORC2, Morc family CW-type zinc finger 2	CMT2Z	Dominant
SETX, senataxin	AOA2, ALS4	Dominant for ALS4
PRDM12, PR/SET domain 12	HSAN8	Recessive
SYNE1, spectrin repeat nuclear envelope 1	Motor neuropathy	Recessive
EGR2, early growth response 2	CMT1D/CMT4E	Dominant/recessive
SOX10	PCWH	Dominant

ALS, *Amyotrophic Lateral Sclerosis*, AOA, *Ataxia with oculomotor apraxia*, AR, *Autosomal Recessive*, CMT, *Charcot-Marie Tooth*, HSAN, *Hereditary Sensory and Autonomic Neuropathy*.

of *Dnmt1* in mice was not reported to cause neurodegeneration [80,81]. Accordingly, there was some reduction in DNA methylation, but there also appears to be functions of DNMT1 outside of S phase, such as binding of DNMT1 to heterochromatin. Moreover, the mutant protein accumulates in cytoplasmic aggresomes that may impact protein homeostatic mechanisms in neurons. Methylation profiling has identified genome-wide changes in methylation patterns caused by *DNMT1* mutations. Neuron-specific knockout mice have been made for *Dnmt1* and homozygous knockout does impair neuronal viability [80,81].

3.4.2 LMNA

LMNA encodes isoforms of scaffolding proteins called lamins, a major component of a nuclear envelope complex that is connected to the actin cytoskeleton, which has been termed the LINC (linker of nucleoskeleton and cytoskeleton) complex. Lamins form the 2D matrix composed of intermediate filaments that constitute the nuclear lamin on the inner side of the nuclear envelope. A single homozygous mutation in *LMNA*, Arg298Cys, was found to be associated with all members of Algerian families with autosomal recessive CMT2 [82–84]. This mutation caused a severe neuropathy, with early onset, symmetrical muscle weakness and wasting, foot deformities, and walking difficulties associated with reduced or absent tendon reflexes. *Lmna*-deficient mice also develop a peripheral neuropathy, with loss of large myelinated fibers. However, a homozygous R298C knockin mouse model did not display neuropathy and surprisingly found induction of *Pmp22* but with no apparent demyelinating neuropathy [82,85].

The *LMNA* gene is mutated in a variety of other genetic disorders affecting muscle (Emery–Dreifuss muscular dystrophy), lipodystrophies, and premature aging (Hutchinson–Gilford progeria). Studies of fibroblasts from such patients show changes in heterochromatin-associated histone modification patterns (H3K9me3, H4K20me3) and EZH2 and H3K27me3 are down regulated [86,87]. However, neuropathy is associated with highly site-specific amino acid substitutions in

LMNA such as R298C. These findings indicate that LMNA may have distinct functional domains for tissue-specific protein interactions, but the mechanism by which this mutation causes neuropathy is unclear.

3.4.3 SYNE1

Other LINC components, SYNE1 and SYNE2 (also known as nesprin proteins), are spectrin repeat proteins that are transmembrane proteins in the nuclear envelope that can bind actin. Mutations of *SYNE1* (nesprin 1) have been identified in motor neuropathies, and like *LMNA*, also in Emery–Dreifuss muscular dystrophy [86,88]. *SYNE1* mutations themselves can cause pure cerebellar ataxia along with cognitive decline, or sometimes copresent with motor neuropathy [89]. The phenotype of motor neuropathy suggests more neuronal defects, but studies of oligodendrocytes have shown that expression of LINC components are developmentally regulated, and silencing of *Syne1* resulted in chromatin changes and impaired myelination [90]. It was proposed that the LINC complex forms a mechanotransduction pathway that conveys signals to the nucleus in oligodendrocytes, and it is possible that it plays a similar role in Schwann cells.

3.4.4 MED25

MED25 is a component of the mediator complex that is essential to the transcriptional control of most RNA polymerase II-dependent genes. Along with other cofactors, MED25 is recruited as a part of a preinitiation complex assembly that serves as a physical bridge between the promoter and the regulatory elements of a gene. This structural formation is considered to integrate multiple information sources into a signaling hub leading to the fine-tuning of gene expression. Interestingly, a homozygous p.A335V mutation in the MED25 gene was found to be linked to the highly specific neuropathy in an extended Costa Rican family, largely afflicted with autosomal recessively inherited CMT2B2 [91]. This mutation could lead to loss of binding specificity by SH3 domain proteins such as ABL and LCK proteins. However, the relationship between MED25 and these potential binding partners is not well characterized. It is also likely that there are additional unknown interacting partners of MED25. Given the interaction of SOX10 with the mediator complex described earlier, one might expect the phenotype to be demyelinating although the clinical signs are more consistent with an axonal neuropathy. However, as noted, Schwann cell dysfunction can lead to axonal degeneration. Utilizing mouse and rat models, it was noted that MED25 and PMP22 dosage are correlated suggesting that the expression of PMP22 may be dependent on the expression level of MED25 [91]. In addition, nerve injury experiments suggest that expression levels in Schwann cells are regulated by axonal signals.

3.4.5 SETX

Heterozygous mutations of the *SETX* (senataxin) gene cause hereditary motor neuropathy with pyramidal features (also known as ALS4) [92]. Recessive *SETX* mutations cause ataxia with oculomotor apraxia (AOA2). This gene shares homology with the yeast Sen1 protein, which has RNA helicase activity at the C-terminus domain. Senataxin is involved in transcription termination at RNA polymerase pausing sites [93], and is known to localize to R loops (RNA-DNA hybrids that arise during transcriptional pausing caused by DNA damage). Transcriptional profiling of patient fibroblasts overexpressing SETX mutants, blood samples from affected patients, and *Setx*-deficient mice revealed dysregulation of genes involved in RNA metabolism and DNA damage/repair [94]. However, the mouse knockout does not recapitulate the AOA2 phenotype, so recent studies of iPSC-derived neural progenitors have been developed as a model system [95] that displays increased sensitivity to DNA damage. The high degree of similarity of Sen1 has enabled the demonstration that *SETX* mutants cause termination read through and thereby alter transcription [96,97]. Interestingly, both SETX and SMN (survival of motor neurons), which is mutated in spinal muscular atrophy, interact with the C-terminal domain of RNA polymerase II, which is regulated by symmetric demethylation of Arg1810 by PRMT5 (protein arginine methyltransferase 5) [98]. Therefore, this transcription termination pathway appears to be related to multiple neurodegenerative disorders.

Given the association of SETX with R loops, a very recent study used cells from a patient with the ALS4 L389S mutation to determine if R loop prevalence was reduced, and cells appeared to have a diminished level of R loops [99]. Expression profiling of patient fibroblasts identified activation of the TGF beta pathway, which was confirmed in analysis of patient spinal cords. Moreover, it was found that an R loop exists within the promoter of the *BAMBI* gene, which encodes a decoy receptor that inhibits activation of the TGF beta pathway. The *SETX* mutation modestly reduces *BAMBI* expression, which is associated with increased DNA methylation of the *BAMBI* promoter. Mechanistic studies suggest that R loops normally inhibit methylation by DNMT1, which may explain how *SETX* mutations affect regulation of *BAMBI* and other genes in motor neurons. Another study provides evidence that SETX plays a role in double strand break repair when breaks occur within transcribed regions, and that SETX recruits RAD51 to prevent inappropriate joining of ends to distal break sites [100].

3.4.6 MORC2

Dominant *MORC2* mutations were recently classified as CMT2Z due to the associated axonal neuropathy, but some cases show a severe spinal muscular atrophy-like disease [101,102]. *MORC2* mutations appear to be relatively common among axonal neuropathies, and in one study, it was the second most prevalent cause of CMT2, after *MFN2* mutations [101]. The MORC2 cofactor is recruited as part of the human silencing hub (HUSH) complex, which deposits repressive H3K9me3 by histone methyltransferase SETDB1 and subsequently silences its target genes. MORC2's role was recently elucidated through a CRISPR-Cas9 screen in HeLa cultures using a reporter that is normally silenced by the HUSH complex. Loss of MORC2 function resulted in transcriptional derepression with concomitant loss of H3K9me3. MORC2 regulates the condensation of heterochromatin in response to DNA damage as well as transcriptional repression. Interestingly, MORC2 and other HUSH subunits turned up in a screen of proteins that repress transcription and transposition of the LINE-1 transposable elements [103].

One mutation is within MORC2's ATPase domain p.R252W, which is a novel linkage to a H3K9 methylation complex. In vitro assays indicated that this mutant leads to hyper-repression by the HUSH complex, and therefore it is considered that gain-of-function mechanism leads to increased and/or inappropriate silencing. Accordingly, expression of this *MORC2* mutant in a neuroblastoma cell line led to increased repression of a number of genes, including a large number of zinc finger (ZFN) transcription factor genes [104]. MORC2 is also expressed in Schwann cells [102], thus it is possible that Schwann cell dysfunction caused by MORC2 loss-of-function leads to axon degeneration. Some *MORC2* mutations lie within the ATPase domain and can affect dimerization dynamics [105], and it was found that repression activity of MORC2 mutants is inversely correlated with ATPase activity.

3.4.7 PRDM12

Recessive mutations in *PRDM12* cause a hereditary sensory neuropathy with insensitivity to pain (inflammation, hot/cold), and analysis of biopsies of cutaneous nerves and skin revealed loss of nociceptive axons: small myelinated Aδ fibers and unmyelinated C fibers [106]. PRDM12 is involved in specifying spinal interneuron and nociceptor neurogenesis, and PRDM12 acts as a repressor of the *Dbx1* and *Nkx6* transcription factor genes [107,108]. PRDM12 is expressed in progenitors of nociceptor sensory neurons, and mutations interfere with histone modification [106]. There is a functional homolog in Drosophila, Hamlet, which is involved in the development of nociceptor neurons [107], and several of the human mutations occur in highly conserved residues. PRDM12 has a SET methyltransferase domain along with three zinc fingers and a polyalanine stretch. Although it lacks intrinsic histone methylase activity, it interacts with the G9A/EHMT2 methyltransferase that creates the repressive histone H3K9me2 modification. Some mutations apparently affect interaction with G9A/EHMT2, but the effects of other mutations remain to be identified.

3.5 Novel mechanisms for CMT mutations

Although most genes mutated in CMT encode proteins with diverse localizations and cellular processes, there is some evidence that diverse CMT-associated mutations may have common effects on DNA damage pathways. One marker of DNA damage responses is phosphorylation of the variant histone, H2AX, which is targeted by several kinases, phosphoinositide-3-kinase-related protein kinase (PIKK), ataxia telangiectasia mutated (ATM), ATM- and RAD3-related (ATR), and DNA-dependent protein kinase (DNA-PKs), which all are involved in DNA repair or checkpoints. A genome-wide siRNA screen in HeLa cells for increased H2AX phosphorylation [109] identified several genes involved in CMT, including demyelinating (*EGR2, PMP22, GJB1, MPZ, NDRG1*) and axonal (*LMNA, HSPB1, NEFL*) forms, along with the expected known DNA repair proteins. This raises the intriguing possibility that these proteins could regulate DNA repair, and the increasing axonal degeneration with age may reflect, in part, the accumulation of DNA damage in Schwann cells and neurons.

3.6 Summary

The study of peripheral neuropathies has been a major success in identification of genetic causes. It is anticipated that the number of genetic causes will continue to grow, particularly if syndromic conditions that exhibit peripheral neuropathy are included [109]. Some of the major neuropathies are directly related to control of transcription (e.g., CMT1A), so it seems quite possible that epigenetic therapies could be applied to such conditions. Overall, characterization of the role of chromatin-modifying enzymes can be complicated due to the potential diverse roles of these complexes.

Acknowledgments

We thank Holly Hung and Ki Ma for providing electron microscopy images, and Steve Scherer, Michael Shy, and Chris Klein for critical comments. PD was supported by NIH training grant T32GM007215.

References

[1] Kidd GJ, Ohno N, Trapp BD. Biology of Schwann cells. Handb Clin Neurol 2013;115:55–79.
[2] Nave KA, Werner HB. Myelination of the nervous system: mechanisms and functions. Annu Rev Cell Dev Biol 2014;30:503–33.
[3] Reilly MM, Murphy SM, Laurá M. Charcot-Marie-Tooth disease. J Peripher Nerv Syst 2011;16:1–14.
[4] Saporta MA, Shy ME. Inherited peripheral neuropathies. Neurol Clin 2013;31:597–619.
[5] Saporta AS, Sottile SL, Miller LJ, Feely SM, Siskind CE, Shy ME. Charcot-Marie-Tooth disease subtypes and genetic testing strategies. Ann Neurol 2011;69:22–33.
[6] Murphy SM, Laura M, Fawcett K, Pandraud A, Liu YT, Davidson GL, Rossor AM, Polke JM, Castleman V, Manji H, et al. Charcot-Marie-Tooth disease: frequency of genetic subtypes and guidelines for genetic testing. J Neurol Neurosurg Psychiatry 2012;83:706–10.
[7] Magyar J, Martini R, Ruelicke T, Aguzzi A, Adlkofer K, Dembic Z, Zielasek J, Toyka K, Suter U. Impaired differentiation of Schwann cells in transgenic mice with increased PMP22 gene dosage. J Neurosci 1996; 16:5351–60.
[8] Huxley C, Passage E, Manson A, Putzu G, Figarella-Branger D, Pellissier J, Fontés M. Construction of a mouse model of Charcot-Marie-Tooth disease type 1A by pronuclear injection of human YAC DNA. Hum Mol Genet 1996;5:563–9.
[9] Huxley C, Passage E, Robertson A, Youl B, Huston S, Manson A, Sabéran-Djoniedi D, Figarella-Branger D, Pellissier J, Thomas P, et al. Correlation between varying levels of PMP22 expression and the degree of demyelination and reduction in nerve conduction velocity in transgenic mice. Hum Mol Genet 1998;7:449–58.
[10] Robertson A, Perea J, McGuigan A, King R, Muddle J, Gabreëls-Festen A, Thomas P, Huxley C. Comparison of a new PMP22 transgenic mouse line with other mouse models and human patients with CMT1A. J Anat 2002;200:377–90.
[11] Sereda M, Griffiths I, Pühlhofer A, Stewart H, Rossner M, Zimmerman F, Magyar J, Schneider A, Hund E, Meinck H, et al. A transgenic rat model of Charcot-Marie-Tooth disease. Neuron 1996;16:1049–60.
[12] Passage E, Norreel J, Noack-Fraissignes P, Sanguedolce V, Pizant J, Thirion X, Robaglia-Schlupp A, Pellissier J, Fontés M. Ascorbic acid treatment corrects the phenotype of a mouse model of Charcot-Marie-Tooth disease. Nat Med 2004;10:396–401.
[13] Perea J, Robertson A, Tolmachova T, Muddle J, King R, Ponsford S, Thomas P, Huxley C. Induced myelination and demyelination in a conditional mouse model of Charcot-Marie-Tooth disease type 1A. Hum Mol Genet 2001;10:1007–18.
[14] Sereda MW, Meyer zu Horste G, Suter U, Uzma N, Nave KA. Therapeutic administration of progesterone antagonist in a model of Charcot-Marie-Tooth disease (CMT-1A). Nat Med 2003;9:1533–7.
[15] Timmerman V, Strickland AV, Züchner S. Genetics of Charcot-Marie-Tooth (CMT) disease within the frame of the human genome project success. Genes 2014;5:13–32.
[16] Rossor AM, Tomaselli PJ, Reilly MM. Recent advances in the genetic neuropathies. Curr Opin Neurol 2016;29:537–48.

[17] Fridman V, Bundy B, Reilly MM, Pareyson D, Bacon C, Burns J, Day J, Feely S, Finkel RS, Grider T, et al. CMT subtypes and disease burden in patients enrolled in the inherited neuropathies consortium natural history study: a cross-sectional analysis. J Neurol Neurosurg Psychiatry 2015;86:873–8.
[18] Rossor AM, Polke JM, Houlden H, Reilly MM. Clinical implications of genetic advances in Charcot-Marie-Tooth disease. Nat Rev Neurol 2013;9:562–71.
[19] Jacob C. Chromatin-remodeling enzymes in control of Schwann cell development, maintenance and plasticity. Curr Opin Neurobiol 2017;47:24–30.
[20] Weider M, Reiprich S, Wegner M. Sox appeal – Sox10 attracts epigenetic and transcriptional regulators in myelinating glia. Biol Chem 2013;394:1583–93.
[21] Jessen KR, Mirsky R, Lloyd AC. Schwann cells: development and role in nerve repair. Cold Spring Harb Perspect Biol 2015;7:a020487.
[22] Grigoryan T, Birchmeier W. Molecular signaling mechanisms of axon-glia communication in the peripheral nervous system. Bioessays 2015;37:502–13.
[23] Svaren J, Meijer D. The molecular machinery of myelin gene transcription in Schwann cells. Glia 2008;56:1541–51.
[24] Meijer D, Svaren J. Specification of macroglia by transcription factors: Schwann cells. In: Alvarez-Buylla A, Rowitch D, editors. Patterning and cell type specification in the developing CNS and PNS comprehensive developmental neuroscience, vol. 1. Elsevier; 2013. p. 759–70.
[25] Stolt CC, Wegner M. Schwann cells and their transcriptional network: evolution of key regulators of peripheral myelination. Brain Res 2015;1641(Pt A):101–10.
[26] Reiprich S, Kriesch J, Schreiner S, Wegner M. Activation of Krox20 gene expression by Sox10 in myelinating Schwann cells. J Neurochem 2010;112:744–54.
[27] He Y, Kim JY, Dupree J, Tewari A, Melendez-Vasquez C, Svaren J, Casaccia P. Yy1 as a molecular link between neuregulin and transcriptional modulation of peripheral myelination. Nat Neurosci 2010;13:1472–80.
[28] Kao SC, Wu H, Xie J, Chang CP, Ranish JA, Graef IA, Crabtree GR. Calcineurin/NFAT signaling is required for neuregulin-regulated Schwann cell differentiation. Science 2009;323:651–4.
[29] Lopez-Anido C, Poitelon Y, Gopinath C, Moran JJ, Ma KH, Law WD, Antonellis A, Feltri ML, Svaren J. Tead1 regulates the expression of peripheral myelin protein 22 during Schwann cell development. Hum Mol Genet 2016;25:3055–69.
[30] Ghislain J, Charnay P. Control of myelination in Schwann cells: a krox20 cis-regulatory element integrates Oct6, Brn2 and Sox10 activities. EMBO Rep 2006;7:52–8.
[31] Finzsch M, Schreiner S, Kichko T, Reeh P, Tamm E, Bösl M, Meijer D, Wegner M. Sox10 is required for Schwann cell identity and progression beyond the immature Schwann cell stage. J Cell Biol 2010;189:701–12.
[32] Decker L, Desmarquet-Trin-Dinh C, Taillebourg E, Ghislain J, Vallat JM, Charnay P. Peripheral myelin maintenance is a dynamic process requiring constant Krox20 expression. J Neurosci 2006;26:9771–9.
[33] Warner LE, Mancias P, Butler IJ, McDonald CM, Keppen L, Koob KG, Lupski JR. Mutations in the early growth response 2 (EGR2) gene are associated with hereditary myelinopathies. Nat Genet 1998;18:382–4.
[34] Warner LE, Svaren J, Milbrandt J, Lupski JR. Functional consequences of mutations in the early growth response 2 gene (EGR2) correlate with severity of human myelinopathies. Hum Mol Genet 1999;8:1245–51.
[35] Boerkoel CF, Takashima H, Bacino CA, Daentl D, Lupski JR. EGR2 mutation R359W causes a spectrum of Dejerine-Sottas neuropathy. Neurogenetics 2001;3:153–7.
[36] Chen Y, Wang H, Yoon SO, Xu X, Hottiger MO, Svaren J, Nave KA, Kim HA, Olson EN, Lu QR. HDAC-mediated deacetylation of NF-κB is critical for Schwann cell myelination. Nat Neurosci 2011;14:437–41.

[37] Jacob C, Christen CN, Pereira JA, Somandin C, Baggiolini A, Lötscher P, Ozçelik M, Tricaud N, Meijer D, Yamaguchi T, et al. HDAC1 and HDAC2 control the transcriptional program of myelination and the survival of Schwann cells. Nat Neurosci 2011;14:429–36.

[38] Jacob C, Lötscher P, Engler S, Baggiolini A, Varum Tavares S, Brügger V, John N, Büchmann-Møller S, Snider PL, Conway SJ, et al. HDAC1 and HDAC2 control the specification of neural crest cells into peripheral glia. J Neurosci 2014;34:6112–22.

[39] Srinivasan R, Mager GM, Ward RM, Mayer J, Svaren J. NAB2 represses transcription by interacting with the CHD4 subunit of the NuRD complex. J Biol Chem 2006;281:15129–37.

[40] Russo MW, Sevetson BR, Milbrandt J. Identification of NAB1, a repressor of NGFI-A and Krox20 mediated transcription. Proc.Natl.Acad.Sci.(USA) 1995;92:6873–7.

[41] Svaren J, Sevetson BR, Golda T, Stanton JJ, Swirnoff AH, Milbrandt J. Novel mutants of NAB corepressors enhance activation by Egr transactivators. EMBO J 1998;17:6010–9.

[42] Baloh RH, Strickland A, Ryu E, Le N, Fahrner T, Yang M, Nagarajan R, Milbrandt J. Congenital hypomyelinating neuropathy with lethal conduction failure in mice carrying the Egr2 I268N mutation. J Neurosci 2009;29:2312–21.

[43] Le N, Nagarajan R, Wang JY, Svaren J, LaPash C, Araki T, Schmidt RE, Milbrandt J. Nab proteins are essential for peripheral nervous system myelination. Nat Neurosci 2005;8:932–40.

[44] Desmazieres A, Decker L, Vallat JM, Charnay P, Gilardi-Hebenstreit P. Disruption of Krox20-Nab interaction in the mouse leads to peripheral neuropathy with biphasic evolution. J Neurosci 2008;28:5891–900.

[45] Hung HA, Kohnken R, Svaren J. The nucleosome remodeling and deacetylase chromatin remodeling (NuRD) complex is required for peripheral nerve myelination. J Neurosci 2012;32:1517–27.

[46] Le N, Nagarajan R, Wang JY, Araki T, Schmidt RE, Milbrandt J. Analysis of congenital hypomyelinating Egr2Lo/Lo nerves identifies Sox2 as an inhibitor of Schwann cell differentiation and myelination. Proc Natl Acad Sci U S A 2005;102:2596–601.

[47] Sevetson B, Svaren J, Milbrandt J. A novel activation function for NAB proteins in EGR-dependent transcription of the luteinizing hormone beta gene. J Biol Chem 2000;275:9749–57.

[48] Srinivasan R, Sun G, Keles S, Jones EA, Jang SW, Krueger C, Moran JJ, Svaren J. Genome-wide analysis of EGR2/SOX10 binding in myelinating peripheral nerve. Nucleic Acids Res 2012;40:6449–60.

[49] Wang Z, Zang C, Cui K, Schones DE, Barski A, Peng W, Zhao K. Genome-wide mapping of HATs and HDACs reveals distinct functions in active and inactive genes. Cell 2009;138:1019–31.

[50] Brügger V, Duman M, Bochud M, Münger E, Heller M, Ruff S, Jacob C. Delaying histone deacetylase response to injury accelerates conversion into repair Schwann cells and nerve regeneration. Nat Commun 2017;8:14272.

[51] Quintes S, Brinkmann BG, Ebert M, Fröb F, Kungl T, Arlt FA, Tarabykin V, Huylebroeck D, Meijer D, Suter U, et al. Zeb2 is essential for Schwann cell differentiation, myelination and nerve repair. Nat Neurosci 2016;19:1050–9.

[52] Wu LM, Wang J, Conidi A, Zhao C, Wang H, Ford Z, Zhang L, Zweier C, Ayee BG, Maurel P, et al. Zeb2 recruits HDAC-NuRD to inhibit Notch and controls Schwann cell differentiation and remyelination. Nat Neurosci 2016;19:1060–72.

[53] Brügger V, Engler S, Pereira JA, Ruff S, Horn M, Welzl H, Münger E, Vaquié A, Sidiropoulos PN, Egger B, et al. HDAC1/2-Dependent P0 expression maintains paranodal and nodal integrity independently of myelin stability through interactions with neurofascins. PLoS Biol 2015;13:e1002258.

[54] Marathe HG, Mehta G, Zhang X, Datar I, Mehrotra A, Yeung KC, de la Serna IL. SWI/SNF enzymes promote SOX10- mediated activation of myelin gene expression. PLoS One 2013;8:e69037.

[55] Weider M, Küspert M, Bischof M, Vogl MR, Hornig J, Loy K, Kosian T, Müller J, Hillgärtner S, Tamm ER, et al. Chromatin-remodeling factor Brg1 is required for Schwann cell differentiation and myelination. Dev Cell 2012;23:193–201.

[56] Limpert AS, Bai S, Narayan M, Wu J, Yoon SO, Carter BD, Lu QR. NF-κB forms a complex with the chromatin remodeler BRG1 to regulate Schwann cell differentiation. J Neurosci 2013;33:2388–97.
[57] Vogl MR, Reiprich S, Küspert M, Kosian T, Schrewe H, Nave KA, Wegner M. Sox10 cooperates with the mediator subunit 12 during terminal differentiation of myelinating glia. J Neurosci 2013;33:6679–90.
[58] Kagey MH, Newman JJ, Bilodeau S, Zhan Y, Orlando DA, van Berkum NL, Ebmeier CC, Goossens J, Rahl PB, Levine SS, et al. Mediator and cohesin connect gene expression and chromatin architecture. Nature 2010;467:430–5.
[59] Martini R, Zielasek J, Toyka KV, Giese KP, Schachner M. Protein zero (P0)-deficient mice show myelin degeneration in peripheral nerves characteristic of inherited human neuropathies. Nat Genet 1995;11: 281–6.
[60] Shy ME. HMSN related to MPZ (P0) mutations. In: Dyck PJ, Thomas PK, editors. Peripheral neuropathy. W.B. Saunders; 2005. p. 1681–706.
[61] Grandis M, Vigo T, Passalacqua M, Jain M, Scazzola S, La Padula V, Brucal M, Benvenuto F, Nobbio L, Cadoni A, et al. Different cellular and molecular mechanisms for early and late-onset myelin protein zero mutations. Hum Mol Genet 2008;17:1877–89.
[62] Wrabetz L, Feltri ML, Quattrini A, Imperiale D, Previtali S, D'Antonio M, Martini R, Yin X, Trapp BD, Zhou L, et al. P(0) glycoprotein overexpression causes congenital hypomyelination of peripheral nerves. J Cell Biol 2000;148:1021–34.
[63] Feely S, Saade D, Shy ME. Duplication of myelin protein zero causing early onset Charcot-Marie-Tooth disease Type 1B. J Peripher Nerv Syst 2017;22:281.
[64] Suter U, Snipes GJ, Schoener-Scott R, Welcher AA, Pareek S, Lupski JR, Murphy RA, Shooter EM, Patel PI. Regulation of tissue-specific expression of alternative peripheral myelin protein-22 (PMP22) gene transcripts by two promoters. J Biol Chem 1994;269:25795–808.
[65] Maier M, Castagner F, Berger P, Suter U. Distinct elements of the peripheral myelin protein 22 (PMP22) promoter regulate expression in Schwann cells and sensory neurons. Mol Cell Neurosci 2003;24:803–17.
[66] Jones EA, Lopez-Anido C, Srinivasan R, Krueger C, Chang LW, Nagarajan R, Svaren J. Regulation of the PMP22 gene through an intronic enhancer. J Neurosci 2011;31:4242–50.
[67] Jones EA, Brewer MH, Srinivasan R, Krueger C, Sun G, Charney KN, Keles S, Antonellis A, Svaren J. Distal enhancers upstream of the Charcot-Marie-Tooth type 1A disease gene PMP22. Hum Mol Genet 2012;21:1581–91.
[68] Hung HA, Sun G, Keles S, Svaren J. Dynamic regulation of Schwann cell enhancers after peripheral nerve injury. J Biol Chem 2015;290:6937–50.
[69] Weterman MA, van Ruissen F, de Wissel M, Bordewijk L, Samijn JP, van der Pol WL, Meggouh F, Baas F. Copy number variation upstream of PMP22 in Charcot-Marie-Tooth disease. Eur J Hum Genet 2010;18: 421–8.
[70] Zhang F, Seeman P, Liu P, Weterman MA, Gonzaga-Jauregui C, Towne CF, Batish SD, De Vriendt E, De Jonghe P, Rautenstrauss B, et al. Mechanisms for nonrecurrent genomic rearrangements associated with CMT1A or HNPP: rare CNVs as a cause for missing heritability. Am J Hum Genet 2010;86:892–903.
[71] Roy A, Lin YN, Agno JE, DeMayo FJ, Matzuk MM. Tektin 3 is required for progressive sperm motility in mice. Mol Reprod Dev 2009;76:453–9.
[72] Sereda MW, Nave KA. Animal models of Charcot-Marie-Tooth disease type 1A. NeuroMolecular Med 2006;8:205–16.
[73] Fang J, Qian JJ, Yi S, Harding TC, Tu GH, VanRoey M, Jooss K. Stable antibody expression at therapeutic levels using the 2A peptide. Nat Biotechnol 2005;23:584–90.
[74] Houlden H, Girard M, Cockerell C, Ingram D, Wood NW, Goossens M, Walker RW, Reilly MM. Connexin 32 promoter P2 mutations: a mechanism of peripheral nerve dysfunction. Ann Neurol 2004;56:730–4.

[75] Tomaselli PJ, Rossor AM, Horga A, Jaunmuktane Z, Carr A, Saveri P, Piscosquito G, Pareyson D, Laura M, Blake JC, et al. Mutations in noncoding regions of GJB1 are a major cause of X-linked CMT. Neurology 2017;88:1445–53.
[76] Klein CJ, Botuyan MV, Wu Y, Ward CJ, Nicholson GA, Hammans S, Hojo K, Yamanishi H, Karpf AR, Wallace DC, et al. Mutations in DNMT1 cause hereditary sensory neuropathy with dementia and hearing loss. Nat Genet 2011;43:595–600.
[77] Klein CJ, Bird T, Ertekin-Taner N, Lincoln S, Hjorth R, Wu Y, Kwok J, Mer G, Dyck PJ, Nicholson GA. DNMT1 mutation hot spot causes varied phenotypes of HSAN1 with dementia and hearing loss. Neurology 2013;80:824–8.
[78] Sun Z, Wu Y, Ordog T, Baheti S, Nie J, Duan X, Hojo K, Kocher JP, Dyck PJ, Klein CJ. Aberrant signature methylome by DNMT1 hot spot mutation in hereditary sensory and autonomic neuropathy 1E. Epigenetics 2014;9:1184–93.
[79] Baets J, Duan X, Wu Y, Smith G, Seeley WW, Mademan I, McGrath NM, Beadell NC, Khoury J, Botuyan MV, et al. Defects of mutant DNMT1 are linked to a spectrum of neurological disorders. Brain 2015;138:845–61.
[80] Golshani P, Hutnick L, Schweizer F, Fan G. Conditional Dnmt1 deletion in dorsal forebrain disrupts development of somatosensory barrel cortex and thalamocortical long-term potentiation. Thalamus Relat Syst 2005;3:227–33.
[81] Hutnick LK, Golshani P, Namihira M, Xue Z, Matynia A, Yang XW, Silva AJ, Schweizer FE, Fan G. DNA hypomethylation restricted to the murine forebrain induces cortical degeneration and impairs postnatal neuronal maturation. Hum Mol Genet 2009;18:2875–88.
[82] De Sandre-Giovannoli A, Chaouch M, Kozlov S, Vallat JM, Tazir M, Kassouri N, Szepetowski P, Hammadouche T, Vandenberghe A, Stewart CL, et al. Homozygous defects in LMNA, encoding lamin A/C nuclear-envelope proteins, cause autosomal recessive axonal neuropathy in human (Charcot-Marie-Tooth disorder type 2) and mouse. Am J Hum Genet 2002;70:726–36.
[83] Tazir M, Azzedine H, Assami S, Sindou P, Nouioua S, Zemmouri R, Hamadouche T, Chaouch M, Feingold J, Vallat JM, et al. Phenotypic variability in autosomal recessive axonal Charcot-Marie-Tooth disease due to the R298C mutation in lamin A/C. Brain 2004;127:154–63.
[84] Hamadouche T, Poitelon Y, Genin E, Chaouch M, Tazir M, Kassouri N, Nouioua S, Chaouch A, Boccaccio I, Benhassine T, et al. Founder effect and estimation of the age of the c.892C>T (p.Arg298Cys) mutation in LMNA associated to Charcot-Marie-Tooth subtype CMT2B1 in families from North Western Africa. Ann Hum Genet 2008;72:590–7.
[85] Poitelon Y, Kozlov S, Devaux J, Vallat JM, Jamon M, Roubertoux P, Rabarimeriarijaona S, Baudot C, Hamadouche T, Stewart CL, et al. Behavioral and molecular exploration of the AR-CMT2A mouse model Lmna (R298C/R298C). NeuroMolecular Med 2012;14:40–52.
[86] Janin A, Bauer D, Ratti F, Millat G, Méjat A. Nuclear envelopathies: a complex LINC between nuclear envelope and pathology. Orphanet J Rare Dis 2017;12:147.
[87] McCord RP, Nazario-Toole A, Zhang H, Chines PS, Zhan Y, Erdos MR, Collins FS, Dekker J, Cao K. Correlated alterations in genome organization, histone methylation, and DNA-lamin A/C interactions in Hutchinson-Gilford progeria syndrome. Genome Res 2013;23:260–9.
[88] Wiethoff S, Hersheson J, Bettencourt C, Wood NW, Houlden H. Heterogeneity in clinical features and disease severity in ataxia-associated SYNE1 mutations. J Neurol 2016;263:1503–10.
[89] Izumi Y, Miyamoto R, Morino H, Yoshizawa A, Nishinaka K, Udaka F, Kameyama M, Maruyama H, Kawakami H. Cerebellar ataxia with SYNE1 mutation accompanying motor neuron disease. Neurology 2013;80:600–1.
[90] Hernandez M, Patzig J, Mayoral SR, Costa KD, Chan JR, Casaccia P. Mechanostimulation promotes nuclear and epigenetic changes in oligodendrocytes. J Neurosci 2016;36:806–13.

[91] Leal A, Huehne K, Bauer F, Sticht H, Berger P, Suter U, Morera B, Del Valle G, Lupski JR, Ekici A, et al. Identification of the variant Ala335Val of MED25 as responsible for CMT2B2: molecular data, functional studies of the SH3 recognition motif and correlation between wild-type MED25 and PMP22 RNA levels in CMT1A animal models. Neurogenetics 2009;10:375–6.

[92] Drew AP, Zhu D, Kidambi A, Ly C, Tey S, Brewer MH, Ahmad-Annuar A, Nicholson GA, Kennerson ML. Improved inherited peripheral neuropathy genetic diagnosis by whole-exome sequencing. Mol Genet Genomic Med 2015;3:143–54.

[93] Skourti-Stathaki K, Proudfoot NJ, Gromak N. Human senataxin resolves RNA/DNA hybrids formed at transcriptional pause sites to promote Xrn2-dependent termination. Mol Cell 2011;42:794–805.

[94] Fogel BL, Cho E, Wahnich A, Gao F, Becherel OJ, Wang X, Fike F, Chen L, Criscuolo C, De Michele G, et al. Mutation of senataxin alters disease-specific transcriptional networks in patients with ataxia with oculomotor apraxia type 2. Hum Mol Genet 2014;23:4758–69.

[95] Becherel OJ, Sun J, Yeo AJ, Nayler S, Fogel BL, Gao F, Coppola G, Criscuolo C, De Michele G, Wolvetang E, et al. A new model to study neurodegeneration in ataxia oculomotor apraxia type 2. Hum Mol Genet 2015;24:5759–74.

[96] Chen X, Müller U, Sundling KE, Brow DA. *Saccharomyces cerevisiae* Sen1 as a model for the study of mutations in human Senataxin that elicit cerebellar ataxia. Genetics 2014;198:577–90.

[97] Chen X, Poorey K, Carver MN, Müller U, Bekiranov S, Auble DT, Brow DA. Transcriptomes of six mutants in the Sen1 pathway reveal combinatorial control of transcription termination across the *Saccharomyces cerevisiae* genome. PLoS Genet 2017;13:e1006863.

[98] Zhao DY, Gish G, Braunschweig U, Li Y, Ni Z, Schmitges FW, Zhong G, Liu K, Li W, Moffat J, et al. SMN and symmetric arginine dimethylation of RNA polymerase II C-terminal domain control termination. Nature 2016;529:48–53.

[99] Grunseich C, Wang IX, Watts JA, Burdick JT, Guber RD, Zhu Z, Bruzel A, Lanman T, Chen K, Schindler AB, et al. Senataxin mutation reveals how R-loops promote transcription by blocking DNA methylation at gene promoters. Mol Cell 2018;69:426–37. e427.

[100] Cohen S, Puget N, Lin YL, Clouaire T, Aguirrebengoa M, Rocher V, Pasero P, Canitrot Y, Legube G. Senataxin resolves RNA:DNA hybrids forming at DNA double-strand breaks to prevent translocations. Nat Commun 2018;9:533.

[101] Ando M, Okamoto Y, Yoshimura A, Yuan JH, Hiramatsu Y, Higuchi Y, Hashiguchi A, Mitsui J, Ishiura H, Fukumura S, et al. Clinical and mutational spectrum of Charcot-Marie-Tooth disease type 2Z caused by MORC2 variants in Japan. Eur J Neurol 2017;24:1274–82.

[102] Sevilla T, Lupo V, Martínez-Rubio D, Sancho P, Sivera R, Chumillas MJ, García-Romero M, Pascual-Pascual SI, Muelas N, Dopazo J, et al. Mutations in the MORC2 gene cause axonal Charcot-Marie-Tooth disease. Brain 2016;139:62–72.

[103] Liu N, Lee CH, Swigut T, Grow E, Gu B, Bassik MC, Wysocka J. Selective silencing of euchromatic L1s revealed by genome-wide screens for L1 regulators. Nature 2018;553:228–32.

[104] Tchasovnikarova IA, Timms RT, Douse CH, Roberts RC, Dougan G, Kingston RE, Modis Y, Lehner PJ. Hyperactivation of HUSH complex function by Charcot-Marie-Tooth disease mutation in MORC2. Nat Genet 2017;49:1035–44.

[105] Douse CH, Bloor S, Liu Y, Shamin M, Tchasovnikarova IA, Timms RT, Lehner PJ, Modis Y. Neuropathic MORC2 mutations perturb GHKL ATPase dimerization dynamics and epigenetic silencing by multiple structural mechanisms. Nat Commun 2018;9:651.

[106] Chen YC, Auer-Grumbach M, Matsukawa S, Zitzelsberger M, Themistocleous AC, Strom TM, Samara C, Moore AW, Cho LT, Young GT, et al. Transcriptional regulator PRDM12 is essential for human pain perception. Nat Genet 2015;47:803–8.

[107] Nagy V, Cole T, Van Campenhout C, Khoung TM, Leung C, Vermeiren S, Novatchkova M, Wenzel D, Cikes D, Polyansky AA, et al. The evolutionarily conserved transcription factor PRDM12 controls sensory neuron development and pain perception. Cell Cycle 2015;14:1799–808.

[108] Gard C, Gonzalez Curto G, Frarma YE, Chollet E, Duval N, Auzié V, Auradé F, Vigier L, Relaix F, Pierani A, et al. Pax3- and Pax7-mediated Dbx1 regulation orchestrates the patterning of intermediate spinal interneurons. Dev Biol 2017;432(1):24–33.

[109] Rossor AM, Carr AS, Devine H, Chandrashekar H, Pelayo-Negro AL, Pareyson D, Shy ME, Scherer SS, Reilly MM. Peripheral neuropathy in complex inherited diseases: an approach to diagnosis. J Neurol Neurosurg Psychiatry 2017;88(10):846–63.

CHAPTER

4 Epigenetic mechanisms in Huntington's disease

Elizabeth A. Thomas

Department of Neuroscience, The Scripps Research Institute, La Jolla, CA, United States

Chapter Outline

4.1 Introduction ... **74**
4.2 Huntington's disease ... **74**
4.2.1 Neuropathology of HD ... 75
4.3 Transcriptional dysregulation in HD ... **75**
4.4 Altered epigenetic marks in HD ... **76**
4.4.1 Histone modifications ... 76
4.4.1.1 Histone acetylation ... 77
4.4.1.2 Histone acetylation alterations in HD ... 80
4.4.1.3 Histone methylation ... 80
4.4.1.4 Histone methylation changes in HD ... 81
4.4.1.5 Histone phosphorylation ... 82
4.4.1.6 Histone phosphorylation and HD ... 82
4.4.1.7 Histone ubiquitination ... 82
4.4.1.8 Altered histone ubiquitination in HD ... 83
4.4.2 DNA methylation ... 83
4.4.3 DNA methylation changes in HD ... 84
4.4.3.1 Global DNA methylation changes ... 84
4.4.3.2 Gene-specific DNA methylation changes ... 85
4.4.3.3 Implicating DNA methylation enzymes ... 85
4.5 Epigenetic-based therapies ... **86**
4.5.1 HDAC inhibitors as a treatment for HD ... 86
4.5.2 Methylation-inhibiting drugs ... 87
4.6 Concluding remarks ... **87**
4.7 Abbreviations ... **87**
References ... **88**

Chromatin Signaling and Neurological Disorders. https://doi.org/10.1016/B978-0-12-813796-3.00004-3

4.1 Introduction

Increasing evidence from the literature implicates epigenetic factors in the pathological mechanisms of Huntington's disease (HD), a devastating, genetic neurodegenerative disease, which is ultimately fatal. Much of this evidence surrounds the disease protein, Huntingtin (Htt) and its associations with several epigenetic events, including histone modifications (i.e., acetylation, methylation, and ubiquitination), and DNA methylation [1–6]. These changes are involved in the regulation of transcriptional processes, which are known to be disrupted in HD and lead to ensuing disease symptoms. This chapter will highlight the most recent findings linking epigenetic mechanisms to the pathology of HD in cell and animals models of HD, as well as in HD patients, and the therapeutic implications of these findings.

4.2 Huntington's disease

Huntington's disease is an autosomal-dominant neurodegenerative disorder caused by an expansion of a cytosine–adenine–guanine (CAG) repeat region in the coding region of the Huntingtin (*HTT*) gene [7]. It is one of a family of neurodegenerative disorders caused by CAG repeat expansions, including dentatorubral-pallidoluysian atrophy, spinobulbar muscular atrophy, and the spinocerebellar ataxias. In all of the CAG repeat diseases, there is a striking threshold effect of the minimal polyglutamine length to cause disease. A CAG repeat expansion greater than a threshold of 40 causes HD, while lengths below 35 repeats are generally considered nonpathological; 36–39 repeat lengths show incomplete penetrance [8,9]. HD is typically diagnosed during adulthood; however, 5%–10% of cases show juvenile onset [10,11]. Juvenile cases are caused by CAG mutations of greater than 60 repeats. Death typically follows 15–20 years after diagnosis from complications such as cardiac failure or aspiration pneumonia [12].

HD occurs at about a rate of 10 cases per 100,000, with a 1–3% new mutation rate due to instability of the CAG repeat region [9]. In the United States, there are ~30,000 HD patients and another 200,000 are at a genetic risk of getting the disease. The frequency of disease is much higher in the Venezuela villages along the shores of Lake Maracaibo [13]. These villages have the highest concentration of HD in the world and the largest family living with the disease [14]. The founder of this family lived in the early 1800s. Her family tree encompasses over 18,000 individuals spanning 10 generations, with more than 14,000 of these individuals currently living today.

The clinical features of HD are a combination of movement dysfunction, cognitive impairment, and psychiatric manifestations [15]. Chorea is the most characteristic movement symptom of HD and consists of involuntary, abnormal movements, which appear unpredictably in all the parts of the body. Cognitive impairment usually precedes motor symptoms or can occur during the course of the disease. Psychiatric symptoms including anxiety, social withdrawal, depression, and impulsivity can often be present, which significantly contribute to the overall disability [15].

Since the identification of the HD gene in 1993, there have been enormous advancements in the diagnosis and understanding of the molecular and pathophysiological features of the disorder. However, there is currently no available satisfactory treatment, nor cure for this disease. Tetrabenazine, and its longer-acting deuterated form, are U.S. Food and Drug Administration-approved agents for the symptomatic management of HD. Tetrabenazine binds reversibly to the type 2 vesicular monoamine transporters and has been shown to inhibit monoamine uptake in presynaptic vesicles, resulting in monoamine depletion [16]. Although it is useful in managing HD-related chorea, tetrabenazine is associated with numerous adverse effects and several drug–drug interactions [17]. Hence, there is a critical need for improved therapeutics to treat HD.

4.2.1 Neuropathology of HD

HD pathology results from expression of the Htt protein with an expanded polyglutamine domain. Although the normal Htt protein exists primarily as a soluble, cytoplasmic protein [18], the polyglutamine-expanded mutant Htt (mHtt) and its N-terminal fragments accumulate in the nucleus and form insoluble protein aggregates [19]. Although the nature of these intranuclear inclusions remains controversial (pathogenic or protective), they represent a pathological hallmark for HD [20].

Although the Htt/mHtt protein is expressed ubiquitously throughout the brain and body, the most striking neuropathological changes are observed predominantly in the striatum. It has been shown that the severity of striatal pathology is correlated with the degree of motor and cognitive impairments [21,22], suggesting that striatal degeneration plays a central role in HD. Neuronal death in HD proceeds regionally within the striatum as the disease progresses. The caudate nucleus exhibits the first signs of neurodegeneration, followed by putamen and finally nucleus accumbens, which degenerates only in the late stages of disease [23]. The medium spiny projection neurons selectively degenerate, while interneurons are relatively spared [24,25]. In addition to the substantial striatal degeneration, neuronal death and atrophy also occur throughout other brain structures, including cortex and hippocampus.

4.3 Transcriptional dysregulation in HD

The exact mechanisms of how the mHtt protein leads to the degeneration of neurons and neuropathological manifestations is not entirely clear [26,27]. However, transcriptional dysregulation has been a key feature long implicated in HD pathology. Aberrant transcriptional alterations occur early in the disease course of illness and have been observed not only in multiple cellular and animal models of HD, but also in HD patients [1,5,27,28]. Early studies in mouse models showed changes in dopamine receptor mRNA and protein levels that occur before the disease onset. These findings were consistent with positron emission tomography studies showing a decrease in dopamine D1 and D2 receptors in gene-positive but clinically asymptomatic HD patients [29,30]. Subsequently, cDNA microarray analyses performed on HD mouse models allowed thousands of genes to be monitored at one time, which provided a global genomic view of gene expression dysfunction in HD. From these analyses, several groups of genes were shown to be altered in expression in HD versus normal animals, including those associated with synaptic function, neuronal structure, stress response, and inflammation [31–34]. These changes were reproducibly observed in multiple HD mouse models, as well as in transcriptome-wide studies from human brain tissue samples [31,32]. More recent studies using an updated technology, RNA sequencing, has also corroborated these older findings [35,36].

Interestingly, more than 80% of striatal-enriched genes (genes with higher relative expression in the striatum when compared to other brain regions) were found to be decreased in an HD mouse model and in the caudate of HD patients [37]. Downregulation of novel striatal-enriched genes involved in vesicle transport and trafficking, tryptophan metabolism, and neuroinflammation has also been identified in both HD mouse striatum and caudate from HD patients [38], implicating a particular transcriptional imbalance in the striatum, which might account for the extensive degeneration observed in this region of the brain.

Altogether these observations strongly support that transcriptional dysregulation is an important process in HD pathogenesis. Several different mechanisms can account for the observed gene expression changes, including interactions of mHtt with specific transcription factors, interference of mHtt with the core transcriptional machinery and direct binding of mHtt to DNA. However, accumulating evidence supports a major role for epigenetic disturbances, such as histone modifications and DNA methylation changes, contributing to the observed alterations in gene expression [1,3–5].

4.4 Altered epigenetic marks in HD

The term "epigenetics" encompasses any inheritable change in gene activity not associated with a change in DNA sequence [39]. Epigenetic gene regulation typically involves postreplication and posttranslational modifications of DNA and histones that lead to lasting changes in chromatin structure, which results in alterations of gene transcription [40]. There is a diversity of epigenetic marks that govern gene regulation, many of which operate in context-dependent manners to control gene expression. These modifications include alterations in chemical modifications of DNA (methylation and hydroxymethylation), posttranslational modification of histones (acetylation, methylation, ubiquitylation, phosphorylation, SUMOylation, ADP ribosylation, etc.), as well as interactions of noncoding RNAs with chromatin [41]. There can also be complex interplay between these different epigenetic marks, for example, between DNA methylation and repressive histone modifications related to gene silencing. Later, I will discuss the evidence for the involvement of histone modifications and DNA methylation alterations in HD pathology and pathogenesis. Several cell culture and mouse models have been developed for HD research, which differ largely in the Htt fragments expressed, lengths of the CAG repeat mutations, and species used. A summary of the most common cell and mouse HD models used in the epigenetic studies described later can be found in Table 4.1.

4.4.1 Histone modifications

The basic unit of chromatin is the nucleosome, which consists of 147 base pairs of DNA wrapped 1.6 times around an octamer of four core histone proteins, H2A, H2B, H3, and H4 [42,43]. The amino-terminal tails of these core histone proteins contain amino acid residues that are sites for acetylation, methylation, phosphorylation, and ubiquitination (Fig. 4.1A), which alter the interactions of histones with DNA and determine the transcriptionally active status of the chromatin [40]. This specific patterns of posttranslationally modified histones are often referred to as the "histone code," and corresponds to the activation or repression of gene expression [44]. Posttranslational modifications of histones occur on several different residues (Fig. 4.1A), with some residues being sites for more than one type of modification. For example, lysine (K) residues can be either acetylated or methylated. Different types of modifications can also occur on more than one amino acid residue. For example, methylation can occur on K or arginine (R) residues. Additionally, residues of histone tails can accommodate more than one mark, being monomethylated, dimethylated, or trimethylated, making their functional status even more complex. New techniques that can detect histone modifications on a

Table 4.1 A list of the cell culture and mouse models used for HD research in studies reviewed in this chapter

Mouse model	Htt fragment	CAG repeat	Lifespan	References
R6/1	67 a.a.	115	9 months	(Mangiarini et al. 1996)
R6/2	67 a.a.	150	12 weeks	(Mangiarini et al. 1996)
N171-82Q	171 a.a.	82	22 weeks	(Schilling et al. 2001)
CAG140KI	Full length	140	18–24 months	(Wheeler et al. 1999)
HdhQ111KI	Full length	111	18 months	(Trettel et al. 2000)
YAC128	Full length	128	18–24 months	(Hodgson et al. 1999)
Cell model	**Htt fragment**	**Species**	**CAG repeat**	**References**
STHdhQ111	Full length	Mouse	111	(Trettel et al. 2000)
Fibroblasts	Full length	Human	41–44	(Coriell Institute)
Primary neurons	Varied	Mouse	Varied	(Biagioli et al. 2015, Ratovitski et al. 2015, Pan et al. 2016)

The Huntingtin (Htt) fragment refers to amino acids (a.a.) of the N-terminal part of the protein is expressed in the given model. KI, *knock-in.*
Biagioli, M., F. et al., (2015). "Htt CAG repeat expansion confers pleiotropic gains of mutant huntingtin function in chromatin regulation." Hum Mol Genet 24: 2442–2457. Hodgson, J. G. et al. (1999). "A YAC mouse model for Huntington's disease with full-length mutant huntingtin, cytoplasmic toxicity, and selective striatal neurodegeneration." Neuron 23: 181–192. Mangiarini, L., K. et al., (1996). "Exon 1 of the HD gene with an expanded CAG repeat is sufficient to cause a progressive neurological phenotype in transgenic mice." Cell 87: 493–506. Pan, Y. et al., (2016). "Inhibition of DNA Methyltransferases Blocks Mutant Huntingtin-Induced Neurotoxicity." Sci Rep 6: 31,022.
Ratovitski, T., N. et al., (2015). "PRMT5-mediated symmetric arginine dimethylation is attenuated by mutant huntingtin and is impaired in Huntington's disease (HD)." Cell Cycle 14(11): 1716–1729. Schilling, G., H. et al., (2001). "Distinct behavioral and neuropathological abnormalities in transgenic mouse models of HD and DRPLA." Neurobiol Dis 8: 405–418. Trettel, F. et al., (2000). "Dominant phenotypes produced by the HD mutation in STHdh(Q111) striatal cells." Hum Mol Genet 9: 2799–2809. Wheeler, V. C. et al., (1999). "Length-dependent gametic CAG repeat instability in the Huntington's disease knock-in mouse." Hum Mol Genet 8: 115–122.

global level are expected to improve our understanding of the complex roles of histone modifications on regulation of gene expression.

4.4.1.1 Histone acetylation

Histone acetylation, the transfer of an acetyl group from acetyl coenzyme A to the lysine side chain in the acceptor histone, is one of the best-studied histone posttranslational modifications. This process is modulated by the actions of two opposing enzyme families, histone acetyltransferases (HATs), and histone deacetylases (HDACs) [40,44]. The association between histone acetylation and increased transcription has been known for many years, whereby, an increase in HAT activity promotes acetylation of histone proteins leading to a more open chromatin configuration and ensuing increases in gene transcription. Alternatively, increased HDAC activity removes the acetyl groups from histones, resulting in condensation of chromatin structure and ensuing repression of gene expression (Fig. 4.2). Acetylation of histones is a relatively transient mark, which is vital for precise temporal transcriptional

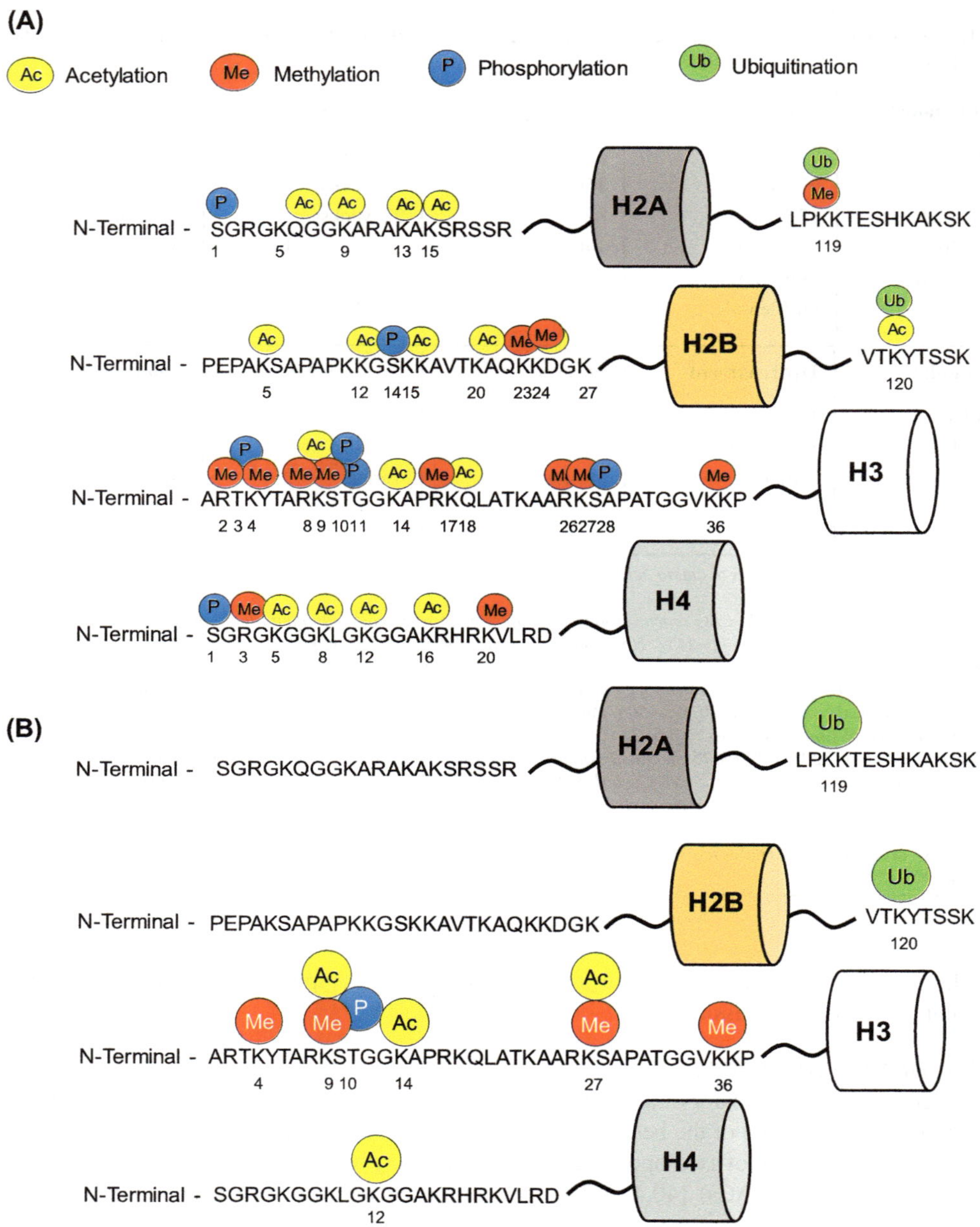

FIGURE 4.1 Different posttranslational modifications present on histones.

(A) The most commonly modified residues on histone proteins (H2A, H2B, H3, and H4) are shown. (B) Histone marks that have been shown to be altered in Huntington's disease model systems. *Ac*, acetylation, *M*, methylation, *P*, phosphorylation, *Ub*, ubiquitination.

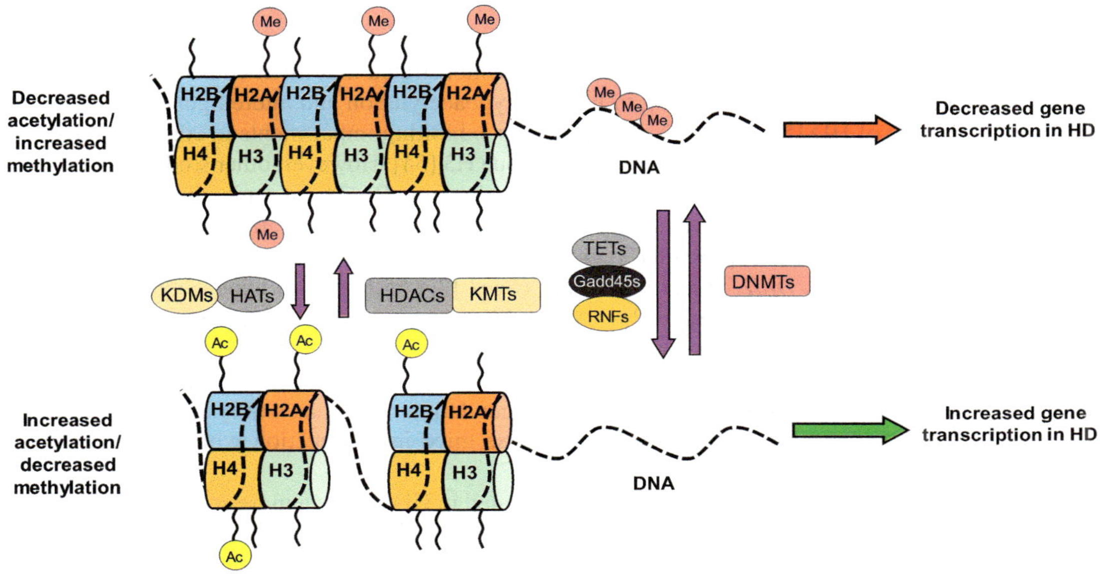

FIGURE 4.2 Schematic depiction of epigenetic factors that regulate condensed and open chromatin and have been implicated in HD.

The dynamic state of histone acetylation/deacetylation is regulated by histone acetyltransferase (*HAT*) and histone deacetylase (*HDAC*) enzymes, whereby histone methylation is governed broadly by lysine methyltransferases (KMTs) and lysine demethylases (KDMs). DNA methylation is governed by DNA methyltransferases (DNMTs), whereby DNA demethylation can be achieved via the actions of ten-eleven translocation (TET) enzymes, Growth arrest, and DNA-damage-inducible 45 (Gadd45) enzymes or RING finger protein 4 (Rnf4) proteins.

control. However, this process is likely to be more complex, involving an array of other DNA-binding and chromatin-related proteins.

The HAT family of proteins comprises five diverse clusters that are based on structural and functional similarities. These include the Gcn5-related N-acetyltransferase superfamily members, MYST proteins, p300/CREB-binding protein (CBP) HATs, general transcription factor HATs, and the steroid receptor coactivators/nuclear receptor coactivator family [45]. In contrast, the HDAC enzymes represent a more related family of proteins with sequence and structural similarities [46]. In humans, the HDAC family of enzymes has 18 subtypes, which are divided into four main classes, however, only classes I, II, and IV are particularly implicated in chromatin regulation [47]. Importantly, HDAC proteins mostly exist in large multiprotein complexes, and evidence suggests that most, if not all, HDAC enzymes require interaction with other HDACs or proteins for optimal function [48,49]. HDAC-interacting proteins serve important roles in recruiting HDACs to their chromatin targets given that HDAC proteins lack a DNA-binding motif [50].

4.4.1.2 Histone acetylation alterations in HD

The landmark studies providing the initial evidence between mHtt and histone acetylation, came from Leslie Thompson, Joan Steffan, and colleagues, who demonstrated that mHtt could bind the histone acetyltransferase domain of CBP, leading to histone hypoacetylation and altered gene expression in cell and fly models of HD [51–53]. HD mouse models exhibiting impairments in cognitive function also showed global decreased levels of acetylated histone H3 [54]. However, other studies did not find global hypoacetylation of histones H2B, H3, or H4 in brains from R6/2 transgenic mice, and other HD mouse models [3,55,56]. Rather, it was found that decreased histone acetylation occurred at specific gene loci, in particular at the promoter regions of those genes found to be downregulated in expression in HD, such as *Drd2, Penk1, Actb,* and *Grin1* [3,4,55,57], with hypoacetylation of both histones H3 and H4 being implicated in R6/2 mice [57]. Later studies using genome-wide methods identified hundreds of loci that were hypoacetylated specifically on histone 3 at lysines 9 and 14 (H3K9, H3K14) in the striatum of transgenic R6/2 mice [58] (Fig. 4.1B). These findings were supported by another study that also demonstrated hypoacetylation of histone 4 at lysine 12 (H4K12) in the chromatin in the cortex and hippocampus of these mice in conjunction with gene expression changes [59].

Other specific histone marks have also been implicated in HD. Using a more recent method, RNA sequencing and chromatin-immunoprecipitation followed by massively parallel sequencing (ChIP-seq), it was shown that downregulated genes in HD mouse striatum were associated with selective decreases in acetylation of histone 3 at lysine 27 (H3K27), a mark of active enhancers [60] (Fig. 4.1B). These studies showed that decreased acetylation of H3K27 was associated with lowered RNA polymerase II binding at regions enriched in downregulated genes in HD mouse striatum [60].

4.4.1.3 Histone methylation

Although the effects of histone acetylation on gene expression are more straightforward, the implications of histone methylation are less clear, as methylation can cause both gene activation and silencing events [61]. Methylation of histones occurs via enzymes called lysine and arginine methyltransferases (KMT and PRMT, respectively), with S-adenosyl-L-methionine serving as the methyl donor [62]. Histone methylation involves the transfer of one to three methyl groups, thus resulting in mono-, di-, or trimethylated lysines, and mono- or dimethylated arginines. The histone methyltransferase group of enzymes is large, with more than 50 members identified in humans [63]. There are two broad families of enzymes divided on the basis of their catalytic domain sequence: the "DOT1-like proteins" and the "SET domain-containing proteins," which are further divided into four families, which include SET1, SET2, suppressor of variegation 39, and PR/SET domain 2 [63].

Different from the HAT enzymes, histone methyltransferase enzymes are specific for the lysine or arginine residues that they modify. Further specificity comes from the fact that many enzymes are specific only for particular residues [45,64]. For example, SET1, MLL, and SMYD3 are histone methyltransferases that catalyze methylation of histone H3 at K4 in mammalian cells [65]. Similarly, SUV39H1, G9a, GLP, and SET domain bifurcated 1 (SETDB1) catalyze methylation of histone H3 at K9. G9a and polycomb group enzymes such as EZH2 are histone methyltransferases that catalyze methylation of histone H3 at K27. Methylation at either H3K9 and H3K27 promotes heterochromatin formation leading to gene silencing [64]. In contrast, trimethylation of histone H3 on lysine 4 (H3K4me3) or lysine 36 (H3K36me3) increases gene expression [61]. Histone methyltransferases are also components of large, multiprotein nuclear complexes that contain other histone-modifying enzymes and other regulatory proteins including HATs, HDACs, and DNA methyltransferases.

Histone demethylation involves the removal of methyl groups on histone proteins, and this occurs via the actions of histone lysine demethylases (KDMs). The discovery of histone demethylase enzymes demonstrates that histone methylation is not a permanent modification, as was previously hypothesized. To date, two classes of KDM have been described: the amine-oxidase type lysine-specific demethylases 1 and 2 (LSD1 and 2; also known as KDM1A and B, respectively) and the Jumonji C (JmjC) domain-containing histone demethylases [63]. The latter consists of a group that contains over 30 members and can be divided, based on the JmjC-domain homology, into seven subfamilies (KDMs 2–8) [63]. Although the downstream effects of these enzymes are not entirely clear, inhibition of histone demethylases is considered to lead to histone remethylation at specific residues that could be important for chromatin dynamics and gene expression.

4.4.1.4 Histone methylation changes in HD

Although early studies focused on histone acetylation, later findings implicated altered histone methylation in HD. These studies focused more on individual methylation marks, in particular, di- and trimethylation of lysine residues on histone 3 (Fig. 4.1B). Trimethylation of lysine 4 (H3K4me3), a mark of active promoters, was found to be decreased at the promoter regions of downregulated genes in the striatum and cortex of HD R6/2 mice, a finding that was also observed in postmortem brain from HD patients [66]. In addition, H3K4me3 occupancy was found at other regulatory sites, downstream of the transcription start site, suggesting this mark is involved in complex transcriptional regulation in HD [66]. In other studies using cultured cells expressing mHtt with various CAG sizes, a correlation was found between CAG-repeat-dependent gene expression changes and H3K4me3 levels [67]. Given the nature of these models, such results suggested that methylation of H3K4 may be an important early event in gene expression control. Methylation of H3K4 has been investigated in neuronal nuclei extracted from prefrontal cortex of HD patients using ChIP-seq, although these levels were not highly correlated with gene expression changes [68].

In contrast, methylation marks associated with decreased gene expression were found to be increased in HD models. Histone H3K9 dimethylation (H3K9me2) and trimethylation (H3K9me3) were found to be increased in brain tissue from R6/2 and N171-82Q mice as well as in HD patients [4,69,70]. This H3K9 hypermethylation was considered to be due to an increase in the expression of its specific histone lysine methyltransferase, SETDB1 [70]. Interestingly, SETDB1 has also been shown to methylate nonhistone proteins, such as upstream binding factor (UBF), which is important for the regulation of ribosomal DNA transcription [71]. In that same study, it was further shown that UBF trimethylation, specifically at K232/254, by SETDB1 caused disruption of transcription in HD ST*Hdh*Q111 striatal cells [71]. A related methyltransferase SETD2, which methylates histone 3 at K36 (H3K36), has also been implicated in HD pathogenesis, whereby the Htt protein was found to interact with SETD2 in a polyglutamine-dependent manner [72,73].

Methylation of arginine residues on histones by the action of PRMTs may also be involved in HD. Htt and mHtt proteins have been linked to increased activity of PRMT5, an enzyme mediating symmetrical dimethylation of arginine [74]. These interactions led to increased methylation of R on histones H2A and H4, in primary cortical neurons transfected with mutant full-length Htt, resulting in detrimental effects to the cells [74]. These toxic effects could be reversed by knocking down the arginine demethylase, Jumonji C domain-containing protein 6 (JMJD6), the only enzyme known to erase dimethyl arginine marks from histones [74]. However, in another study, overexpression of a lysine demethylase, KDM5D, and ensuing decreases in K4 methylation on histone H3 (H3K4), was found to improve metabolic deficits in HD ST*Hdh*Q111 cells [75].

4.4.1.5 Histone phosphorylation

Phosphorylation of histones occurs at serine, threonine, and tyrosine residues. The extent of this mark and its ensuing effects on gene transcription are less well studied and likely to be complex. There are eight characterized phosphorylation sites on the core canonical histones H2A, H2B, H3, and H4 (Fig. 4.1), which have been linked to specific kinase enzymes. These events occur in several pathways including DNA damage, mitosis/meiosis, apoptosis, and nuclear hormone signaling. Several hundred protein kinases exist in mammals and are classified into distinct superfamilies. Protein phosphatases are primarily responsible for dephosphorylation of histones, and these proteins can be grouped into 104 distinct enzyme families that all catalyze the same hydrolysis reaction. Phosphatases are classified by substrate specificity and sequence homology of its catalytic domain. Phosphorylation of histones is diverse and complex with several different kinases being able to phosphorylate the same amino acid residues [76].

Probably, the most important feature of histone phosphorylation is that it plays an important role in the regulation of combinatorial posttranslational modifications [77]. An extensive cross talk exists between phosphorylation and other posttranslational modifications (e.g., Kme), which together regulate various biological processes, including not only gene transcription, but also DNA repair and cell cycle progression [76].

4.4.1.6 Histone phosphorylation and HD

A few studies have examined histone phosphorylation in HD. Yazawa and colleagues were the first group to report aberrant phosphorylation of histone H3 in brain tissue of patients with HD, as well as other polyglutamine diseases [78]. Later studies showed that histone H3 phosphorylation specifically at serine 10 was decreased in primary striatal neurons transfected with the mutated form of the *HTT* gene in conjunction with decreased expression of c-fos [79] (Fig. 4.1B). Subsequently, those same authors demonstrated that mitogen- and stress-activated kinase (MSK-1), a nuclear protein kinase involved in chromatin remodeling through histone H3 phosphorylation, is deficient in the striatum of HD patients and in HD R6/2 mice. Restoring MSK-1 expression in cultured striatal cells prevented neuronal dysfunction and death induced by mHtt [79].

4.4.1.7 Histone ubiquitination

Although less well studied than the other posttranslational modifications, histone ubiquitination also represents an additional opportunity for regulating the epigenome. In contrast to polyubiquitination that marks a protein for proteasomal degradation, monoubiquitination of histones is associated with the transcriptional control of gene expression and the DNA damage response. In this context, histones are monoubiquitinated by the addition of a single 8.5 kDa ubiquitin molecule to specific lysine residues on histone tails (Fig. 4.1). Histones are the most abundantly monoubiquitinated conjugates in the nucleus of mammalian cells. Histone H2A is monoubiquitylated (H2Aub1) at K119 by the Polycomb repressor complex PRC1. H2Aub1 is associated with silencing of developmental genes. Although H2A ubiquitylation leads to transcriptional repression, histone H2B monoubiquitylated (H2Bub1) at K120 is typically associated with active genes. H2Bub1 is a prerequisite step to histone H3 methylation at K4 and K79, and more importantly, H2Bub1 regulates general chromatin structure by ensuring accurate organization of histone proteins.

The process of ubiquitination requires the sequential activities of three enzymes: ubiquitin activating enzymes, ubiquitin-conjugating enzymes, and ubiquitin ligase enzymes, which comprise large families, several of which specifically target histone 2A and 2B proteins [80] (Fig. 4.1). Similar to other histone posttranslational modifications, H2A and H2B ubiquitination is reversible, and deubiquitination occurs via a class of thiol proteases known as deubiquitylating enzymes [80].

4.4.1.8 Altered histone ubiquitination in HD

Altered histone monoubiquitylation mediated by mHtt has been associated with the transcriptional dysregulation observed in HD [81], although evidence suggests that this process is complex. Genome-wide increases in histone H2A ubiquitination at K119 have been shown in mouse models of HD; whereas, ubiquitination of histone H2B at K120 was reduced in the R6/2 mouse brain [81,82] (Fig. 4.1B). Reduction in H2A ubiquitination was shown to reverse transcriptional repression and inhibit methylation of H3K9 in cell culture, whereby reduction in H2B ubiquitination induces transcriptional repression and inhibits methylation of H3K4 [82]. Other studies have specifically implicated a role for the polyubiquitin gene, Ubc, in these effects [83].

4.4.2 DNA methylation

DNA methylation is one of the oldest known, and most intensely studied, epigenetic modifications in mammals and has been shown to be an essential regulator of gene expression and gene silencing [84] (Fig. 4.2). Initially, DNA methylation of promoters was considered to primarily control gene expression; however, DNA methylation of distal regulatory sites has also been found to have an important role in gene regulation [85]. DNA methylation involves the addition of a methyl group at the 5′ position on the pyrimidine ring of cytosines, creating 5-methylcytosine (5-mC) [84]. Although this process typically occurs at cytosine–phosphate–guanine (CpG) islands, non-CpG methylation might also be important for transcriptional regulation [86]. Furthermore, in addition to cytosines, guanine and adenine nucleotides can also be methylated, resulting in 7-methylguanine (7-mG) and 3-methyladenine (3-mA), respectively. In more recent years, another DNA modification has been discovered, 5-hydroxymethylcytosine (5-hmC), which is an oxidation product of 5-mC and also a potential epigenetic modifier that might have important roles in neurodegenerative diseases [87].

DNA methylation is controlled by a class of enzymes called DNA methyltransferases (DNMTs) [88], of which five members have been identified in mammals: DNMT1, DNMT2, DNMT3a, DNMT3b, and DNMT3L. However, DNMT2 and DNMT3L are not considered to function as cytosine methyltransferases, with DNMT3L being a catalytically inactive member, and DNMT2 considered to play a role in methylation of RNA [89]. It is believed that DNA methylation patterns initiated during embryogenesis and development occur via de novo DNMTs (DNMT3a and DNMT3b), which catalyze the methyl transfer onto a "naked" or unmethylated DNA template [90]. DNMT1, on the other hand, is the "maintenance" methylation enzyme, which copies DNA marks from a hemimethylated template and which has major importance preserving methylation patterns in adult tissues, including postmitotic cells in the brain. DNMTs are expressed throughout neural development and considered to play important roles in neuronal survival and plasticity [91].

DNA methylation was initially considered to be unidirectional and irreversible; however, evidence now indicates that this is not the case. Although it is possible to reverse DNA methylation in replicating cells by passive demethylation via blocking DNMT1 activity during DNA synthesis, this mechanism would not work in postmitotic, differentiated cells. Alternatively, several studies support active DNA demethylation [92–94]. The ten-eleven translocation (TET) family enzymes can oxidize 5-mC to 5-hmC, a first step in active DNA demethylation. The 5-hmC form can then be actively reverted to cytosine through iterative oxidation and thymine DNA glycosylase-mediated removal mechanisms. Other genes encoding proteins associated with DNA demethylation include Methyl-CpG-binding domain protein 3 (Mbd3), Growth arrest, and DNA-damage-inducible 45 (Gadd45) and RING finger protein 4 (Rnf4) [94–96] (Fig. 4.2).

4.4.3 DNA methylation changes in HD

4.4.3.1 Global DNA methylation changes

Growing literature supports an involvement of DNA methylation changes in HD. Given the known functions of DNA methylation on gene expression, changes in DNA methylation clearly represent a relevant means to alter gene expression and ensuing neuronal dysfunction in this disease [75,97]. Studies have reported changes in DNA methylation in association with expression of mHtt protein in different HD model systems, as well as in human HD brain (Table 4.2). Using reduced representation bisulfite sequencing, Ng and colleagues found altered patterns of DNA methylation in HD ST*Hdh*Q111 cells compared to wild-type STHdhQ7 cells (WTQ7) cells [97]. Changes in DNA methylation were found at promoter, proximal, and distal regulatory regions, with some sites increasing in methylation and others decreasing [97]. Furthermore, these authors found that methylation changes at CpG-rich regions, which were largely located near the transcription start sites, were inversely correlated with gene expression and were enriched in categories of genes related to developmental processes, neuron migration, signal transduction, and cell differentiation. Similar

Table 4.2 DNA methylation changes related to Huntington's disease (HD)

Methylation change	Source	HD model	Type	Ref #
5-mC	Fibroblasts	Human	Genome-wide	[75]
5-mC	STHdhQ111 cells	Cells	Genome-wide	[97]
5-mC	Cortex	Human	*HES4* promoter	[68]
5-mC	Cortex	Human	*HTT promoter*	[98]
5-mC/5-hmC	Putamen	Human	*ADORA2A* promoter	[99]
5-mC/5-hmC	Striatum	R6/1, R6/2 mice	*Adora2a* promoter	[99]
5-hmC	Striatum and cortex	YAC128 mice	Genome-wide	[100]
7-mG	Brain	R6/2, CAG140 KI mice	Genome-wide	[101]
7-mG	Cortex	Human	Genome-wide	[101]

5-hmC, *5-hydroxymethylcytosine;* 5-mC, *5-methylcytosine;* 7-mG, *7-methyguanine;* ADORA2a, *adenosine receptor 2A;* HES4, *hairy/enhancer of split 4;* HTT, *Huntington's gene.*

results were found in studies using an array method in human fibroblasts from normal and HD patients [75]. However, using a similar array method, only minimal changes in global DNA methylation were detected in postmortem brain from HD patients [98], suggesting complex regulation in the human brain on a global level.

Dysregulation of other methylated marks has also been studied on a global level in HD, and these may also contribute to the known transcriptional pathology of the disease. Genome-wide analysis of the 5-hmC epigenetic mark showed lower levels in the striatum and cortex of YAC128 transgenic mice compared to wild-type (WT) controls, with the corresponding genes being related to neurogenesis, neuronal function, and survival [100]. An HPLC-based method also detected differential levels of 7-mG in DNA samples from brains of two different HD mouse models, as well as in caudate DNA from HD patients [101]. Overall, it is clear that neither hyper- nor hypomethylation is desired across the board for all genes, which highlights the need to investigate gene and CpG sequence-specific changes.

4.4.3.2 Gene-specific DNA methylation changes

In human HD cortex, a genome-wide mapping study of the transcriptional mark, histone H3 trimethylated on lysine 4 (H3K4me3), led to the identification of an important site-specific DNA methylation change on the hairy and enhancer of split 4 (*HES4*) gene promoter [68]. Loss of H3K4me3 at CpG-rich sequences on the *HES4* promoter was associated with excessive DNA methylation and altered expression of *HES4* and its target genes. Moreover, hypermethylation of the *HES4* promoter was correlated with measures of striatal degeneration and age of onset in HD patients [68]. Other studies in postmortem brain from HD patients found that aberrant methylation of the *HTT* promoter was correlated to the age of disease onset, and further suggested that DNA methylation contributes, in part, to tissue-specific *HTT* transcription [98].

Studies have examined both 5-mC and 5-hmC contents in the 5′UTR region of the adenosine A2A receptor gene (*ADORA2A*) in the putamen of HD patients and in the striatum of R6/1 and R6/2 mice [99]. Their findings suggested that altered methylation patterns of the *ADORA2A* gene are linked to the pathological decrease of *ADORA2A* expression levels found in HD [99].

4.4.3.3 Implicating DNA methylation enzymes

Studies have begun to explore the possible causes for DNA methylation abnormalities in HD by measuring the expression of genes encoding DNMT enzymes in HD models. One study found that *Dnmt1* expression was decreased twofold in HD ST*Hdh*Q111 striatal cells compared to Q7 WT cells [97]. Lowered *Dnmt1* expression in HD is supported by other studies demonstrating lower levels of *Dnmt1* in striatum and cortex of N171-82Q transgenic mice compared to littermate controls. Another study using human postmortem brain found that *DNMT1* and *DNMT3A* were differentially coexpressed in a combined cohort of HD and Alzheimer's disease patients [102]. More recent studies showed that knockdown of *DNMT3A* or *DNMT1* protected neurons against mHtt-induced toxicity, further implicating a requirement for DNMTs in mHtt-triggered neuronal death [103].

Decreased expression of the putative DNA demethylase gene, Growth arrest, and DNA-damage-inducible 45a (*Gadd45a*) has also been reported in striatum of R6/2 transgenic mice [104] and muscle from N171-82Q transgenic mice [75]. In contrast, *Gadd45g*, was found to be increased in HD ST*Hdh*Q111 cells [97]. Finally, *Rnf4*, another gene associated with DNA demethylation, was also found to be differentially expression in HD mice from microarray studies [75,104].

4.5 Epigenetic-based therapies

4.5.1 HDAC inhibitors as a treatment for HD

Given the extensive literature supporting histone acetylation disturbances in HD, it is not surprising that HDAC inhibitors have emerged as a relevant disease-modifying therapeutic approach for HD. The most straightforward consequence of HDAC inhibition is the elevation of histone acetylation, thereby promoting an open chromatin structure and facilitation of gene transcription (Fig. 4.2). Accordingly, studies have demonstrated that different types of HDAC inhibitors elevate histone acetylation on both the local and/or global levels [57,105–107]. Consistent with this mechanism, microarray studies have demonstrated that HDAC inhibitors do indeed affect the expression of select groups of genes in mouse models of HD [57,108]. However, contrary to early predictions, HDAC inhibitors do not cause global upregulation of gene transcription, rather, previous expression profiling studies have demonstrated a relatively small number of genes (2–10%) modulated by HDAC inhibitors [108].

With regards to effects on HD phenotypes, nonselective HDAC inhibitors, such as suberoylanilide hydroxamic acid (SAHA), phenylbutyrate, and sodium butyrate, have been shown to be beneficial in cell [51,109], Drosophila [52], and mouse models of HD [56,110,111]. However, during the past decade, there has been great progress in identifying isotype-selective HDAC inhibitors (reviewed in Refs. [112–116]), and these are considered to be more beneficial for targeting disease symptoms and minimizing harmful side effects in HD [117].

Extensive research has been carried out using a novel class of HDAC inhibitors, which selectively target HDACs 1 and/or 3 [105,118,119]. These HDAC1/HDAC3-targeting compounds showed beneficial effects in reducing eye neurodegeneration in HD Drosophila and improving metabolic function in mutant HD striatal cells [120]. *In vivo* studies showed that HDAC1/3 inhibitors improved motor deficits and cognitive decline in R6/2 and N171-82Q transgenic mouse models [57,75,121]. Inhibitors specifically targeting HDAC3 were also found to elicit beneficial effects in HD mouse models from multiple research groups [75,118,122,123]. Importantly, these selective compounds have been shown to exhibit low *in vitro* and *in vivo* toxicity [57,120,121]. HDAC3 inhibition showing beneficial effects in neurodegenerative diseases is consistent with studies demonstrating that genetic knockdown of HDAC3 expression protected against neuronal cell death, while overexpression of HDAC3 showed detrimental effects to neurons [124,125].

HDAC6-selective inhibitors have also been developed and may represent a promising therapeutic approach against HD, as well as other neurodegenerative diseases. Genetic knockdown of HDAC6 has been described to have different effects in HD mouse models. In one study using the R6/2 mouse model, HDAC6 knockdown did not modify disease progression [126]. However, another study using the related R6/1 described no effect on motor deficits but an exacerbation of some disease phenotypes was observed with HDAC6 deletion [127].

Overall, these findings demonstrate that various HDAC inhibitors are effective in suppressing pathogenic symptoms in various HD models, with HDAC3-selective compounds exhibiting some of the strongest effects. These studies support the use of selective HDAC inhibitors as disease-modifying therapeutics for HD.

4.5.2 Methylation-inhibiting drugs

Methyltransferase inhibitors also have potential as a therapeutic strategy for HD. Such compounds might exert their effects by interacting with histone acetylation. These events have been investigated by studying guanine cytosine (GC)-binding anthracyclines, such as mithramycin A and chromomycin, which are compounds with antibiotic and anticancer properties that interact with the minor groove of DNA and inhibit binding of GC-rich-binding transcription factors [128]. In studies on R6/2 HD mice, mithramycin treatment attenuated the hypermethylation of histone H4, thereby extending survival and improving motor phenotypes in these mice [69]. Other studies showed a significant shift in the balance between methylation and acetylation of histones in treated HD mice, resulting in greater acetylation of histone H3 at lysine 9 thereby promoting gene transcription [4]. These findings were correlated with extended survival, improved motor phenotypes, and increased brain volume in transgenic R6/2 mice [4]. Mithramycin treatment also restored the expression of the methyltransferase SETDB1, a key enzyme that regulates histone methylation as mentioned earlier [70].

4.6 Concluding remarks

As reviewed earlier, alterations in gene expression and the epigenome are important contributors to early mechanisms underlying HD pathogenesis and ensuing pathology. Although many correlations between individual epigenetic marks and gene expression changes were observed in HD, it is still unclear to what extent epigenetic changes are causative to the known transcriptional abnormalities. These findings demonstrate that alterations in the epigenome in HD are more complex than previously appreciated and highlight the importance of understanding epigenetic "signatures," as opposed to individual epigenetic marks when considering biomarker identification and/or therapeutics for HD. Moreover, a number of additional epigenetic loci have yet to be fully explored in HD. It is likely that combinations of epigenetic marks converge on selective loci leading to enhancing or reduced transcription in HD.

4.7 Abbreviations

HD	Huntington's disease
Htt	Huntingtin
CAG	cytosine–adenine–guanine
HAT	histone acetyltransferase
HDAC	histone deacetylase
DNMT	DNA methyltransferase
CBP	CREB-binding protein
KMTs	lysine methyltransferases
KDMs	lysine demethylases
SETDB1	SET domain bifurcated 1
ChIP-seq	chromatin-immunoprecipitation sequencing

References

[1] Cha JH. Transcriptional dysregulation in Huntington's disease. Trends Neurosci 2000;23(9):387–92.
[2] Lee J, Hwang YJ, Kim KY, Kowall NW, Ryu H. Epigenetic mechanisms of neurodegeneration in Huntington's disease. Neurotherapeutics 2013;10(4):664–76.
[3] Sadri-Vakili G, Bouzou B, Benn CL, Kim MO, Chawla P, Overland RP, Glajch KE, Xia E, Qiu Z, Hersch SM, Clark TW, Yohrling GJ, Cha JH. Histones associated with downregulated genes are hypo-acetylated in Huntington's disease models. Hum Mol Genet 2007;16(11):1293–306.
[4] Stack EC, Del Signore SJ, Luthi-Carter R, Soh BY, Goldstein DR, Matson S, Goodrich S, Markey AL, Cormier K, Hagerty SW, Smith K, Ryu H, Ferrante RJ. Modulation of nucleosome dynamics in Huntington's disease. Hum Mol Genet 2007;16(10):1164–75.
[5] Sugars KL, Rubinsztein DC. Transcriptional abnormalities in Huntington disease. Trends Genet 2003; 19(5):233–8.
[6] Thomas EA. Involvement of HDAC1 and HDAC3 in the pathology of polyglutamine disorders: therapeutic implications for selective HDAC1/HDAC3 inhibitors. Pharmaceuticals 2014;7(6):634–61.
[7] Group HSDCR. A novel gene containing a trinucleotide repeat that is expanded and unstable on Huntington's disease chromosomes. The Huntington's Disease Collaborative Research Group. Cell 1993;72(6): 971–83.
[8] Kremer B, Goldberg P, Andrew SE, Theilmann J, Telenius H, Zeisler J, Squitieri F, Lin B, Bassett A, Almqvist E, et al. A worldwide study of the Huntington's disease mutation. The sensitivity and specificity of measuring CAG repeats. N Engl J Med 1994;330(20):1401–6.
[9] Myers RH, MacDonald ME, Koroshetz WJ, Duyao MP, Ambrose CM, Taylor SA, Barnes G, Srinidhi J, Lin CS, Whaley WL, et al. De novo expansion of a (CAG)n repeat in sporadic Huntington's disease. Nat Genet 1993;5(2):168–73.
[10] Nance MA, Myers RH. Juvenile onset Huntington's disease–clinical and research perspectives. Ment Retard Dev Disabil Res Rev 2001;7(3):153–7.
[11] van Dijk JG, van der Velde EA, Roos RA, Bruyn GW. Juvenile Huntington disease. Hum Genet 1986;73(3): 235–9.
[12] Heemskerk AW, Roos RA. Aspiration pneumonia and death in Huntington's disease. PLoS Curr 2012;4: RRN1293.
[13] Penney Jr JB, Young AB, Shoulson I, Starosta-Rubenstein S, Snodgrass SR, Sanchez-Ramos J, Ramos-Arroyo M, Gomez F, Penchaszadeh G, Alvir J, et al. Huntington's disease in Venezuela: 7 years of follow-up on symptomatic and asymptomatic individuals. Mov Disord 1990;5(2):93–9.
[14] Wexler NS, Lorimer J, Porter J, Gomez F, Moskowitz C, Shackell E, Marder K, Penchaszadeh G, Roberts SA, Gayan J, Brocklebank D, Cherny SS, Cardon LR, Gray J, Dlouhy SR, Wiktorski S, Hodes ME, Conneally PM, Penney JB, Gusella J, Cha JH, Irizarry M, Rosas D, Hersch S, Hollingsworth Z, MacDonald M, Young AB, Andresen JM, Housman DE, De Young MM, Bonilla E, Stillings T, Negrette A, Snodgrass SR, Martinez-Jaurrieta MD, Ramos-Arroyo MA, Bickham J, Ramos JS, Marshall F, Shoulson I, Rey GJ, Feigin A, Arnheim N, Acevedo-Cruz A, Acosta L, Alvir J, Fischbeck K, Thompson LM, Young A, Dure L, O'Brien CJ, Paulsen J, Brickman A, Krch D, Peery S, Hogarth P, Higgins Jr DS, Landwehrmeyer B. Venezuelan kindreds reveal that genetic and environmental factors modulate Huntington's disease age of onset. Proc Natl Acad Sci U S A 2004;101(10):3498–503.
[15] Sturrock A, Leavitt BR. The clinical and genetic features of Huntington disease. J Geriatr Psychiatr Neurol 2010;23(4):243–59.
[16] Mestre T, Ferreira J, Coelho MM, Rosa M, Sampaio C. Therapeutic interventions for symptomatic treatment in Huntington's disease. Cochrane Database Syst Rev 2009;3:CD006456.

[17] Guay DR. Tetrabenazine, a monoamine-depleting drug used in the treatment of hyperkinetic movement disorders. Am J Geriatr Pharmacother 2010;8(4):331−73.

[18] DiFiglia M, Sapp E, Chase K, Schwarz C, Meloni A, Young C, Martin E, Vonsattel JP, Carraway R, Reeves SA. Huntingtin is a cytoplasmic protein associated with vesicles human and rat brain neurons. Neuron 1995;14:1075−81.

[19] Davies SW, Turmaine M, Cozens BA, DiFiglia M, Sharp AH, Ross CA, Scherzinger E, Wanker EE, Mangiarini L, Bates GP. Formation of neuronal intranuclear inclusions underlies the neurological dysfunction in mice transgenic for the HD mutation. Cell 1997;90(3):537−48.

[20] Zoghbi HY, Orr HT. Glutamine repeats and neurodegeneration. Annu Rev Neurosci 2000;23:217−47.

[21] Bamford KA, Caine ED, Kido DK, Plassche WM, Shoulson I. Clinical-pathologic correlation in Huntington's disease: a neuropsychological and computed tomography study. Neurology 1989;39(6): 796−801.

[22] Rosenblatt A, Abbott MH, Gourley LM, Troncoso JC, Margolis RL, Brandt J, Ross CA. Predictors of neuropathological severity in 100 patients with Huntington's disease. Ann Neurol 2003;54(4): 488−93.

[23] Vonsattel JP, Myers RH, Stevens TJ, Ferrante RJ, Bird ED, Richardson EPJ. Neuropathological classification of Huntington's disease. J Neuropathol Exp Neurol 1985;44:559−77.

[24] Cicchetti F, Gould PV, Parent A. Sparing of striatal neurons coexpressing calretinin and substance P (NK1) receptor in Huntington's disease. Brain Res 1996;730(1−2):232−7.

[25] Ferrante RJ, Kowall NW, Beal MF, Richardson Jr EP, Bird ED, Martin JB. Selective sparing of a class of striatal neurons in Huntington's disease. Science 1985;230(4725):561−3.

[26] Ross CA. Huntington's disease: new paths to pathogenesis. Cell 2004;118(1):4−7.

[27] Thomas EA. Striatal specificity of gene expression dysregulation in Huntington's disease. J Neurosci Res 2006;84(6):1151−64.

[28] Okazawa H. Polyglutamine diseases: a transcription disorder? Cell Mol Life Sci 2003;60(7):1427−39.

[29] Augood SJ, Faull RL, Emson PC. Dopamine D1 and D2 receptor gene expression in the striatum in Huntington's disease. Ann Neurol 1997;42:215−21.

[30] Weeks RA, Piccini P, Harding AE, Brooks DJ. Striatal D1 and D2 dopamine receptor loss in asymptomatic mutation carriers of Huntington's disease. Ann Neurol 1996;40(1):49−54.

[31] Hodges A, Strand AD, Aragaki AK, Kuhn A, Sengstag T, Hughes G, Elliston LA, Hartog C, Goldstein DR, Thu D, Hollingsworth ZR, Collin F, Synek B, Holmans PA, Young AB, Wexler NS, Delorenzi M, Kooperberg C, Augood SJ, Faull RL, Olson JM, Jones L, Luthi-Carter R. Regional and cellular gene expression changes in human Huntington's disease brain. Hum Mol Genet 2006;15(6):965−77.

[32] Kuhn A, Goldstein DR, Hodges A, Strand AD, Sengstag T, Kooperberg C, Becanovic K, Pouladi MA, Sathasivam K, Cha JH, Hannan AJ, Hayden MR, Leavitt BR, Dunnett SB, Ferrante RJ, Albin R, Shelbourne P, Delorenzi M, Augood SJ, Faull RL, Olson JM, Bates GP, Jones L, Luthi-Carter R. Mutant huntingtin's effects on striatal gene expression in mice recapitulate changes observed in human Huntington's disease brain and do not differ with mutant huntingtin length or wild-type huntingtin dosage. Hum Mol Genet 2007;16:1845−61.

[33] Luthi-Carter R, Hanson SA, Strand AD, Bergstrom DA, Chun W, Peters NL, Woods AM, Chan EY, Kooperberg C, Krainc D, Young AB, Tapscott SJ, Olson JM. Dysregulation of gene expression in the R6/2 model of polyglutamine disease: parallel changes in muscle and brain. Hum Mol Genet 2002;11(17): 1911−26.

[34] Luthi-Carter R, Strand A, Peters NL, Solano SM, Hollingsworth ZR, Menon AS, Frey AS, Spektor BS, Penney EB, Schilling G, Ross CA, Borchelt DR, Tapscott SJ, Young AB, Cha JH, Olson JM. Decreased expression of striatal signaling genes in a mouse model of Huntington's disease. Hum Mol Genet 2000;9(9): 1259−71.

[35] Hensman Moss DJ, Flower MD, Lo KK, Miller JR, van Ommen GB, t Hoen PA, Stone TC, Guinee A, Langbehn DR, Jones L, Plagnol V, van Roon-Mom WM, Holmans P, Tabrizi SJ. Huntington's disease blood and brain show a common gene expression pattern and share an immune signature with Alzheimer's disease. Sci Rep 2017;7:44849.
[36] Miller JR, Lo KK, Andre R, Hensman Moss DJ, Trager U, Stone TC, Jones L, Holmans P, Plagnol V, Tabrizi SJ. RNA-Seq of Huntington's disease patient myeloid cells reveals innate transcriptional dysregulation associated with proinflammatory pathway activation. Hum Mol Genet 2016;25(14):2893–904.
[37] Desplats PA, Kass KE, Gilmartin T, Stanwood GD, Woodward EL, Head SR, Sutcliffe JG, Thomas EA. Selective deficits in the expression of striatal-enriched mRNAs in Huntington's disease. J Neurochem 2006; 96(3):743–57.
[38] Mazarei G, Neal SJ, Becanovic K, Luthi-Carter R, Simpson EM, Leavitt BR. Expression analysis of novel striatal-enriched genes in Huntington disease. Hum Mol Genet 2010;19(4):609–22.
[39] Feinberg AP. Phenotypic plasticity and the epigenetics of human disease. Nature 2007;447(7143):433–40.
[40] Kornberg RD, Lorch Y. Twenty-five years of the nucleosome, fundamental particle of the eukaryote chromosome. Cell 1999;98(3):285–94.
[41] Kanduri C. Long noncoding RNA and epigenomics. Adv Exp Med Biol 2011;722:174–95.
[42] Luger K, Mader AW, Richmond RK, Sargent DF, Richmond TJ. Crystal structure of the nucleosome core particle at 2.8 A resolution. Nature 1997;389(6648):251–60.
[43] Quina AS, Buschbeck M, Di Croce L. Chromatin structure and epigenetics. Biochem Pharmacol 2006; 72(11):1563–9.
[44] Strahl BD, Allis CD. The language of covalent histone modifications. Nature 2000;403(6765):41–5.
[45] An W. Histone acetylation and methylation: combinatorial players for transcriptional regulation. Subcell Biochem 2007;41:351–69.
[46] Gregoretti IV, Lee YM, Goodson HV. Molecular evolution of the histone deacetylase family: functional implications of phylogenetic analysis. J Mol Biol 2004;338(1):17–31.
[47] Xu WS, Parmigiani RB, Marks PA. Histone deacetylase inhibitors: molecular mechanisms of action. Oncogene 2007;26(37):5541–52.
[48] Adcock IM, Ford P, Ito K, Barnes PJ. Epigenetics and airways disease. Respir Res 2006;7:21.
[49] Hildmann C, Riester D, Schwienhorst A. Histone deacetylases–an important class of cellular regulators with a variety of functions. Appl Microbiol Biotechnol 2007;75(3):487–97.
[50] Reichert N, Choukrallah MA, Matthias P. Multiple roles of class I HDACs in proliferation, differentiation, and development. Cell Mol Life Sci 2012;69(13):2173–87.
[51] McCampbell A, Taye AA, Whitty L, Penney E, Steffan JS, Fischbeck KH. Histone deacetylase inhibitors reduce polyglutamine toxicity. Proc Natl Acad Sci U S A 2001;98(26):15179–84.
[52] Steffan JS, Bodai L, Pallos J, Poelman M, McCampbell A, Apostol BL, Kazantsev A, Schmidt E, Zhu YZ, Greenwald M, Kurokawa R, Housman DE, Jackson GR, Marsh JL, Thompson LM. Histone deacetylase inhibitors arrest polyglutamine-dependent neurodegeneration in Drosophila. Nature 2001;413(6857): 739–43.
[53] Steffan JS, Kazantsev A, Spasic-Boskovic O, Greenwald M, Zhu YZ, Gohler H, Wanker EE, Bates GP, Housman DE, Thompson LM. The Huntington's disease protein interacts with p53 and CREB-binding protein and represses transcription. Proc Natl Acad Sci U S A 2000;97(12):6763–8.
[54] Giralt A, Puigdellivol M, Carreton O, Paoletti P, Valero J, Parra-Damas A, Saura CA, Alberch J, Gines S. Long-term memory deficits in Huntington's disease are associated with reduced CBP histone acetylase activity. Hum Mol Genet 2012;21(6):1203–16.
[55] Guiretti D, Sempere A, Lopez-Atalaya JP, Ferrer-Montiel A, Barco A, Valor LM. Specific promoter deacetylation of histone H3 is conserved across mouse models of Huntington's disease in the absence of bulk changes. Neurobiol Dis 2016;89:190–201.

[56] Hockly E, Richon VM, Woodman B, Smith DL, Zhou X, Rosa E, Sathasivam K, Ghazi-Noori S, Mahal A, Lowden PA, Steffan JS, Marsh JL, Thompson LM, Lewis CM, Marks PA, Bates GP. Suberoylanilide hydroxamic acid, a histone deacetylase inhibitor, ameliorates motor deficits in a mouse model of Huntington's disease. Proc Natl Acad Sci U S A 2003;100(4):2041–6.

[57] Thomas EA, Coppola G, Desplats PA, Tang B, Soragni E, Burnett R, Gao F, Fitzgerald KM, Borok JF, Herman D, Geschwind DH, Gottesfeld JM. The HDAC inhibitor, 4b, ameliorates the disease phenotype and transcriptional abnormalities in Huntington's disease transgenic mice. Proc Natl Acad Sci U S A 2008; 105(40):15564–9.

[58] McFarland KN, Das S, Sun TT, Leyfer D, Xia E, Sangrey GR, Kuhn A, Luthi-Carter R, Clark TW, Sadri-Vakili G, Cha JH. Genome-wide histone acetylation is altered in a transgenic mouse model of Huntington's disease. PLoS One 2012;7(7):e41423.

[59] Valor LM, Guiretti D, Lopez-Atalaya JP, Barco A. Genomic landscape of transcriptional and epigenetic dysregulation in early onset polyglutamine disease. J Neurosci 2013;33(25):10471–82.

[60] Achour M, Le Gras S, Keime C, Parmentier F, Lejeune FX, Boutillier AL, Neri C, Davidson I, Merienne K. Neuronal identity genes regulated by super-enhancers are preferentially down-regulated in the striatum of Huntington's disease mice. Hum Mol Genet 2015;24(12):3481–96.

[61] Rice JC, Allis CD. Histone methylation versus histone acetylation: new insights into epigenetic regulation. Curr Opin Cell Biol 2001;13(3):263–73.

[62] Morales Y, Caceres T, May K, Hevel JM. Biochemistry and regulation of the protein arginine methyltransferases (PRMTs). Arch Biochem Biophys 2016;590:138–52.

[63] Morera L, Lubbert M, Jung M. Targeting histone methyltransferases and demethylases in clinical trials for cancer therapy. Clin Epigenet 2016;8:57.

[64] Greer EL, Beese-Sims SE, Brookes E, Spadafora R, Zhu Y, Rothbart SB, Aristizabal-Corrales D, Chen S, Badeaux AI, Jin Q, Wang W, Strahl BD, Colaiacovo MP, Shi Y. A histone methylation network regulates transgenerational epigenetic memory in *C. elegans*. Cell Rep 2014;7(1):113–26.

[65] Alam H, Gu B, Lee MG. Histone methylation modifiers in cellular signaling pathways. Cell Mol Life Sci 2015;72(23):4577–92.

[66] Vashishtha M, Ng CW, Yildirim F, Gipson TA, Kratter IH, Bodai L, Song W, Lau A, Labadorf A, Vogel-Ciernia A, Troncosco J, Ross CA, Bates GP, Krainc D, Sadri-Vakili G, Finkbeiner S, Marsh JL, Housman DE, Fraenkel E, Thompson LM. Targeting H3K4 trimethylation in Huntington disease. Proc Natl Acad Sci U S A 2013;110(32):E3027–36.

[67] Biagioli M, Ferrari F, Mendenhall EM, Zhang Y, Erdin S, Vijayvargia R, Vallabh SM, Solomos N, Manavalan P, Ragavendran A, Ozsolak F, Lee JM, Talkowski ME, Gusella JF, Macdonald ME, Park PJ, Seong IS. Htt CAG repeat expansion confers pleiotropic gains of mutant huntingtin function in chromatin regulation. Hum Mol Genet 2015;24(9):2442–57.

[68] Bai G, Cheung I, Shulha HP, Coelho JE, Li P, Dong X, Jakovcevski M, Wang Y, Grigorenko A, Jiang Y, Hoss A, Patel K, Zheng M, Rogaev E, Myers RH, Weng Z, Akbarian S, Chen JF. Epigenetic dysregulation of hairy and enhancer of split 4 (HES4) is associated with striatal degeneration in postmortem Huntington brains. Hum Mol Genet 2015;24(5):1441–56.

[69] Ferrante RJ, Ryu H, Kubilus JK, D'Mello S, Sugars KL, Lee J, Lu P, Smith K, Browne S, Beal MF, Kristal BS, Stavrovskaya IG, Hewett S, Rubinsztein DC, Langley B, Ratan RR. Chemotherapy for the brain: the antitumor antibiotic mithramycin prolongs survival in a mouse model of Huntington's disease. J Neurosci 2004;24(46):10335–42.

[70] Ryu H, Lee J, Hagerty SW, Soh BY, McAlpin SE, Cormier KA, Smith KM, Ferrante RJ. ESET/SETDB1 gene expression and histone H3 (K9) trimethylation in Huntington's disease. Proc Natl Acad Sci U S A 2006;103(50):19176–81.

[71] Hwang YJ, Han D, Kim KY, Min SJ, Kowall NW, Yang L, Lee J, Kim Y, Ryu H. ESET methylates UBF at K232/254 and regulates nucleolar heterochromatin plasticity and rDNA transcription. Nucleic Acids Res 2014;42(3):1628–43.

[72] Faber PW, Barnes GT, Srinidhi J, Chen J, Gusella JF, MacDonald ME. Huntingtin interacts with a family of WW domain proteins. Hum Mol Genet 1998;7(9):1463–74.

[73] Sun XJ, Wei J, Wu XY, Hu M, Wang L, Wang HH, Zhang QH, Chen SJ, Huang QH, Chen Z. Identification and characterization of a novel human histone H3 lysine 36-specific methyltransferase. J Biol Chem 2005; 280(42):35261–71.

[74] Ratovitski T, Arbez N, Stewart JC, Chighladze E, Ross CA. PRMT5- mediated symmetric arginine dimethylation is attenuated by mutant huntingtin and is impaired in Huntington's disease (HD). Cell Cycle 2015;14(11):1716–29.

[75] Jia H, Morris CD, Williams RM, Loring JF, Thomas EA. HDAC inhibition imparts beneficial transgenerational effects in Huntington's disease mice via altered DNA and histone methylation. Proc Natl Acad Sci U S A 2015;112(1):E56–64.

[76] Banerjee T, Chakravarti D. A peek into the complex realm of histone phosphorylation. Mol Cell Biol 2011; 31(24):4858–73.

[77] Sawicka A, Seiser C. Sensing core histone phosphorylation - a matter of perfect timing. Biochim Biophys Acta 2014;1839(8):711–8.

[78] Yazawa I, Hazeki N, Nakase H, Kanazawa I, Tanaka M. Histone H3 is aberrantly phosphorylated in glutamine-repeat diseases. Biochem Biophys Res Commun 2003;302(1):144–9.

[79] Roze E, Betuing S, Deyts C, Marcon E, Brami-Cherrier K, Pages C, Humbert S, Merienne K, Caboche J. Mitogen- and stress-activated protein kinase-1 deficiency is involved in expanded-huntingtin-induced transcriptional dysregulation and striatal death. FASEB J 2008;22(4):1083–93.

[80] Fuchs G, Oren M. Writing and reading H2B monoubiquitylation. Biochim Biophys Acta 2014;1839(8): 694–701.

[81] McFarland KN, Das S, Sun TT, Leyfer D, Kim MO, Xia E, Sangrey GR, Kuhn A, Luthi-Carter R, Clark TW, Sadri-Vakili G, Cha JH. Genome-wide increase in histone H2A ubiquitylation in a mouse model of Huntington's disease. J Huntingtons Dis 2013;2(3):263–77.

[82] Kim MO, Chawla P, Overland RP, Xia E, Sadri-Vakili G, Cha JH. Altered histone monoubiquitylation mediated by mutant huntingtin induces transcriptional dysregulation. J Neurosci 2008;28(15):3947–57.

[83] Bett JS, Benn CL, Ryu KY, Kopito RR, Bates GP. The polyubiquitin Ubc gene modulates histone H2A monoubiquitylation in the R6/2 mouse model of Huntington's disease. J Cell Mol Med 2009;13(8b): 2645–57.

[84] Razin A, Riggs AD. DNA methylation and gene function. Science 1980;210(4470):604–10.

[85] Stadler MB, Murr R, Burger L, Ivanek R, Lienert F, Scholer A, van Nimwegen E, Wirbelauer C, Oakeley EJ, Gaidatzis D, Tiwari VK, Schubeler D. DNA-binding factors shape the mouse methylome at distal regulatory regions. Nature 2011;480(7378):490–5.

[86] Guo JU, Su Y, Shin JH, Shin J, Li H, Xie B, Zhong C, Hu S, Le T, Fan G, Zhu H, Chang Q, Gao Y, Ming GL, Song H. Distribution, recognition and regulation of non-CpG methylation in the adult mammalian brain. Nat Neurosci 2014;17(2):215–22.

[87] Sherwani SI, Khan HA. Role of 5-hydroxymethylcytosine in neurodegeneration. Gene 2015;570(1):17–24.

[88] Wu JC, Santi DV. On the mechanism and inhibition of DNA cytosine methyltransferases. Prog Clin Biol Res 1985;198:119–29.

[89] Goll MG, Kirpekar F, Maggert KA, Yoder JA, Hsieh CL, Zhang X, Golic KG, Jacobsen SE, Bestor TH. Methylation of tRNAAsp by the DNA methyltransferase homolog Dnmt2. Science 2006;311(5759):395–8.

[90] Okano M, Bell DW, Haber DA, Li E. DNA methyltransferases Dnmt3a and Dnmt3b are essential for de novo methylation and mammalian development. Cell 1999;99(3):247–57.

[91] Klose RJ, Bird AP. Genomic DNA methylation: the mark and its mediators. Trends Biochem Sci 2006; 31(2):89–97.

[92] Carey N, Marques CJ, Reik W. DNA demethylases: a new epigenetic frontier in drug discovery. Drug Discov Today 2011;16(15–16):683–90.

[93] Kohli RM, Zhang Y. TET enzymes, TDG and the dynamics of DNA demethylation. Nature 2013; 502(7472):472–9.

[94] Schafer A. Gadd45 proteins: key players of repair-mediated DNA demethylation. Adv Exp Med Biol 2013; 793:35–50.

[95] Brown SE, Suderman MJ, Hallett M, Szyf M. DNA demethylation induced by the methyl-CpG-binding domain protein MBD3. Gene 2008;420(2):99–106.

[96] Hu XV, Rodrigues TM, Tao H, Baker RK, Miraglia L, Orth AP, Lyons GE, Schultz PG, Wu X. Identification of RING finger protein 4 (RNF4) as a modulator of DNA demethylation through a functional genomics screen. Proc Natl Acad Sci U S A 2010;107(34):15087–92.

[97] Ng CW, Yildirim F, Yap YS, Dalin S, Matthews BJ, Velez PJ, Labadorf A, Housman DE, Fraenkel E. Extensive changes in DNA methylation are associated with expression of mutant huntingtin. Proc Natl Acad Sci U S A 2013;110(6):2354–9.

[98] De Souza RA, Islam SA, McEwen LM, Mathelier A, Hill A, Mah SM, Wasserman WW, Kobor MS, Leavitt BR. DNA methylation profiling in human Huntington's disease brain. Hum Mol Genet 2016;25(10): 2013–30.

[99] Villar-Menendez I, Blanch M, Tyebji S, Pereira-Veiga T, Albasanz JL, Martin M, Ferrer I, Perez-Navarro E, Barrachina M. Increased 5-methylcytosine and decreased 5-hydroxymethylcytosine levels are associated with reduced striatal A2AR levels in Huntington's disease. NeuroMolecular Med 2013;15(2):295–309.

[100] Wang F, Yang Y, Lin X, Wang JQ, Wu YS, Xie W, Wang D, Zhu S, Liao YQ, Sun Q, Yang YG, Luo HR, Guo C, Han C, Tang TS. Genome-wide loss of 5-hmC is a novel epigenetic feature of Huntington's disease. Hum Mol Genet 2013;22(18):3641–53.

[101] Thomas B, Matson S, Chopra V, Sun L, Sharma S, Hersch S, Rosas HD, Scherzer C, Ferrante R, Matson W. A novel method for detecting 7-methyl guanine reveals aberrant methylation levels in Huntington disease. Anal Biochem 2013;436(2):112–20.

[102] Narayanan M, Huynh JL, Wang K, Yang X, Yoo S, McElwee J, Zhang B, Zhang C, Lamb JR, Xie T, Suver C, Molony C, Melquist S, Johnson AD, Fan G, Stone DJ, Schadt EE, Casaccia P, Emilsson V, Zhu J. Common dysregulation network in the human prefrontal cortex underlies two neurodegenerative diseases. Mol Syst Biol 2014;10:743.

[103] Pan Y, Daito T, Sasaki Y, Chung YH, Xing X, Pondugula S, Swamidass SJ, Wang T, Kim AH, Yano H. Inhibition of DNA methyltransferases blocks mutant huntingtin-induced neurotoxicity. Sci Rep 2016;6: 31022.

[104] Tang B, Seredenina T, Coppola G, Kuhn A, Geschwind DH, Luthi-Carter R, Thomas EA. Gene expression profiling of R6/2 transgenic mice with different CAG repeat lengths reveals genes associated with disease onset and progression in Huntington's disease. Neurobiol Dis 2011;42(3):459–67.

[105] Chou CJ, Herman DM, Gottesfeld JM. Pimelic diphenylamide 106 is a slow, tight-binding inhibitor of class I histone deacetylases. J Biol Chem 2008;283(51):35402–9.

[106] Herman D, Jenssen K, Burnett R, Soragni E, Perlman SL, Gottesfeld JM. Histone deacetylase inhibitors reverse gene silencing in Friedreich's ataxia. Nat Chem Biol 2006;2(10):551–8.

[107] Quinti L, Chopra V, Rotili D, Valente S, Amore A, Franci G, Meade S, Valenza M, Altucci L, Maxwell MM, Cattaneo E, Hersch S, Mai A, Kazantsev A. Evaluation of histone deacetylases as drug targets in Huntington's disease models. Study of HDACs in brain tissues from R6/2 and CAG140 knock-in HD mouse models and human patients and in a neuronal HD cell model. PLoS Curr 2010;2.

[108] Peart MJ, Smyth GK, van Laar RK, Bowtell DD, Richon VM, Marks PA, Holloway AJ, Johnstone RW. Identification and functional significance of genes regulated by structurally different histone deacetylase inhibitors. Proc Natl Acad Sci U S A 2005;102(10):3697–702.
[109] Nucifora Jr FC, Sasaki M, Peters MF, Huang H, Cooper JK, Yamada M, Takahashi H, Tsuji S, Troncoso J, Dawson VL, Dawson TM, Ross CA. Interference by huntingtin and atrophin-1 with cbp-mediated transcription leading to cellular toxicity. Science 2001;291(5512):2423–8.
[110] Ferrante RJ, Kubilus JK, Lee J, Ryu H, Beesen A, Zucker B, Smith K, Kowall NW, Ratan RR, Luthi-Carter R, Hersch SM. Histone deacetylase inhibition by sodium butyrate chemotherapy ameliorates the neurodegenerative phenotype in Huntington's disease mice. J Neurosci 2003;23(28):9418–27.
[111] Gardian G, Browne SE, Choi DK, Klivenyi P, Gregorio J, Kubilus JK, Ryu H, Langley B, Ratan RR, Ferrante RJ, Beal MF. Neuroprotective effects of phenylbutyrate in the N171-82Q transgenic mouse model of Huntington's disease. J Biol Chem 2005;280(1):556–63.
[112] Harrison IF, Dexter DT. Epigenetic targeting of histone deacetylase: therapeutic potential in Parkinson's disease? Pharmacol Ther 2013;140(1):34–52.
[113] Kalin JH, Bergman JA. Development and therapeutic implications of selective histone deacetylase 6 inhibitors. J Med Chem 2013;56(16):6297–313.
[114] Pan H, Cao J, Xu W. Selective histone deacetylase inhibitors. Anti Cancer Agents Med Chem 2012;12(3): 247–70.
[115] Rajak H, Singh A, Dewangan PK, Patel V, Jain DK, Tiwari SK, Veerasamy R, Sharma PC. Peptide based macrocycles: selective histone deacetylase inhibitors with antiproliferative activity. Curr Med Chem 2013; 20(14):1887–903.
[116] Soragni E, Xu C, Plasterer HL, Jacques V, Rusche JR, Gottesfeld JM. Rationale for the development of 2-aminobenzamide histone deacetylase inhibitors as therapeutics for Friedreich ataxia. J Child Neurol 2012; 27(9):1164–73.
[117] Balasubramanian S, Verner E, Buggy JJ. Isoform-specific histone deacetylase inhibitors: the next step? Cancer Lett 2009;280(2):211–21.
[118] Jia H, Kast RJ, Steffan JS, Thomas EA. Selective histone deacetylase (HDAC) inhibition imparts beneficial effects in Huntington's disease mice: implications for the ubiquitin-proteasomal and autophagy systems. Hum Mol Genet 2012a;21(24):5280–93.
[119] Xu C, Soragni E, Chou CJ, Herman D, Plasterer HL, Rusche JR, Gottesfeld JM. Chemical probes identify a role for histone deacetylase 3 in Friedreich's ataxia gene silencing. Chem Biol 2009;16(9): 980–9.
[120] Jia H, Pallos J, Jacques V, Lau A, Tang B, Cooper A, Syed A, Purcell J, Chen Y, Sharma S, Sangrey GR, Darnell SB, Plasterer H, Sadri-Vakili G, Gottesfeld JM, Thompson LM, Rusche JR, Marsh JL, Thomas EA. Histone deacetylase inhibitors targeting HDAC3 and HDAC1 ameliorate polyglutamine-elicited phenotypes in Huntington's disease model systems. In: CHDI 7th annual Huntington's disease therapeutics conference: a forum for drug discovery & development; 2012.
[121] Jia H, Pallos J, Jacques V, Lau A, Tang B, Cooper A, Syed A, Purcell J, Chen Y, Sharma S, Sangrey GR, Darnell SB, Plasterer H, Sadri-Vakili G, Gottesfeld JM, Thompson LM, Rusche JR, Marsh JL, Thomas EA. Histone deacetylase (HDAC) inhibitors targeting HDAC3 and HDAC1 ameliorate polyglutamine-elicited phenotypes in model systems of Huntington's disease. Neurobiol Dis 2012c;46(2):351–61.
[122] Jia H, Wang Y, Morris CD, Jacques V, Gottesfeld JM, Rusche JR, Thomas EA. The effects of pharmacological inhibition of histone deacetylase 3 (HDAC3) in Huntington's disease mice. PLoS One 2016; 11(3):e0152498.
[123] Suelves N, Kirkham-McCarthy L, Lahue RS, Gines S. A selective inhibitor of histone deacetylase 3 prevents cognitive deficits and suppresses striatal CAG repeat expansions in Huntington's disease mice. Sci Rep 2017;7(1):6082.

[124] Bardai FH, D'Mello SR. Selective toxicity by HDAC3 in neurons: regulation by Akt and GSK3beta. J Neurosci 2011;31(5):1746–51.

[125] Bardai FH, Price V, Zaayman M, Wang L, D'Mello SR. Histone deacetylase (HDAC1) is a molecular switch between neuronal survival and death. J Biol Chem 2012;287(42):35444–53.

[126] Bobrowska A, Paganetti P, Matthias P, Bates GP. Hdac6 knock-out increases tubulin acetylation but does not modify disease progression in the R6/2 mouse model of Huntington's disease. PLoS One 2011;6(6): e20696.

[127] Ragot A, Pietropaolo S, Vincent J, Delage P, Zhang H, Allinquant B, Leinekugel X, Fischer A, Cho YH. Genetic deletion of the Histone Deacetylase 6 exacerbates selected behavioral deficits in the R6/1 mouse model for Huntington's disease. Brain Behav 2015;5(9):e00361.

[128] Piekarski M, Jelinska A. Anthracyclines still prove effective in anticancer therapy. Mini Rev Med Chem 2013;13(5):627–34.

CHAPTER

5 The epigenetics of multiple sclerosis

Tove Christensen[1], Christian Muchardt[2,3]

Department of Biomedicine, Aarhus University, Bartholins Allé 6, DK-8000 Aarhus C, Denmark[1]; Institut Pasteur, Département de Biologie du Développement et Cellules Souches, Unité de Régulation Epigénétique, Paris, France[2]; CNRS UMR 3738, Paris, France[3]

Chapter Outline

5.1 Multiple sclerosis, the knowns and the unknowns 98
5.1.1 A chronic progressive disease of the central nervous system 98
5.1.2 The genetics of MS 98
5.1.3 Gender bias and parent-of-origin effect 98
5.1.4 A role for environmental factors 99
5.1.5 A viral component in MS 99
5.2 MS as an epigenetic disorder 100
5.2.1 Nongenetic causes of MS and their link to chromatin and transcription 100
5.2.1.1 Vitamin D and the vitamin D receptor 100
5.2.1.2 Reactivation of HERVs 100
5.2.2 DNA and histone modifications are footprints of transcriptional regulation 101
5.3 DNA and histone modifications linked to MS 101
5.3.1 DNA methylation 102
5.3.1.1 Imprinting 102
5.3.1.2 Differential methylation in blood cells and in CNS 103
5.3.1.3 Cytosine hydroxymethylation 103
5.3.2 A possibly reduced efficiency of the H3K9me/HP1 axis of transcriptional repression in MS 103
5.3.2.1 Reduced recruitment of HP1 at HERVs and proinflammatory genes in patients with MS 104
5.3.2.2 Peptidylarginine deiminases interfere with the H3K9me/HP1 axis of transcriptional repression 104
5.3.2.3 The H3K9me/HP1 axis: a central player in the onset of MS? 106
5.4 Epigenetics beyond transcription 108
5.4.1 Exosomal miRNA silencing 109
5.4.2 Microbiota 109
5.4.3 Environment: pollutants that may interfere with silencing machineries 110

Chromatin Signaling and Neurological Disorders. https://doi.org/10.1016/B978-0-12-813796-3.00005-5

5.5 Conclusions 110
Acknowledgments 111
References 111

5.1 Multiple sclerosis, the knowns and the unknowns

5.1.1 A chronic progressive disease of the central nervous system

Multiple sclerosis (MS) is a complex immune-mediated, inflammatory, and demyelinating disease of the central nervous system (CNS) [1]. In young adults, MS is the most prevalent progressive CNS disease and is associated with considerable morbidity, including chronic physical disability [2]. The pathology encompasses areas of inflammatory demyelination that spread through the CNS. These lesions are known as plaques.

The disease course may take several forms: most patients with MS have a disease course of exacerbations followed by remission (relapsing-remitting (RRMS)), while others display continuous progression (primary progressive). In most cases, RRMS develops to secondary progressive MS (SPMS) [3,4]. At early phases of the disease, patients are without spreading of neurological symptoms in both time and space and clinical onset commonly presents as an acute or subacute episode of neurological disturbance due to a single CNS lesion, diagnosed as clinically isolated syndrome, which often, but not always, progresses to MS [5].

5.1.2 The genetics of MS

The genetic susceptibility for MS is shown in the familial recurrence rate of approximately 15% with an age-adjusted risk for siblings (3%), parents (2%), and children (2%) that is higher than for second-degree and third-degree relatives (reviewed in Ref. [3]). Concordance among identical twins is about 25%, while concordance among dizygotic twins is 3% [6]. Thus, the MS risk for individuals with an affected family member increases roughly in proportion to the genetic similarity between themselves and the proband [7].

Among the genes contributing to MS susceptibility (OMIM 126200), the most important are variations in the human leukocyte antigen (HLA) genes from the major histocompatibility complex (MHC) [8]. Subsequently, over the recent decades, results from accumulating genome-wide association studies have facilitated the definition of more than 100 non-MHC risk loci that however still explain less than one-third of the heritability [9].

5.1.3 Gender bias and parent-of-origin effect

In MS epidemiology, there is also a "gender bias." Among patients with MS, women outnumber men, and the female-to-male ratio is apparently increasing, as is the overall prevalence, as demonstrated for example by studies in Denmark (female-to-male ratio 1.31 for 1950, but 2.02 for 2005 [10]), Canada (1.9 for 1931, but 3.2 for 1980 [11]), and Sweden (1.70 for patients born in the 1930s, but 2.67 for the 1980s [12]). The gender bias is further augmented by a parent-of-origin effect. In a large Canadian cohort, studies on half-siblings and avuncular pairs suggested that the maternal route was favored in disease transmission, with maternal half-siblings of patients with MS being at significantly higher risk for developing MS compared to paternal half-siblings [13,14]. Maternal parent-of-origin effects have also been suggested to lower the age of onset of MS in affected female offspring [15,16].

5.1.4 A role for environmental factors

The non-Mendelian inheritance of MS susceptibility suggests that genetic differences between individuals are not solely responsible for the onset of the disease and thus the current paradigm for causality comprises multiple factors, associating hereditary predisposition with environmental factors and, to some extent, lifestyle. Among these factors, the most firmly established are low vitamin D levels (exposure to sunlight/the latitude gradient, and/or diet), adolescent obesity, and adolescent Epstein–Barr virus (EBV) infection.

One of several strong indicators of an interplay between hereditary factors and environment in MS pathogenesis is the finding of a latitude gradient. This gradient shapes as a significant increase in incidence and prevalence of MS with distance to the Equator, both in the northern and southern hemispheres [17,18]. Thus, the highest MS prevalence is found for people of Northern European/Scandinavian descent [17,19], although there are exceptions, such as in Italy [18], or in certain sporadic epidemiological clusters [20]. The impact of environment on this issue is further nuanced by an apparent decrease over recent years of the latitude-gradient effect, for example, in studies of US veterans [21] and by increasing prevalence in historically low-risk areas such as South America [22]. Furthermore, studies of migration have indicated that the age of migrants affects the risk of MS, in that preadolescents migrating from a low-risk to a high-risk country/region "acquire" the higher risk [23,24]. As a direct connection with lifestyle, it should also be noted that increased exposure to tobacco smoke and environmental pollutants have been linked with disease onset [25]. Childhood obesity may also predispose for MS, with recent observational/retrospective studies from Scandinavia and the United States having indicated that obesity (high BMI) in early life increases the risk [26–28]. With the caveat of confounding factors in such studies, a recent Mendelian randomization approach to assess causality between early obesity and MS risk has substantiated this risk factor [29,30].

5.1.5 A viral component in MS

Disease development in MS is also impacted by exposure to a number of viral or bacterial infections, and there is a clear correlation between phases of relapse and the occurrence of nonspecific infectious syndromes [31–33]. A specific example is the association of MS risk with EBV infection when it occurs during puberty rather than early childhood. EBV is a B-lymphotropic member of the Herpesviridae, a family of ubiquitous large DNA viruses causing persistent, latent infections. EBV is associated with certain lymphoproliferative and immune disorders. Most commonly, both the primary infection (usually in early childhood) and subsequent latency are asymptomatic, but a number of large studies have demonstrated that almost all MS patients are EBV-positive and often have a history of infectious mononucleosis (often manifested when primary EBV infection occurs during puberty). Similarly, high levels of anti-EBV antibodies increase the risk of MS [34–38].

Finally, a specific viral–MS relationship spanning the bridge between genetic susceptibility and environmental factors is the association of increased expression of certain human endogenous retroviruses (HERVs) with MS, as will be further discussed later. HERVs are vestigial retroviral sequences, inserted in the human genome both before and during primate evolution. Many of these loci retain intact promoter activity, some encode proteins, and in rare cases, sufficiently competent loci may complement each other to form secreted particles [39,40]. At least one HERV locus is classified as a non-MHC-locus contributing to MS risk: the HERV-H/Fc1 locus on chromosome X [41,42].

5.2 MS as an epigenetic disorder

The challenge in deciphering the etiology of MS lies in untangling the complex web of interactions between pathogenic pathways in this polygenetic, multifactorial disease. Fresh insight into this complexity may come from characterizing transcriptional events underlying the cellular and systemic symptoms of MS. Along these lines, in the following sections, we will discuss how some of these symptoms may be linked to chromatin structure and its effect on transcription, and we will describe several mechanisms of chromatin-mediated gene silencing that may be altered in patients with MS.

5.2.1 Nongenetic causes of MS and their link to chromatin and transcription

Many suggested nongenetic causal factors in MS share, an intriguing connection with transcriptional regulation. Latitude affects exposure to sunlight and consequently availability of vitamin D, a variable that directly impacts on the activity of the transcription factor VDR (vitamin D receptor). Similarly, pollutants, whether exposure comes from cigarette smoke, food, or other sources, frequently stimulate cellular stress signaling pathways, resulting in transcriptional activation of defense genes, including many cytokines [25,43]. Finally, transcriptional activity of HERVs is a likely manifestation of reduced chromatin-dependent gene silencing.

5.2.1.1 Vitamin D and the vitamin D receptor

Studies of vitamin D levels as well as vitamin D intake have established that deficiency of this vitamin contributes to MS risk [44,45]. Inversely, higher serum levels of 25-hydroxyvitamin D (25(OH)D_3) is associated with lower MS relapse risk [46]. Vitamin D is the ligand of the nuclear receptor VDR. This transcription factor dimerizes with retinoid X receptor (RXR), another nuclear receptor. When unliganded, this dimer associates with histone deacetylase activities to restrict chromatin accessibility and prevent gene activation. In the presence of the agonist, the VDR/RXR complex is stabilized, and becomes preferentially localized to the nucleus where it mostly promotes transcriptional activation by recruiting histone acetylases to enhancers and promoters of target genes (for a review, see Ref. [47]). Liganded VDR/RXR can also repress transcription, possibly by interfering with other promoter-bound transcription factors, and preventing them from recruiting coactivators (see for example [48]). In this context, we note that vitamin D-stimulated binding of VDR/RXR is observed inside the HIST1 cluster of histone genes, embedded in the extended human MHC on chromosome 6 [49,50] and it can be speculated that this VDR binding interval participates in the normal regulation of both MHC [51] and histone proteins upon sufficiency of vitamin D.

5.2.1.2 Reactivation of HERVs

HERVs originate from ancient retroviral infections, which have entered the germ line, and thus are present in the genome of all host cells in subsequent generations. As a consequence of multiple infections and possibly reinfection events, these sequences are present in thousands of loci and multiple copies in the human genome [40,52]. In the germline, the repeated nature of these sequences allows for their transcriptional repression by Argonaute proteins and piRNAs, a subtype of microRNAs [53,54]. Whether a similar repression mechanism exists in somatic cells is still controversial; however, it is clear that HERVs [55], like other imperfect repeats, such as alpha satellites or satellite two sequences [56], are kept in check by chromatin-mediated repression mechanisms. The transcriptional activity of HERVs therefore strongly suggests that in patients with MS, some chromatin-dependent repression mechanisms are defective [57].

5.2.2 DNA and histone modifications are footprints of transcriptional regulation

To unravel the possible role played by chromatin in the onset of MS, it is important to appreciate the intimate link uniting chromatin and transcriptional regulation. Within the nucleus, the DNA is associated with histones, and together, they constitute the chromatin. Chromatin is an obstacle for transcription. Consequently, initiation of transcription is largely an enzymatic process where transcription factors recruited by DNA motifs gather enzymes modifying the histones to create a transcription-friendly environment at promoters (for reviews, see Refs. [58,59]). Without going into details, the enzymes frequently referred to as transcription cofactors will either methylate the DNA, introduce posttranslational modifications (PTM) on histones, or "remodel" the nucleosomes, that is, modify the contact between histones and DNA while consuming adenosine triphosphate (ATP). In other words, tampering with chromatin is the essence of transcriptional regulation. This also means that the chromatin carries the footprint of transcription, and it is now possible to gather information on the function of a given locus from the histone modifications present at the site. The "chromatin states" as they have been defined by the "Epigenome Roadmap Consortium" are an example of how histone modifications can be translated into functional information [60]. By mapping approximately 30 different histone modifications and combining them with data on DNA methylation, DNAse accessibility, and local transcriptional activity, the consortium has defined a limited number of "states" allowing the prediction of whether a DNA locus is a promoter, an enhancer, a transcribed region, and so on. Importantly, these chromatin states are not defined by a single histone modification, but by the presence—in defined proportions—of several modifications. For example, an enhancer region will be rich in histone H3 lysine 27 acetylation (H3K27ac) and lysine four monomethylation (H3K4me1), while poor in lysine four trimethylation (H3K4me3), providing a clear illustration of the concept of the "histone code" as it was defined by Jenuwein and Allis in 2001 [61].

The term "epigenetic" frequently used to allude to the chromatin-borne information *stricto sensu* supposes transmission from one generation to the other. As we will see later, there is very little evidence for transgenerational transmission of this information that, however, remains fascinating by the long-term effects it may have throughout the lifetime of individuals.

5.3 DNA and histone modifications linked to MS

Our knowledge of changes in chromatin-borne information in association with MS is still in its infancy. Examining histone modifications on a genome-wide scale requires significant amounts of live patient material, and therefore available studies frequently restrict themselves to peripheral blood that may not be the most informative tissue. Analysis of DNA methylation is more manageable because DNA is very stable and methylation can be detected by chemical approaches. Consequently, this modification is currently the best characterized in MS.

5.3.1 DNA methylation

By default, DNA CpG dinucleotides are methylated at position 5 of the pyrimidine ring of cytosine (5-mC). DNA methylation favors C to T mutations, and CpG dinucleotides are therefore statistically underrepresented, particularly outside coding and regulatory regions. Inside coding regions, CpG dinucleotides may be maintained because of the selective pressure on codon integrity, but also because DNA methylation appears to be required for transcriptional elongation or mRNA maturation [62–64]. Promoter sequences are frequently enriched in CpG dinucleotides and, in fact, 60% of all human genes use CpG-rich regions (CpG islands or CGIs) as promoters. These regions are largely unmethylated, but their methylation in the context of imprinting, during development, and in cancer results in transcriptional repression (for a review, see Ref. [65]). It is within these regions that connections between MS and DNA methylation have been investigated.

5.3.1.1 Imprinting

Genomic imprinting consists of CpG methylation marks on the DNA of one or the other parent that persist through the wave of genome-wide DNA demethylation occurring after fertilization. Genomic imprinting in mammals results in the expression of the alleles of a given gene being dependent on their parental origin. In mammals, this phenomenon affects a few hundred genes [66]. It is one of the rare true epigenetic events documented as transgenerational, and it generates patterns of transmission difficult to untangle.

As discussed earlier, several studies have suggested a parent-of-origin effect in MS. Although parent-of-origin effects can easily be confounded by environmental and *in utero* effects, they may also be the reflection of imprinting. Maternally transmitted HLA-DRB1*1501 haplotypes have been associated with a lower age of onset of MS of affected female offspring [16,38]. Whether this parent-of-origin effect is due to imprinting has yet to be demonstrated and to date, no gene directly linked to MS has been confirmed as imprinted. Yet, several genes involved in inflammation are imprinted. One example, *DLK1* encodes an epidermal growth factor (EGF)-repeat-containing protein similar to molecules involved in the Notch signaling pathway. Several studies provide evidence for the involvement of DLK1 in immune system development and function, and genetic studies have suggested a causal role of DLK1 in inflammatory diseases [67]. Imprinting also affects expression of noncoding RNAs, and several imprinted domains contain clusters of miRNAs, many of which are also imprinted [68]. Although miRNAs are suspected actors in MS [69], a role for these miRNAs in the onset of MS has yet to be demonstrated.

Finally, we need to introduce the notion of metabolic imprinting. Nutrition and environment of the pregnant mother in the perinatal period can lead to epigenetic changes in the child that persist into adulthood. A textbook example of this comes from epidemiologic studies of survivors of the Dutch famine of 1944-45, revealing that low-calorie availability just before birth can influence body mass index in the adult (for a review, see Ref. [70]). In the case of MS, it is the exposure of the mother to sunlight and the subsequent availability of vitamin D for the child that has been suggested to play a role in protecting against MS [71]. This hypothesis is in part supported by later studies showing that prenatal and early life vitamin D deficiency in rats affects the adult brain and causes aberrant expression of several proteins also affected in patients with MS [72].

5.3.1.2 Differential methylation in blood cells and in CNS

Although differences in genomic imprinting in patients with MS is still speculative, several studies have reported differences in levels of DNA methylation between patients with MS and healthy controls. Based mainly on Human Methylation 450K arrays, these studies have shown that a relatively large series of CGIs are differentially methylated, although it is difficult to discern a general pattern: analysis of patient $CD8^+$ T cells leads the authors to suggest extensive CGI methylation associated with MS [73], while other studies on brain tissues suggest more specific changes, with MS being associated with gene-specific hyper- or hypomethylation [74,75]. In parallel, a study examining global levels of DNA methylation (not only CGIs) using an antimethyl cytosine antibody reports a two-third global 5-mC loss in normal appearing white matter (NAWM) of the brain [76]. In contrast, and rather counterintuitively, methylation levels at repetitive elements such as Alu, LINE-1, and SAT-α are increased in blood samples from patients with MS [77]. Regarding endogenous retroviruses, hypomethylation at some loci has been reported for systemic lupus erythematosus [78], but no study has addressed this in patients with MS.

5.3.1.3 Cytosine hydroxymethylation

Hydroxymethyl cytosine, an additional DNA modification originating from 5-mC oxidation by the Ten-eleven translocation 1−3 (TET 1−3) proteins, has recently attracted much attention. The TET enzymes sequentially convert 5-mC into 5-hydroxymethylcytosine (5-hmC), 5-formylcytosine (5-fC), and 5-carboxymethylcytosine (5-caC). These are intermediates in active DNA demethylation processes, which lead to removal of inadvertent 5-mC by the base-excision repair mechanism (for a review, see Ref. [79]). However, 5-fC, 5-caC, and 5-hmC may be more than mere 5-mC oxidation intermediates, particularly 5-hmC which is highly stable and abundant in specific tissues such as the brain. Recent work has indeed shown that 5-hmC formation renders the DNA methylome more dynamic and ductile under specific environmental stimuli. This is particularly evident in tissues characterized by poor regenerative potential such as the brain [80]. Thus, the 5-hmC modification arises as an additional opportunity to explain MS-related methylome changes. Currently, a link between 5-hmC reduction and MS has been observed in peripheral blood mononuclear cell (PBMCs) from patients with MS, in parallel with a diminished expression of the related methylcytosine-hydroxylase TET2 [81]. A reduced expression of TET enzymes was also observed in MS brain tissue in demyelinated hippocampus [74].

5.3.2 A possibly reduced efficiency of the H3K9me/HP1 axis of transcriptional repression in MS

At the surface of the nucleosome, the tail of histone H3 harbors two "alanine arginine lysine serine (ARKS)" motifs that are both methylated on the lysine (K), phosphorylated on the serine (S), and methylated or citrullinated on the arginine (R) residues. The two lysines (H3K9 and H3K27) within these ARKS motifs function as repressive marks when methylated, and are recognized by conserved proteins domains of which the chromodomains are probably the most studied. The chromodomain proteins recognizing methylated H3K27 are members of the Polycomb group, all subunits of large complexes mainly involved in gene silencing during development. Methylated H3K9 (H3K9me) is recognized by another series of chromatin readers whereof HP1 proteins are the best characterized. In mammals, this family of chromodomain proteins includes HP1α, HP1β, and HP1γ. Although the role of these proteins in the repression of HERVs is still debated [82], several lines of evidence suggest a possibly weakened H3K9me/HP1 axis in patients with MS.

5.3.2.1 Reduced recruitment of HP1 at HERVs and proinflammatory genes in patients with MS

H3K9me marks are enriched at DNA repeats present within pericentromeric regions, but also at other repeats such as Sat2 [56]. Expression of HERVs is also largely kept in check by H3K9me [55]. Interestingly, the H3K9me/HP1 axis also participates in transcriptional repression of a series of proinflammatory cytokine and interferon-responsive genes in the absence of stimulation and may be an important component in the control of inflammation (Fig. 5.1A, and [83,84]). Other studies have further shown that HP1 proteins are involved in the stable resilencing/tolerization of proinflammatory promoters after acute phases of inflammation [85,86].

The mechanisms allowing proinflammatory signaling to overcome HP1-mediated repression at innate immune genes is relatively well characterized. At these genes, the binding of HP1 proteins to H3K9me is antagonized by H3S10 phosphorylation combined with H3K14 acetylation (Fig. 5.1B and C, and [87–89]). Histone H3S10 phosphorylation is catalyzed by IκB kinase α (IKKα), an NF-κB pathway component [90] and by mitogen- and stress-activated protein kinase (MSK1), a kinase immediately downstream of extracellular signal–regulated kinase and p38 in the mitogen-activated protein (MAP) kinase signal transduction cascade [91]. IKK-α and MSK1 also phosphorylate the HP1γ isoform directly [92,93]. Thus, both the NF-κB and MAP kinase pathways are rigged with histone/HP1 kinases allowing them to overcome HP1-mediated repression at proinflammatory genes.

As MS pathogenesis combines an inflammatory condition with an increased expression of HERVs (see for example [41,94]), we investigated recruitment of HP1α at some of the HERVs and at the TNFα promoter. In all cases, we observed a reduced recruitment of this protein [95].

5.3.2.2 Peptidylarginine deiminases interfere with the H3K9me/HP1 axis of transcriptional repression

A reduced recruitment of the transcriptional repressor HP1α at promoters of activated inflammation-related genes and HERVs was an expected result, leaving open the question of the mechanism causing the eviction of the HP1 protein. In this context, our attention was drawn to the possible effect of H3R8 citrullination on HP1 binding to H3K9me. Citrullination is a process whereby an arginine residue is converted to the nonstandard amino acid citrulline [96]. This PTM is carried out by a family of peptidylarginine deiminases, with PADI2 and PADI4 being the most common isoforms found in the brain [97]. These peptidyl arginine deiminases (PADIs) also happen to be the only isoforms entering the nucleus [97,98]. Therein, one of their substrates is histone H3, and conversion of H3R8 into a citrulline destroys the HP1 binding site formed by methylated H3K9 (Fig. 5.1D). We also observed increased expression of the PADI4 gene in PBMCs in approximately one-third of the patients we examined [95].

This observation is to be viewed in parallel with the accumulating evidence for increased citrullination in MS. PADI2, which also modifies histone H3, accumulates in NAWM of patients with MS in correlation with a decreased methylation of the promoter of the gene [76]. Excess citrullination has

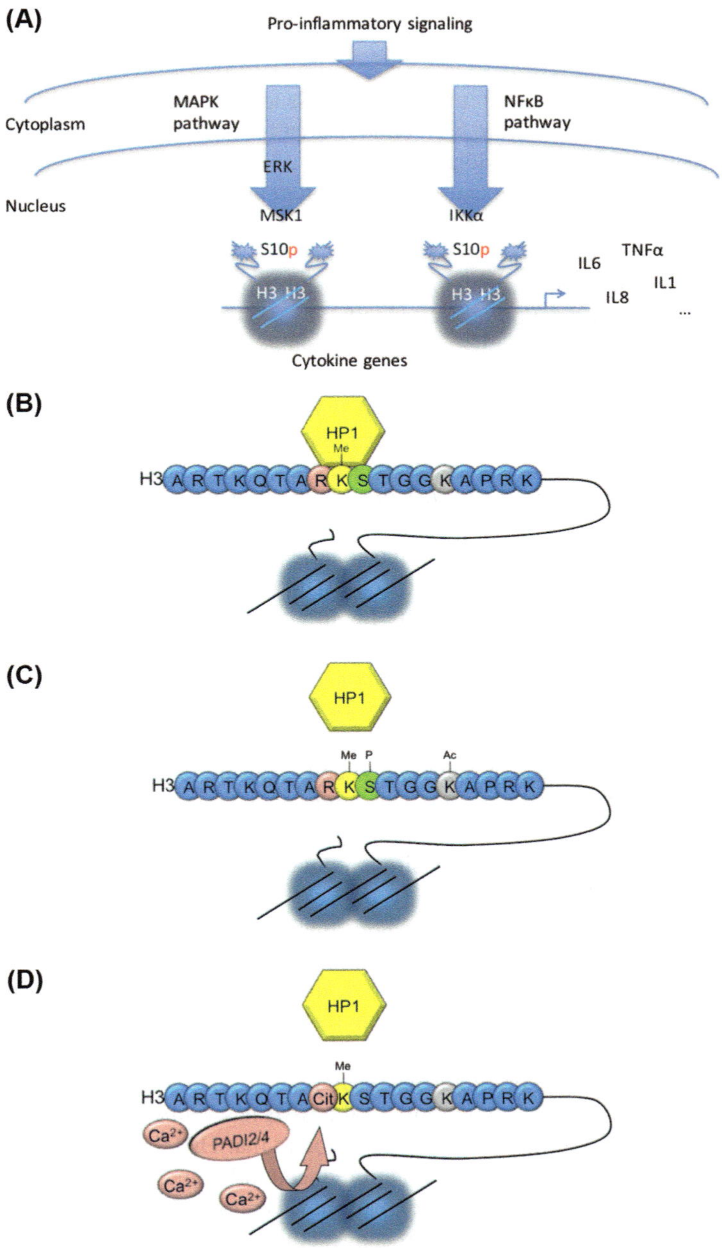

FIGURE 5.1

Proinflammatory signaling to chromatin. (A) Activation of the MAP kinase pathway and/or the NFkB pathway results in activation of kinases (MSK1 and IKKα, respectively) that phosphorylates histone H3 at serine 10 (H3S10). This phosphorylation event participates in the transcriptional activation of several cytokine genes, including TNFα, IL6, IL8, and IL1. (B) HP1 proteins bind to the tail of histone H3 when it is methylated on lysine 9 (H3K9me). (C) This binding is disrupted by H3S10 phosphorylation and subsequent acetylation of H3K14. (D) Citrullination (Cit) of histone H3R8 in the presence of calcium also disrupts binding of HP1 proteins.

been reported in the CNS in postmortem MS brain tissue [99,100], and more recently, increased citrullinated glial fibrillary acidic protein was found in areas of ongoing demyelination and myelin loss in active and chronically active MS lesions [101].

Citrullination has been suggested to cause partial protein unfolding and increased solvent exposure, possibly increasing immunogenicity [102–104]. In parallel, citrullination of histone H1 has been shown to favor chromatin accessibility [105], while citrullination of the transcription factor ELK-1 facilitates its phosphorylation and is part of the activation of the c-Fos promoter [106]. This later study also shows that PADI4 accumulates at specific sites on chromatin, engendering citrullination events that are unlikely to be detected by global approaches such as immunocytochemistry.

These observations suggest that citrullination functions as a regulator of protein–protein interaction and gene activity in healthy cells, and that deregulated expression or activity of PADI enzymes result in both immunological and transcriptional issues.

5.3.2.3 The H3K9me/HP1 axis: a central player in the onset of MS?

One of the remaining questions is whether weakening of silencing machineries such as the H3K9me/HP1 axis could have a systemic impact and elicit a complex disorder such as MS. Inactivation of HP1γ in the mouse has provided some hints on that issue. The mutation is embryonically lethal, but it is possible to isolate mouse embryonic fibroblasts for in vitro culture. Stimulation of these cells with phorbol 12-myristate 13-acetate (PMA), an activator of protein kinase-C frequently used to mimic a proinflammatory signal, followed by transcriptome analysis, revealed that inactivation of HP1γ results in both increased amplitude and decreased specificity of the inflammatory response [93]. One manifestation of this is an exacerbated (mostly down-) regulation of genes involved in the formation of the extracellular matrix (collagens), and in matrix- and cell–cell interactions (in particular, the vascular cell adhesion protein VCAM-1) (Fig. 5.2A). This study also showed a role for HP1 in the negative regulation of pathogen-associated molecular pattern (PAMP) receptors (Fig. 5.2B). Finally, interferon-inducible genes, and genes involved in the positive regulation of cell proliferation were stimulated by PMA only in the absence of HP1γ [93].

Although this is still very speculative, we note that there are several similarities between this HP1$\gamma^{-/-}$ phenotype and the manifestations observed in tissues from patients with MS, including the high expression of proinflammatory cytokine genes, the increased levels of PAMP receptors in the absence of any apparent microbial stimulus (for a review, see Ref. [107]), and the modified expression of genes involved in extracellular matrix to receptor interactions that are essential in the processes by which T-cells penetrate the blood–brain barrier (BBB).

The impact of HP1γ inactivation on integrity of the BBB has yet to be examined, but in a different barrier, that is, the gut epithelium of guinea pigs, we found that HP1γ is abundantly expressed in the proliferative compartment, consistent with a role in reduced inflammatory response and rapid renewal observed in this tissue [93]. A similar expression pattern was reported earlier for HP1α in mouse colon and intestine [108]. Together, these observations suggest that a defective HP1-mediated transcriptional regulation may impact cells outside, within, and possibly beyond the BBB, generating molecular events favoring the onset of an autoimmune condition (see model in Fig. 5.3).

(A)

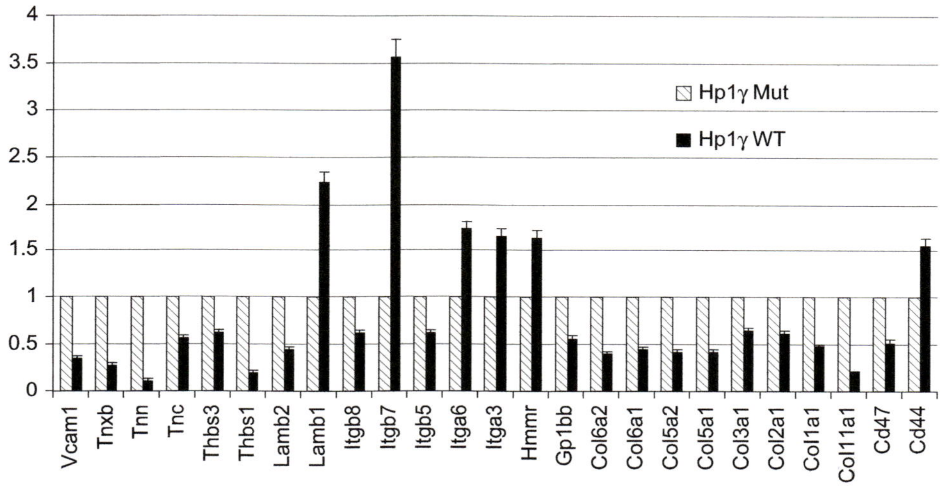

(B)

Gene Symbol	Regulation	Fold-Change	P-Value	Gene Name/function	Aliases and Synonyms
Nlrp10	down	8.30	2.33E-47	NLR family, pyrin domain containing 10	Nalp10, Pynod
Lbp	down	7.95	4.92E-25	lipopolysaccharide binding protein	Bpifd2
Nod1	down	6.02	2.62E-30	nucleotide-binding oligomerization domain containing 1	Card4, Nlrc1
C3	down	4.35	7.51E-14	complement component 3	Plp
Tlr2	down	3.95	7.31E-63	toll-like receptor 2	Ly105
Ticam1	down	2.15	8.87E-19	toll-like receptor adaptor molecule 1	Tlr3, Trif
Sigirr	up	2.02	1.18E-11	inhibitor of TLR-induced cytokines	Sigirr, Tir8

FIGURE 5.2

Inactivation of HP1γ results in modified expression of cell–cell interaction and PAMP receptor genes. (A) Mouse fibroblasts either HP1γ null (Mut) or restored with wild-type HP1γ (WT) were stimulated with PMA, an activator of protein kinase C. After 1 h, RNA was collected and derived cDNA analyzed by next-generation sequencing (NGS). Graph shows genes belonging to the Kyoto encyclopedia of genes and genomes (KEGG) pathway "ECM-Receptor interaction" and altered more than 1.5-fold in their expression with a *P*-value $<.05$ when comparing HP1γ Mut with HP1γ WT cells in the presence of PMA. (B) As in (A), mouse fibroblasts either HP1γ Mut or HP1γ WT were stimulated with PMA. After 1 h, RNA was collected and derived cDNA analyzed by NGS. Table lists genes encoding known PAMP receptors or regulators thereof, altered more than 2-fold in their expression with a *P*-value $< 5.10–10$ when comparing HP1γ Mut with HP1γ WT cells in the presence of PMA. GEO Data accession: GSE60283. *P*-value calculated with Student's *t*-test.

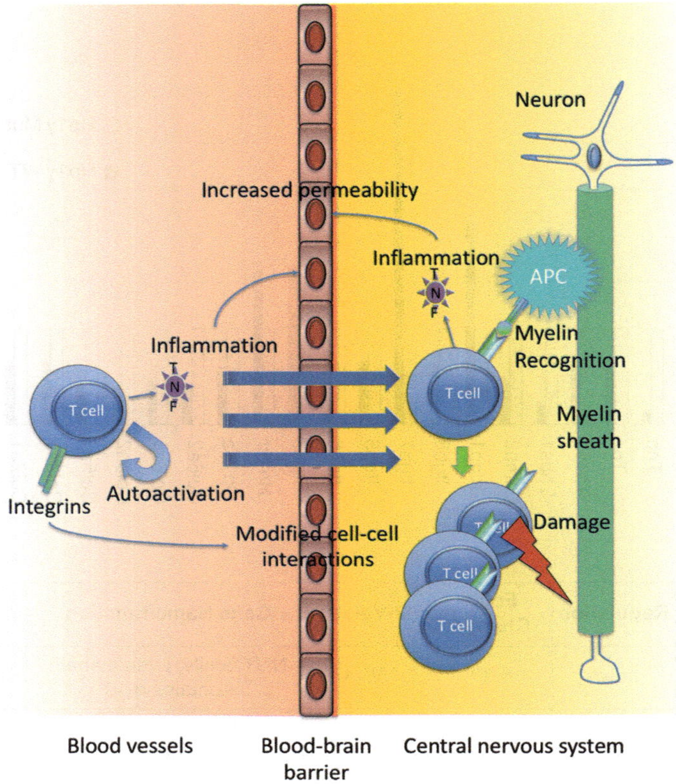

FIGURE 5.3

Some hypothetical consequences of reduced HP1-mediated transcriptional regulation. The illustration is T-cell based, but could also be envisioned with B cells. Outside the blood–brain barrier (BBB), T cells and other cell populations have an exacerbated reactivity to proinflammatory stimuli, generating inflammation that in turn weakens the BBB). These cells also express a modified repertoire of integrins that changes their adhesion to and ability to penetrate through the BBB endothelium. Cells of the BBB may also be burdened by inefficient HP1-mediated repression, possibly affecting cell–cell interaction within the endothelium. Inside the CNS, the increased influx of immune cells also include cells of the monocyte/macrophage lineage contributing to the antigen-presenting cell (APC) repertoire, and increases the probability for entry of T or B cell clones expressing either antimyelin T-cell receptor (TCR) or immunoglobulins, respectively. Their encounter with the antigen results in amplification of the antimyelin load.

5.4 Epigenetics beyond transcription

In the previous paragraphs, much of our attention focused on transcription factors and cofactors on chromatin. Later, we will address some of the mechanisms providing possible direct links between environmental stimuli and chromatin.

5.4.1 Exosomal miRNA silencing

MicroRNAs (miRNAs) are small noncoding RNAs (about 22 nucleotides) regulating a large number of biological processes by blocking translation or inducing degradation of mRNAs via base pairing to their untranslated or coding regions [109]. MicroRNAs are found intracellularly but are also abundantly present in exosomes that are cell-derived vesicles shed by almost all cell types [110]. This allows for transmission of miRNAs from mother to child via breast milk, and it is in this context that they may be considered as epigenetic factors. Breastmilk is the essential postnatal nutrition at the very beginning of extrauterine life. Several lines of evidence also document that breastfeeding lowers the risk of diseases with an autoimmune component, such as type 1 diabetes, celiac disease, and inflammatory bowel diseases [111]. Breastfeeding is also discussed as possibly lowering the risk of MS: the absence of infant breastfeeding is associated with an increased risk of pediatric-onset MS [112], and males breastfed for less than 4 months also have increased MS risk [113].

The possible prophylactic effects of breastmilk are complex. In the short term, it provides a direct immune defense against pathogens through immune cells and molecules such as immunoglobulins and lysozymes [114,115]. However, it also appears to participate in the shaping of the developing immune system with long-term consequences for the health of the child and the adult he or she will become [114–116]. Exosomal miRNAs may play an active role in this process, and their absence may cause functional alterations in the pool of hematopoietic stem cells. Such a mechanism may be at the root of the recent successes obtained with MS therapy in the form of immunoablation and autologous hematopoietic stem cell transplantation that seem to inhibit disease activity and to prevent progression in patients with RRMS [117].

5.4.2 Microbiota

The microbiota plays a fundamental role in the induction, and the training of the host immune system. In return, the immune system maintains the symbiotic relationship of the host with the commensal microorganisms that reside within the gut. When this system operates optimally, it allows the combination of protective responses against pathogens with tolerance to innocuous antigens. However, overuse of antibiotics and changes in diet are suspected to favor a microbiota that lacks the resilience and diversity required to establish balanced immune responses [118]. This phenomenon may contribute to some of the autoimmune and inflammatory disorders prevalent in high-income countries. Accumulating data further suggest a potential communication between the gut microbiota and the brain. The reasoning behind this gut–mind axis is that the development of a healthy, functional brain is guided by environmental cues, many of which originate from the microbiome. These cues affect phases of neurogeneration that are highly relevant for MS, including the formation of the BBB, and myelination [119]. A recent study has substantiated a possible role for the gut microbiota in the onset of MS. It shows that microbiota from patients with MS is more apt to induce autoimmunity than that of healthy controls when transplanted to a transgenic mouse model of spontaneous brain autoimmunity [120]. Although the mechanisms involved are still unclear, these observations have important implications not only for the understanding of the pathogenesis but also for the therapy and possibly prevention of MS.

5.4.3 Environment: pollutants that may interfere with silencing machineries

Environmental pollutants have repeatedly been suspected to play a role in the onset of MS [121]. A recent study suggests that air pollution, in particular, may trigger phases of relapse and explain in part the apparent seasonal variations of these phases [122]. The organic pollutants do indeed deserve attention in this respect because some have known effects on chromatin structure. For example, the unsaturated aldehyde acrolein, present in tobacco smoke and urban air pollution, preferentially binds to methylated CpG-rich DNA and causes G to A and G to T mutations that may interfere with transcriptional repression [123]. Similarly, air enriched in aluminum, manganese, nickel, zinc, arsenic, lead, and iron was shown to cause increased H3K9 acetylation in a cohort of industrial workers [124]. Finally, several elements such as arsenic, nickel, and cadmium are suspected to induce hypomethylation of DNA [125]. Whether the concentrations of these pollutants present in the environment are sufficient to dysregulate the immune system via epigenetic mechanisms remains to be established.

5.5 Conclusions

Numerous high-quality studies have clearly established a genetic component in MS. Yet pinpointing a specific mutation, or even a number of mutations allowing to predict MS risk with reasonable confidence is not possible at this stage. Examination of the chromatin from patients with MS has provided some new tracks to follow, including altered DNA methylation and deficient HP1 binding, both predicting inappropriate gene activation. Moreover, when it comes to explain the cause of such chromatin anomalies, the environment is the primary suspect.

However, the environment is shared by many, while only few suffer from MS, the difference between affected and unaffected is probably lying in genetic susceptibility. This leaves us in a loop that emphasizes the need for a better understanding of the links between genetics and epigenetics. Presently, we understand a DNA mutation when it affects a codon or a splice site, or modifies a promoter. The consequences of mutations further away from genes are more difficult to decipher.

The field of chromatin research is currently building maps of the topological structure of the genome, defining chromatin loops, and sites of attachment of the chromatin to the nuclear lamina. Others are mapping distant enhancers by cataloging histone modifications and monitoring the production of enhancer RNAs (eRNAs). Combining the two maps (loops and enhancers) will progressively allow identifying the genes in proximity to the enhancers in the 3D nucleus, and possibly permit understanding of the functional impact of intergenic and intronic DNA mutations. Such approaches are already implemented in the field of cancer (see for example [126]).

For MS, the limitation of the available data is that they are derived from studies of healthy tissues only, or cancer cell lines that have interest mostly for cancer and basic research. Should one want to identify an "epimutation" associated with MS, one would need to carry out genome-wide chromatin immunoprecipitations with 10 or more different chromatin activators and silencers to feed a "chromatin state" algorithm: each in triplicate, and for a sufficiently high number of patients for the data to be significant. Subsequently, repeat this with healthy controls. So, unfortunately, the currently available techniques require unreasonable amounts of patient material, human resources, and funding.

Possibly, the rapidly evolving single-cell approaches may allow us to overcome this limitation. Alternatively, approaches based on the detection of eRNAs could be a powerful source of information on the epigenetic landscape of patients with MS [127]. With these tools at hand, it may also be possible to identify the histone and DNA modifiers at play on the chromatin, and inhibitors of those enzymes may be the MS drugs of tomorrow.

Acknowledgments

We thank Frederique Michel, Yevheniia Porozhan, and Jacob Seeler for critical reading of the manuscript.

References

[1] Hafler DA. Multiple sclerosis. J Clin Invest 2004;113:788–94. https://doi.org/10.1172/JCI21357.
[2] Hempel S, Fu N, Estrada E, Chen A, Miake-Lye I, Beroes J, et al. Modifiable risk factors in the progression of multiple sclerosis: a systematic review of the epidemiology and treatment. 2015.
[3] Compston A, Coles A. Multiple sclerosis. Lancet 2002;359:1221–31. https://doi.org/10.1016/S0140-6736(02)08220-X.
[4] Dendrou CA, Fugger L, Friese MA. Immunopathology of multiple sclerosis. Nat Rev Immunol 2015;15: 545–58. https://doi.org/10.1038/nri3871.
[5] Miller D, Barkhof F, Montalban X, Thompson A, Filippi M. Clinically isolated syndromes suggestive of multiple sclerosis, part I: natural history, pathogenesis, diagnosis, and prognosis. Lancet Neurol 2005;4: 281–8. https://doi.org/10.1016/S1474-4422(05)70071-5.
[6] Hansen T, Skytthe A, Stenager E, Petersen HC, Brønnum-Hansen H, Kyvik KO. Concordance for multiple sclerosis in Danish twins: an update of a nationwide study. Mult Scler 2005;11:504–10. https://doi.org/10.1191/1352458505ms1220oa.
[7] Goodin DS. The epidemiology of multiple sclerosis: insights to disease pathogenesis. 1st ed. Elsevier B.V; 2014. https://doi.org/10.1016/B978-0-444-52001-2.00010-8.
[8] Jersild C, Svejgaard A, Fog T. HL-A antigens and multiple sclerosis. Lancet 1972;1:1240–1.
[9] Gourraud P-A, Harbo HF, Hauser SL, Baranzini SE. The genetics of multiple sclerosis: an up-to-date review. Immunol Rev 2012;248:87–103. https://doi.org/10.1111/j.1600-065X.2012.01134.x.
[10] Bentzen J, Flachs EM, Stenager E, Brønnum-Hansen H, Koch-Henriksen N. Prevalence of multiple sclerosis in Denmark 1950–2005. Mult Scler 2010;16:520–5. https://doi.org/10.1177/1352458510364197.
[11] Orton SM, Herrera BM, Yee IM, Valdar W, Ramagopalan SV, Sadovnick AD, et al. Sex ratio of multiple sclerosis in Canada: a longitudinal study. Lancet Neurol 2006;5:932–6.
[12] Westerlind H, Boström I, Stawiarz L, Landtblom A-M, Almqvist C, Hillert J. New data identify an increasing sex ratio of multiple sclerosis in Sweden. Mult Scler 2014;20:1578–83. https://doi.org/10.1177/1352458514530021.
[13] Herrera BM, Ramagopalan SV, Lincoln MR, Orton SM, Chao MJ, Sadovnick AD, et al. Parent-of-origin effects in MS: observations from avuncular pairs. Neurology 2008;71:799–803. https://doi.org/10.1212/01.wnl.0000312377.50395.00.
[14] Ebers GC, Sadovnick AD, Dyment DA, Yee I, Willer CJ, Risch N. Parent-of-origin effect in multiple sclerosis: observations in half-siblings. Lancet 2004;363:1773–4. https://doi.org/10.1016/S0140-6736(04)16304-6.
[15] Ramagopalan SV, Byrnes JK, Dyment DA, Guimond C, Handunnetthi L, Disanto G, et al. Parent-of-origin of HLA-DRB1∗1501 and age of onset of multiple sclerosis. J Hum Genet 2009;54:547–9. https://doi.org/10.1038/jhg.2009.69.

[16] Chao MJ, Herrera BM, Ramagopalan SV, Deluca G, Handunetthi L, Orton SM, et al. Parent-of-origin effects at the major histocompatibility complex in multiple sclerosis. Hum Mol Genet 2010;19:3679–89. https://doi.org/10.1093/hmg/ddq282.
[17] Kurtzke JF. Epidemiology of multiple sclerosis. Does this really point toward an etiology? Lectio Doctoralis, Neurol. Sci 2000;21:383–403.
[18] Simpson S, Blizzard L, Otahal P, van der Mei I, Taylor B. Latitude is significantly associated with the prevalence of multiple sclerosis: a meta-analysis. J Neurol Neurosurg Psychiatry 2011;82:1132–41. https://doi.org/10.1136/jnnp.2011.240432.
[19] Kurtzke JF. Multiple sclerosis in time and space–geographic clues to cause. J Neurovirol 2000;6(Suppl. 2): S134–40.
[20] Haahr S, Munch M, Christensen T, Møller-Larsen A, Hvas J. Cluster of multiple sclerosis patients from Danish community. Lancet 1997;349:923. https://doi.org/10.1016/S0140-6736(05)62698-0.
[21] Wallin MT, Page WF, Kurtzke JF. Multiple sclerosis in US veterans of the Vietnam era and later military service: race, sex, and geography. Ann Neurol 2004;55:65–71. https://doi.org/10.1002/ana.10788.
[22] Cristiano E, Patrucco L, Miguez J, Giunta D, Correale J, Fiol M, et al. Increasing prevalence of multiple sclerosis in Buenos Aires, Argentina. Mult Scler Relat Disord 2016;9:91–4. https://doi.org/10.1016/j.msard.2016.07.007.
[23] Kurtzke JF, Beebe GW, Norman JE. Epidemiology of multiple sclerosis in US veterans: III. migration and the risk of MS. Neurology 1985;35:672–8.
[24] Ahlgren C, Odén A, Lycke J. A nationwide survey of the prevalence of multiple sclerosis in immigrant populations of Sweden. Mult Scler 2012;18:1099–107. https://doi.org/10.1177/1352458511433062.
[25] Olsson T, Barcellos LF, Alfredsson L. Interactions between genetic, lifestyle and environmental risk factors for multiple sclerosis. Nat Rev Neurol 2017;13:25–36. https://doi.org/10.1038/nrneurol.2016.187.
[26] Munger KL, Bentzen J, Laursen B, Stenager E, Koch-Henriksen N, Sørensen TIA, et al. Childhood body mass index and multiple sclerosis risk: a long-term cohort study. Mult Scler 2013;19:1323–9. https://doi.org/10.1177/1352458513483889.
[27] Wesnes K, Riise T, Casetta I, Drulovic J, Granieri E, Holmøy T, et al. Body size and the risk of multiple sclerosis in Norway and Italy: the EnvIMS study. Mult Scler 2015;21:388–95. https://doi.org/10.1177/1352458514546785.
[28] Munger KL, Chitnis T, Ascherio A. Body size and risk of MS in two cohorts of US women. Neurology 2009;73:1543–50. https://doi.org/10.1212/WNL.0b013e3181c0d6e0.
[29] Ascherio A, Munger KL. Weighing evidence from mendelian randomization-early-life obesity as a causal factor in multiple sclerosis? PLoS Med 2016;13:e1002054. https://doi.org/10.1371/journal.pmed.1002054.
[30] Mokry LE, Ross S, Timpson NJ, Sawcer S, Davey Smith G, Richards JB. Obesity and multiple sclerosis: a mendelian randomization study. PLoS Med 2016;13:e1002053. https://doi.org/10.1371/journal.pmed.1002053.
[31] Libbey JE, Cusick MF, Fujinami RS. Role of pathogens in multiple sclerosis. Int Rev Immunol 2014;33: 266–83. https://doi.org/10.3109/08830185.2013.823422.
[32] Cossu D, Yokoyama K, Hattori N. Conflicting role of Mycobacterium species in multiple sclerosis. Front Neurol 2017;8:216. https://doi.org/10.3389/fneur.2017.00216.
[33] Loebermann M, Winkelmann A, Hartung H-P, Hengel H, Reisinger EC, Zettl UK. Vaccination against infection in patients with multiple sclerosis. Nat Rev Neurol 2012;8:143–51. https://doi.org/10.1038/nrneurol.2012.8.
[34] Haahr S, Koch-Henriksen N, Møller-Larsen A, Eriksen LS, Andersen HM. Increased risk of multiple sclerosis after late Epstein-Barr virus infection: a historical prospective study. Mult Scler 1995;1:73–7. https://doi.org/10.1177/135245859500100203.
[35] Christensen T. The role of EBV in MS pathogenesis. Int MS J 2006;13:52–7.

[36] Ascherio A, Munch M. Epstein-Barr virus and multiple sclerosis. Epidemiology 2000;11:220–4.
[37] Ascherio A, Munger KL. Epstein-barr virus infection and multiple sclerosis: a review. J Neuroimmune Pharmacol 2010;5:271–7. https://doi.org/10.1007/s11481-010-9201-3.
[38] Ramagopalan SV, Dobson R, Meier UC, Giovannoni G. Multiple sclerosis: risk factors, prodromes, and potential causal pathways. Lancet Neurol 2010;9:727–39. https://doi.org/10.1016/S1474-4422(10)70094-6.
[39] Christensen T. Human endogenous retroviruses in neurologic disease. APMIS 2016;124:116–26. https://doi.org/10.1111/apm.12486.
[40] Villesen P, Aagaard L, Wiuf C, Pedersen FS. Identification of endogenous retroviral reading frames in the human genome. Retrovirology 2004;1:32. https://doi.org/10.1186/1742-4690-1-32.
[41] Nexø BA, Christensen T, Frederiksen J, Moller-Larsen A, Oturai AB, Villesen P, et al. The etiology of multiple sclerosis: genetic evidence for the involvement of the human endogenous retrovirus HERV-Fc1. PLoS One 2011;6:e16652. https://doi.org/10.1371/journal.pone.0016652.
[42] de la Hera B, Varadé J, Garcia-Montojo M, Alcina A, Fedetz M, Alloza I, et al. Human endogenous retrovirus HERV-Fc1 association with multiple sclerosis susceptibility: a meta-analysis. PLoS One 2014;9:e90182. https://doi.org/10.1371/journal.pone.0090182.
[43] Brodin P, Davis MM. Human immune system variation. Nat Rev Immunol 2017;17:21–9. https://doi.org/10.1038/nri.2016.125.
[44] Munger KL, Zhang SM, O'Reilly E, Hernán MA, Olek MJ, Willett WC, et al. Vitamin D intake and incidence of multiple sclerosis. Neurology 2004;62:60–5.
[45] Munger KL, Levin LI, Hollis BW, Howard NS, Ascherio A. Serum 25-hydroxyvitamin D levels and risk of multiple sclerosis. JAMA 2006;296:2832–8. https://doi.org/10.1001/jama.296.23.2832.
[46] Simpson S, Taylor B, Blizzard L, Ponsonby A-L, Pittas F, Tremlett H, et al. Higher 25-hydroxyvitamin D is associated with lower relapse risk in multiple sclerosis. Ann Neurol 2010;68:193–203. https://doi.org/10.1002/ana.22043.
[47] Rochel N, Molnár F. Structural aspects of vitamin D endocrinology. Mol Cell Endocrinol 2017;453:22–35. https://doi.org/10.1016/j.mce.2017.02.046.
[48] Turunen MM, Dunlop TW, Carlberg C, Vaisanen S. Selective use of multiple vitamin D response elements underlies the 1 ,25-dihydroxyvitamin D3-mediated negative regulation of the human CYP27B1 gene. Nucleic Acids Res 2007;35:2734–47. https://doi.org/10.1093/nar/gkm179.
[49] Horton R, Wilming L, Rand V, Lovering RC, Bruford EA, Khodiyar VK, et al. Gene map of the extended human MHC. Nat Rev Genet 2004;5:889–99. https://doi.org/10.1038/nrg1489.
[50] Ramagopalan SV, Heger A, Berlanga AJ, Maugeri NJ, Lincoln MR, Burrell A, et al. A ChIP-seq defined genome-wide map of vitamin D receptor binding: associations with disease and evolution. Genome Res 2010;20:1352–60. https://doi.org/10.1101/gr.107920.110.
[51] Unger WWJ, Laban S, Kleijwegt FS, van der Slik AR, Roep BO. Induction of Treg by monocyte-derived DC modulated by vitamin D 3or dexamethasone: differential role for PD-L1. Eur J Immunol 2009;39:3147–59. https://doi.org/10.1002/eji.200839103.
[52] Christensen T. Association of human endogenous retroviruses with multiple sclerosis and possible interactions with herpes viruses. Rev Med Virol 2005;15:179–211. https://doi.org/10.1002/rmv.465.
[53] Reuter M, Berninger P, Chuma S, Shah H, Hosokawa M, Funaya C, et al. Miwi catalysis is required for piRNA amplification-independent LINE1 transposon silencing. Nature 2011;480:264–7. https://doi.org/10.1038/nature10672.
[54] Ha H, Song J, Wang S, Kapusta A, Feschotte C, Chen KC, et al. A comprehensive analysis of piRNAs from adult human testis and their relationship with genes and mobile elements. BMC Genomics 2014;15:545. https://doi.org/10.1186/1471-2164-15-545.

[55] Friedli M, Trono D. The developmental control of transposable elements and the evolution of higher species. Annu Rev Cell Dev Biol 2015;31:429–51. https://doi.org/10.1146/annurev-cellbio-100814-125514.

[56] Wang D, Zhou J, Liu X, Lu D, Shen C, Du Y, et al. Methylation of SUV39H1 by SET7/9 results in heterochromatin relaxation and genome instability. Proc Natl Acad Sci USA 2013;110:5516–21. https://doi.org/10.1073/pnas.1216596110.

[57] Muchardt C. Transcriptional activity of endogenous retroviruses in multiple sclerosis: a sign of deficient chromatin-mediated silencing. Epigenomics 2015;7:1235–7. https://doi.org/10.2217/epi.15.81.

[58] Lee TI, Young RA. Transcriptional regulation and its misregulation in disease. Cell 2013;152:1237–51. https://doi.org/10.1016/j.cell.2013.02.014.

[59] Gibney ER, Nolan CM. Epigenetics and gene expression. Heredity 2010;105:4–13. https://doi.org/10.1038/hdy.2010.54.

[60] Roadmap Epigenomics Consortium, Kundaje A, Meuleman W, Ernst J, Bilenky M, Yen A, et al. Integrative analysis of 111 reference human epigenomes. Nature 2015;518:317–30. https://doi.org/10.1038/nature14248.

[61] Jenuwein T, Allis CD. Translating the histone code. Science Magazine 2001;293:1074–80. https://doi.org/10.1126/science.1063127.

[62] Shukla S, Kavak E, Gregory M, Imashimizu M, Shutinoski B, Kashlev M, et al. CTCF-promoted RNA polymerase II pausing links DNA methylation to splicing. Nature 2011. https://doi.org/10.1038/nature10442.

[63] Veloso A, Kirkconnell KS, Magnuson B, Biewen B, Paulsen MT, Wilson TE, et al. Rate of elongation by RNA polymerase II is associated with specific gene features and epigenetic modifications. Genome Res 2014;24:896–905. https://doi.org/10.1101/gr.171405.113.

[64] Yearim A, Gelfman S, Shayevitch R, Melcer S, Glaich O, Mallm J-P, et al. HP1 is involved in regulating the global impact of DNA methylation on alternative splicing. Cell Rep 2015;10:1122–34. https://doi.org/10.1016/j.celrep.2015.01.038.

[65] Jones PA. Functions of DNA methylation: islands, start sites, gene bodies and beyond. Nat Rev Genet 2012;13:484–92. https://doi.org/10.1038/nrg3230.

[66] Kelsey G, Bartolomei MS. Imprinted genes … and the number is? PLoS Genet 2012;8:e1002601. https://doi.org/10.1371/journal.pgen.1002601.

[67] Ruhrmann S, Stridh P, Kular L, Jagodic M. Genomic imprinting: a missing piece of the multiple sclerosis puzzle? Int J Biochem Cell Biol 2015;67:49–57. https://doi.org/10.1016/j.biocel.2015.05.010.

[68] Zhang Y, Guan D-G, Yang J-H, Shao P, Zhou H, Qu L-H. ncRNAimprint: a comprehensive database of mammalian imprinted noncoding RNAs. RNA 2010;16:1889–901. https://doi.org/10.1261/rna.2226910.

[69] Jagot F, Davoust N. Is it worth considering circulating microRNAs in multiple sclerosis? Front Immunol 2016;7:129. https://doi.org/10.3389/fimmu.2016.00129.

[70] Waterland RA, Garza C. Potential mechanisms of metabolic imprinting that lead to chronic disease. Am J Clin Nutr 1999;69:179–97.

[71] McGrath J. Does "imprinting" with low prenatal vitamin D contribute to the risk of various adult disorders? Med Hypotheses 2001;56:367–71. https://doi.org/10.1054/mehy.2000.1226.

[72] Almeras L, Eyles D, Benech P, Laffite D, Villard C, Patatian A, et al. Developmental vitamin D deficiency alters brain protein expression in the adult rat: implications for neuropsychiatric disorders. Proteomics 2007;7:769–80. https://doi.org/10.1002/pmic.200600392.

[73] Bos SD, Page CM, Andreassen BK, Elboudwarej E, Gustavsen MW, Briggs F, et al. Genome-wide DNA methylation profiles indicate CD8+ T cell hypermethylation in multiple sclerosis. PLoS One 2015;10:e0117403–16. https://doi.org/10.1371/journal.pone.0117403.

[74] Chomyk AM, Volsko C, Tripathi A, Deckard SA, Trapp BD, Fox RJ, et al. DNA methylation in demyelinated multiple sclerosis hippocampus. Sci Rep 2017:1–10. https://doi.org/10.1038/s41598-017-08623-5.
[75] Huynh JL, Casaccia P. Epigenetic mechanisms in multiple sclerosis: implications for pathogenesis and treatment. Lancet Neurol 2013;12:195–206. https://doi.org/10.1016/S1474-4422(12)70309-5.
[76] Mastronardi FG, Noor A, Wood DD, Paton T, Moscarello MA. Peptidyl argininedeiminase 2 CpG island in multiple sclerosis white matter is hypomethylated. J Neurosci Res 2007;85:2006–16. https://doi.org/10.1002/jnr.21329.
[77] Neven KY, Piola M, Angelici L, Cortini F, Fenoglio C, Galimberti D, et al. Repetitive element hypermethylation in multiple sclerosis patients. BMC Genet 2016;17:84. https://doi.org/10.1186/s12863-016-0395-0.
[78] Nakkuntod J, Sukkapan P, Avihingsanon Y, Mutirangura A, Hirankarn N. DNA methylation of human endogenous retrovirus in systemic lupus erythematosus. J Hum Genet 2013;58:241–9. https://doi.org/10.1038/jhg.2013.6.
[79] Rasmussen KD, Helin K. Role of TET enzymes in DNA methylation, development, and cancer. Genes Dev 2016;30:733–50. https://doi.org/10.1101/gad.276568.115.
[80] Tognini P, Napoli D, Pizzorusso T. Dynamic DNA methylation in the brain: a new epigenetic mark for experience-dependent plasticity. Front Cell Neurosci 2015;9:331. https://doi.org/10.3389/fncel.2015.00331.
[81] Calabrese R, Valentini E, Ciccarone F, Guastafierro T, Bacalini MG, Ricigliano VAG, et al. TET2 gene expression and 5-hydroxymethylcytosine level in multiple sclerosis peripheral blood cells. Biochim Biophys Acta 2014;1842:1130–6. https://doi.org/10.1016/j.bbadis.2014.04.010.
[82] Maksakova IA, Goyal P, Bullwinkel J, Brown JP, Bilenky M, Mager DL, et al. H3K9me3-binding proteins are dispensable for SETDB1/H3K9me3-dependent retroviral silencing. Epigenet Chromatin 2011;4:12. https://doi.org/10.1186/1756-8935-4-12.
[83] Lavigne M, Eskeland R, Azebi S, Saint-André V, Jang SM, Batsché E, et al. Interaction of HP1 and Brg1/Brm with the globular domain of histone H3 is required for HP1-mediated repression. PLoS Genet 2009;5:e1000769. https://doi.org/10.1371/journal.pgen.1000769.
[84] Jang SM, Azebi S, Soubigou G, Muchardt C. DYRK1A phoshorylates histone H3 to differentially regulate the binding of HP1 isoforms and antagonize HP1-mediated transcriptional repression. EMBO Rep 2014;15:686–94. https://doi.org/10.15252/embr.201338356.
[85] El Gazzar M, Yoza BK, Hu JY-Q, Cousart SL, McCall CE. Epigenetic silencing of tumor necrosis factor alpha during endotoxin tolerance. J Biol Chem 2007;282:26857–64. https://doi.org/10.1074/jbc.M704584200.
[86] Chen X, El Gazzar M, Yoza BK, McCall CE. The NF-kappaB factor RelB and histone H3 lysine methyltransferase G9a directly interact to generate epigenetic silencing in endotoxin tolerance. J Biol Chem 2009;284:27857–65. https://doi.org/10.1074/jbc.M109.000950.
[87] Mateescu B, England P, Halgand F, Yaniv M, Muchardt C. Tethering of HP1 proteins to chromatin is relieved by phosphoacetylation of histone H3. EMBO Rep 2004;5:490–6. https://doi.org/10.1038/sj.embor.7400139.
[88] Fischle W, Tseng BS, Dormann HL, Ueberheide BM, Garcia BA, Shabanowitz J, et al. Regulation of HP1-chromatin binding by histone H3 methylation and phosphorylation. Nature 2005;438:1116–22. https://doi.org/10.1038/nature04219.
[89] Mateescu B, Bourachot B, Rachez C, Ogryzko V, Muchardt C. Regulation of an inducible promoter by an HP1beta-HP1gamma switch. EMBO Rep 2008;9:267–72. https://doi.org/10.1038/embor.2008.1.
[90] Yamamoto Y, Verma UN, Prajapati S, Kwak Y-T, Gaynor RB. Histone H3 phosphorylation by IKK-alpha is critical for cytokine-induced gene expression. Nature 2003;423:655–9. https://doi.org/10.1038/nature01576.

[91] Soloaga A, Thomson S, Wiggin GR, Rampersaud N, Dyson MH, Hazzalin CA, et al. MSK2 and MSK1 mediate the mitogen- and stress-induced phosphorylation of histone H3 and HMG-14. EMBO J 2008;22: 2788–97. https://doi.org/10.1093/emboj/cdg273.

[92] Thorne JL, Ouboussad L, Lefevre PF. Heterochromatin protein 1 gamma and IκB kinase alpha interdependence during tumour necrosis factor gene transcription elongation in activated macrophages. Nucleic Acids Res 2012;40:7676–89. https://doi.org/10.1093/nar/gks509.

[93] Harouz H, Rachez C, Meijer BM, Marteyn B, Donnadieu F, Cammas F, et al. Shigella flexneri targets the HP1γ subcode through the phosphothreonine lyase OspF. EMBO J 2014;33:2606–22. https://doi.org/10.15252/embj.201489244.

[94] Brudek T, Christensen T, Aagaard L, Petersen T, Hansen HJ, Moller-Larsen A. B cells and monocytes from patients with active multiple sclerosis exhibit increased surface expression of both HERV-H Env and HERV-W Env, accompanied by increased seroreactivity. Retrovirology 2009;6:104. https://doi.org/10.1186/1742-4690-6-104.

[95] Sharma P, Azebi S, England P, Christensen T, Moller-Larsen A, Petersen T, et al. Citrullination of histone H3 interferes with HP1-mediated transcriptional repression. PLoS Genet 2012;8:e1002934. https://doi.org/10.1371/journal.pgen.1002934.

[96] Beniac DR, Wood DD, Palaniyar N, Ottensmeyer FP, Moscarello MA, Harauz G. Cryoelectron microscopy of protein-lipid complexes of human myelin basic protein charge isomers differing in degree of citrullination. J Struct Biol 2000;129:80–95. https://doi.org/10.1006/jsbi.1999.4200.

[97] Vossenaar ER, Zendman AJW, van Venrooij WJ, Pruijn GJM. PAD, a growing family of citrullinating enzymes: genes, features and involvement in disease. Bioessays 2003;25:1106–18. https://doi.org/10.1002/bies.10357.

[98] Jang B, Shin H-Y, Choi J-K, Nguyen DPT, Jeong B-H, Ishigami A, et al. Subcellular localization of peptidylarginine deiminase 2 and citrullinated proteins in brains of scrapie-infected mice: nuclear localization of PAD2 and membrane fraction-enriched citrullinated proteins. J Neuropathol Exp Neurol 2011;70: 116–24. https://doi.org/10.1097/NEN.0b013e318207559e.

[99] Mastronardi FG, Wood DD, Mei J, Raijmakers R, Tseveleki V, Dosch H-M, et al. Increased citrullination of histone H3 in multiple sclerosis brain and animal models of demyelination: a role for tumor necrosis factor-induced peptidylarginine deiminase 4 translocation. J Neurosci 2006;26:11387–96. https://doi.org/10.1523/JNEUROSCI.3349-06.2006.

[100] Nicholas AP, Sambandam T, Echols JD, Tourtellotte WW. Increased citrullinated glial fibrillary acidic protein in secondary progressive multiple sclerosis. J Comp Neurol 2004;473:128–36. https://doi.org/10.1002/cne.20102.

[101] Bradford CM, Ramos I, Cross AK, Haddock G, McQuaid S, Nicholas AP, et al. Localisation of citrullinated proteins in normal appearing white matter and lesions in the central nervous system in multiple sclerosis. J Neuroimmunol 2014;273:85–95. https://doi.org/10.1016/j.jneuroim.2014.05.007.

[102] Wood DD, Moscarello MA. The isolation, characterization, and lipid-aggregating properties of a citrulline containing myelin basic protein. J Biol Chem 1989;264:5121–7.

[103] Musse AA, Boggs JM, Harauz G. Deimination of membrane-bound myelin basic protein in multiple sclerosis exposes an immunodominant epitope. Proc Natl Acad Sci U S A 2006;103:4422–7. https://doi.org/10.1073/pnas.0509158103.

[104] James EA, Moustakas AK, Bui J, Papadopoulos GK, Bondinas G, Buckner JH, et al. HLA-DR1001 presents "altered-self" peptides derived from joint-associated proteins by accepting citrulline in three of its binding pockets. Arthritis Rheum 2010;62:2909–18. https://doi.org/10.1002/art.27594.

[105] Christophorou MA, Castelo-Branco G, Halley-Stott RP, Oliveira CS, Loos R, Radzisheuskaya A, et al. Citrullination regulates pluripotency and histone H1 binding to chromatin. Nature 2014;507:104–8. https://doi.org/10.1038/nature12942.

[106] Zhang X, Gamble MJ, Stadler S, Cherrington BD, Causey CP, Thompson PR, et al. Genome-wide analysis reveals PADI4 cooperates with Elk-1 to activate c-Fos expression in breast cancer cells. PLoS Genet 2011; 7:e1002112. https://doi.org/10.1371/journal.pgen.1002112.

[107] Hernández-Pedro NY, Vosoughi R, Freedman MS, Espinosa-Ramirez G, de la Cruz VP, Pineda B, et al. Therapy of MS. Clin Neurol Neurosurg 2010;112:365–85. https://doi.org/10.1016/j.clineuro.2010.03.010.

[108] Gould KA, Dove WF. Action of Min and Mom1 on neoplasia in ectopic intestinal grafts. Cell Growth Differ 1996;7:1361–8.

[109] Bartel DP. MicroRNAs: target recognition and regulatory functions. Cell 2009;136:215–33. https://doi.org/10.1016/j.cell.2009.01.002.

[110] Kim KM, Abdelmohsen K, Mustapic M, Kapogiannis D, Gorospe M. RNA in extracellular vesicles. Wiley Interdiscip Rev RNA 2017;8. https://doi.org/10.1002/wrna.1413.

[111] Schack-Nielsen L, Michaelsen KF. Breast feeding and future health. Curr Opin Clin Nutr Metab Care 2006; 9:289–96. https://doi.org/10.1097/01.mco.0000222114.84159.79.

[112] Brenton JN, Engel CE, Sohn M-W, Goldman MD. Breastfeeding during infancy is associated with a lower future risk of pediatric multiple sclerosis. Pediatr Neurol 2017;77:67–72. https://doi.org/10.1016/j.pediatrneurol.2017.09.007.

[113] G. Ragnedda, S. Leoni, M. Parpinel, I. Casetta, T. Riise, K.-M. Myhr, et al., Reduced duration of breast-feeding is associated with a higher risk of multiple sclerosis in both Italian and Norwegian adult males: the EnvIMS study, J Neurol. 262 (2015) 1271–1277. doi:10.1007/s00415-015-7704-9.

[114] Vieira Borba V, Sharif K, Shoenfeld Y. Breastfeeding and autoimmunity: programing health from the beginning. Am J Reprod Immunol 2017;8. e12778–11. https://doi.org/10.1111/aji.12778.

[115] Alsaweed M, Hartmann P, Geddes D, Kakulas F. MicroRNAs in breastmilk and the lactating breast: potential immunoprotectors and developmental regulators for the infant and the mother. Ijerph 2015;12: 13981–4020. https://doi.org/10.3390/ijerph121113981.

[116] Zempleni J, Aguilar-Lozano A, Sadri M, Sukreet S, Manca S, Wu D, et al. Biological activities of extracellular vesicles and their cargos from bovine and human milk in humans and implications for infants. J Nutr 2017;147:3–10. https://doi.org/10.3945/jn.116.238949.

[117] Atkins HL, Bowman M, Allan D, Anstee G, Arnold DL, Bar-Or A, et al. Immunoablation and autologous haemopoietic stem-cell transplantation for aggressive multiple sclerosis: a multicentre single-group phase 2 trial. Lancet 2016;388:576–85. https://doi.org/10.1016/S0140-6736(16)30169-6.

[118] Belkaid Y, Harrison OJ. Homeostatic immunity and the microbiota. Immunity 2017;46:562–76. https://doi.org/10.1016/j.immuni.2017.04.008.

[119] Sharon G, Sampson TR, Geschwind DH, Mazmanian SK. The central nervous system and the gut microbiome. Cell 2016;167:915–32. https://doi.org/10.1016/j.cell.2016.10.027.

[120] Berer K, Gerdes LA, Cekanaviciute E, Jia X, Xiao L, Xia Z, et al. Gut microbiota from multiple sclerosis patients enables spontaneous autoimmune encephalomyelitis in mice. Proc Natl Acad Sci USA 2017;114: 10719–24. https://doi.org/10.1073/pnas.1711233114.

[121] Parrón T, Requena M, Hernández AF, Alarcón R. Association between environmental exposure to pesticides and neurodegenerative diseases. Toxicol Appl Pharmacol 2011;256:379–85. https://doi.org/10.1016/j.taap.2011.05.006.

[122] Roux J, Bard D, Le Pabic E, Segala C, Reis J, Ongagna J-C, et al. Air pollution by particulate matter PM10 may trigger multiple sclerosis relapses. Environ Res 2017;156:404–10. https://doi.org/10.1016/j.envres.2017.03.049.

[123] Wang H-T, Weng M-W, Chen W-C, Yobin M, Pan J, Chung F-L, et al. Effect of CpG methylation at different sequence context on acrolein- and BPDE-DNA binding and mutagenesis. Carcinogenesis 2013; 34:220–7. https://doi.org/10.1093/carcin/bgs323.

[124] Cantone L, Nordio F, Hou L, Apostoli P, Bonzini M, Tarantini L, et al. Inhalable metal-rich air particles and histone H3K4 dimethylation and H3K9 acetylation in a cross-sectional study of steel workers. Environ Health Perspect 2011;119:964–9. https://doi.org/10.1289/ehp.1002955.

[125] Hou L, Zhang X, Wang D, Baccarelli A. Environmental chemical exposures and human epigenetics. Int J Epidemiol 2012;41:79–105. https://doi.org/10.1093/ije/dyr154.

[126] Babaei S, Akhtar W, de Jong J, Reinders M, de Ridder J. 3D hotspots of recurrent retroviral insertions reveal long-range interactions with cancer genes. Nat Commun 2015;6:1–10. https://doi.org/10.1038/ncomms7381.

[127] Dukler N, Booth GT, Huang Y-F, Tippens N, Waters CT, Danko CG, et al. Nascent RNA sequencing reveals a dynamic global transcriptional response at genes and enhancers to the natural medicinal compound celastrol. Genome Res 2017;27:1816–29. https://doi.org/10.1101/gr.222935.117.

CHAPTER

Alterations in epigenetic regulation contribute to neurodegeneration of ataxia-telangiectasia

6

Jiali Li[1,2,3,4], Karl Herrup[5,6]

National Institute on Drug Dependence, Peking University, Beijing, China[1]; PKU/McGovern Institute for Brain Research, Peking University, Beijing, China[2]; Key Laboratory of Animal Models and Human Disease Mechanisms of the Chinese Academy of Sciences & Yunnan Province, Kunming Institute of Zoology, Chinese Academy of Sciences, Kunming, Yunnan, China[3]; CAS Center for Excellence in Animal Evolution and Genetics, Chinese Academy of Sciences, Kunming, Yunnan, China[4]; Division of Life Science and the State Key Laboratory of Molecular Neuroscience, The Hong Kong University of Science and Technology, Kowloon, Hong Kong[5]; Department of Neurobiology, University of Pittsburgh School of Medicine, Pittsburgh, Pennsylvania, United States[6]

Chapter Outline

6.1 Decreased level of histone acetylation induced by nuclear accumulation of HDAC4 drives A-T neurodegeneration 120
6.2 Dysfunction of polycomb repressive complex 2 involved in A-T neurodegeneration 123
6.3 Selective loss of 5-hmC is associated with purkinje cell vulnerability in A-T brain 124
6.4 TETs-mediated DNA oxidation regulates ATM/ATR-dependent DDR 126
6.5 Conclusion 128
6.6 Future perspective 128
Acknowledgments 129
References 129

Ataxia-telangiectasia (A-T) is a multisystem disease characterized by neurodegeneration in the central nervous system (CNS) [1–5]. The earliest and most profound neuropathology involves the Purkinje and granule cells of the cerebellum. A-T is caused by mutation of the A-T mutated (ATM) gene, which is ubiquitously expressed throughout development and encodes a serine/threonine protein kinase of the phosphatidylinositol-3 kinase-related kinase family [6]. A-T is classified as a rare neurodegenerative disease, mainly impacting on cerebellum integrity and functioning, resulting in a progressive deterioration of motor functional capabilities [1]. Moreover, as there is currently no effective treatment to cure or even slow down the rate of cerebellar neurodegeneration, A-T significantly decreases life quality of those suffering from it. The best-known function of ATM is to ensure the integrity of the genome. After DNA damage, ATM activates cell cycle checkpoints, arresting progression of the cycle until DNA repair is complete [7]. Its role in maintaining the health and survival of neurons is complex. Consistent with its function in DNA-damage repair [8], ATM protects postmitotic neurons from degeneration by suppressing the cell cycle [9–11]. Although non-neurological phenotypes including

Chromatin Signaling and Neurological Disorders. https://doi.org/10.1016/B978-0-12-813796-3.00006-7

immune system defects, germ cell defects, hypersensitivity to ionizing radiation, and increased susceptibility to tumorigenesis are found, the origins of the CNS motor phenotypes of A-T are the most poorly understood. Yet, the pathway(s) that leads from these defects to neuronal cell loss and the other classical neurological phenotypes remains unknown.

Epigenetic systems, heritable changes in gene expression that do not involve coding sequence modifications, include DNA methylation, histone modifications, chromatin remodeling, and noncoding RNA regulation. These types of changes have been proposed to be responsible for controlling the expression and function of genes and have emerged as important mediators of development and aging [12–19]. Several diseases are known to have a multifactorial origin, depending not only on genetic but also on environmental factors. They are called "complex disorders" and include cardiovascular disease, cancer, diabetes, as well as neuropsychiatric and neurodegenerative diseases [20–26]. The neurological disorder most intensely studied regarding epigenetic changes is Rett syndrome; patients with Rett syndrome have neurodevelopmental defects associated with mutations in MeCP2, which encodes the methyl-CpG-binding protein 2, that binds to methylated DNA. Other mental retardation disorders are also linked to the disruption of genes involved in epigenetic mechanisms; such disorders include alpha thalassaemia/mental retardation X-linked syndrome (Chapter 11), Rubinstein–Taybi syndrome, and Coffin–Lowry syndrome [26]. Due to epigenetic variability among different neuronal cell types and epigenetic instability during development and maturation of neurons, whether specific epigenetic systems are involved in the vulnerability of ATM-deficient Purkinje cell with a large and pale nucleus will be helpful to understand A-T neurodegeneration.

In this chapter, we place emphasis on addressing abnormal changes in multiple epigenetic regulations, which include nuclear accumulation of histone deacetylase 4 (HDAC4), elevated enhancer of zeste-2 (EZH2)-dependent H3K27me3, selective loss of 5-hydroxymethlycytosine (5-hmC) in ATM-deficient neurons and their roles in regulating cell cycle reentry and Purkinje cell degenerative process in A-T pathological conditions.

6.1 Decreased level of histone acetylation induced by nuclear accumulation of HDAC4 drives A-T neurodegeneration

Histone modification is a major epigenetic regulation mechanism that plays a critical role in many processes [27–29]. The class IIa HDAC4 usually shuttles between the nucleus and cytoplasm in most cell types, and serves as a nuclear corepressor that regulates bone and muscle development and maturation. HDAC4 has normally a cytoplasmic distribution in neurons. Loss of HDAC4 cytoplasmic distribution induces neuronal cell death [30–32]. The function of HDAC4 is essentially regulated by its subcellular localization and the formation of multisubunit complexes with other proteins. HDAC4 subcellular distribution is depended on its phosphorylation by CaMK II and IV and the dephosphorylation carried by the PP2A phosphatase. It has been well documented that HDAC4 is abundant in the brain [33–35]. HDAC4-null mice display an early postnatal atrophy of the cerebellum with surviving Purkinje cells that are notably reduced in dendritic complexity [36]. Unexpectedly, *Atm* deficiency causes nuclear accumulation of HDAC4 in neurons and this shift in HDAC location promotes neurodegeneration [37]. Nuclear HDAC4 suppresses myocyte-specific *enhancer factor 2A* (MEF2A)- and cAMP *response element-binding protein* (CREB)-dependent transcription of neuronal survival genes; blocking HDAC4 nuclear accumulation rescues the degenerative changes found in *Atm*-deficient neurons (Fig. 6.1, left). Significantly, this rescue activity could be achieved by the

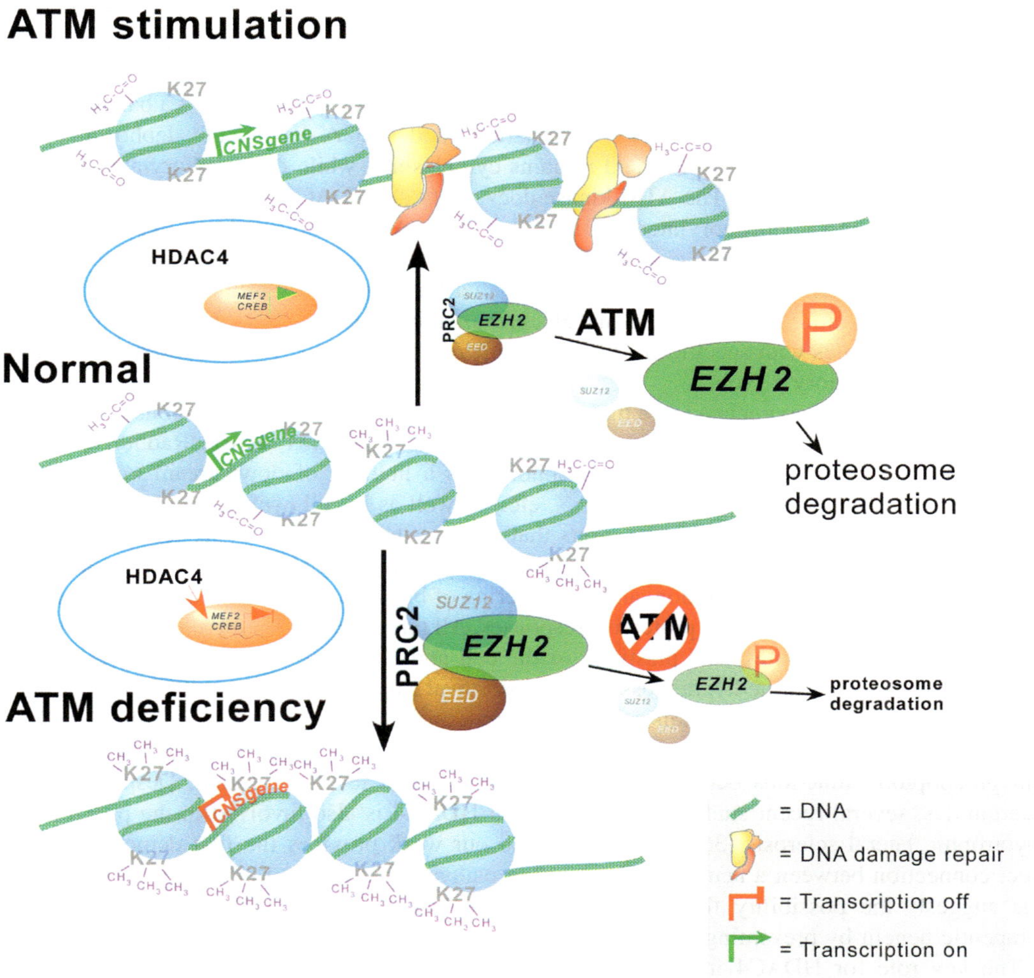

FIGURE 6.1

Diagram of epigenetic basis in A-T neurodegeneration.

overexpression of HDAC4 mutants that were localized to the neuronal cytoplasm, but other HDAC4 mutants, localized to the cell nucleus, not only were incapable of rescuing the *Atm*-deficient phenotype, they actually compromised the phenotype of the wild type. These findings suggest that the A-T phenotype results in part from a loss of cytoplasmic HDAC4 function. To remain cytoplasmic, HDAC4 must be phosphorylated and bound to 14-3-3. The HDAC4 phosphatase, PP2A, is downregulated by ATM-mediated phosphorylation of its A subunit (PR65). In ATM deficiency, enhanced PP2A activity leads to HDAC4 dephosphorylation and nuclear accumulation. Whether nuclear translocation and cytoplasmic depletion of HDAC4 are key factors leading to the degeneration of neurons in A-T remain unknown. In our work, a major new molecular function of ATM was identified. Through its phosphorylation of the A subunit of the PP2A phosphatase protein, ATM normally promotes the nuclear export of HDAC4. The regulation of this translocation is indirect, and our model for how it occurs is diagrammed in Fig. 6.1. HDAC4 phosphorylation is a balance between CaMKII- and CaMKIV-mediated phosphorylation and PP2A-dependent dephosphorylation. Phosphorylated HDAC4 binds 14-3-3 and remains cytoplasmic, while dephosphorylated HDAC4 is free and moves to the nucleus. The ATM kinase does not target HDAC4 itself or either of the CaMK proteins. Rather, ATM phosphorylates S401 of the PP2A-A regulatory subunit. This disrupts the binding of the PP2A holoenzyme to its HDAC4 substrate, inhibiting the dephosphorylation reaction. Once in the nucleus, HDAC4 has many effects that alter the cell biology of the neuron. It lowers the global level of acetylation on histones H3 and H4, presumably condensing the chromatin structure, and it physically associates with specific sites in the regulatory regions of several important neurotrophic and cell cycle regulatory proteins.

ATM has long been appreciated as a major component of the DNA-damage response (DDR) of the cell where it is activated by binding to the MRN complex of proteins (Mre11/Rad50/Nbs1). Once activated by autophosphorylation, ATM phosphorylates a number of downstream targets that are involved in the DDR pathway (Nbs1, γ-H2AX, etc.) as well as in cell cycle arrest (p53, Chk2, etc.). These are believed to be protective functions that prevent the cell from trying to divide when the integrity of its genome is compromised in some way. In neurons, the global loss of histone acetylation, the loss of CREB-induced stimulation, and the lowering of brain-derived *neurotrophic factor* (BDNF) levels might also be envisioned as protective responses. All three changes would dampen the structural growth (dendrites and spines) and plasticity of the cell and its synapses allowing damage repair to proceed unhindered. Admittedly, the loss of the prosurvival MEF2A protein and the failure of the cell cycle suppression do not easily fit this scheme. Perhaps, analogous to the situation with the cell cycle suppressor and proapoptotic functions of p53, the response to HDAC4 nuclear localization is graded with proapoptotic functions occurring only at more elevated levels of signal. Despite remaining uncertainties, several recent studies have shown that HDAC4 is also involved in the progression of amyotrophic lateral sclerosis [36,37a,37b,37c], and our work provides the first demonstration of a direct connection between a neuronal degeneration pathway and the dysfunction of ATM signaling. This suggests the possibility that preventing the nuclear accumulation of HDAC4 could have therapeutic benefit by preventing or delaying the neurological symptoms of A-T.

The key role for HDAC4 in the neuronal response to decreased levels of ATM activity was unexpected but fits well with its apparent functions in other tissues. During muscle and bone development, HDAC4 acts as a transcriptional corepressor that interacts with MEF2-related transcription factors and the N-CoR, BCoR, and CtBP corepressors [32]. For example, in response to neural activity, muscle transcriptional activity reprograms through an HDAC4-regulated MEF2-dependent mechanism [38]. MEF2- and CREB-dependent transcription is suppressed when nuclear accumulation of class II HDACs occurs [31,39,40]. In neurons, evidence as to whether HDAC4 directly regulates MEF2 and CREB-dependent gene expression, especially during development, is less conclusive. In the adult nervous system, however, the family of HDAC proteins is believed to regulate learning and memory as well as neuron survival by altering gene expression through chromatin remodeling [41–45].

This finding is also the first to report a direct association of HDAC4 with the chromatin of any cell. Most current models of HDAC4 are based on a sequestration or "buffering" of proteins such as MEF2A. The buffering of these transcription factors decreases their effective concentration at the promoters of key target genes [40,46]. Our ChIP and ChIPseq data suggest a modification of these models. We find that in the absence of ATM inactivation of PP2A, nuclear HDAC4 can directly associate with chromatin, and while it may also still be associated with MEF2A and/or CREB, it can be described as a full participant in transcriptional suppression. Indeed, it is possible to suggest from the patterns seen in the ChIPseq data that the suppression associated with nuclear HDAC4 could be due as much to its direct DNA or chromatin binding as to its more indirect effects caused by histone deacetylation. Collectively, these data suggest that the neuronal degeneration observed in A-T disease could be a direct consequence of the associated changes in gene expression, in addition to the accumulated burden of chronic unrepaired DNA damage.

These observations, as well as those from other studies, suggest that in brain a complex set of dysregulations underlies the full spectrum of neurological abnormalities that are observed in A-T. Although the deficiencies in the ATM-mediated DDR [1,8,47] undoubtedly play their part, the existence of an additional or alternative pathway could provide a new insight on how the loss of neuronal ATM function leads to the spectrum of neurological deficiencies found in A-T.

6.2 Dysfunction of polycomb repressive complex 2 involved in A-T neurodegeneration

Although nuclear accumulation of HDAC4 plays a critical role in neurodegeneration of A-T [37], it remains possible that other epigenetic systems are also involved in this neurodegenerative process. The polycomb group (PcG) family of proteins are transcriptional repressors that epigenetically modify chromatin and participate in the establishment and maintenance of cell fates [48–56]. They are important factors in both stem cell self-renewal and tumorigenesis as well as in brain development [57,58]. In mammals, there are two distinct polycomb protein complexes—PRC1 (polycomb repressive complex 1) and PRC2—each with different protein constituents [59]. EZH2 functions as part of the PRC2 complex, which also includes SU(Z)12 (suppressor of zeste 12) and EED (embryonic ectoderm development) [60,61]. EZH2 catalyzes the addition of three methyl groups to lysine 27 of histone H3 (H3K27) at target gene promoters, leading to gene silencing. Several studies reported that EZH2 function depends on its phosphorylation status [55,62–66]. Two recent articles showed that Cdk1-mediated EZH2 phosphorylation increased H3K27 methylation [55,67]. However, one study reported that Akt-mediated EZH2 phosphorylation prevented H3K27 methylation [63]. In our study, surprisingly, ATM plays an essential part of the normal homeostatic process that suppresses EZH2 (Fig. 6.1, right). In both A-T patients and in A-T mouse models, the physical association of EZH2 with histone H3, specifically H3K27, is substantially increased by reduced ATM signaling. This leads to hypermethylation, creating the trimethyl form known as H3K27me3. Normally, ATM-mediated phosphorylation of EZH2 prevents its binding to H3K27 as well as the formation of the PRC2 complex.

Through phosphorylation of EZH2, ATM normally attenuates PRC2 formation and EZH2 binding to H3K27. Phosphorylation of EZH2 by ATM leads to lower stability. In the absence of ATM, EZH2 displays increased stability and enhanced binding to H3K27 and PRC2 formation. When this function fails, the resulting hypermethylation of H3K27me3 causes suppression of multiple brain-specific micro-RNAs such as miR-124 and-134 as well as cell cycle reentry. Significantly, knocking down EZH2 in *Atm*-deficient neurons prevents cell cycle reentry and neuronal death. Meanwhile, the observation that ATM activation by DNA damage leads to a substantial reduction in histone methylation suggests a broader perspective. Enhanced histone methylation is generally associated with a more compact form of chromatin. DNA damage, by enhancing ATM activity, would thus lead to a genome-wide reduction in chromatin condensation. This opening of the chromatin might be expected to aid in the access of the various DNA repair protein to the injured double helix and suggests a new and unexpected way in which ATM helps in the DNA repair process. The genome-wide analysis of H3K27me3 further represents the first unbiased, high-resolution mapping of H3K27me3 occupancy in wild type and *Atm*$^{-/-}$ mouse brain. Moreover, it reveals that the EZH2-induced methylation of H3 is accompanied by a dramatic increase in the sites with which it associates in the genome and a correlated shift in the expression of a variety of cell cycle and neurotrophic genes. Lastly, alteration in EZH2-H3K27me3-mediated epigenetic regulation has a potential to affect ATM deficiency—induced increase in neurodegeneration, cell cycle reentry, suppression of neuronal genes, and induction of the apoptotic process. The direct participation of the polycomb *group* (PcG) *proteins* complex in this degenerative phenotype is confirmed by the finding that EZH2 knockdown can prevent cell cycle reentry and neurodegeneration in *Atm*$^{-/-}$ neurons both *in vitro* and *in vivo*. Knockdown of EZH2 partially rescues motor functional disabilities found in *Atm*$^{-/-}$ mice. These findings suggest that the cause of the elevated H3K27me3 in A-T and *Atm*$^{-/-}$ mouse brain is, in major part, mediated through EZH2 rather than EZH1. Moreover, the activity of EZH2 must be consistently suppressed in neurons in order for them to avoid reentrance into cell cycle and thus initiate a pathway that results in neurodegeneration [68].

With these results, an expanded and integrated epigenetic regulatory model for neurodegenerative process in A-T is proposed, which is due to both dysfunction of EZH2 and nuclear accumulation of HDAC4. In this model, the epigenetic regulation of cell cycle and neuronal genes expression is not only dependent, in part, on histone acetylation but also relies on the PRC2-mediated H3K27me3 signaling. Identification of a role for ATM in Polycomb signaling will enhance our understanding of molecular underlying dynamic changes of histone methylation in A-T neurodegenerative process as well as tumorigenesis and will provide therapeutic targets for novel treatment by identifying novel epigenetic mechanisms (Fig. 6.1).

6.3 Selective loss of 5-hmC is associated with purkinje cell vulnerability in A-T brain

Neuropathological and imaging studies suggest that cerebellar neurons are particular targets of A-T—most especially cerebellar Purkinje cells (PCs). Although a high requirement for energy metabolism may contribute to PC vulnerability [69], the same might well be applied to other neuronal cell types in the CNS and thus the mechanism of PC vulnerability in ATM deficiency remains unknown. Studies have demonstrated that nuclear accumulation of HDAC4 [37] and stabilization of EZH2 [68] contribute to A-T neurodegeneration by altering the histone acetylation and methylation status of the genome, respectively. As the genome of all neurons carries the identical *ATM* mutation, the question arises as to whether there are additional epigenetic factors, such as DNA methylation, that play a role in driving specific disease mechanisms.

It has been reported that TET1 regulates the expression of CNS activity-dependent genes and memory formation independently of its catalytic activity [70,71]. The abundance of all three ten-eleven translocation dioxygenase (TET) proteins as well as 5-hmC in animal and human brains further suggests important roles of TET-dependent conversion of 5-mC to 5-hmC in neuronal health and survival [72]. This raises the question of the normal role of TET1 and 5-hmC in the physiology of the neuron. PCs are among the largest cells in the brain. PC nuclei are large and pale, which are characterized by the majority of heterochromatin detectable at the ultrastructural level on the surrounding of the large, centrally located nucleolus. The levels of 5-hmC in PC are much higher than in granule cell nuclei, suggesting that 5-hmC may have a particularly important role in PC function [73].

ATM is ubiquitously expressed throughout the organism and is normally found in many regions of the brain. Although ATM is best known for its role in the DDR, the consequences of the loss of ATM function for the brain are multifaceted. It is activated in response to DNA damage in neurons just as it is in other cell types; yet, defects in the DDR cannot fully account for the specific targeting of PCs in the A-T brain.

Elucidating factors controlling development of neurodegeneration in ATM deficiency is crucial for understanding the mechanism of A-T neurological symptoms. In particular, factors that selectively promote Purkinje cell vulnerability are of key interest as potential targets for A-T. The recent discovery of the role of the TET family in the control of genomic DNA modifications has opened a new and potentially valuable window into the regulation of gene expression in health and in disease. Although all three TET family members catalyze the conversion of 5-mC to 5-hmC, each TET-dependent 5-hmC production shows distinct regulation in different brain functions [74,75]. Yet, 5-hmC is not simply an intermediate of the demethylation reaction; it has its own role in the regulation of chromatin dynamics and gene expression. Immunocytochemistry suggests that 5-hmC is found in many cell types, including PCs. Yet, compared to PCs, the density of staining is much greater in cerebellar and hippocampal granule cells as well as cortical neurons [76]. It seems almost paradoxical therefore that in human A-T and *Atm*$^{-/-}$ mouse models, there is a significant loss of 5-hmC in Purkinje neurons but not in other neuronal cell types. Perhaps, it is precisely because the levels are low that loss of ATM-dependent TET1 activity and the consequent decline in PC 5-hmC levels are deleterious for the cell.

The involvement of DNA demethylation in A-T patients is further demonstrated by the observation of abnormal alterations in TET1-mediated DNA hydroxymethylation, which are also selectively found in ATM-deficient PCs, providing a valuable evidence for understanding PC vulnerability. hMeDIP-seq data reveal a significant genome-wide shift in the 5-hmC landscape in A-T cerebellar cortex but not in samples of frontal cortex. This is particularly noteworthy as it can be assessed by the proportion of sequencing reads overlapping repeated elements in the genome, a direct measure that is independent of algorithms to identify enriched regions, for example. These changes in 5-hmC marks, while widely distributed across the genome, do not appear to be random. The comparison of the 5-hmC and 5-mC DIP-seq data in conjunction with the immunocytochemistry is revealing that the involvement of DNA demethylation in the neurodegeneration of A-T is demonstrated by abnormal alterations in TET1-mediated DNA hydroxymethylation, which is also selectively found in ATM-deficient PCs, thus providing a valuable evidence for understanding PC vulnerability.

It suggests that when TET1 activity is reduced (as in ATM deficiency), not only to the levels of 5-hmC go down, but there is a concomitant increase in the levels of 5-mC, suggesting that there is an accumulation of substrate upstream of the disabled TET1 as well as a loss of the 5-hmC product. In tabulating the "ontology" of the genes whose 5-hmC levels fall at the same time as their 5-mC levels rise, we find it noteworthy that many genes whose function is critical for cerebellar development were identified. These results suggest that ATM regulation of 5-hmC levels may also influence developmental functions such as fate specification. The hMeDIP-seq analysis further shows a high correlation between 5-hmC enrichments and dysregulation of neuronal differentiation-specific functional genes found in ATM deficiency. One potential role for such alterations in TET1-mediated 5-hmC enrichment is the cell-type specificity observed when ATM is deficient. Overexpression of TET1 wild type, but not inactivated kinase ATM mutant, not only increases 5-hmC level but also prevents PC degeneration as well as behavioral deficits in *Atm*$^{-/-}$ mice. The link between selective loss of 5-hmC and the appearance of PC vulnerability sheds new light on our understanding of the defects in signaling in the A-T brain. Manipulation of TET1 activity directly affects the subsequent DNA-damage signaling, cell cycle reentry, and cell death. Meanwhile, several evidences suggest that TET1 activity links to the degenerative process in PCs as well as behavioral deficits in *Atm*$^{-/-}$ mice. This work thus suggests that in ATM deficiency, loss of 5-hmC contributes to a PC-enriched epigenetic alteration that deregulates chromatin structure and alters gene expression as well as DDR.

Interestingly, our latest unpublished data show that selective loss of 5-hmC was also found in the frontal cortical and hippocampal neurons in Alzheimer's disease human samples. Taken together, these findings further support the concept that, while it might not be a cause of the PC vulnerability in A-T, the selective loss of 5-hmC could play, in conjunction with the alterations in other epigenetic regulations, a key role in unique degeneration of PCs. Moreover, it also indicates that the involvement of 5-hmC in the selective pathology of other neurodegenerative disorders will indicate common and particular signatures that might aid in differential diagnosis as biomarkers or in the discovery of novel therapeutic targets.

6.4 TETs-mediated DNA oxidation regulates ATM/ATR-dependent DDR

Maintaining genomic integrity and chromosome stability requires an accurate DDR and a variety of machineries of DNA repair. DDR is a highly complex process involving a plethora of sensors, regulators, and effectors. A variety of DNA repair machineries have evolved to respond to DNA damage. Although eukaryotic cells have developed a plethora of enzymatic tools to capably remodel and repair DNA lesions, an understanding of how their activities are tightly regulated in a temporal and spatial fashion to optimize repair and prevent unnecessary and potentially deleterious alterations in the DNA structure remains incomplete [77]. ATM and ATR, two of the PI3 protein kinase family members, are well-documented regulators of DNA-damage signaling. Studies have shown that ATM- and ATR-dependent DDRs play a key role in DNA repair and genome integrity [78–80]. Although cross talk between ATM- and ATR-dependent signaling pathways has been found in DDR, the ATM/CHK2 (checkpoint *kinase* 2) signaling pathway is mainly activated after DNA double-strand breaks whereas the ATR/CHK1 module responds primarily to DNA single-strand breaks [81–83].

The distinct functions of three TET family dioxygenases are partially supported by several independent studies using complementary approaches in which each TET dioxygenase was demonstrated to be necessary for distinct functions [84,85]. However, the specific activity of TET dioxygenases remains unknown in many biological processes, especially in DDR. A recent study reported that TET2 might be required for damage-associated 5-hmC foci in HeLa cells [86], but whether the three TET enzymes play distinct roles in DDR is unknown.

In ATM deficiency, there is a genome-wide shift of 5-hmC "marks" at both regulatory elements and repeat sequences in cerebellar cortex, but not in frontal cortex. Chromatin status is known to have a substantial impact on the process of DNA-damage signaling and repair. Our findings suggest that DNA damage leads to alteration in TET1-mediated conversion of 5-mC to 5-hmC, and that manipulation of

TET1 activity directly affects DNA-damage signaling. This scenario is consistent with prior reports that ATM-dependent phosphorylation of KAP1, a heterochromatin formation factor, is involved in the slower kinetics of DNA repair in or near areas of heterochromatin [87]. As expected, TET1-mediated conversion of 5-mC to 5-hmC responds to DNA damage in an ATM-dependent manner [76]. Therefore, just as ATM acts in multiple independent ways to regulate the histone code, the same may be true for its effects on the regulation of DNA demethylation. This suggests that TET1-mediated active DNA demethylation in ATM-dependent DDR possibly contributes to DNA repair and it could be a universal phenotype (Fig. 6.2, left).

In contrast to regulation of TET1 by ATM [76], TET3 is a new ATR target, and involved in ATR-dependent DNA-damage signaling. Loss of TET3 activity leads to a decrease in DDR-induced 5-hmC production and increases DDR-induced cell death; whereas, overexpression of wild-type TET3, but not catalytic domain mutant, increases DDR-induced 5-hmC production and reduces DDR-induced cell death. The mechanism for this regulatory function of TET3 in DDR may include its unknown role in facilitating DDR-induced ATR signaling pathway, and we provide evidence that loss of TET3 activity results in an increase in phosphorylation of ATR, CHK1, p53, γ-H2AX, and PARP-1. Overexpression of the nonphosphorylatable TET3 mutant leads to a decrease in phosphorylation of ATR, p53, γ-H2AX, and PARP-1. Evidence from this and previous studies suggest that DDR-induced phosphorylation of ATR, CHK1, p53, PARP-1, and γ-H2AX regulate the speed and accuracy of DDR and DNA repair. Abnormal phosphorylation of these sensors and regulators as well as defects in their recruitment to DNA break sites lead to inappropriate DNA repair processes and genome instability. It is possible that in the absence of TET dioxygenases, aberrant 5-hmC products of stalled replication forks may prevent recruitment of DNA-damage sensors and regulators. Notably, this finding suggests

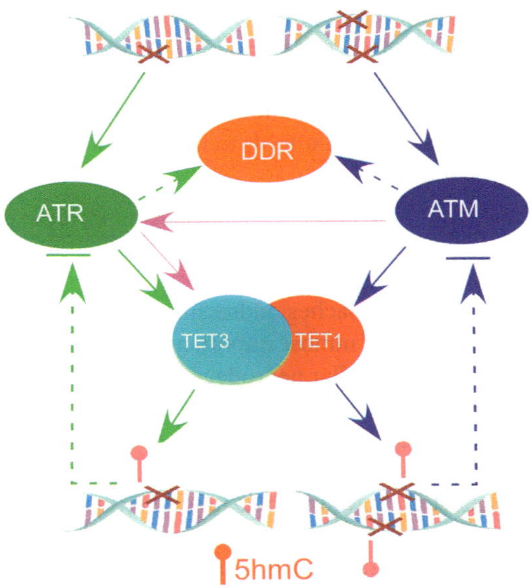

FIGURE 6.2

Diagram TET enzymes-mediated DNA demethylation in DNA-damage response and repair.

that dynamic changes in TET3-mediated 5-hmC production in DDR could be a previously unknown regulatory system in DNA-damage signaling and repair. ATR-dependent DNA-damage response mainly regulates TET3-mediated conversion of 5-mC to 5-hmC [88].

The cross talk between DDR and TET dioxygenases-mediated active DNA demethylation in maintaining genomic stability, and integrity highlights the diversity of functional states and signaling potential associated with dynamic DNA modifications in the DDR. Hence, these findings suggest that in addition to its well-characterized and important role in actively directing epigenetic regulation during development and aging [89–91], TET1/TET3-mediated DNA demethylation implicitly regulates ATM/ATR-dependent DNA-damage signaling and DNA repair.

These data support a model that incorporates distinct roles for TET family dioxygenases in the presence of DDR. One unique pathway involving TET1/TET3 and ATM/ATR are required for DDR leading to activation of ATM/ATR signaling pathway and conversion of 5-mC to 5-hmC, which act in a coordinated way as sensors and regulators for DNA repair and genomic stability. These data suggest that although it is important for TET dioxygenases to deliver active DNA demethylation to maintain genomic stability and regulate transcriptional activity, TET dioxygenases have also evolved the ability to carefully exploit aspects of DNA-damage signaling to differentially regulate DNA-damage events. Increased understanding of regulation of unknown functions of TET proteins in DNA-damage signaling may provide new valuable insights to the process of developing highly specific machinery of DNA repair and genomic integrity.

6.5 Conclusion

The studies discussed here have described the characterization of abnormal HDAC4 function and EZH2-mediated H3K27me3 in ATM-deficient neurons by using genome-wide transcriptome and epigenome datasets from both *Atm*$^{-/-}$ mice and A-T patients. These data reveal patterns and associations that suggest a complex combinatorial mechanism for epigenetic regulation of A-T neurodegeneration. Understanding the contribution of epigenetic changes to neurodegenerative progression of A-T will give us better molecular tools to improve prognosis, diagnosis, and therapeutic intervention for A-T neurodegeneration.

6.6 Future perspective

A systematic genome-wide, multiple approaches addressing the role of the epigenetic changes in the development and progression of A-T is of utmost importance, and integrates both classical as well as entirely distinct, novel views on A-T cerebellar neurodegeneration. Such an approach should integrate genetic and epigenetic variables, preferably in prospectively collected samples with different *ATM* mutations. Examining the possible association between aberrant epigenetic regulation and, for example, Purkinje cell loss would not only decipher to what extent these neuropathological hallmarks of A-T are associated with epigenetic alterations but, moreover, it will enable us to identify whether these epigenetic alterations are causally involved in neuronal cell cycle reentry as well as Purkinje cell loss or rather represent a consequence of their formation (e.g., hydroxymethylation changes preceding or following defective DDR, respectively). The integration of genomic, epigenomic, transcriptomic, and other "omic" data modalities is crucial for improving our understanding of the role of the epigenetic systems in the pathogenesis of A-T. It still remains to be determined which epigenetic system plays a critical role in the degenerative process of ATM-deficient cerebellar neurons, and how closely neurological phenotypes found in $Atm^{-/-}$ mice resemble the cerebellar degeneration of A-T patients. It is clear that further work is necessary to verify and determine the functional connections of multiple epigenetic systems (i.e., histone methylation and histone acetylation) and defects in DDR. Moreover, as important epigenetic systems for many biological processes, it will be intriguing to see whether the same patterns of DNA methylation and micro-RNAs regulation will apply, or whether distinct mechanisms will be uncovered in A-T neurodegeneration. However, for accurate explanation of A-T neurodegeneration, discrepancies between mouse model and human A-T neurological symptoms may be problematic. Although technically challenging, it is important that studies delineating the dynamic role of alteration in specific epigenetic system in contributing to A-T neurodegeneration must be well performed. For some applications, such as high-throughput screening for HDAC4- and EZH2-specific inhibitors in ATM deficiency cells as well as $Atm^{-/-}$ mice, it may be critical for exploring the therapeutic epigenetic targets in curing A-T diseases. However, in this respect, the use of epigenetic drugs such as HDAC4 or EZH2 inhibitors will require substantial progress in the specificity and safety, and holds great prospects for addressing some of key questions on the mechanisms of brain functions and their side effects. Hopefully, with these compelling projects going well, we will enable the identification of novel epigenetic signatures to be used to diagnose, prevent, attenuate, or possibly reverse the pathophysiology of A-T.

Acknowledgments

We are grateful to Kim Heiman Chow for critical reading the manuscript and editorial assistance. This work was supported by the Chinese Academy of Sciences (Y406541141 and 1100050210) to J.L. and the National Science Foundation of China (NSFC 81471313 and 91649119) to J.L.

References

[1] Biton S, Barzilai A, Shiloh Y. The neurological phenotype of ataxia-telangiectasia: solving a persistent puzzle. DNA Repair 2008;7:1028–38.

[2] Boder E. Ataxia-telangiectasia: some historic, clinical and pathologic observations. Birth Defects Original Article Series 1975;11:255–70.

[3] Bundey S. Clinical and genetic features of ataxia-telangiectasia. Int J Radiat Biol 1994;66:S23–9.

[4] Lavin MF, Shiloh Y. Ataxia-telangiectasia: a multifaceted genetic disorder associated with defective signal transduction. Curr Opin Immunol 1996;8:459–64.

[5] Sedgwick RP, Boder E. Progressive ataxia in childhood with particular reference to ataxia-telangiectasia. Neurology 1960;10:705–15.

[6] Savitsky K, Bar-Shira A, Gilad S, Rotman G, Ziv Y, Vanagaite L, Tagle DA, Smith S, Uziel T, Sfez S, et al. A single ataxia telangiectasia gene with a product similar to PI-3 kinase. Science 1995;268:1749–53.

[7] Bakkenist CJ, Kastan MB. DNA damage activates ATM through intermolecular autophosphorylation and dimer dissociation. Nature 2003;421:499–506.

[8] Dar I, Biton S, Shiloh Y, Barzilai A. Analysis of the ataxia telangiectasia mutated-mediated DNA damage response in murine cerebellar neurons. J Neurosci 2006;26:7767–74.

[9] Li, J., Chen, J., Vinters, H.V., Gatti, R.A., and Herrup, K. Stable brain ATM message and residual kinase-active ATM protein in ataxia-telangiectasia. J Neurosci 31, 7568-7577.
[10] Rimkus SA, Katzenberger RJ, Trinh AT, Dodson GE, Tibbetts RS, Wassarman DA. Mutations in String/CDC25 inhibit cell cycle re-entry and neurodegeneration in a Drosophila model of Ataxia telangiectasia. Genes Dev 2008;22:1205−20.
[11] Yang Y, Herrup K. Loss of neuronal cell cycle control in ataxia-telangiectasia: a unified disease mechanism. J Neurosci 2005;25:2522−9.
[12] Abel T, Zukin RS. Epigenetic targets of HDAC inhibition in neurodegenerative and psychiatric disorders. Curr Opin Pharmacol 2008;8:57−64.
[13] Agis-Balboa RC, Pavelka Z, Kerimoglu C, Fischer A. Loss of HDAC5 impairs memory function: implications for alzheimer's disease. J Alzheimers Dis 2012;33(1):35−44.
[14] Alagiakrishnan K, Gill SS, Fagarasanu A. Genetics and epigenetics of Alzheimer's disease. Postgrad Med 2012;88:522−9.
[15] Bakulski KM, Rozek LS, Dolinoy DC, Paulson HL, Hu H. Alzheimer's disease and environmental exposure to lead: the epidemiologic evidence and potential role of epigenetics. Curr Alzheimer Res 2012;9:563−73.
[16] Barbash S, Soreq H. Threshold-independent meta-analysis of Alzheimer's disease transcriptomes shows progressive changes in hippocampal functions, epigenetics and microRNA regulation. Curr Alzheimer Res 2012;9:425−35.
[17] Bihaqi SW, Schumacher A, Maloney B, Lahiri DK, Zawia NH. Do epigenetic pathways initiate late onset Alzheimer disease (LOAD): towards a new paradigm. Curr Alzheimer Res 2012;9:574−88.
[18] Coppieters N, Dragunow M. Epigenetics in Alzheimer's disease: a focus on DNA modifications. Curr Pharmaceut Des 2011;17:3398−412.
[19] Deaton AM, Bird A. CpG islands and the regulation of transcription. Genes Dev 2011;25:1010−22.
[20] Kinney SR, Pradhan S. Ten eleven translocation enzymes and 5-hydroxymethylation in Mammalian development and cancer. Adv Exp Med Biol 2013;754:57−79.
[21] Konsoula Z, Barile FA. Epigenetic histone acetylation and deacetylation mechanisms in experimental models of neurodegenerative disorders. J Pharmacol Toxicol Methods 2012;66(3):215−20.
[22] Kudo Y, Tateishi K, Yamamoto K, Yamamoto S, Asaoka Y, Ijichi H, Nagae G, Yoshida H, Aburatani H, Koike K. Loss of 5-hydroxymethylcytosine is accompanied with malignant cellular transformation. Cancer Sci 2012;103:670−6.
[23] Kwok JB. Role of epigenetics in Alzheimer's and Parkinson's disease. Epigenomics 2010;2:671−82.
[24] Maekawa M, Watanabe Y. Epigenetics: relations to disease and laboratory findings. Curr Med Chem 2007; 14:2642−53.
[25] Marques SC, Oliveira CR, Pereira CM, Outeiro TF. Epigenetics in neurodegeneration: a new layer of complexity. Prog Neuro-Psychopharmacol Biol Psychiatry 2011;35:348−55.
[26] Urdinguio RG, Sanchez-Mut JV, Esteller M. Epigenetic mechanisms in neurological diseases: genes, syndromes, and therapies. Lancet Neurol 2009;8:1056−72.
[27] Chouliaras L, Rutten BP, Kenis G, Peerbooms O, Visser PJ, Verhey F, van Os J, Steinbusch HW, van den Hove DL. Epigenetic regulation in the pathophysiology of Alzheimer's disease. Prog Neurobiol 2010;90: 498−510.
[28] Fagiolini M, Jensen CL, Champagne FA. Epigenetic influences on brain development and plasticity. Curr Opin Neurobiol 2009;19:207−12.
[29] Ma DK, Marchetto MC, Guo JU, Ming GL, Gage FH, Song H. Epigenetic choreographers of neurogenesis in the adult mammalian brain. Nat Neurosci 2010;13:1338−44.
[30] Bolger TA, Yao TP. Intracellular trafficking of histone deacetylase 4 regulates neuronal cell death. J Neurosci 2005;25:9544−53.
[31] Chen B, Cepko CL. HDAC4 regulates neuronal survival in normal and diseased retinas. Science 2009;323:256−9.

[32] Youn HD, Grozinger CM, Liu JO. Calcium regulates transcriptional repression of myocyte enhancer factor 2 by histone deacetylase 4. J Biol Chem 2000;275:22563–7.
[33] Darcy MJ, Calvin K, Cavnar K, Ouimet CC. Regional and subcellular distribution of HDAC4 in mouse brain. J Comp Neurol 2010;518:722–40.
[34] Grozinger CM, Hassig CA, Schreiber SL. Three proteins define a class of human histone deacetylases related to yeast Hda1p. Proc Natl Acad Sci U S A 1999;96:4868–73.
[35] Wang AH, Bertos NR, Vezmar M, Pelletier N, Crosato M, Heng HH, Th'ng J, Han J, Yang XJ. HDAC4, a human histone deacetylase related to yeast HDA1, is a transcriptional corepressor. Mol Cell Biol 1999;19: 7816–27.
[36] Majdzadeh N, Wang L, Morrison BE, Bassel-Duby R, Olson EN, D'Mello SR. HDAC4 inhibits cell-cycle progression and protects neurons from cell death. Dev Neurobiol 2008;68:1076–92.
[37] Li J, Chen J, Ricupero CL, Hart RP, Schwartz MS, Kusnecov A, Herrup K. Nuclear accumulation of HDAC4 in ATM deficiency promotes neurodegeneration in ataxia telangiectasia. Nat Med 2012;18:783–90.
[37a] Williams SR, Aldred MA, V.M. Der Kaloustian, Halal F, Gowans G, McLeod DR, Zondag S, Toriello HV, Magenis RE, Elsea SH. Haploinsufficiency of HDAC4 causes brachydactyly mental retardation syndrome, with brachydactyly type E, developmental delays, and behavioral problems. Am J Hum Genet 2010;87(2): 219–28.
[37b] Cohen TJ, Barrientos T, Hartman ZC, Garvey SM, Cox GA, Yao TP. The deacetylase HDAC4 controls myocyte enhancing factor-2-dependent structural gene expression in response to neural activity. FASEB J 2009;23(1):99–106.
[37c] D'Mello SR. Histone deacetylases as targets for the treatment of human neurodegenerative diseases. Drug News Perspect 2009;22:513–24.
[38] Cohen TJ, Barrientos T, Hartman ZC, Garvey SM, Cox GA, Yao TP. The deacetylase HDAC4 controls myocyte enhancing factor-2-dependent structural gene expression in response to neural activity. FASEB J 2009;23:99–106.
[39] Flavell SW, Kim TK, Gray JM, Harmin DA, Hemberg M, Hong EJ, Markenscoff-Papadimitriou E, Bear DM, Greenberg ME. Genome-wide analysis of MEF2 transcriptional program reveals synaptic target genes and neuronal activity-dependent polyadenylation site selection. Neuron 2008;60:1022–38.
[40] Miska EA, Karlsson C, Langley E, Nielsen SJ, Pines J, Kouzarides T. HDAC4 deacetylase associates with and represses the MEF2 transcription factor. EMBO J 1999;18:5099–107.
[41] Donmez G, Wang D, Cohen DE, Guarente L. SIRT1 suppresses beta-amyloid production by activating the alpha-secretase gene ADAM10. Cell 2010;142:320–32.
[42] Gao J, Wang WY, Mao YW, Graff J, Guan JS, Pan L, Mak G, Kim D, Su SC, Tsai LH. A novel pathway regulates memory and plasticity via SIRT1 and miR-134. Nature 2010;466(7310):1105–9.
[43] Levenson JM, O'Riordan KJ, Brown KD, Trinh MA, Molfese DL, Sweatt JD. Regulation of histone acetylation during memory formation in the hippocampus. J Biol Chem 2004;279:40545–59.
[44] Martin SJ, Grimwood PD, Morris RG. Synaptic plasticity and memory: an evaluation of the hypothesis. Annu Rev Neurosci 2000;23:649–711.
[45] Vecsey CG, Hawk JD, Lattal KM, Stein JM, Fabian SA, Attner MA, Cabrera SM, McDonough CB, Brindle PK, Abel T, et al. Histone deacetylase inhibitors enhance memory and synaptic plasticity via CREB: CBP-dependent transcriptional activation. J Neurosci 2007;27:6128–40.
[46] Chan JK, Sun L, Yang XJ, Zhu G, Wu Z. Functional characterization of an amino-terminal region of HDAC4 that possesses MEF2 binding and transcriptional repressive activity. J Biol Chem 2003;278:23515–21.
[47] Eilam R, Peter Y, Elson A, Rotman G, Shiloh Y, Groner Y, Segal M. Selective loss of dopaminergic nigro-striatal neurons in brains of Atm-deficient mice. Proc Natl Acad Sci U S A 1998;95:12653–6.
[48] Cao R, Zhang Y. The functions of E(Z)/EZH2-mediated methylation of lysine 27 in histone H3. Curr Opin Genet Dev 2004a;14:155–64.

[49] Francis NJ, Follmer NE, Simon MD, Aghia G, Butler JD. Polycomb proteins remain bound to chromatin and DNA during DNA replication in vitro. Cell 2009;137:110–22.
[50] Juan AH, Derfoul A, Feng X, Ryall JG, Dell'Orso S, Pasut A, Zare H, Simone JM, Rudnicki MA, Sartorelli V. Polycomb EZH2 controls self-renewal and safeguards the transcriptional identity of skeletal muscle stem cells. Genes Dev 2011;25:789–94.
[51] Pirrotta V. PcG complexes and chromatin silencing. Curr Opin Genet Dev 1997;7:249–58.
[52] Sawarkar, R., and Paro, R. Interpretation of developmental signaling at chromatin: the Polycomb perspective. Dev Cell 19, 651-661.
[53] Surface, L.E., Thornton, S.R., and Boyer, L.A. Polycomb group proteins set the stage for early lineage commitment. Cell stem cell 7, 288-298.
[54] Vire E, Brenner C, Deplus R, Blanchon L, Fraga M, Didelot C, Morey L, Van Eynde A, Bernard D, Vanderwinden JM, et al. The Polycomb group protein EZH2 directly controls DNA methylation. Nature 2006;439:871–4.
[55] Wei Y, Chen YH, Li LY, Lang J, Yeh SP, Shi B, Yang CC, Yang JY, Lin CY, Lai CC, et al. CDK1-dependent phosphorylation of EZH2 suppresses methylation of H3K27 and promotes osteogenic differentiation of human mesenchymal stem cells. Nat Cell Biol 2011;13:87–94.
[56] Wu Z, Lee ST, Qiao Y, Li Z, Lee PL, Lee YJ, Jiang X, Tan J, Aau M, Lim CZ, et al. Polycomb protein EZH2 regulates cancer cell fate decision in response to DNA damage. Cell Death Differ 2011;18(11):1771–9.
[57] Hirabayashi Y, Suzki N, Tsuboi M, Endo TA, Toyoda T, Shinga J, Koseki H, Vidal M, Gotoh Y. Polycomb limits the neurogenic competence of neural precursor cells to promote astrogenic fate transition. Neuron 2009;63:600–13.
[58] Sauvageau, M., and Sauvageau, G. Polycomb group proteins: multi-faceted regulators of somatic stem cells and cancer. Cell stem cell 7, 299-313.
[59] Margueron R, Reinberg D. The Polycomb complex PRC2 and its mark in life. Nature 2011;469:343–9.
[60] Cao R, Zhang Y. SUZ12 is required for both the histone methyltransferase activity and the silencing function of the EED-EZH2 complex. Mol Cell 2004b;15:57–67.
[61] Sewalt RG, van der Vlag J, Gunster MJ, Hamer KM, den Blaauwen JL, Satijn DP, Hendrix T, van Driel R, Otte AP. Characterization of interactions between the mammalian polycomb-group proteins Enx1/EZH2 and EED suggests the existence of different mammalian polycomb-group protein complexes. Mol Cell Biol 1998;18:3586–95.
[62] Caretti G, Palacios D, Sartorelli V, Puri PL. Phosphoryl-EZH-ion. Cell stem cell 2011;8:262–5.
[63] Cha TL, Zhou BP, Xia W, Wu Y, Yang CC, Chen CT, Ping B, Otte AP, Hung MC. Akt-mediated phosphorylation of EZH2 suppresses methylation of lysine 27 in histone H3. Science 2005;310:306–10.
[64] Kaneko S, Li G, Son J, Xu CF, Margueron R, Neubert TA, Reinberg D. Phosphorylation of the PRC2 component Ezh2 is cell cycle-regulated and up-regulates its binding to ncRNA. Genes Dev 2010;24:2615–20.
[65] Wu SC, Zhang Y. CDK1-mediated phosphorylation of Ezh2 regulates its stability. J Biol Chem 2011; 286(32):28511–9.
[66] Zeng X, Chen S, Huang H. Phosphorylation of EZH2 by CDK1 and CDK2: a possible regulatory mechanism of transmission of the H3K27me3 epigenetic mark through cell divisions. Cell Cycle 2011;10:579–83.
[67] Chen S, Bohrer LR, Rai AN, Pan Y, Gan L, Zhou X, Bagchi A, Simon JA, Huang H. Cyclin-dependent kinases regulate epigenetic gene silencing through phosphorylation of EZH2. Nat Cell Biol 2010;12: 1108–14.
[68] Li J, Hart RP, Mallimo EM, Swerdel MR, Kusnecov AW, Herrup K. EZH2-mediated H3K27 trimethylation mediates neurodegeneration in ataxia-telangiectasia. Nat Neurosci 2013;16:1745–53.
[69] Chakrabarti L, Zahra R, Jackson SM, Kazemi-Esfarjani P, Sopher BL, Mason AG, Toneff T, Ryu S, Shaffer S, Kansy JW, et al. Mitochondrial dysfunction in NnaD mutant flies and Purkinje cell degeneration mice reveals a role for Nna proteins in neuronal bioenergetics. Neuron 2010;66:835–47.

[70] Kaas GA, Zhong C, Eason DE, Ross DL, Vachhani RV, Ming GL, King JR, Song H, Sweatt JD. TET1 controls CNS 5-methylcytosine hydroxylation, active DNA demethylation, gene transcription, and memory formation. Neuron 2013;79:1086–93.

[71] Rudenko A, Dawlaty MM, Seo J, Cheng AW, Meng J, Le T, Faull KF, Jaenisch R, Tsai LH. Tet1 is critical for neuronal activity-regulated gene expression and memory extinction. Neuron 2013;79:1109–22.

[72] Xu Y, Wu F, Tan L, Kong L, Xiong L, Deng J, Barbera AJ, Zheng L, Zhang H, Huang S, et al. Genome-wide regulation of 5hmC, 5mC, and gene expression by Tet1 hydroxylase in mouse embryonic stem cells. Mol Cell 2011;42:451–64.

[73] Kriaucionis S, Heintz N. The nuclear DNA base 5-hydroxymethylcytosine is present in Purkinje neurons and the brain. Science 2009;324:929–30.

[74] West AE. Cocaine shapes chromatin landscapes via Tet1. Nat Neurosci 2015;18:478–80.

[75] Yu H, Su Y, Shin J, Zhong C, Guo JU, Weng YL, Gao F, Geschwind DH, Coppola G, Ming GL, et al. Tet3 regulates synaptic transmission and homeostatic plasticity via DNA oxidation and repair. Nat Neurosci 2015a;18(6):836–43.

[76] Jiang D, Zhang Y, Hart RP, Chen J, Herrup K, Li J. Alteration in 5-hydroxymethylcytosine-mediated epigenetic regulation leads to Purkinje cell vulnerability in ATM deficiency. Brain 2015;138(Pt 12):3520–36.

[77] Ciccia A, Elledge SJ. The DNA damage response: making it safe to play with knives. Mol Cell 2010;40: 179–204.

[78] Abraham AT, Zhou X, Hecht SM. Metallobleomycin-mediated cleavage of DNA not involving a threading-intercalation mechanism. J Am Chem Soc 2001;123:5167–75.

[79] Harper JW, Elledge SJ. The DNA damage response: ten years after. Mol Cell 2007;28:739–45.

[80] Shiloh Y. ATM and related protein kinases: safeguarding genome integrity. Nat Rev Canc 2003;3:155–68.

[81] Falck J, Coates J, Jackson SP. Conserved modes of recruitment of ATM, ATR and DNA-PKcs to sites of DNA damage. Nature 2005;434:605–11.

[82] Reinhardt HC, Yaffe MB. Kinases that control the cell cycle in response to DNA damage: Chk1, Chk2, and MK2. Curr Opin Cell Biol 2009;21:245–55.

[83] Shiloh Y, Ziv Y. The ATM protein kinase: regulating the cellular response to genotoxic stress, and more. Nat Rev Mol Cell Biol 2013;14:197–210.

[84] Feng J, Shao N, Szulwach KE, Vialou V, Huynh J, Zhong C, Le T, Ferguson D, Cahill ME, Li Y, et al. Role of Tet1 and 5-hydroxymethylcytosine in cocaine action. Nat Neurosci 2015;18:536–44.

[85] Yu H, Su Y, Shin J, Zhong C, Guo JU, Weng YL, Gao F, Geschwind DH, Coppola G, Ming GL, et al. Tet3 regulates synaptic transmission and homeostatic plasticity via DNA oxidation and repair. Nat Neurosci 2015b;18:836–43.

[86] Kafer GR, Li X, Horii T, Suetake I, Tajima S, Hatada I, Carlton PM. 5-Hydroxymethylcytosine marks sites of DNA damage and promotes genome stability. Cell Rep 2016;14:1283–92.

[87] Goodarzi AA, Jeggo P, Lobrich M. The influence of heterochromatin on DNA double strand break repair: getting the strong, silent type to relax. DNA Repair 2010;9:1273–82.

[88] Jiang D, Wei S, Chen F, Zhang Y, Li J. TET3-mediated DNA oxidation promotes ATR-dependent DNA damage response. EMBO Rep 2017;18:781–96.

[89] Beerman I, Rossi DJ. Epigenetic control of stem cell potential during homeostasis, aging, and disease. Cell stem cell 2015;16:613–25.

[90] Koh KP, Yabuuchi A, Rao S, Huang Y, Cunniff K, Nardone J, Laiho A, Tahiliani M, Sommer CA, Mostoslavsky G, et al. Tet1 and Tet2 regulate 5-hydroxymethylcytosine production and cell lineage specification in mouse embryonic stem cells. Cell stem cell 2011;8:200–13.

[91] Pujadas E, Feinberg AP. Regulated noise in the epigenetic landscape of development and disease. Cell 2012; 148:1123–31.

CHAPTER

Cockayne syndrome

7

Luca Proietti-De-Santis[1], Vincent Laugel[2], Giorgio Prantera[3]

Laboratorio di Genetica Molecolare dell'Invecchiamento, Dipartimento di Scienze Ecologiche e Biologiche, Università della Tuscia, Viterbo, Italy[1]; Laboratoire de Génétique Médicale - INSERM U1112, Institut de Génétique Médicale d'Alsace (IGMA), Faculté de médecine de Strasbourg, Strasbourg, France[2]; Laboratorio di Epigenetica, Dipartimento di Scienze Ecologiche e Biologiche, Università della Tuscia, Viterbo, Italy[3]

Chapter outline

7.1 Clinical phenotypes **135**
7.1.1 Classical (moderate) type I cockayne syndrome 136
7.1.2 Early-onset (severe) subtypes 138
7.1.3 Late-onset subtypes 139
7.2 Genetics **140**
7.3 CSA and CSB proteins **140**
7.4 Cellular and molecular aspects **140**
7.5 The molecular basis of neurodegeneration **141**
7.6 Concluding remarks **146**
References **147**

7.1 Clinical phenotypes

Cockayne syndrome (CS) is a rare, multisystem autosomal recessive condition, which was first described by E. Cockayne in 1936 as a form of cachectic dwarfism [1]. Clinical diagnosis requires three major criteria (progressive growth failure, mental retardation, and microcephaly) together with two minor criteria identified among the following: cataracts or pigmentary retinopathy, sensorineural hearing loss, dental abnormalities, loss of subcutaneous fat, and skin photosensitivity [2–4]. The symptoms of this condition are always progressive and always worsen with time. CS encompasses several severity subtypes that draw a continuous clinical spectrum from the most severe prenatal subtype (also called COFS syndrome) to the mildest adult-onset subtype. Despite this continuous distribution of clinical symptoms without a clear threshold, it remains relevant for diagnostic purposes to distinguish classical (moderate) type I CS, severe early-onset subtypes (prenatal subtype or COFS syndrome and neonatal subtype or type II), and mild late-onset subtypes (adult-onset or type III CS) [5,6]. The incidence of CS was evaluated at 2.7 cases per million births in Western Europe [7] but this

Chromatin Signaling and Neurological Disorders. https://doi.org/10.1016/B978-0-12-813796-3.00007-9

incidence is probably underestimated, as CS remains largely underdiagnosed due to the rarity of the disease and to the wide variety of symptoms.

7.1.1 Classical (moderate) type I cockayne syndrome

The moderate type I CS was first recognized among the various subgroups of CS. Usually, type I CS patients hardly show any symptom at birth and progressively develop the cardinal symptoms of the disease. Type I CS patients usually live up to their teens or twenties. Postnatal growth failure, delayed developmental milestones, and progressive microcephaly are usually the first hallmarks of the disease, occurring for most patients before the end of the first year of life. Weight, height, and head circumference tend to reach an early plateau, which progressively and severely deviates from the normal range. The pattern of these growth curves is correlated with the overall severity of the disease. The loss of subcutaneous fat is particularly significant in CS patients leading to a cachectic phenotype with severe sunken eyes, which is one of the most typical clinical characteristics of the condition. Feeding difficulties and limited oral intake are constant features in CS patients: optimal nutritional care should aim at preventing weight loss, overfeeding is constantly poorly tolerated, and tube feeding is often required in the late stages of the evolution of type I patients.

The mechanisms of this severe growth restriction are not clearly elucidated. A direct effect of transcriptional defects [8] or mitochondrial defects [9] on cell proliferation in CS was suggested. An alternative or complementary explanation postulates that the growth restriction might be an adaptive mechanism to the DNA repair defect of CS cells to limit the production of reactive oxygen species (ROS) through a downregulation of the IGF—GH axis [10]. The use of growth hormone (GH) in CS patients remains controversial. Several dietary interventions have been promoted to improve the clinical course of the disease: caloric restriction [11] or high-fat diet [12] has proven partly effective to rescue the metabolic imbalance in CS mouse models and improve their survival but it is unclear if these interventions can be safely applied to CS children. At present, clinical data are still lacking to validate either of these hypotheses.

Combined neurological manifestations constantly occur in type I CS patients. Mental deficiency is present at various levels in this subgroup and shows an unusual pattern with a typical relative preservation of communication skills. Cerebellar ataxia and dysarthria together with pyramidal spasticity rapidly limit the motor functions and autonomy of the patients. Extrapyramidal akinesia and in some cases resting tremor further impair the motility as the condition worsens. Peripheral neuropathy is also present but is usually nonvisible at the clinical level except for the late abolition of deep tendon reflexes. Type I CS patients are usually able to learn to walk but lose this ability later on. Cognitive decline is present in the late stages of the disease and then leads to the loss of communication abilities. Seizures occur in a minority of CS patients but seem more frequent than in the general population.

In line with these clinical signs, brain imaging (Fig. 7.1) typically shows hypomyelination and progressive atrophy of the white matter, progressive cerebellar atrophy, bilateral calcification of the basal ganglia (mostly putamina), and dentate nuclei [13]. The cortical regions are better preserved and global atrophy is observed only in the late stages of the disease. A particular sensitivity of Purkinje cells to the oxidative stress damage was put forward to account for the early involvement of the cerebellar vermis. The defective production of myelin was linked to defective transcription capacities in highly specialized cells, such as CS oligodendrocytes, but any formal evidence is

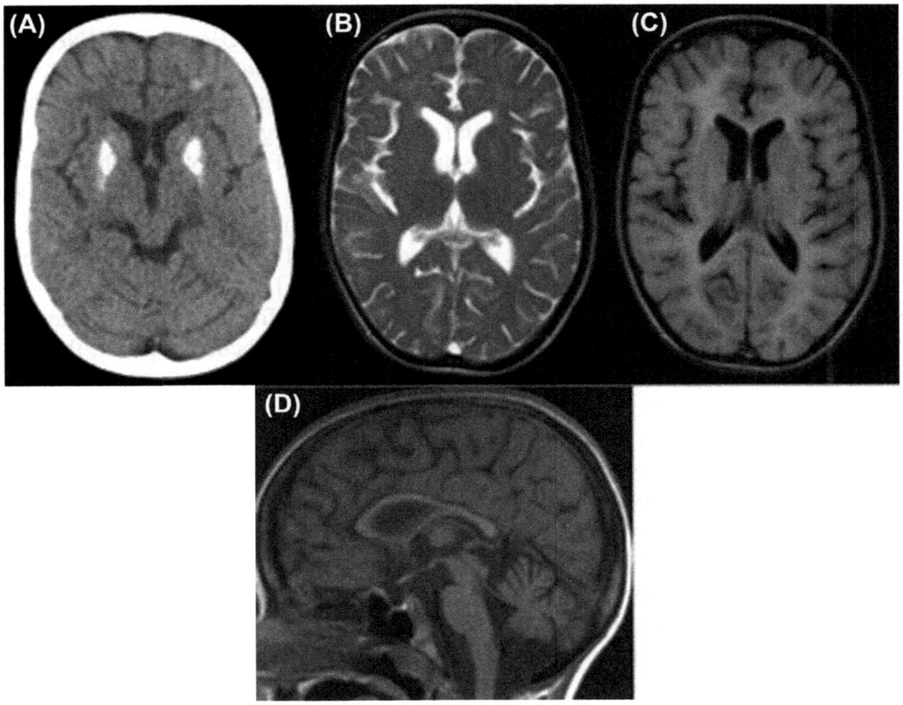

FIGURE 7.1

Brain CT scan (A), axial T2-weighted MRI (B), axial FLAIR MRI (C), and sagittal MRI (D) of a type I CS patient showing calcifications of the putamina, hypomyelination of the white matter, and cerebellar hypoplasia.

still lacking. Vascular (capillary) damage was also observed in some postmortem examinations [14]. Nerve conduction studies show progressive demyelinating neuropathy with secondary axonal involvement [15].

Both progressive cataracts and pigmentary retinopathy may impair visual function. Sensorineural hearing loss is a frequent feature of type I CS patients that sometimes leads to complete deafness. Hearing devices or cochlear implants should be implemented whenever necessary to usefully maintain the communication capacities of CS patients despite their cognitive impairment. The progressive neuropathy, pigmentary retinopathy, and sensorineural hearing loss often show a parallel and steadily worsening course, which suggests a common degenerative process. These measurable and steadily decreasing markers could be used in clinical trials to assess the positive effect of putative therapeutic molecules.

Dental anomalies are frequent in type I CS patients [16]. Frequent dental cavities were mentioned from the first reports on CS and were considered as one of the minor diagnostic criteria of the disease. Enamel hypoplasia is actually the main cause underlying frequent cavities in CS and is much more reliable and specific for diagnostic purposes. Abnormalities in the number and shape of the teeth are also very frequently observed.

Cutaneous photosensitivity was described only in later reports obviously resulting as a key element of the link between the pathophysiology of CS and a defective repair of UV-induced DNA lesions. This cutaneous photosensitivity is almost constantly found by skin phototesting when performed (standardized UV exposure of a small skin area) but is only clinically reported in about half of type I CS patients, depending on the level of sun exposure and skin subtype. Pigmentary macules can accumulate with age and are therefore more prominent in late-onset subtypes of CS. Both the photosensitivity and the skin pigmentation however can in no way be compared to the symptoms observed in Xeroderma pigmentosum (XP) patients. Furthermore, in any cohort of CS patients, no skin cancer proneness was reported. CSB suppression showed significant anticancer effects [17–19], which might counteract the risk of carcinogenesis because of the accumulation of unrepaired DNA lesions.

Hyperuricemia, elevated blood pressure, and diabetes are more frequent in CS patients than in the general population of the same age. Together with the loss of the subcutaneous fat, these symptoms further substantiated the hypothesis of premature aging in CS as they also occur later in life in normal aging. However, not all the previously described symptoms fit into the process of aging and therefore CS can hardly be considered as a full-blown progeroid syndrome but rather as a segmental progeroid condition. In addition, it is not yet known whether the mechanisms of hyperuricemia, elevated blood pressure, and diabetes, in CS fully mimic those occurring in normal aging. Renal symptoms may also occur in the late stages of CS: proteinuria, nephrotic syndrome, and progressive renal failure may accompany the final stages of CS natural course. The relationship between renal symptoms and the potential nephrotic reduction linked to premature aging are not yet fully understood. Elevated liver enzymes and elevated CSF protein levels are also common biomarkers of CS and are linked to normal aging. Aspiration pneumonia, renal failure, and acute liver failure are among the main causes of death in type I CS.

7.1.2 Early-onset (severe) subtypes

Cerebro-oculo-facio-skeletal (COFS) syndrome has been initially reported as a distinct clinical entity characterized by severe congenital microcephaly, cataracts, microphthalmia, prominent metopic suture, kyphosis, and arthrogryposis. It was originally described in an isolated population in Manitoba and subsequently identified in other populations worldwide. It was later demonstrated that COFS patients shared mutations in CSB or other nucleotide excision repair (NER) genes (*XPB*, *XPG*, *ERCC1*, *XPF*). COFS syndrome is now considered to be allelic to CS and constitutes the most severe end of the CS clinical spectrum. Specific diagnostic criteria were elaborated to improve the diagnosis of this prenatal form of CS and require the association of arthrogryposis, congenital microcephaly, congenital cataracts and/or microphthalmia, severe developmental delay, and severe pre- and postnatal growth failure [20,21]. COFS patients usually show very little developmental progress, and their life expectancy is usually very short. Most of them do not show the same degree of subcutaneous fat as in CS. In most cases, their short life expectancy does not allow enough time to develop all the classical symptoms of type I CS thus rendering the diagnosis difficult. COFS syndrome appears mostly as a very severe neurodevelopmental disorder as compared to the more neurodegenerative pattern of later-onset subtypes of CS: it is hypothesized that the underlying mechanisms may be the same for COFS as for the later-onset subtypes of CS, but that these pathological processes occur already during the prenatal period in COFS patients leading to already severe neurodevelopmental symptoms at birth.

Type II (severe early-onset) CS is overlapping with COFS syndrome and is characterized by congenital symptoms without arthrogryposis or the cardinal symptoms of COFS. Type II CS patients show congenital microcephaly, severe developmental delay, and severe growth failure already from the first weeks of life. Poor feeding, weak cry, axial hypotonia, and peripheral hypertonia are frequent in the neonatal period. Congenital cataracts are common in this group and require surgical intervention. Type II CS patients show very little developmental achievements and are usually unable to sit if unaided. Contrarily to later-onset subtypes of CS, type II CS patients have very limited communication capacities and no language. Tube feeding is usually necessary from the first weeks or months of life. Early brain imaging may show subcortical calcifications rather than calcifications in the basal ganglia. Skin photosensitivity is usually of little concern in these patients, as they are often less exposed to the sun. However, UV sensitivity can be demonstrated when formally challenged in clinical photo testing. Life expectancy is limited to a few years in most cases, and respiratory failure is the main cause of death in this subgroup of CS. In some cases, type II CS patients show a prolonged survival in spite of the early onset of very severe symptoms [22]. This situation may be caused by different specific mutations, but the exact mechanism remains elusive. In this situation, type II CS patients progressively acquire the same symptoms as type I CS patients, with progressive ataxia, spasticity, peripheral neuropathy, retinopathy, hearing loss, dental abnormalities, and cachexia. Depending on their life duration, type II CS patients show a combination of neurodevelopmental and neurodegenerative symptoms.

7.1.3 Late-onset subtypes

Type III (mild) CS patients usually show the first symptoms of the disease after 2 years of life. Growth failure and psychomotor retardation remain the first signs of the disease, as in type I, but they occur at later ages and are less severe. Growth curves usually do not reach a plateau but remain below the third percentile. These type III CS patients eventually develop all the neurological and extraneurological symptoms described for type I CS, mostly after age 10. Their life expectancy is usually longer than 30 years. Some very mild patients even reach a normal height and weight, a normal intellectual level but show late-onset cerebellar ataxia and secondary cognitive decline: this adult-onset subgroup of CS can be named type IV CS and often represents a very difficult diagnostic challenge [23]. These patients, however, underline that the mutational status of CS genes should be investigated in cases of early dementia or adult-onset progressive ataxia even in the absence of the classical diagnostic criteria of CS. It is in these late-onset subgroups that the segmental progeroid symptoms of CS are prominent. Progressive hearing loss, loss of subcutaneous fat, increased incidence of diabetes, cognitive decline, and elevated blood pressure are particularly frequent in these groups. Type III and type IV CS show purely neurodegenerative symptoms of the CS spectrum.

Finally, it was shown that some patients showing isolated cutaneous sun sensitivity, with no other symptoms, could also be bearing pathogenic mutations in *CSA* or *CSB* and that their fibroblasts could show the same DNA repair defects in the NER pathway as classical CS patients [24–26]. These patients were diagnosed with UV-sensitive syndrome (UVSS). However, none of these patients had a long-term clinical follow-up, and it is unclear whether these patients belong to a distinct clinical entity and will never develop any other symptoms or whether they will show mild symptoms of CS at very late stages.

7.2 Genetics

CS is an autosomal recessive disorder, caused by mutations falling into two different complementation groups: (1) CS-A, identified by a mutation in the *ERCC8* gene located at locus 5q12–q31 and encoding the protein named CSA; and (2) CS-B, identified by a mutation in the *ERCC6* gene at locus 10q11 and encoding the protein named CSB. Mutations in *ERCC6* gene account for about 70%–75% of CS cases [5,6]. CSA mutations predominantly cause the type I CS phenotype, while the CSB ones are mainly associated to the type II CS phenotype [6]. However, no obvious genotype–phenotype correlation can be associated to either gene; moreover, no mutational hot spot exists, while missense, nonsense, and frameshift mutations are distributed along the entire gene.

A number of patients exhibit the combined phenotype CS/Xeroderma pigmentosum (XP/CS) [27,28]. These patients present mutations in *XPB* (*ercc3*), *XPD* (*ercc2*), and *XPG* genes and show the unfortunate combination of severe photosensitivity and increased skin cancer risk typical of XP patients, along with the serious somatic and neurological manifestations of CS patients. The degenerative disorder COFS also proved to be a particularly severe variant of CS, caused by mutations in CSB. The UVSS is considered to be part of the spectrum of CS, but with mild features mainly related to photosensitivity.

7.3 CSA and CSB proteins

CSA belongs to the WD40 repeat family of proteins exhibiting both structural and regulatory roles. CSA binds to DDB1 and is able to assemble an E3 ligase complex with Cul4 and Rbx1 [29]. CSA is involved in the ubiquitination/degradation of proteins including CSB, RNA polymerase II, p53, and ATF3 [30–33].

CSB is instead a member of the SWI2/SNF2 family of chromatin remodeling complexes [34] as it contains helicase motifs and a DNA-stimulated ATPase activity associated with nucleosome remodeling [35,36]. However, CSB does not show classical helicase activity [37]. Chromatin remodeling is a dynamic ATP-dependent process necessary for the highly condensed chromatin to become accessible to the factors involved in the DNA metabolic activities, such as replication, transcription, repair, and recombination. Importantly, chromatin remodeling events are also critical for the maintenance of the epigenomic patterns of both multipoint adult stem and differentiated cells in a tissue microenvironment. By interacting with a series of transcription and excision repair proteins, CSB may promote DNA repair and transcription complexes formation at initiation sites. Further and interestingly, CSB was found to interact with the mitochondrial transcription factor TFAM (transcription factor A, mitochondrial) [38], thus suggesting a possible critical CSB role for mitochondrial DNA metabolism.

CSB is also essential to various processes of DNA repair including base excision repair [39], interstrand crosslink repair [40], and DNA double-strand break repair [41,42].

7.4 Cellular and molecular aspects

CS defects were initially associated to transcription elongation impairment because of a deficiency in transcription-coupled nucleotide excision repair (TC-NER), and the DNA repair pathway aimed to the removal of DNA bulky adducts located on the transcribed strand of active genes [43,44]. A deficiency in TC-NER leads to the stalling of RNA polymerase in front of the not repaired lesion, in turn triggering genomic instability and apoptotic cell death. During TC-NER, CSA and CSB proteins first participate in the removal of the stalled RNA polymerase, by inducing its ubiquitin-dependent degradation, and then recruit NER proteins to remove the DNA damage [45,46].

However, the DNA repair defects alone cannot explain the wide clinical spectrum of CS patients and the multiple pathological symptoms. Thus, in the last years, CS became an ever-complicated molecular puzzle. Defective repair of UV-induced lesions or of other adducts may explain photosensitivity, but not other clinical aspects such as cachexia and neurodegeneration. This strongly argues that CS proteins have additional roles in processes other than the transcription-coupled repair (TCR) pathway [17,18,47–49]. Consistently with this hypothesis, CSB protein proved to be itself a functional transcription factor [50] interacting with and stimulating the transcriptional complexes of all the three RNA polymerases (I, II, and III) [51–57]. Moreover, CSA and CSB proteins were recently demonstrated to possess an additional regulatory key role in the turnover of p53 and ATF3 transcription factors by promoting their ubiquitination and degradation [32,33].

7.5 The molecular basis of neurodegeneration

How CS proteins function in NER and BER (base excision repair) could account for neurodegenerative features was previously thoroughly reviewed [58,59]. In particular, the consideration of the oxidative nature of the main DNA damage accumulation in the brain indicates that the role of CS proteins in BER may be more significant for neurodegeneration than its role in NER. Thus, here we mainly focused on the role played by CS proteins in chromatin remodeling that is required for transcription activation, and how eventually, the dysregulation of these processes may account for neurodegeneration.

Even though XP, which also impairs the NER pathway, confers neurological defects, these differ from those associated with CS. Indeed, XP is primarily a neurodegenerative disease owing to the degeneration of neurons, whereas CS appears to be more a neurodevelopmental disease because of the neurologic features, such as early microcephaly, cerebellar hypoplasia, and hypomyelination. Moreover, the early onset of the neurological symptoms, which appears either at birth or during early childhood, suggests that CS proteins may play crucial roles in the transcriptional programs underpinning the plasticity and the homeostasis of the central nervous system during both perinatal and postnatal life. It is reasonable to assume that typical CS features, such as microcephaly, are caused by defects in the postnatal development of prefrontal cortex due to a lack of postnatal neurogenesis, at least for what concerns CS type I patients.

Similarly, the dysmyelination and demyelination features of CS patients may arise from a deficiency in neurogenesis in the subcortical white matter. Moreover, defective neurogenesis, particularly in the cerebellum, could account for the cerebellar degenerative features of CS patients. Interestingly, in support of these considerations, suppression of CSB was shown to affect the neuronal differentiation capability of human neural progenitor cells [49]. More specifically, CSB suppression led to a highly reduced expression of neuronal markers, such as MAP2 and NEUROD1, accompanied by impaired cell polarization and neuritogenesis. Very interestingly, by chromatin immunoprecipitation (ChIP), it was shown that CSB suppression affected the recruitment of the transcriptional initiation complex on the promoters of the above genes as indicated by the abrogated occupancy of both RNA polymerase II and TFIIB. ChIP assays also revealed the presence of CSB at the *MAP2* promoter site that was further enriched during differentiation. Analysis of chromatin modifications in CSB-suppressed cells showed a highly reduced extent of the active chromatin marker H3K9 acetylation around the *MAP2* promoter region. Moreover, the reduced level of H3K9 acetylation observed in CSB-suppressed cells consistently correlated with the impaired recruitment of the p300 histone acetyltransferase at the *MAP2* promoter. It is well known that the acetyltransferase p300 is an essential cofactor in adult neurogenesis [60]. It should be also noticed that during differentiation, BAF (BRG1-associated factor) chromatin remodeling complex is positioned at the *MAP2* promoter and is required for its activation. Consistently, in CSB-suppressed cells, BAF is not properly recruited at the *MAP2* promoter. In addition, the activation of *MAP2* promoter during neural differentiation correlates with the loss of HDAC3 deacetylase activity in normal cells, but not in CSB-suppressed cells [49]. Thus, the synchrony between the CSB binding to *MAP2* promoter and the recruitment of BAF implies a role for CSB in BAF recruitment. This is partly supported by the observed interaction of CSB with some BAF complex subunits [61]. Indeed, by coupling TAP (tandem affinity purification)-tag technology and mass spectrometry, we identified CSB-associated proteins, including four prominent chromatin remodeling factors (SMARCA1, 2, 4, and 5), which belong to the SWI/SNF family of proteins. The association of some SMARCA proteins with WINAC complex is also required for the coactivation of estrogen-responsive promoters and vitamin D-coupled transcription regulation. Deficiencies in hormonal response may thus account for shunted growth and skeletal abnormalities frequently observed in CS patients.

As underlined earlier, CSB associates with members of the SWI/SNF family of chromatin remodeling proteins [34,36]. These proteins belong to neural progenitor-specific chromatin remodeling complex (npBAF complex) and neuron-specific chromatin remodeling complex (nBAF complex) that are essential for the self-renewal/proliferative capacity of neural stem cells and for the transition from the proliferative state of neural stem cells to the postmitotic state of neurons, respectively [62]. Furthermore, the nBAF complex, together with calcium-responsive transactivator, regulates the activity of genes essential for dendrite growth [63]. Copurification of CSB with components of npBAF and nBAF complexes suggests that CSB plays a crucial role to preserve the functional integrity of different brain cell types by finely coordinating transcription through chromatin remodeling activities. This role may involve not only MAP2 and NEUROD1, but also a large plethora of genes involved in these processes [64].

Alternatively, CSB, by interacting with p300, could cause histone acetylation changes, thereby facilitating the recruitment of essential factors to BAF. Both hypotheses, not mutually exclusive, confer an important role to CSB in the signaling mechanisms linking neuronal differentiation to specific gene expression. The demonstration of MAP2 transcriptional regulation by CSB indicates that a functional *CSB* gene is required for the efficient differentiation of neural progenitor cells into neurons. In CSB-suppressed cells, the impaired neuronal differentiation, coupled with reduced expression of the neuronal-specific marker MAP2, strongly indicates that *CSB* gene product is pivotal for the neuronal differentiation process.

Along this line, Wong et colleagues [64] demonstrated that reprogramming of CS fibroblasts to neuron-like cells is inefficient unless an exogenous *CSB* gene was introduced. Moreover, CSB-depleted neuroblastoma cells show defects in gene expression programs required for neuronal differentiation, and fail to differentiate and extend neurites. Similarly, neuron-like cells cannot be maintained without CSB.

Furthermore, transcriptome analyses performed in CS neurons derived from human CS-induced pluripotent stem cells revealed that several pathways related to neuronal function and communications were dysregulated in CSB neurons that accordingly showed a reduced synaptogenesis [65], as already observed with a different approach in Csb-KO mice [64]. Reduced synaptogenesis was shown to correlate with impaired transcription of the GH/IGF-1 signaling pathway in CSB neurons. This feature can be at the basis of several neurologic features exhibited by CS patients, such as reduced survival of Purkinje cells in neonatal cerebella, reduction of brain size, hypomyelination, and loss of granule cells.

Interestingly, reduced GH and IGF-1 transcription is also described in double deficient ($Xpa^{-/-}$; $Csb^{-/-}$) murine model that in fact displays growth retardation, progressive cachexia, and neurodegeneration [66]. In contrast, $Csb^{-/-}$ mice instead fail to display several CS human neurological features [67].

Therefore, the impairment of the GH/IGF-1 signaling pathway in CSB may account for both the neurodegenerative and the progeroid features exhibited by CS patients.

These findings consistently explain the molecular basis of the neurological symptoms reported in CS patients, at least as far as CS-B patients are concerned. Even if, as mentioned earlier, the neurodevelopmental defects cannot be directly attributed to a defect in TCR-NER, it is however possible that other neurological and sensorial aspects of CS patients, including cataracts and cerebellar atrophy, might be induced by an additive effect between the defective neurogenesis and the TCR defect.

In some CS patients, neurological defects appear only after birth. An intriguing question thus arises: why CS proteins are more prominently crucial for the maintenance of the integrity of the nervous system in extrauterine life? Strikingly, patients with the classical form (type I CS) display normal intrauterine body and brain growth with developmental crisis only after birth in the first or second year of life. This allowed us to hypothesize that *CSA* and *CSB* gene products could be dispensable during the fetal stage but become essential starting from postnatal development. Manifestation of CS patients developmental defects after birth indicates that CS proteins may perform essential functions during the extrauterine life when the child is no longer protected by the controlled environment of mother's womb in terms of angiogenic factors and gaseous oxygen. Oxygen homeostasis may well represent one of the Achilles' heel of CS patients during extrauterine life. In agreement, this hypothesis is supported by the impairment in CSB-deficient cells of the hypoxia-inducible factor 1 (HIF-1)-dependent transcriptional response to hypoxia. Oxygen deprivation (hypoxia) occurs in tissues when O_2 supply fails to meet the cellular demand. Hypoxia may occur in physiological as well as in pathological conditions. HIF-1 is a major regulator of oxygen homeostasis within the cells [68]. As a transcription factor, it regulates the expression of dozens of genes involved in homeostasis maintenance following oxygen concentration changes. This vast array of genes controls multiple cellular functions such as angiogenesis, glucose metabolism, and cell survival [69]. The transcriptional induction is mediated by HIF-1 binding to the hypoxia-responsive element at the promoter of HIF-responsive genes. HIF-1 is a heterodimer composed by the inducible HIF-1α subunit and the constitutively expressed HIF-1β subunit [68]. HIF-1-mediated transcriptional activation requires the recruitment of transcriptional cofactors, such as the acetyltransferase p300, to the promoter of target genes [70].

Importantly, whether not mitigated by HIF-1-induced adaptive responses, hypoxic stress results in p53 stabilization [71]. Consequently, p53 becomes active, promoting the transcription not only of cell cycle-regulator genes, such as *CDKN1A* gene product p21, but also of the genes involved in the apoptotic cascade, such as *BAX* (BCL2-associated X protein), thus initiating cell death. Most p53 downstream effects depend on its intrinsic function as a transcription factor, which contributes to the regulation of an expanding spectrum of cellular processes. p53 target genes share a consensus response element proximal to the transcription start site [72]. Besides the transactivating function of genes harboring the p53-responsive element in the promoter, it recently emerged that p53 can also trans-repress genes without necessarily binding to their promoters [73]. Transrepression by p53 may occur through competition with other transcription factors for coactivators.

Some studies, for instance, demonstrated that p53 and HIF-1 compete for the coactivator p300 [74,75] thus antagonizing each other until the integration of different signals reaches its balance and the fate of the cell is determined. Similarly, p53 inhibits the p300-dependent activation of the *TFF2* gene by sequestering p300/CBP away from its promoter [76]. We recently showed that the p53 repressive activity becomes exacerbated by CSB deprivation. Indeed, we showed that p300 and CSB compete for p53 binding, with the latter showing a stronger affinity. Consequently, p53 transcriptional activity is negatively modulated by CSB [74]. In contrast, in the absence of CSB, an increased binding of p53 to p300 occurs causing the stabilization of p53 and the activation of its target genes, including those involved in the apoptotic commitment. Therefore, the overactivation of p53 response is toxic because this protein titrates away essential transcription factors such as p300.

This finding might have great implications for neurodegeneration in CS. Actually, the oxygen tension is lower in brain than in other organs and tissues. Thus, the optimal activation of HIF-1 pathway is, therefore, crucial for the functional integrity of brain cell types including neural stem and progenitor cells. Alternatively, in metabolically active brain regions, hypoxia, if not adequately mitigated, may stimulate mitochondrial-dependent reactive oxygen species (ROS) production [77] that in turn can lead to DNA damage accumulation, eventually resulting in either senescence or cell death. Accordingly, in CSB-suppressed neural progenitor cells, grown under hypoxic conditions, we observed an increased fraction of senescent cells with exacerbation of neurogenesis defects. Neurodegeneration therefore may also arise not only from a defective differentiation of the neural stem cells but also from an accelerated depletion of the neural stem cell niche. Therefore, neurodegeneration may arise either from a failed differentiation of the neural stem cells in neurons or from a failure of the brain repair activity as a consequence of the depletion of the stem cell niche.

Interestingly, the most affected neurons are located in the basal ganglia and in the hippocampus where oxygen requirement is maximum. Moreover, CSB was found to interact with the mitochondrial transcription factor TFAM [38], thus implying an important function of CSB in mitochondrial DNA metabolism. Indeed, CSA and CSB proteins localize in mitochondria where they participate in the repair of mitochondrial DNA oxidative damage and in the maintenance of mitochondria functional integrity. These observations add a new dimension to the panoply of complex cellular functions in which CS proteins are involved [78,79]. In agreement, CS patients exhibit alterations of mitochondrial functions [9,80–83]. However, the scenario may be even more complex. It was recently described that the hyperactivation of p53 also represses PGC-1α, a master inducer of mitochondrial biogenesis and function involved in the defense against ROS. Therefore, the high levels of p53, as observed in CS cells may give rise to a vicious cycle where inefficient repair of DNA damage and/or decreased efficiency of p53 degradation trigger elevated levels of p53, that disturb HIF adaptation and repress PGC-1α pathway, damaging mitochondrial functionality and enhancing ROS levels that in turn increase DNA damage and so on (Fig. 7.2).

Intriguingly, the clinical manifestations of CS resemble those of mitochondrial diseases, characterized by the presence of ataxia, seizures, myopathy, hearing loss, neuropathy, cerebellar atrophy,

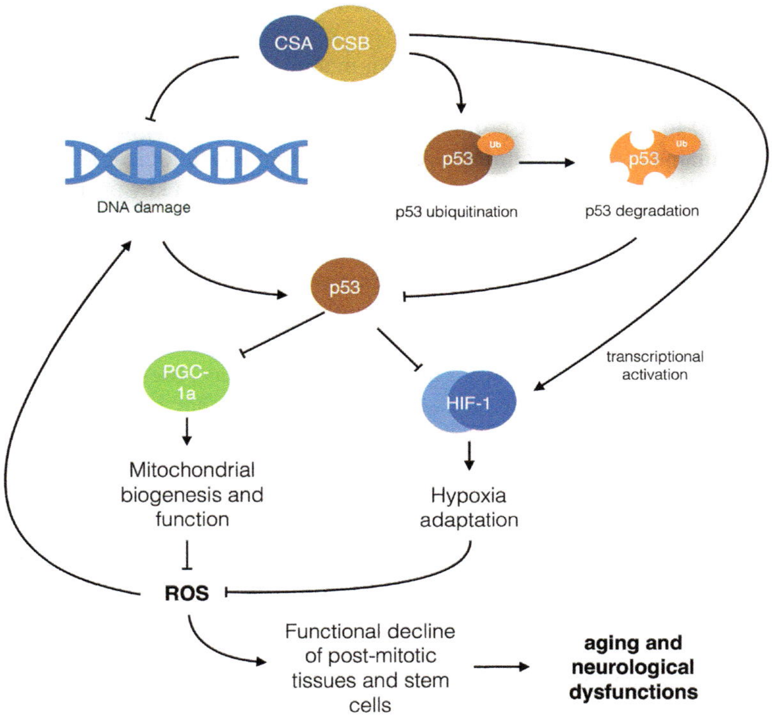

FIGURE 7.2

The vicious cycle. CSA and CSB play multiple functions in the context of the cell. On one side, following DNA damage, they first participate to the removal of the RNA polymerase, stalled in front of damage, by inducing its ubiquitination-dependent degradation. Then, they recruit NER proteins to remove the DNA damage. On the other side, CSA and CSB also promote p53 protein downregulation by ubiquitination-dependent degradation. As a further level of surveillance, these proteins induce the activation of prosurvival pathways in response to cellular stresses, including hypoxia. In their absence, a vicious cycle may be triggered where increased p53 levels, due to inefficient repair of DNA damage and/or decreased efficiency of p53 degradation, disturb hypoxia adaptation and repress PGC-1α pathway, damaging mitochondrial functionality and enhancing ROS levels that in turn increase DNA damage, which in turn further induces p53, thus restarting this detrimental vicious cycle.

basal ganglia pathology, and leukodystrophy [83]. A recent paper demonstrated that CSB prevents the accumulation of damaged mitochondria by promoting mitochondrial autophagy [67]. CSB also promotes transcriptional elongation in mitochondria through a crosstalk with mitochondrial proteins [38]. The protection of mitochondrial DNA from excessive generation of ROS and RNS (reactive nitrogen species) is crucial for the integrity of metabolically highly active brain cells. Therefore, loss of CSA and CSB proteins and the consequent enhanced DNA damage accumulation is expected to lead to deleterious effects in terms of neuronal loss in addition to the impact on neurogenesis. Consistently, a recent analysis revealed that somatic mutations accumulated slowly but inexorably with age in the neurons of normal human brain, a phenomenon termed *genosenium*, which occurs even more rapidly in Cockayne syndrome [84].

CSB has a double essential function in the down modulation of p53 activity: (1) as we have seen earlier, it interacts with p53, causing the release of the essential factor p300 from p53 and the redistribution of p300 to HIF-1; (2) it promotes the ubiquitination-dependent degradation of p53, in synergy with CSA, maintaining the cellular level of p53 within physiological levels. Indeed, CSB and CSA proteins physically interact with p53 in a complex that involves a key posttranslational regulator of p53, MDM2 [32]. Accordingly, in CSA- and CSB-deficient cells, p53 level increases due to its insufficient ubiquitination and degradation [32]. *Csa*$^{-/-}$ and *Csb*$^{-/-}$ mice show elevated p53 immunoreactivity in the brain [85]. As neural precursors are concerned, p53 regulates neural precursor function in the olfactory bulb of mice at both proliferation and differentiation levels [86]. In cultures from p53-null mice, a higher number of neurospheres resulted (increased proliferation of neural stem cells) and an increased differentiation toward the neural lineage was observed [86]. As p53 is elevated in brains of *Csa*$^{-/-}$ and *Csb*$^{-/-}$ mice, it is logical to predict that an opposite neural precursor phenotype to the one observed in p53-knockout mice would be observed if the regulation of p53 by Csa and Csb is relevant for neural precursor function. The defective self-renewal observed in *Csb*$^{-/-}$ neural precursors is consistent with this prediction [49]. In addition, *Csa*$^{-/-}$ and *Csb*$^{-/-}$ mice reproduce the major aspects of CS neuropathology, albeit in a mild form, which may explain the absence of macroscopic neuropathological and obvious neurological deficits associated with CS in man. The discrepancies between human and rodents may follow from differences in the rate of production and type of DNA lesions caused by endogenous metabolites, and from the shorter lifespan of mice. Indeed, complete abolishment of NER functions, as obtained in *Csb*$^{-/-}$; *Xpa*$^{-/-}$ mice, strongly affects neuronal maintenance and survival, and more closely mimics the neurological abnormalities exhibited by CS patients [87].

7.6 Concluding remarks

CS is a devastating neurodevelopmental disorder, with growth abnormalities and progeroid features. A deep understanding of the mechanistic basis of CS neurologic disease is urgently needed. Although investigations into the mechanisms of TC-NER continue to produce important insights into the role played by CS proteins in DNA repair, pioneering studies on CS abnormalities in either basal or activated transcription started to provide a better explanation of many neurologic aspects of CS. Homeostasis of the brain tissue requires the replenishment of neurons as well as the protection of various brain cell types against endogenous and exogenous cellular stresses, including oxidative DNA damage. CSA and CSB proteins play a multifaceted role in DNA repair and in the activation of transcriptional programs regulating cellular stress response (oxidative attack and hypoxia) and neural differentiation. Consequently, reduced or absent functionality of either of these proteins may result in defective neurogenesis and also in a reduced protection of the functional integrity of progenitor and differentiated neural cells. All together these defects contribute to the dramatic complex phenotype exhibited by CS patients. Moreover, this picture may account for the large heterogeneity of the phenotypes displayed by CS patients in terms of both severity and time of onset. In the future, molecular mechanistic studies will certainly generate further important insights, which will contribute to the full understanding of all aspects of CS and eventually to the opening of new therapeutic strategies.

References

[1] Cockayne EA. Dwarfism with retinal atrophy and deafness. Arch Dis Child 1936;11(61):1–8.

[2] Nance MA, Berry SA. Cockayne syndrome: review of 140 cases. Am J Med Genet 1992;42(1):68–84.

[3] Laugel V, Dalloz C, Tobias ES, Tolmie JL, Martin-Coignard D, Drouin-Garraud V, Valayannopoulos V, Sarasin A, Dollfus H. Cerebro-oculo-facio-skeletal syndrome: three additional cases with CSB mutations, new diagnostic criteria and an approach to investigation. J Med Genet 2008;45(9):564–71.

[4] Wilson BT, Stark Z, Sutton RE, Danda S, Ekbote AV, Elsayed SM, Gibson L, Goodship JA, Jackson AP, Keng WT, King MD, McCann E, Motojima T, Murray JE, Omata T, Pilz D, Pope K, Sugita K, White SM, Wilson IJ. The Cockayne Syndrome Natural History (CoSyNH) study: clinical findings in 102 individuals and recommendations for care. Genet Med 2016;18(5):483–93.

[5] Natale V. A comprehensive description of the severity groups in Cockayne syndrome. Am J Med Genet 2011;155A(5):1081–95.

[6] Laugel V. Cockayne syndrome: the expanding clinical and mutational spectrum. Mech Ageing Dev 2013; 134(5–6):161–70.

[7] Kleijer WJ, Laugel V, Berneburg M, Nardo T, Fawcett H, Gratchev A, Jaspers NG, Sarasin A, Stefanini M, Lehmann AR. Incidence of DNA repair deficiency disorders in western Europe: xeroderma pigmentosum, Cockayne syndrome and trichothiodystrophy. DNA Repair 2008;7(5):744–50.

[8] Vélez-Cruz R, Egly JM. Cockayne syndrome group B (CSB) protein: at the crossroads of transcriptional networks. Mech Ageing Dev 2013;134(5–6):234–42.

[9] Chatre L, Biard DS, Sarasin A, Ricchetti M. Reversal of mitochondrial defects with CSB-dependent serine protease inhibitors in patient cells of the progeroid Cockayne syndrome. Proc Natl Acad Sci USA 2015; 112(22):E2910–9.

[10] Hinkal G, Donehower LA. How does suppression of IGF-1 signaling by DNA damage affect aging and longevity? Mech Ageing Dev 2008;129(5):243–53.

[11] Vermeij WP, Dollé ME, Reiling E, Jaarsma D, Payan-Gomez C, Bombardieri CR, Wu H, Roks AJ, Botter SM, van der Eerden BC, Youssef SA, Kuiper RV, Nagarajah B, van Oostrom CT, Brandt RM, Barnhoorn S, Imholz S, Pennings JL, de Bruin A, Gyenis Á, Pothof J, Vijg J, van Steeg H, Hoeijmakers JH. Restricted diet delays accelerated ageing and genomic stress in DNA-repair-deficient mice. Nature 2016; 537(7620):427–31.

[12] Scheibye-Knudsen M, Mitchell SJ, Fang EF, Iyama T, Ward T, Wang J, Dunn CA, Singh N, Veith S, Hasan-Olive MM, Mangerich A, Wilson MA, Mattson MP, Bergersen LH, Cogger VC, Warren A, Le Couteur DG, Moaddel R, Wilson 3rd DM, Croteau DL, de Cabo R, Bohr VA. A high-fat diet and NAD(+) activate Sirt1 to rescue premature aging in cockayne syndrome. Cell Metabol 2014;20(5):840–55.

[13] Koob M, Laugel V, Durand M, Fothergill H, Dalloz C, Sauvanaud F, Dollfus H, Namer IJ, Dietemann JL. Neuroimaging in cockayne syndrome. AJNR Am J Neuroradiol 2010;31(9):1623–30.

[14] Rapin I, Weidenheim K, Lindenbaum Y, Rosenbaum P, Merchant SN, Krishna S, Dickson DW. Cockayne syndrome in adults: review with clinical and pathologic study of a new case. J Child Neurol 2006;21(11): 991–1006.

[15] Gitiaux C, Blin-Rochemaure N, Hully M, Echaniz-Laguna A, Calmels N, Bahi-Buisson N, Desguerre I, Dabaj I, Wehbi S, Quijano-Roy S, Laugel V. Progressive demyelinating neuropathy correlates with clinical severity in Cockayne syndrome. Clin Neurophysiol 2015;126(7):1435–9.

[16] Bloch-Zupan A, Rousseaux M, Laugel V, Schmittbuhl M, Mathis R, Desforges E, Koob M, Zaloszyc A, Dollfus H, Laugel V. A possible cranio-oro-facial phenotype in Cockayne syndrome. Orphanet J Rare Dis 2013;8:9.

[17] Caputo M, Frontini M, Velez-Cruz R, Nicolai S, Prantera G, Proietti-De-Santis L. The CSB repair factor is overexpressed in cancer cells, increases apoptotic resistance, and promotes tumor growth. DNA Repair 2013; 12(4):293–9.

[18] Caputo M, Balzerano A, Arisi I, D'Onofrio M, Brandi R, Bongiorni S, Brancorsini S, Frontini M, Proietti-De-Santis L. CSB ablation induced apoptosis is mediated by increased endoplasmic reticulum stress response. PLoS One 2017;12(3):e0172399.

[19] Proietti-De-Santis L, Balzerano A, Prantera G. CSB: An emerging actionable target for cancer therapy. Trends Cancer 2018;4(3):172–5.

[20] Graham Jr JR, Greenberg CR, Anyane-Yeboa K, Jaspers NGT, Raam A, Coleman DM, Ziffer DW, Busch D. COFS syndrome caused by defective nucleotide excision repair. Am J Hum Genet 1998;63:A33.

[21] Meira LB, Graham Jr JM, Greenberg CR, Busch DB, Doughty AT, Ziffer DW, Coleman DM, Savre-Train I, Friedberg EC. Manitoba aboriginal kindred with original cerebro-oculo- facio-skeletal syndrome has a mutation in the Cockayne syndrome group B (CSB) gene. Am J Hum Genet 2000;66(4):1221–8.

[22] Schalk A, Greff G, Drouot N, Obringer C, Dollfus H, Laugel V, Chelly J, Calmels N. Deep intronic variation in splicing regulatory element of the ERCC8 gene associated with severe but long-term survival Cockayne syndrome. Eur J Hum Genet 2018;26(4):527–36.

[23] Hashimoto S, Suga T, Kudo E, Ihn H, Uchino M, Tateishi S. Adult-onset neurological degeneration in a patient with Cockayne syndrome and a null mutation in the CSB gene. J Invest Dermatol 2008;128(6): 1597–9.

[24] Nardo T, Oneda R, Spivak G, Vaz B, Mortier L, Thomas P, Orioli D, Laugel V, Stary A, Hanawalt PC, Sarasin A, Stefanini M. A UV-sensitive syndrome patient with a specific CSA mutation reveals separable roles for CSA in response to UV and oxidative DNA damage. Proc Natl Acad Sci USA 2009;106(15): 6209–14.

[25] Nakazawa Y, Sasaki K, Mitsutake N, Matsuse M, Shimada M, Nardo T, Takahashi Y, Ohyama K, Ito K, Mishima H, Nomura M, Kinoshita A, Ono S, Takenaka K, Masuyama R, Kudo T, Slor H, Utani A, Tateishi S, Yamashita S, Stefanini M, Lehmann AR, Yoshiura K, Ogi T. Mutations in UVSSA cause UV-sensitive syndrome and impair RNA polymerase IIo processing in transcription-coupled nucleotide-excision repair. Nat Genet 2012;44(5):586–92.

[26] Zhang X, Horibata K, Saijo M, Ishigami C, Ukai A, Kanno S, Tahara H, Neilan EG, Honma M, Nohmi T, Yasui A, Tanaka K. Mutations in UVSSA cause UV-sensitive syndrome and destabilize ERCC6 in transcription-coupled DNA repair. Nat Genet 2012;44(5):593–7.

[27] Natale V, Raquer H. Xeroderma pigmentosum-Cockayne syndrome complex. Orphanet J Rare Dis 2017; 12(1):65.

[28] Calmels N, Greff G, Obringer C, Kempf N, Gasnier C, Tarabeux J, Miguet M, Baujat G, Bessis D, Bretones P, Cavau A, Digeon B, Doco-Fenzy M, Doray B, Feillet F, Gardeazabal J, Gener B, Julia S, Llano-Rivas I, Mazur A, Michot C, Renaldo-Robin F, Rossi M, Sabouraud P, Keren B, Depienne C, Muller J, Mandel JL, Laugel V. Uncommon nucleotide excision repair phenotypes revealed by targeted high-throughput sequencing. Orphanet J Rare Dis 2016;11:26.

[29] Fischer ES, Scrima A, Böhm K, Matsumoto S, Lingaraju GM, Faty M, Yasuda T, Cavadini S, Wakasugi M, Hanaoka F, Iwai S, Gut H, Sugasawa K, Thomä NH. The molecular basis of CRL4DDB2/CSA ubiquitin ligase architecture, targeting, and activation. Cell 2011;147(5):1024–39.

[30] Groisman R, Polanowska J, Kuraoka I, Sawada J, Saijo M, Drapkin R, Kisselev AF, Tanaka K, Nakatani Y. The ubiquitin ligase activity in the DDB2 and CSA complexes is differentially regulated by the COP9 signalosome in response to DNA damage. Cell 2003;113(3):357–67.

[31] Fousteri M, Vermeulen W, van Zeeland AA, Mullenders LH. Cockayne syndrome A and B proteins differentially regulate recruitment of chromatin remodeling and repair factors to stalled RNA polymerase II in vivo. Mol Cell 2006;23(4):471–82.

[32] Latini P, Frontini M, Caputo M, Gregan J, Cipak L, Filippi S, Kumar V, Vélez-Cruz R, Stefanini M, Proietti-De-Santis L. CSA and CSB proteins interact with p53 and regulate its Mdm2-dependent ubiquitination. Cell Cycle 2011;10(21):3719–30.

[33] Epanchintsev A, Costanzo F, Rauschendorf MA, Caputo M, Ye T, Donnio LM, Proietti-de-Santis L, Coin F, Laugel V, Egly JM. Cockayne's syndrome A and B proteins regulate transcription arrest after genotoxic stress by promoting ATF3 degradation. Mol Cell 2017;68(6):1054–66.
[34] Matson SW, Bean DW, George JW. DNA helicases: enzymes with essential roles in all aspects of DNA metabolism. Bioessays 1994;16(1):13–22.
[35] Cho I, Tsai PF, Lake RJ, Basheer A, Fan HY. ATP-dependent chromatin remodeling by Cockayne syndrome protein B and NAP1-like histone chaperones is required for efficient transcription-coupled DNA repair. PLoS Genet 2013;9(4):e1003407.
[36] Citterio E, Van Den Boom V, Schnitzler G, Kanaar R, Bonte E, Kingston RE, Hoeijmakers JH, Vermeulen W. ATP-dependent chromatin remodeling by the Cockayne syndrome B DNA repair-transcription-coupling factor. Mol Cell Biol 2000;20(20):7643–53.
[37] Selby CP, Sancar A. Human transcription-repair coupling factor CSB/ERCC6 is a DNA-stimulated ATPase but is not a helicase and does not disrupt the ternary transcription complex of stalled RNA polymerase II. J Biol Chem 1997;272(3):1885–90.
[38] Berquist BR, Canugovi C, Sykora P, Wilson 3rd DM, Bohr VA. Human Cockayne syndrome B protein reciprocally communicates with mitochondrial proteins and promotes transcriptional elongation. Nucleic Acids Res 2012;40(17):8392–405.
[39] Stevnsner T, Muftuoglu M, Aamann MD, Bohr VA. The role of Cockayne Syndrome group B (CSB) protein in base excision repair and aging. Mech Ageing Dev 2008;129(7–8):441–8.
[40] Iyama T, Lee SY, Berquist BR, Gileadi O, Bohr VA, Seidman MM, McHugh PJ, Wilson 3rd DM. CSB interacts with SNM1A and promotes DNA interstrand crosslink processing. Nucleic Acids Res 2015;43(1): 247–58.
[41] Batenburg NL, Thompson EL, Hendrickson EA, Zhu XD. Cockayne syndrome group B protein regulates DNA double-strand break repair and checkpoint activation. EMBO J 2015;34(10):1399–416.
[42] Wei L, Nakajima S, Böhm S, Bernstein KA, Shen Z, Tsang M, Levine AS, Lan L. DNA damage during the G0/G1 phase triggers RNA-templated, Cockayne syndrome B-dependent homologous recombination. Proc Natl Acad Sci USA 2015;112(27):E3495–504.
[43] Mayne LV, Lehmann AR. Failure of RNA synthesis to recover after UV irradiation: an early defect in cells from individuals with Cockayne's syndrome and xeroderma pigmentosum. Cancer Res 1982;42(4): 1473–8.
[44] Troelstra C, van Gool A, de Wit J, Vermeulen W, Bootsma D, Hoeijmakers JH. ERCC6, a member of a subfamily of putative helicases, is involved in Cockayne's syndrome and preferential repair of active genes. Cell 1992;71(6):939–53.
[45] Wilson 3rd DM, Bohr VA. Special issue on the segmental progeria Cockayne syndrome. Mech Ageing Dev 2013;134(5–6):159–60.
[46] Groisman R, Kuraoka I, Chevallier O, Gaye N, Magnaldo T, Tanaka K, Kisselev AF, Harel-Bellan A, Nakatani Y. CSA-dependent degradation of CSB by the ubiquitin-proteasome pathway establishes a link between complementation factors of the Cockayne syndrome. Genes Dev 2006;20(11):1429–34.
[47] Sarasin A, Stary A. New insights for understanding the transcription-coupled repair pathway. DNA Repair 2007;6(2):265–9.
[48] Laposa RR, Huang EJ, Cleaver JE. Increased apoptosis, p53 up-regulation, and cerebellar neuronal degeneration in repair-deficient Cockayne syndrome mice. Proc Natl Acad Sci USA 2007;104(4): 1389–94.
[49] Ciaffardini F, Nicolai S, Caputo M, Canu G, Paccosi E, Costantino M, Frontini M, Balajee AS, Proietti-De-Santis L. The cockayne syndrome B protein is essential for neuronal differentiation and neuritogenesis. Cell Death Dis 2014;5:e1268.

[50] Kristensen U, Epanchintsev A, Rauschendorf MA, Laugel V, Stevnsner T, Bohr VA, Coin F, Egly JM. Regulatory interplay of Cockayne syndrome B ATPase and stress-response gene ATF3 following genotoxic stress. Proc Natl Acad Sci USA 2013;110(25):E2261−70.
[51] Selby CP, Sancar A. Cockayne syndrome group B protein enhances elongation by RNA polymerase II. Proc Natl Acad Sci USA 1997;94(21):11205−9.
[52] Balajee AS, May A, Dianov GL, Friedberg EC, Bohr VA. Reduced RNA polymerase II transcription in intact and permeabilized Cockayne syndrome group B cells. Proc Natl Acad Sci USA 1997;94(9): 4306−11.
[53] Tantin D, Kansal A, Carey M. Recruitment of the putative transcription-repair coupling factor CSB/ERCC6 to RNA polymerase II elongation complexes. Mol Cell Biol 1997;17(12):6803−14.
[54] van Gool AJ, Citterio E, Rademakers S, van Os R, Vermeulen W, Constantinou A, Egly JM, Bootsma D, Hoeijmakers JH. The Cockayne syndrome B protein, involved in transcription-coupled DNA repair, resides in an RNA polymerase II-containing complex. EMBO J 1997;16(19):5955−65.
[55] Bradsher J, Auriol J, Proietti de Santis L, Iben S, Vonesch JL, Grummt I, Egly JM. CSB is a component of RNA pol I transcription. Mol Cell 2002;10(4):819−29.
[56] Yuan X, Feng W, Imhof A, Grummt I, Zhou Y. Activation of RNA polymerase I transcription by cockayne syndrome group B protein and histone methyltransferase G9a. Mol Cell 2007;27(4):585−95.
[57] Proietti-De-Santis L, Drané P, Egly JM. Cockayne syndrome B protein regulates the transcriptional program after UV irradiation. EMBO J 2006;25(9):1915−23.
[58] McKinnon PJ. DNA repair deficiency and neurological disease. Nat Rev Neurosci February 2009;10(2): 100−12. https://doi.org/10.1038/nrn2559. Epub 2009 Jan 15. Review. Erratum in: Nat Rev Neurosci 2009; 10(3):242.
[59] Aamann MD, Muftuoglu M, Bohr VA, Stevnsner T. Multiple interaction partners for Cockayne syndrome proteins: implications for genome and transcriptome maintenance. Mech Ageing Dev 2013;134(5−6): 212−24.
[60] Chatterjee S, Mizar P, Cassel R, Neidl R, Selvi BR, Mohankrishna DV, Vedamurthy BM, Schneider A, Bousiges O, Mathis C, Cassel JC, Eswaramoorthy M, Kundu TK, Boutillier AL. A novel activator of CBP/p300 acetyltransferases promotes neurogenesis and extends memory duration in adult mice. J Neurosci 2013; 33(26):10698−712.
[61] Nicolai S, Filippi S, Caputo M, Cipak L, Gregan J, Ammerer G, Frontini M, Willems D, Prantera G, Balajee AS, Proietti-De-Santis L. Identification of novel proteins Co purifying with cockayne syndrome group B (CSB) reveals potential roles for CSB in RNA metabolism and chromatin dynamics. PLoS One 2015;10(6):e0128558.
[62] Yoo AS, Crabtree GR. ATP-dependent chromatin remodeling in neural development. Curr Opin Neurobiol 2009;19(2):120−6.
[63] Wu JI, Lessard J, Olave IA, Qiu Z, Ghosh A, Graef IA, Crabtree GR. Regulation of dendritic development by neuron-specific chromatin remodeling complexes. Neuron 2007;56(1):94−108.
[64] Wong JC, Visanji NP, Dabek MK, Laposa RR, Hazrati LN. Dendritic spine density is altered in a mouse model of Cockayne syndrome. Neuropathol Appl Neurobiol 2013;39(4):437−40.
[65] Vessoni AT, Herai RH, Karpiak JV, Leal AM, Trujillo CA, Quinet A, Agnez Lima LF, Menck CF, Muotri AR. Cockayne syndrome-derived neurons display reduced synapse density and altered neural network synchrony. Hum Mol Genet 2016;25(7):1271−80.
[66] van der Pluijm I, Garinis GA, Brandt RM, Gorgels TG, Wijnhoven SW, Diderich KE, de Wit J, Mitchell JR, van Oostrom C, Beems R, Niedernhofer LJ, Velasco S, Friedberg EC, Tanaka K, van Steeg H, Hoeijmakers JH, van der Horst GT. Impaired genome maintenance suppresses the growth hormone−insulin-like growth factor 1 axis in mice with Cockayne syndrome. PLoS Biol 2007;5(1):e2. Erratum in: PLoS Biol 2008;6(11):e304.

[67] Scheibye-Knudsen M, Ramamoorthy M, Sykora P, Maynard S, Lin PC, Minor RK, Wilson 3rd DM, Cooper M, Spencer R, de Cabo R, Croteau DL, Bohr VA. Cockayne syndrome group B protein prevents the accumulation of damaged mitochondria by promoting mitochondrial autophagy. J Exp Med 2012;209(4): 855–69.

[68] Semenza GL. Targeting HIF-1 for cancer therapy. Nat Rev Canc 2003;3(10):721–32.

[69] Pugh CW, Ratcliffe PJ. Regulation of angiogenesis by hypoxia: role of the HIF system. Nat Med 2003;9(6): 677–84.

[70] Arany Z, Huang LE, Eckner R, Bhattacharya S, Jiang C, Goldberg MA, Bunn HF, Livingston DM. An essential role for p300/CBP in the cellular response to hypoxia. Proc Natl Acad Sci USA 1996;93(23): 12969–73.

[71] Harris AL. Hypoxia–a key regulatory factor in tumour growth. Nat Rev Canc 2002;2(1):38–47.

[72] Beckerman R, Prives C. Transcriptional regulation by p53. Cold Spring Harb Perspect Biol 2010;2(8): a000935.

[73] Manni I, Mazzaro G, Gurtner A, Mantovani R, Haugwitz U, Krause K, Engeland K, Sacchi A, Soddu S, Piaggio G. NF-Y mediates the transcriptional inhibition of the cyclin B1, cyclin B2, and cdc25C promoters upon induced G2 arrest. J Biol Chem 2001;276(8):5570–6.

[74] Filippi S, Latini P, Frontini M, Palitti F, Egly JM, Proietti-De-Santis L. CSB protein is (a direct target of HIF-1 and) a critical mediator of the hypoxic response. EMBO J 2008;27(19):2545–56.

[75] Frontini M, Proietti-De-Santis L. Cockayne syndrome B protein (CSB): linking p53, HIF-1 and p300 to robustness, lifespan, cancer and cell fate decisions. Cell Cycle 2009;8(5):693–6.

[76] Tu SP, Chi AL, Ai W, Takaishi S, Dubeykovskaya Z, Quante M, Fox JG, Wang TC. p53 inhibition of AP1-dependent TFF2 expression induces apoptosis and inhibits cell migration in gastric cancer cells. Am J Physiol Gastrointest Liver Physiol 2009;297(2):G385–96.

[77] Guzy RD, Schumacker PT. Oxygen sensing by mitochondria at complex III: the paradox of increased reactive oxygen species during hypoxia. Exp Physiol 2006;91(5):807–19.

[78] Aamann MD, Sorensen MM, Hvitby C, Berquist BR, Muftuoglu M, Tian J, de Souza-Pinto NC, Scheibye-Knudsen M, Wilson 3rd DM, Stevnsner T, Bohr VA. Cockayne syndrome group B protein promotes mitochondrial DNA stability by supporting the DNA repair association with the mitochondrial membrane. FASEB J 2010;24(7):2334–46.

[79] Kamenisch Y, Berneburg M. Mitochondrial CSA and CSB: protein interactions and protection from ageing associated DNA mutations. Mech Ageing Dev 2013;134(5–6):270–4.

[80] Cleaver JE, Brennan-Minnella AM, Swanson RA, Fong KW, Chen J, Chou KM, Chen YW, Revet I, Bezrookove V. Mitochondrial reactive oxygen species are scavenged by Cockayne syndrome B protein in human fibroblasts without nuclear DNA damage. Proc Natl Acad Sci USA 2014;111(37):13487–92.

[81] Osenbroch PØ, Auk-Emblem P, Halsne R, Strand J, Forstrøm RJ, van der Pluijm I, Eide L. Accumulation of mitochondrial DNA damage and bioenergetic dysfunction in CSB defective cells. FEBS J 2009;276(10): 2811–21.

[82] Pascucci B, Lemma T, Iorio E, Giovannini S, Vaz B, Iavarone I, Calcagnile A, Narciso L, Degan P, Podo F, Roginskya V, Janjic BM, Van Houten B, Stefanini M, Dogliotti E, D'Errico M. An altered redox balance mediates the hypersensitivity of Cockayne syndrome primary fibroblasts to oxidative stress. Aging Cell 2012;11(3):520–9.

[83] Scheibye-Knudsen M, Croteau DL, Bohr VA. Mitochondrial deficiency in Cockayne syndrome. Mech Ageing Dev 2013;134(5–6):275–83.

[84] Lodato MA, Rodin RE, Bohrson CL, Coulter ME, Barton AR, Kwon M, Sherman MA, Vitzthum CM, Luquette LJ, Yandava CN, Yang P, Chittenden TW, Hatem NE, Ryu SC, Woodworth MB, Park PJ, Walsh CA. Aging and neurodegeneration are associated with increased mutations in single human neurons. Science 2018;359(6375):555–9.

[85] Jaarsma D, van der Pluijm I, de Waard MC, Haasdijk ED, Brandt R, Vermeij M, Rijksen Y, Maas A, van Steeg H, Hoeijmakers JH, van der Horst GT. Age-related neuronal degeneration: complementary roles of nucleotide excision repair and transcription-coupled repair in preventing neuropathology. PLoS Genet 2011; 7(12):e1002405.

[86] Armesilla-Diaz A, Bragado P, Del Valle I, Cuevas E, Lazaro I, Martin C, Cigudosa JC, Silva A. p53 regulates the self-renewal and differentiation of neural precursors. Neuroscience 2009;158(4):1378–89.

[87] Jaarsma D, van der Pluijm I, van der Horst GT, Hoeijmakers JH. Cockayne syndrome pathogenesis: lessons from mouse models. Mech Ageing Dev 2013;134(5–6):180–95.

CHAPTER

8 Epigenetic processes in Alzheimer's disease

Isabel Castanho, Katie Lunnon

University of Exeter Medical School, University of Exeter, Devon, United Kingdom

Chapter Outline

8.1 Alzheimer's disease: a need for new drug targets 153
8.2 Alzheimer's disease: the genomic era 155
8.3 An additional layer of information: Alzheimer's disease from an epigenetic perspective 156
8.3.1 DNA modifications 157
8.3.2 Histone modifications 160
8.3.3 Regulatory RNA–based mechanisms 162
8.3.4 Epigenetic signatures as blood biomarkers 164
8.4 Modeling Alzheimer's disease: mouse models as powerful tools 166
8.5 Current challenges and future directions 169
8.6 Final considerations 171
References 172

8.1 Alzheimer's disease: a need for new drug targets

Alzheimer's disease (AD), first described by the German psychiatrist Alois Alzheimer in 1898 [1,2], is the most common type of dementia. AD is characterized by memory loss, cognitive decline, behavioral changes, and, ultimately, death [3], which are consequences of chronic progressive neurodegeneration and synaptic loss [4,5]. AD affects more than 5 million Americans and about 26 million people worldwide [6]. The high prevalence of AD is not only devastating for affected families but also represents a high economic burden for the nation as the cost of caring for patients with AD or other dementias was estimated to be $259 billion in 2017 in the United States alone [7]. The neuropathology of AD is believed to start decades before the manifestation of clinical symptoms [8]. Although potentially promising new drugs are currently being tested in clinical trials, at present, there are no available disease-modifying treatments, and the existing medications simply alleviate the symptoms of AD [7].

Patients affected by AD display brain atrophy, which is restricted to the cortical and limbic regions, accompanied by enlargement of the ventricles, as a consequence of differential neurodegeneration in these brain regions compared with other brain areas [3] (Fig. 8.1). Moreover, AD brains are characterized by two key histopathologic hallmarks of disease neuropathology:

Chromatin Signaling and Neurological Disorders. https://doi.org/10.1016/B978-0-12-813796-3.00008-0

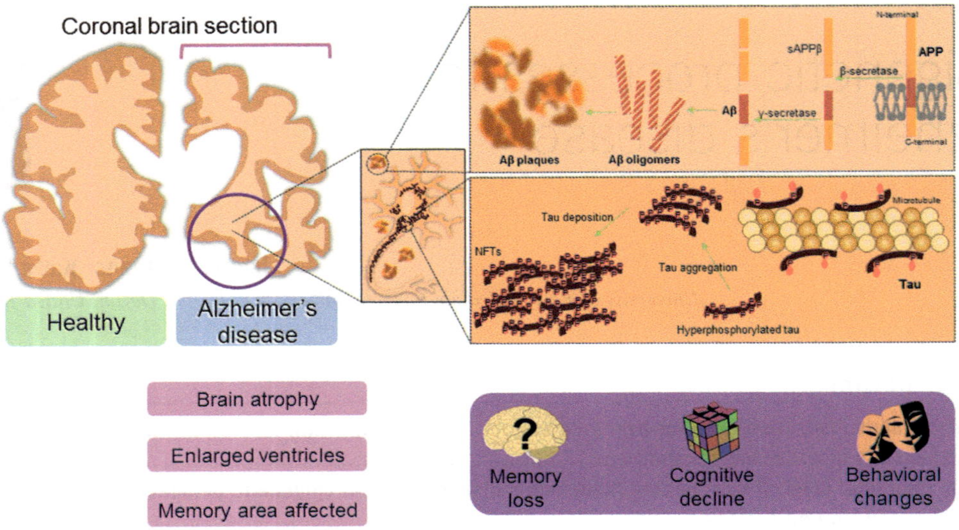

FIGURE 8.1 Pathologic hallmarks of AD.

On a microscopic level, AD is characterized by the presence of extracellular amyloid beta (Aβ) plaques and intracellular neurofibrillary tangles (NFTs). On a macroscopic level the AD-affected brain shows extensive brain atrophy of limbic and cortical regions and ventricular enlargement. *APP*, amyloid precursor protein.

extracellular amyloid beta (Aβ) plaques and intracellular neurofibrillary tangles (NFTs) [9,10]. Amyloid plaques are aggregates of Aβ, a protein fragment that results from the amyloidogenic cleavage of amyloid precursor protein (APP), a transmembrane protein suggested to be involved in synapse formation [9]. The tangles are deposits of hyperphosphorylated and aggregated microtubule-associated protein tau (MAPT), a protein with important functions in neurons, particularly in axonal growth and transport [10]. The microscopic presence of these hallmarks is used for the definitive postmortem diagnosis of AD. Other characteristic features of AD include neuronal cell death and the activation of astrocytes and microglia [3]. Notably, some regions of the human brain are particularly affected by AD, whereas other brain regions seem to be relatively resistant. The hippocampus and the entorhinal cortex, for example, both involved in memory formation and recall, are particularly affected, from the early stages of disease, while the cerebellum is relatively spared from the pathologic condition [11]. Considering this, AD can be categorized into different stages according to the spread of NFTs, which are known as Braak stages [12,13]. In the initial stages of the disease (I and II), tau pathology starts in the transentorhinal and entorhinal regions, followed by the limbic system and adjoining neocortex in stages III and IV, before later spreading to other neocortical regions in stages V and VI. As the neuropathology precedes clinical symptoms, currently the disease can only be clinically diagnosed relatively late in the disease process, with a definitive diagnosis only being possible post mortem.

8.2 Alzheimer's disease: the genomic era

AD can be divided into two subtypes: familial AD (FAD) and sporadic AD (SAD). FAD is usually early-onset AD (EOAD), generally occurring in individuals <65 years, whereas SAD is mainly late-onset AD (LOAD), affecting people >65 years (Fig. 8.2). FAD accounts for less than 5% of disease incidence and is caused by one or more autosomal dominant mutations in the gene that encodes for APP, or in the genes that encode for presenilin-1 or -2 (*PSEN1* or *PSEN2*), which are components of the γ-secretase enzyme that is involved in the cleavage of APP [14]. Although the role of genetics in SAD is less well defined than that in FAD, a heritable component has been recognized, exemplified by the larger monozygotic twin concordance rate for the disease than the dizygotic twin concordance rate [15,16].

To date, the greatest genetic risk factor identified for SAD is in the apolipoprotein E (*APOE*) gene, more specifically the presence of the ε4 allele [17,18]. However, many common low-penetrance genetic variants have been shown to be involved in the disease, suggesting that SAD may have a polygenic component in which many common variants, each of small effect, are likely to influence disease. In addition to *APOE*, several large genome-wide association studies (GWASs) and subsequent meta-analyses have identified numerous common single nucleotide polymorphisms at

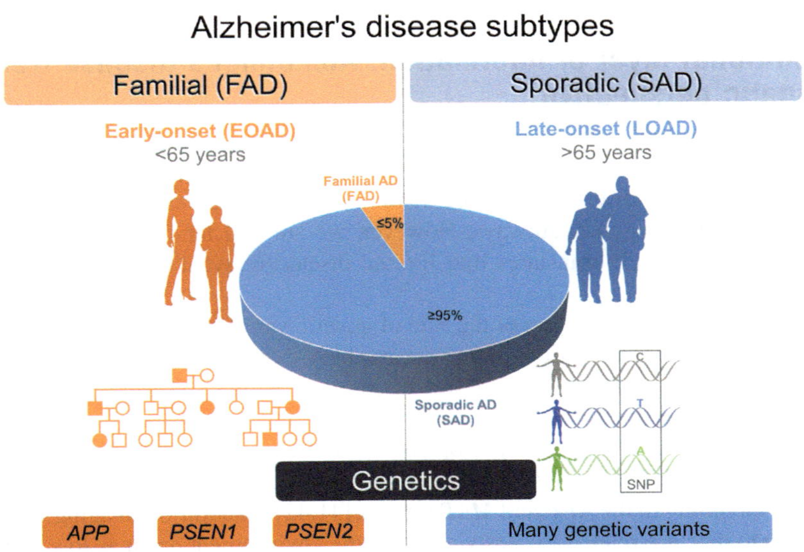

FIGURE 8.2 Alzheimer's disease (AD) subtypes and their genetics.

Familial AD (FAD) accounts for <5% of AD cases and is caused by autosomal dominant mutations in the amyloid precursor protein (APP), PSEN1, or PSEN2 genes. Sporadic AD (SAD) accounts for >95% of AD cases and many common genetic variants have been associated with this type of disease. *SNP*, single nucleotide polymorphism.

specific loci, which infer a relatively small risk for disease. Results from these studies have identified variants in numerous loci, including clusterin (*CLU*), phosphatidylinositol-binding clathrin assembly protein (*PICALM*), complement receptor 1 (*CR1*) [19–24], bridging integrator 1 (*BIN1*), cluster of differentiation 2–associated protein (*CD2AP*), cluster of differentiation 33 (*CD33*), erythropoietin-producing hepatoma receptor A1 (*EPHA1*), and ATP-binding cassette subfamily A member 7 (*ABCA7*) [25,26]. Through advances in genomic sequencing technology, researchers have also identified a number of rarer variants with a potential medium disease risk, including the genes for triggering receptors expressed on myeloid cells 2 protein (*TREM2*) [27,28], ABI family 3 (*ABI3*) [28], and phospholipase D3 (*PLD3*) [29]. Interestingly, overall, the variants identified from genomic studies suggest specific biological pathways are altered in AD, such as synaptic pathways, immune regulation, cholesterol transport and lipid metabolism, endocytosis, ubiquitination, and protein folding [30]. Recently, three large studies (GWAS and meta-analyses) have provided further support for a role of the immune system and lipid metabolism in AD by confirming the association of previously identified loci with AD. More importantly, these reports collectively identified 15 novel loci, of which one, ADAM metallopeptidase domain 10 (*ADAM10*), was common to all three studies [31–33]. Additionally, AD has indeed been shown to have a significant polygenic component [34,35]. Of note, many of the risk variants identified in GWAS actually reside in noncoding regions of the genome [14], possibly within regulatory regions, for which their functional significance is not entirely understood.

8.3 An additional layer of information: Alzheimer's disease from an epigenetic perspective

Although the neuropathologic indications of AD have now been well characterized in postmortem brain tissue from human patients, little is still known about the specific mechanisms involved in the onset and progression of the disease. The aforementioned genetic studies have identified a number of novel molecular mechanisms and pathways. However, we are still far from fully understanding the cause of AD, which ultimately translates into limited disease-relevant targets for developing novel pharmacologic treatments.

Despite the increased understanding of the role of genetic variation in AD in the past few decades, recent studies have highlighted that common variants only explain about 33% of disease incidence [36], with much of the remaining heritability for disease still being unclear. A number of features of AD are suggestive of an epigenetic contribution to the cause of the disease, for example, monozygotic twin discordance [16,37] for both the development of disease [15] and age of onset [38]. Aging, considered the greatest risk factor for AD, has been shown to be associated with significant epigenetic changes [39,40], including in the *APP* and *MAPT* genes [41,42]. In addition to aging, most risk factors identified for AD to date are related to lifestyle and/or the environment, such as low education levels, hypertension, obesity, smoking, depression and/or stress, physical inactivity, social isolation, and diabetes [6,7]. Taken together, this suggests that a combination of both genetic and epigenetic factors is likely involved in AD.

Epigenetic processes that control gene regulation can be divided into three major categories: (1) DNA modifications, (2) histone modifications, and (3) RNA-based mechanisms (Fig. 8.3), which together compose the epigenome. Epigenetic mechanisms regulate gene activation/repression in a

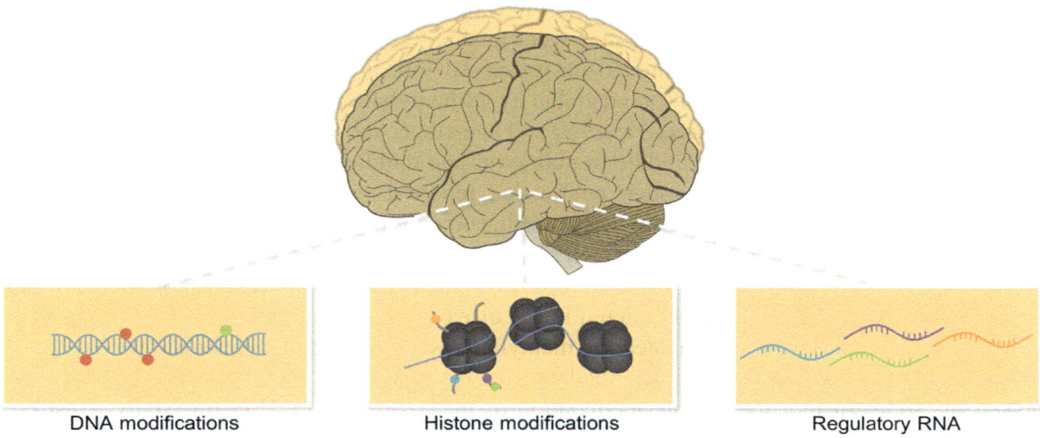

FIGURE 8.3 Epigenetic processes that potentially play a role in Alzheimer's disease.

Epigenetic mechanisms can be divided into three categories: DNA modifications, histone modifications, and RNA-based mechanisms.

dynamic and reversible fashion, independent of genetic variation, and have been suggested to play a role in AD [43–45]. Targeting disease-associated epigenetic changes makes them potentially attractive therapeutic options for the treatment of AD.

8.3.1 DNA modifications

Methylation of DNA at the fifth carbon of the cytosine nucleotide is the most studied epigenetic modification, mainly because of its relative stability as an epigenetic modification and because numerous robust methods have been developed for its evaluation. Accordingly, it has been the primary focus of most AD-related epigenetic studies to date [46]. DNA methylation in vertebrates corresponds to the covalent addition of a methyl group to cytosine, which is established and maintained by DNA methyltransferases (DNMTs), resulting in the formation of 5-methylcytosine (5-mC). It can be demethylated by oxidation into 5-hydroxymethylcytosine (5-hmC) by the action of the ten–eleven translocation (TET) family of enzymes [47].

Many epigenetic studies of AD have used immunoreactive approaches to evaluate global DNA methylation, with varying results. Some studies have identified disease-associated hypomethylation, other studies have reported hypermethylation, and others have found no changes in AD samples [37,48–53]. Perhaps these discrepancies arose because different studies explored distinct brain regions and/or different types of AD. Mastroeni et al. [37] initially investigated brain samples from monozygotic twins discordant for AD, revealing reduced levels of global 5-mC (hypomethylation) in the temporal neocortex in the twin with AD. The same authors later reported hypomethylation in the entorhinal cortex of AD patients when compared with cognitively and neurologically healthy individuals [53]. In a follow-up study, Chouliaras and colleagues [49] also reported reduced levels of 5-mC and 5-hmC in the hippocampus of AD patients when compared with controls. Condliffe et al. [50]

detected decreased global 5-hmC levels in both the entorhinal cortex and cerebellum in AD patients when compared with controls; however, they were unable to detect any differences in 5-mC levels. These results contrast with other studies that observed global hypermethylation or no changes. Bradley-Whitman and Lovell [48] observed global hypermethylation and hyperhydroxymethylation in the hippocampus of patients with preclinical AD and late-stage AD when compared with age-matched controls, a finding replicated in the middle frontal gyrus and middle temporal gyrus of AD patients by Coppieters and colleagues [51]. Another publication reported no differences in 5-mC and 5-hmC levels in the entorhinal cortex in AD cases when compared with controls [52]. It is, however, important to note that immunoassays, particularly immunohistochemistry, are only able to evaluate DNA modifications at a global level; as such, it provides no insight into the exact biological pathways or specific loci that are epigenetically altered in the disease.

A number of studies have investigated the methylation status of specific candidate genes. The first published study assessing epigenetic dysregulation in the context of AD used enzymatic digestion followed by Southern blot to evaluate methylation of the *APP* gene [54]. In this study, West et al. compared an AD patient with a nondemented subject and reported the absence of DNA methylation at a site in the *APP* gene in the temporal lobe of the AD-affected individual. A candidate-based study used pyrosequencing to measure aberrant CpG DNA methylation in the *APP, MAPT,* and glycogen synthase kinase 3 beta (*GSK3B*) genes in postmortem AD brain samples [55]. Interestingly, Iwata et al. suggest that the hypermethylation and hypomethylation changes they identified in *APP* and *MAPT,* respectively, were associated with increased expression of both genes. In another publication, gene-specific DNA methylation of *MAPT* was evaluated by pyrosequencing in FAD brains of patients bearing a mutation in *PSEN1* (which results in higher production and accumulation of Aβ), revealing no changes in DNA methylation in the visual cortex, but a significant decrease in methylation levels in the cerebellum of AD patients when compared with age-matched controls [56]. One study used real-time polymerase chain reaction (RT-PCR) to measure gene-specific changes in DNA methylation in postmortem frontal cortex of AD patients when compared with age-matched controls [57]. The authors assessed the methylation status of the promoter regions for a number of genes, reporting hypomethylation of cyclooxygenase 2 (*PTGS2*) and nuclear transcription factor kappa B (*NF-κB*) and hypermethylation of brain-derived neurotrophic factor (*BDNF*), transcription factor cAMP response element-binding protein (*CREB*), and synaptic protein synaptophysin (*SYP*). DNA methylation changes in transposable elements (repetitive elements) such as Alu sequences and long interspersed nuclear elements have also been suggested to be altered in AD [58]. However, studies exploring this relationship are very limited in number and further investigations are necessary, particularly in brain tissues.

The first study to take a more systematic genome-wide approach used the Illumina 27K methylation array to assess DNA methylation at >27,000 CpG sites in the prefrontal cortex of patients with LOAD and compared the results with control individuals [59]. The promoter region of transmembrane protein 59 (*TMEM59*) was the CpG site that was most strongly associated with AD status, and gene ontology analysis suggested hypermethylation of genes related to transcription and DNA replication and hypomethylation of membrane transporters. When the authors checked the methylation status of specific candidate genes, they also reported hypomethylation in one site of the *PSEN1* gene in AD cases compared with controls.

The successor to the 27K array, the Illumina 450K methylation array, has more recently been used to assess DNA methylation at >485,000 CpG sites. Two independent epigenome-wide association studies (EWAS) were published simultaneously in 2014 and they used this method to compare DNA methylation profiles in postmortem brain samples from AD cases [60,61]. Both studies highlighted neuropathology-associated differential methylation in ankyrin 1 (*ANK1*), ribosomal protein L1 (*RPL13*), rhomboid 5 homolog 2 (*RHBDF2*), and cadherin-related family member 23 (*CDH23*) [62]. Disease-associated *ANK1* hypermethylation was observed in cortical samples from four independent study cohorts, and interestingly, the brain regions most highly affected in the disease (e.g., the entorhinal cortex, superior temporal gyrus, and frontal cortex) showed robust differences, whereas the cerebellum and premortem blood showed no alterations in *ANK1* [61]. The study by De Jager and colleagues [60] also highlighted two additional genes, *ABCA7* and *BIN1*, that overlapped with the previous genetic studies. A follow-up study assessing levels of DNA methylation in known AD risk genes within this data-set highlighted neuropathology-associated differential DNA methylation in sortilin-related receptor 1 (*SORL1*), *ABCA7*, major histocompatibility complex class II DR beta 5 (*HLA-DRB5*), solute carrier family 24 member 4 (*SLC24A4*), and *BIN1* [63]. A follow-up study, using the Lunnon et al. data set, highlighted AD-associated hypermethylation at a CpG site within the *TREM2* gene in the superior temporal gyrus, which was validated in two other cohorts [64]. Other AD EWAS have been published; Watson et al. [65] identified 479 AD-associated differential methylated regions (DMRs) in the superior temporal gyrus, the majority of which were hypermethylated in AD cases compared with nondemented controls and enriched for brain-specific histone signatures and for binding motifs of transcription factors, and found a significant correlation between their data set and the Lunnon et al. data set, while Smith et al. [66] highlighted AD-associated hypermethylation within the homeobox A (*HOXA*) gene cluster in the frontal cortex and superior temporal gyrus, which spanned 48 kb. Another study has shown that *ANK1* DNA methylation changes in the entorhinal cortex seem to be specific to only certain neurodegenerative diseases, with hypermethylation observed in AD and Huntington disease (HD) and to a lesser extent in Parkinson disease (PD). Interestingly, in individuals with vascular dementia (VaD) or dementia with Lewy bodies (DLB), disease-associated hypermethylation was only observed in individuals with coexisting AD [67].

One issue with all the EWAS data sets published to date is that they have utilized sodium bisulfite conversion to enable quantification of DNA methylation levels. This method converts unmodified cytosines to uracil, while both methylated and hydroxymethylated cytosines are protected from the conversion, and the presence of a DNA modification can be determined in downstream analyses as a difference in the sequence between treated and untreated DNA. As such, they cannot differentiate between DNA methylation and hydroxymethylation, and the published EWAS data actually represents a sum of these two modifications. An adaptation to the bisulfite conversion protocol, where the DNA is oxidized before bisulfite treatment, converts 5-hmC to 5-formylcytosine (which is converted to uracil), meaning that data generated from oxidative bisulfite treated DNA gives a measure of "true" DNA methylation (5-mC). Furthermore, by performing bisulfite and oxidative bisulfite treatment in parallel, one can calculate 5-hmC levels by subtracting 5-mC levels from total modifications. In conjunction with array- or sequencing-based assays, this allows for the evaluation of 5-mC and 5-hmC in a genome-wide fashion [68,69], although this method has not yet been applied to the study of AD brain samples. Investigating 5-hmC specifically in the context of dementia is particularly exciting, given that 5-hmC has been shown to be enriched in the brain (particularly in synaptic genes [70]), is found at different levels across different brain regions [68], and is altered during brain development [71]. Only one study has aimed to profile DNA hydroxymethylation at the genome-wide level in AD brains [69]. This study used a biotin tag capturing method followed by high-throughput sequencing and reported both hyperhydroxymethylation and hypohydroxymethylation changes in the dorsolateral prefrontal cortex from patients with a high burden of AD pathology compared with individuals with a low burden of AD neuropathology. The study nominated ~300 loci that were associated with high amyloid burden, 50 loci that were associated with high tau neuropathology, and 4 genes that were associated with both amyloid and tau pathologic conditions: 4-aminobutyrate aminotransferase (*ABAT*), calcium/calmodulin-dependent protein kinase (*CAMK1D*), high-temperature requirement factor A3 (*HTRA3*), and leucine-rich repeat neuronal 1 (*LRRN1*). Using network analyses to explore cohydroxymethylation patterns, Zhao and colleagues identified dysregulation of modules containing genes involved in ion channel activity, neuron differentiation and development, neurotransmitter transport, glutamatergic synapse, and calcium ion–dependent exocytosis of neurotransmitter, to name but a few examples. However, the authors do highlight that the low resolution of the sequencing used in their study means they might not have been able to discriminate 5-hmC from 5-mC [69]. As such, additional studies using the oxidative bisulfite treatment approach are eagerly awaited by the field.

One issue with the Illumina 450K array and its successor, the EPIC array, is that they only cover the nuclear genome. As mitochondrial DNA (mtDNA) can also be subject to methylation and hydroxymethylation, these processes have been hypothesized to be altered with aging and in AD [72], although empirical research into this phenomenon is still in its infancy. Using pyrosequencing, Blanch et al. [73] analyzed and compared mtDNA methylation in the entorhinal cortex, in three mitochondrial loci, of AD cases (Braak stages I–II and III–IV) with control individuals. AD samples showed increased methylation in the D-loop region of mtDNA (a noncoding region containing transcription and replication elements), with higher levels in Braak stages I/II than in III/IV. They also showed decreased methylation in mitochondrially encoded NADH:ubiquinone oxidoreductase core subunit 1 (*MT-ND1*), which codes for a subunit of the NADH dehydrogenase, part of the mitochondrial electron transport chain [73]. In blood samples, a study of mtDNA methylation in the D-loop showed decreased levels in AD patients [74].

8.3.2 Histone modifications

Histones are structural and functional proteins that package and order the DNA into nucleosomes, controlling the transition between active and inactive chromatin states [75]. The nucleosomes are the fundamental units of chromatin and consist of octamers of two molecules of each core histone (H2A, H2B, H3, and H4) [75,76]. The control is achieved as a consequence of posttranslational modifications in the N-terminus tail of the histones, such as phosphorylation, methylation, acetylation, ubiquitination, or sumoylation [77]. The sum of the presence, or absence, of different histone marks influences DNA compaction and accessibility, regulating DNA transcription, replication, repair, and recombination [78]. A number of studies have provided evidence demonstrating the importance of histone modifications in neuronal plasticity, learning, and memory events [79–82]. Although posttranslational histone marks have been suggested to play a role in AD [43–45], only a small number of studies to date have looked into changes in histone modifications in AD, with the vast majority simply assessing global changes. These studies have primarily used immunohistochemistry to compare levels of phosphorylation, methylation, or acetylation in specific histone variants between AD and elderly control donors.

Histone phosphorylation reduces the positive charge of histones and thus their affinity for DNA, presumably facilitating gene expression [83]. Studies of histone phosphorylation that have been described so far in AD have focused on histone H3. Specifically, increases in H3 phosphorylation have been reported; using an enzyme-linked immunosorbent assay (ELISA), Rao et al. [57] found significantly increased global H3 phosphorylation in postmortem frontal cortex from AD patients when compared with age-matched controls. Similarly, another study has shown elevated levels of phosphorylation in serine 10 (H3S10ph) in hippocampal neurons of AD samples compared with age-matched controls [84], although this was not replicated in another report, perhaps because of the assessment of a smaller number of samples [85]. Another study has shown an increase in phosphorylation of the histone H2A variant H2AX at serine 139 (γ-H2AX) in hippocampal astrocytes of AD patients when compared with controls [86].

Similarly, the evaluation of histone methylation has mainly focused on histone H3 posttranslational modification sites, including methylation of histone H3 lysine 9 (H3K9) and lysine 27 (H3K27), which are common sites for gene inactivation [87]. Increased levels of lysine 9 dimethylation (H3K9me2) were observed in postmortem occipital cortex from AD patients when compared with controls [85], and changes in methylation at the same site were also detected in another study in which two monozygotic twins discordant for AD were compared for levels of trimethylation of histone H3 lysine 9 (H3K9me3) [88]. Specifically, Ryu et al. detected an increase in H3K9me3 in the temporal cortex and hippocampus of the twin with AD compared with the twin without AD. A different study that analyzed trimethylation of histone H3 lysine 4 (H3K4me3), which is usually associated with active transcription, showed increased immunoreactivity in the cytoplasm, accompanied by decreased levels in the nuclear compartment in the hippocampus and temporal gyrus with increasing Braak stage [89]. In another study, histone modifications in postmortem frontal cortex were evaluated using mass spectrometry, and it showed a decrease in both histone H2B lysine 108 monomethylation (H2BK108me) and H4 arginine 55 monomethylation (H4R55me) in AD patients when compared with age-matched controls [90]. Histone methylation is maintained by histone methyltransferases (HMTs) and histone demethylases [91]. A histone demethylase inhibitor of lysine-specific demethylase 1 (LSD1), which is responsible for H3K4me2/1 demethylation, has been shown to ameliorate memory deficits and pathology-associated changes in gene expression in a mouse model of accelerated aging and AD [92], and an LSD1 inhibitor is currently in a phase 1 clinical trial for a prospective treatment for AD [93]. Moreover, from studies using rodent models, a role for H3-modifying enzymes in AD has also been proposed, particularly euchromatic histone-lysine N-methyltransferase 2 (G9a or EHMT2, a H3K9 histone methyltransferase), and KDM5A (JARID1A or RBBP2, a H3K4 histone demethylase) [94–97].

Histone acetylation, similar to phosphorylation, reduces histone positivity, weakening the electrostatic attraction between histones and the negatively charged DNA, therefore promoting DNA accessibility [83]. Research in histone acetylation in AD has also mainly focused on histone H3. Although Rao et al. [57] did not identify any changes in levels of H3 acetylation in AD postmortem frontal cortex, as assessed by ELISA, another publication using mass spectrometry reported a reduction in histone H3 lysine 18/lysine 23 acetylation (H3K18acK23ac) in the temporal lobe of AD patients when compared with controls [98]. Furthermore, using immunoblotting, Lithner et al. [85] detected an increase in lysine 14 acetylation (H3K14ac) in postmortem AD brain (occipital cortex) samples. In the same publication, Lithner et al. further reported *in vitro* changes in H3K14ac modulated by the presence of Aβ, specifically reduced H3K14ac after inhibiting Aβ secretion in a human neuroblastoma cell line expressing a mutated form of human APP, and increased H3K14ac in primary rat neurons exposed to soluble Aβ oligomers. Anderson and Turko [90] used mass spectrometry and identified an increase in histone 4 lysine 12/lysine 16 acetylation (H4K12acK16ac) in postmortem frontal cortex from AD patients when compared with age-matched controls, with no changes detected in histone H4 lysine 4/lysine 9 acetylation (H3K4acK9ac) and lysine 8/lysine 12/lysine 16 acetylation (H4K8acK12acK16ac). Moreover, levels of total and acetylated H3 and H4 have also been reported to be increased in postmortem temporal gyrus from AD patients when compared with controls [99]. The first studies undertaking a genome-wide approach to study histone acetylation in AD used chromatin immunoprecipitation sequencing (ChIP-seq) and looked at H3 lysine 27 acetylation (H3K27ac) [100] and H4 lysine 16 acetylation (H4K16ac) [101]. Marzi et al. [100] compared postmortem AD entorhinal cortex samples with controls and detected increased H3K27ac in the *APP*, *PSEN1*, *PSEN2*, and *MAPT* genes. Nativio et al. [101] compared H4K16ac genome-wide profiles in postmortem lateral temporal lobe samples from AD patients with age-matched and younger individuals without dementia, reporting distinct age-related patterns of H4K16ac in AD subjects compared with elderly cognitively normal controls, including changes in genes involved in specific biological functions such as myeloid differentiation, cell death, and Wnt and Ras signal transduction pathways. Histone acetylation is regulated by two families of enzymes: histone acetyltransferases (HATs) and histone deacetylases (HDACs) [83]. Studies using rodent models have provided evidence that HDAC inhibitors promote cognitive functions at the functional and cellular levels, particularly long-term, associative, and spatial memory in physiologic and pathologic conditions [76]. Additionally, a limited number of studies have quantified different HDAC subtypes, comparing protein levels in AD patients and control individuals [102,103]. Specifically, Graff and colleagues used immunohistochemistry and assessed the levels of HDAC2 in the nucleus of neurons from the CA1 region of the hippocampus, reporting an elevation in HDAC2 levels in AD patients with Braak stages I–II, III–IV, and V–VI when compared with nondemented controls [102]. Odagiri et al. [103] used immunohistochemistry and immunoblotting to evaluate protein levels of HDAC6 in the temporal cortex; however, the authors did not find any AD-related differences. Despite the promising preclinical findings from animal studies, further research to better understand pathology-related changes in HDACs and their inhibition as a therapeutic strategy for AD is still needed. In addition, research on the role of HATs and the potential therapeutic value of HAT agonists is required.

Other histone marks have been less well studied. One publication used mass spectrometry to assess histone ubiquitination changes in AD, specifically histone H2B lysine 120 monoubiquitination (H2BK120ub), and reported an increase in H2BK120ub in the frontal cortex of AD patients when compared with age-matched normal donors [90].

Overall, there is growing evidence demonstrating changes in histone modifications in AD brain, and a number of studies hypothesize that targeting these changes could result in promising novel therapeutic strategies. However, a clear understanding about gene-specific changes is still lacking. Furthermore, studies of histone modifications such as ubiquitination and sumoylation, in particular, are still in their infancy. On the other hand, methods for studying histone modifications are rapidly being developed and will certainly be applied to the AD research field in the upcoming years.

8.3.3 Regulatory RNA–based mechanisms

Although most of the genome seems to be transcribed as RNA, $<1.5\%$ of the genome encodes for proteins, meaning intronic and intergenic non–protein-coding genomic regions may hold large amounts of regulatory information [104]. Noncoding RNA (ncRNA) is usually defined as a functionally conserved RNA molecule that is not translated into protein but that modulates gene expression at the transcriptional and posttranscriptional levels instead. According to their size, ncRNA can be classified into short noncoding RNA (sncRNA) and long noncoding RNA (lncRNA). sncRNA can be further divided into three major classes: microRNA (miRNA), short interfering RNA, and piwi-interacting RNAs [105,106].

miRNAs, short 22-nucleotide-long regulatory ncRNA, have gained particular interest in the recent years, particularly because of the increasing evidence implying an important role of miRNAs in brain processes. miRNAs usually act as posttranscriptional repressors through translational inhibition [106]. A role for miRNAs in causing neurodegenerative diseases, including AD, has been suggested [107–119]. Using an RT-PCR-based assay, a study profiled miRNAs in three brain regions (hippocampus, medial frontal gyrus, and cerebellum) from individuals ranging from Braak stages 0–VI [120]. In this report, nine miRNAs (miR-26a, miR-30c, miR-30e-5p, miR-92, miR-145, miR-148a, miR-200c, miR-210, and miR-423) were significantly changed in the hippocampus and/or medial frontal gyrus at Braak stages III–IV and V–VI, but not in the cerebellum. Additionally, in individuals with severe AD (Braak stage V–VI), miR-29a and miR-29b were increased in the medial frontal gyrus, and miR-423 was upregulated in both the hippocampus and medial frontal gyrus when compared with age-matched controls. However, the authors also reported 12 miRNAs that showed significant alterations in the cerebellum (miR-9, miR-132, miR-146b, miR-212, miR-425, miR-27a, miR-27b, miR-34a, miR-100, miR-125b, miR-381, and miR-422a), in addition to the other brain regions studied [120]. The majority of studies published to date have used microarrays to profile miRNAs. Hébert et al. [121] investigated miRNA expression in the anterior temporal cortex from AD patients and compared it with age-matched controls and found changes in 16 miRNAs. These changes included decreased levels of miR-29a and miR-29b-1, which were suggested to be involved in the transcriptional regulation of *APP* and β-secretase 1 (*BACE1*), the latter encoding an enzyme involved in the amyloidogenic cleavage of APP. This finding was contradictory to the observations of Cogswell et al. [120], perhaps because the upregulation of miR-29a and miR-29b reported by Cogswell and colleagues was only detected at the later stages of AD, and variations may occur across the different stages of pathology. Downregulation of miR-29a has been further reported in AD in a different study that investigated the frontal cortex and compared brains from donors with AD, PD, amyotrophic lateral sclerosis, and controls who died from nonneurologic causes [115]. Another potential regulator of *BACE1*, miR-107, has been reported to be decreased in AD in the temporal cortex, from early stages of pathology, as it was identified as being downregulated in individuals with AD and in also those with mild cognitive impairment (MCI) [112,118]. In a different study, Nunez-Iglesias et al. [113] profiled miRNAs and their target messenger RNAs (mRNAs) in the parietal lobe cortex from AD patients and age-matched controls. They identified 48 differentially expressed miRNAs, including downregulation of miR-29b, and 11 differentially expressed mRNAs in AD samples when compared with controls, further demonstrating correlations in expression between miRNAs and their target mRNA. Interestingly, different miRNAs have been reported to be altered in the white matter and gray matter in AD temporal cortex when compared with controls, including changes in miR-212 and miR-424 in the white matter and miR-15/107 and miR-29 paralogs in the gray matter [117]. Two additional studies identified miRNA changes in AD hippocampus [107,111]. Lau et al. identified dysregulation of 35 miRNAs when comparing AD samples with age-matched controls, including downregulation of miR-124-3p, miR-128, miR-129-2-3p, miR-129-5p, miR-132-3p, miR-136-5p, and miR-138-5p and upregulation of miR-27a-3p, miR-92b-3p, miR-142-3p, and miR-200a-3p. For validation, the authors profiled an independent cohort of prefrontal cortex samples, showing dynamic expression patterns in

41 miRNAs with Braak stage. For example, eight miRNAs, including miR-132-3p (the most significantly downregulated miRNA in the hippocampus samples), were found to be upregulated between Braak stages I and II and downregulated between Braak stages II and VI. miR-132-3p was also shown to be downregulated in the temporal gyrus in a third, smaller, independent cohort in the same study [111]. Bekris et al. [107] found changes in miR-15a in postmortem AD hippocampus when compared with controls, which correlated with amyloid plaque score. In a different publication, Patrick et al. [114] explored changes in miRNA in two longitudinal cohort studies of aging and detected AD-associated changes in miR-129 and miR-132 in the dorsolateral prefrontal cortex, consistent with the previous reports [111,120]. The authors also reported AD-associated upregulation of miR-200a, which they suggest warrants further investigation [114]. Of note, there are several potential explanations for the inconsistency in miRNA findings reported across studies, for example, the assessment of a small number of samples, insufficient phenotypic information in some of the studies, differences across distinct brain regions and/or stages of a disease, and/or variation in classification criteria (Table 8.1).

lncRNAs, usually defined as transcripts longer than 200 nucleotides that are not translated into proteins, have also been suggested to play a role in AD pathophysiology, but very few studies have attempted to address this question, with all reports to date only investigating specific target genes using RT-PCR. Faghihi et al. [122] identified a conserved noncoding antisense transcript for *BACE1* (*BACE1*-AS), reporting elevated levels of *BACE1*-AS in several brain regions, namely, the hippocampus, entorhinal cortex, superior frontal gyrus, and cerebellum, of AD-affected individuals when compared with controls. An lncRNA involved in γ-aminobutyric acid B regulation, lncRNA 17A, was found to be upregulated in AD frontal and temporal cortex when compared with control cases, and its overexpression was shown to cause overproduction of Aβ in neuroblastoma cells [123]. Another publication reported upregulation of the lncRNA 51A, which regulates the expression of *SORL1*, in postmortem AD cerebral cortex when compared with nondemented control samples [124].

An emerging field of research in the context of complex disease is that of RNA modifications, particularly N6-methyladenosine (m6A), the addition of a methyl group to the nitrogen-6 position of the adenosine base of mRNA [125]. Although no m6A studies in AD have been yet reported, in the coming years, it is anticipated that studies examining the role of this mechanism in AD will emerge.

8.3.4 Epigenetic signatures as blood biomarkers

At present, AD is clinically diagnosed only when robust symptoms of the disease emerge, which can be decades after the onset of neuropathologic conditions. As such, specific peripheral biomarkers for the disease are highly desired. Biomarkers are sought for improved accuracy of AD diagnosis (diagnostic biomarkers), for earlier diagnosis of AD (predictive biomarkers) and to predict future changes in cognition, and/or to monitor the outcome of therapeutic strategies (prognostic biomarkers). Epigenetic signatures in blood, which correlate with the neuropathologic conditions in the brain, have been proposed to represent promising candidate biomarkers for AD [44,126], particularly given that there is evidence that some epigenetic changes may be conserved across brain and blood [127,128].

Given that DNA methylation is the most widely studied epigenetic process, it is not surprising that the field has mainly focused on identifying blood DNA methylation signatures of the disease. However, results so far are still limited and inconclusive. Most studies to date used pyrosequencing to assess a small number of genomic sites; Chang et al. [129] reported higher *BDNF* methylation and

Table 8.1 Main reports of miRNAs in AD.

miRNA	Status in AD	References
miR-9	Upregulated	[120]
miR-15a	Upregulated	[107]
miR-26a	Upregulated/ downregulated	[120]
miR-27a	Upregulated	[111,120]
miR-27b	Upregulated	[120]
miR-29a	Upregulated Downregulated	[120] [115,121]
miR-29b	Upregulated Downregulated	[120] [113,121]
miR-30c	Downregulated/ upregulated	[120]
miR-30e-5p	Upregulated	[120]
miR-34a	Upregulated	[120]
miR-92	Upregulated	[111,120]
miR-100	Upregulated	[120]
miR-107	Downregulated	[112,118]
124-3p	Downregulated	[111]
miR-125b	Upregulated	[120]
miR-128	Downregulated	[111]
miR-129	Downregulated	[111,114]
miR-132	Upregulated Downregulated	[114,120] [111]
miR-136	Downregulated	[111]
miR-138	Downregulated	[111]
miR-142	Upregulated	[111]
miR-145	Upregulated	[120]
miR-146b	Upregulated	[120]
miR-148a	Upregulated	[120]
miR-200a	Upregulated	[111,114]
miR-200c	Upregulated/ downregulated	[120]
miR-210	Downregulated	[120]
miR-212	Upregulated	[120]
miR-381	Upregulated	[120]
miR-422a	Upregulated	[120]
miR-423	Upregulated	[120]
miR-425	Upregulated	[120]

Most miRNA alterations have been either identified in just a single study or inconsistent across studies. Therefore these findings warrant further exploration.
AD, *Alzheimer's disease;* miRNA, *microRNA.*

Ji et al. [130] found elevated *OPRD1* promoter methylation in AD when compared with healthy controls. Two studies reported no differences in *PIN1* promoter DNA methylation between AD and controls [131,132], in contrast to two other publications that used RT-PCR and found AD-associated decreased *PIN1* promoter methylation in AD when compared with cognitively healthy age-matched controls [133], and in AD-discordant monozygotic twins [134]. Yamazaki and colleagues [135,136] reported no significant differences in the *SLC6A4* and *ABCA7* promoters in AD. Using PCR-based methods, no differences were found in promoter methylation of *PSEN1*, *BACE1*, *MTHFR*, *DNMT1*, *DNMT3A*, *DNMT3B*, *BDNF*, and *SIRT1* in the blood of AD patients compared with that of non-demented controls [132,137]. A study using the Illumina 450K methylation array identified decreased DNA methylation of the *NCAPH2/LMF2* promoter in AD and MCI cases when compared with healthy controls [138]. A systematic review studied the current evidence for an association between dementia and blood DNA methylation, reporting that most studies have assessed a limited number of individuals, with the methods used and results obtained so far being highly heterogeneous, and concluding that further investigations are still needed [126]. Indeed, studies of DNA methylation changes in AD blood samples are expected to be published in the near future, with such analyses currently being undertaken in the European Medical Information Framework – Alzheimer's Disease (EMIF-AD) study [139] and the Alzheimer's Disease Neuroimaging Initiative (ADNI) cohort [140].

Very little attention has been given to blood histone modification profiles in AD, possibly given that these epigenetic marks are less stable than DNA methylation and because the available methods are mainly sequencing based rather than array based. One study has reported increased H4K12ac levels in monocytes from human patients with MCI when compared with controls, but not from patients with AD compared with controls, suggesting this could reflect early changes in AD development [141].

A number of studies, although still in their infancy, have aimed to identify miRNA signatures of AD in blood. Two systematic reviews investigated the relationship between blood miRNA dysregulation and AD diagnosis in previously published studies and suggested a promising role for miRNAs in the detection of AD, but cautioned for the need for further analysis and validation [109,142]. From 52 miRNAs that have been reported to be altered in AD blood across the 18 studies that Wu et al. reviewed, miR-29b, miR-181c miR-15b, miR-146a, and miR-107 were reported as the most consistently dysregulated miRNA markers for AD, with consistent downregulation observed in at least two independent studies, and downregulation of miR-29b, miR-181c, and miR-15b was described as concordant to changes found in cerebrospinal fluid and brain [142]. Both publications highlight the issue of study design differences between different reports, with few common miRNAs being assessed, which makes the interpretation of results much more difficult [109,142].

8.4 Modeling Alzheimer's disease: mouse models as powerful tools

The use of mouse models over the past few decades has enabled researchers to better comprehend diseases at a molecular level, including important pathways affected by and/or involved in pathology; to establish and validate biomarkers; and to identify novel therapeutic targets [143]. Age-related disorders, such as AD-associated changes that usually take decades to manifest in humans, rarely occur spontaneously in animals used for research, mainly because their lifespan may not be long enough to observe them; therefore, the most used animal models are genetically engineered mice that recapitulate many aspects of human diseases. Nonetheless, animal models, in general, and mouse models, in particular, represent valuable tools for the understanding of pathogenic mechanisms in a wide range of diseases, especially those that affect inaccessible tissues such as the brain.

Most mouse models currently used for the study of AD overexpress human *APP*, alone or in conjunction with *PSEN* overexpression, bearing mutations identified in FAD. These mice show some of the characteristics of AD, particularly the formation of Aβ plaques. Tau mouse models are also used, overexpressing human *MAPT* bearing mutations identified in patients with frontotemporal dementia (FTD), which recapitulate NFT neuropathology. Finally, mice that develop plaques and tangles due to overexpressing mutated *APP*, *PSEN*, and *MAPT* are also widely used [143]. It is important to note that these mice do not truly mimic SAD, given that their pathology is driven by the introduction of mutations identified in FAD; however, studying them can still be highly useful, as it provides a means to understand processes that mediate early changes and progression of AD pathology. Furthermore, the use of laboratory animals is highly regulated, including the control of their environment and the general conditions they are exposed to, allowing the exclusion of potential confounders common in human research, such as medication, diet, and potential stressors.

Despite their widespread utility, limited work has been undertaken to assess regulatory genomic changes in AD-relevant mouse models. Most studies carried out to date have focused either on global levels of epigenetic modifications, particularly histone modifications, or on a small number of loci [94,108,144–147]. Francis et al. [144] used immunoblotting and reported differences in hippocampal H4 acetylation levels during associative memory formation in 3- to 4-month-old APP/PSEN1 transgenic mice when compared with littermate controls. In a different mouse model, the 3xTg-AD line that contains three mutations identified in either FAD (Swedish *APP* mutation and *PSEN1* M146V mutation) or FTD (*MAPT* P301L mutation), Marques et al. [145] used ChIP followed by quantitative RT-PCR and detected increased H3 acetylation in the promoter region of *Bace1* in cortex samples from 15-month-old transgenic mice when compared with controls. In a different study using the same mouse model, Walker et al. [94] used immunofluorescence to assess changes in H3 and H4 and found increased H3 and H4 acetylation with age in cultured hippocampal/cortical neurons from 3xTg-AD mice when compared with nontransgenic mice, as well as higher levels of H3 acetylation in nontransgenic neurons treated with Aβ. The same study showed increased H3K9 methylation in 3xTg mice cultured neurons when compared with controls. In addition to increased levels of H3K14ac in postmortem AD brain and in vitro, due to the presence of Aβ discussed earlier, Lithner et al. [85] used immunoblotting and found increased H3K14ac and H3K9me2 in the cortex and hippocampus of Tg2576 transgenic mice (a mouse line expressing *APP* with the Swedish mutation) when compared with controls at a preplaque stage. Mastroeni et al. [89] investigated H3K4me3 changes in 3xTg-AD mice in addition to human AD brain samples and found an age-related loss of H3K4me3 in the hippocampus of 3xTg-AD transgenic mice versus nontransgenic mice when studied at 8, 12, 15, and 19 months of age. However, they observed a different pattern in the transgenic mice than what they observed in human hippocampal samples; for example, the most significant loss in the 3xTg-AD mice was seen in the *cornu ammonis* (CA) 3 subregion of the hippocampus, whereas in human samples, this was seen in the CA1 subregion [89]. Seven histone modifications (H3K4me1, H3K4me3, H3K9me3, H3K27ac, H3K27me3, H3K36me3, and H4K20me1) were profiled in a genome-wide fashion in the hippocampus of CK-p25 mice, which exhibit inducible accumulation of the cyclin dependent kinase 5 activator protein p25 that leads to increased Aβ levels at early stages, as well as neuronal and synaptic loss, followed by cognitive impairment at later stages [148]. Using ChIP-seq and RNA sequencing to measure gene expression, Gjoneska et al. found downregulation of neural plasticity genes and corresponding regulatory regions, as well as upregulation of genes involved in immune and inflammatory pathways and their regulatory regions, in CK-p25 transgenic mice when compared with CK littermate controls. The authors also compared these results with human AD brain data and identified shared functional changes as a consequence of AD-associated neurodegeneration [148].

As mentioned earlier, several studies have explored the effect of HDAC inhibitors on cognitive function in rodent models, including in AD mouse models. In fact, HDAC inhibitors have been shown to improve cognitive deficits, including learning and memory impairment, as well as synaptic dysfunction, in mouse models of AD and neurodegeneration [149–154], suggesting a potential promising therapeutic approach for HDAC inhibitors to be further explored.

The first studies assessing DNA modifications in AD mouse models, particularly DNA methylation, used genome-wide approaches. One study used a custom-designed mouse brain genome-wide promoter DNA methylation array, covering 384 genes, and reported hypermethylation of thromboxane A_2 receptor (*Tbxa2r*), coagulation factor II receptor-like 2 (*F2rl2*), sorbin and SH3 domain containing 3 (*Sorbs3*), and spectrin beta 4 (*Sptbn4*) in the frontal cortex of ~2-month-old APP/PSEN1 transgenic mice when compared with control mice, with a similar trend in 3xTg-AD mice, accompanied by downregulation of the corresponding RNA transcripts and proteins [146]. The authors further explored these findings using pyrosequencing in frontal cortex samples from AD patients and detected increased promoter hypermethylation for the *TBXA2R*, *SORBS3*, and *SPTBN4* genes in AD brains when compared with healthy controls. In a different study, using a different methylation array, methylation profiles in cortex samples from APP/PSEN1 transgenic mice and control mice (11 month old) were also investigated [155]. Cong et al. found hypermethylation in 485 genes, with several differentially methylated genes from the same gene family, for example, 54 genes were olfactory receptor family members, 43 were solute carrier family genes, 18 were interleukin family members, 14 were keratin, 6 were keratin-associated protein family members, 12 were cytochrome P450 family members, and 12 were nuclear receptor subfamily members. Some of the reported top hypermethylated genes included serine (or cysteine) peptidase inhibitor, clade B (ovalbumin), member 10 (*Serpinb10*), transforming growth factor β_1 (*Tgfb1*), sodium channel epithelial 1 alpha subunit (*Scnn1a*), THO Complex 7 (*Thoc7*), opticin (*Optc*), myosin heavy chain 4 (*Myh4*), and prion protein (*Prnp*) [155]. *Mapt* promoter methylation has also been evaluated by pyrosequencing in the hippocampus from J20 mice that overexpress human *APP* bearing the Swedish and Indiana mutations identified in FAD cases. In 5 month-old mice, which display little or no plaque pathology at this time point, Coupland et al. [156] detected lower DNA methylation in J20 transgenic mice when compared with wild-type mice in the *Mapt* promoter. A few studies have assessed global methylation using immunohistochemistry. Cadena-del-Castillo et al. [157] examined levels of 5-mC and 5-hmC in young and old 3xTg-AD mice, reporting an age-associated decline in 5-mC levels (from 3 to 17 month old) in the cerebral cortex and a larger increase in 5-hmC levels in the cortex and hippocampus (dentate gyrus [DG] and CA1) of transgenic mice when compared with control mice. In contrast, in a different study, Lardenoije et al. [158] found an age-associated increase in 5-mC levels in the DG and CA1–2 hippocampal subregions in 3xTg-AD mice (from 5 to 25 month old), with no differences in 5-hmC levels in this mouse model. The same study investigated hippocampal methylation changes in the J20 mouse model, identifying an age-related decrease in 5-mC levels in the DG, a decrease in the ratio of 5-mC and 5-hmC in the DG and CA3, and a negative correlation between plaque load and 5-mC levels in the DG in transgenic mice when compared with controls. Lardenoije et al. [158] also explored changes in APP/PSEN1dE9 mice and a nonhuman primate model, but no significant age-associated alterations were identified.

Regarding studies of miRNA changes in AD mouse models, very limited work has been reported to date. Wang et al. used microarrays to identify changes in 37 miRNAs in 3- and 6-month-old double transgenic APPswe/PSEN1dE9 mice (with the *APP* Swedish and *PSEN1*: deltaE9 mutations) and compared them with age-matched controls in the cerebral cortex, highlighting upregulation of miR-34a, which they also confirmed by RT-PCR [119]. Another study used the same mouse model and a similar microarray approach to profile changes in the brain of 1-, 3-, 6-, and 9-month-old APPswe/PSEN1dE9 mice [116]. By comparing APP/PSEN1 transgenic mice with wild-type littermates, Wang et al. found 8 differentially expressed miRNAs in 1-month-old mice, 14 miRNAs at 3 months, and 30 miRNAs at 6 months, as well as changes in 14 miRNAs in 9-month-old mice; these included consistent upregulation of miR-342-3p at 1, 6, and 9 months of age, upregulation of miR-342-5p at 1 and 3 months, and upregulation of miR-2136, let-7f-5p, miR-431-5p, and miR-491-3p in both 3- and 6-month-old APP/PSEN1 transgenic mice. A pathway analysis revealed Ras-related protein 1 (Rap1) signaling pathway, mitogen-activated protein kinase (MAPK) signaling pathway, TGF-β signaling pathway, and the Wnt signaling pathway to be some of the most significantly enriched pathways regulated by these miRNAs [116].

8.5 Current challenges and future directions

Although epigenetic studies in AD have demonstrated promising findings, more studies, particularly at the genome-wide level, are still required. However, there are a number of challenges and caveats associated with performing such studies, which need to be adequately addressed to ultimately allow for the identification of novel genes and neurobiological pathways altered in the disease, so that it could deliver novel pharmacologic targets.

Different tissues and cell types exhibit unique epigenetic profiles, and therefore it can be quite challenging to disentangle disease-specific changes from nondirectly affected tissues and/or heterogeneous tissues. As discussed earlier, some regions of the brain are more affected than others by AD neuropathology, with specific epigenetic changes reported in distinct brain regions and for specific cell types. In addition, AD is characterized by progressive changes in brain cell composition, including neuronal loss due to neurodegeneration, increased number of activated microglia (microgliosis), and abnormal hypertrophy of astrocytes (reactive astrogliosis), which is particularly true in brain regions highly susceptible to neuropathology, such as the hippocampus [3]. The vast majority of AD epigenetic studies have used bulk brain tissue; therefore, it becomes difficult to draw conclusions about cell-specific AD-associated epigenetic changes. Furthermore, it is plausible that some important disease-associated epigenetic variations might be missed if epigenetic changes in one cell type are masked by opposing changes in a different cell type. The gold standard analysis would use single-cell profiling to investigate epigenomic changes in individual cells; however, the costs associated with performing such experiments with adequate power to detect disease-associated differences are still prohibitive. Nonetheless, epigenomic studies utilizing methods aimed at isolating specific cell populations, such as laser capture microdissection, fluorescence-activated cell sorting (FACS), magnetic affinity cell sorting, and density gradients, are becoming more commonplace. A study used FACS to investigate AD-associated DNA methylation changes in sorted neuronal and nonneuronal (glial) nuclei from postmortem human occipital cortex samples, in addition to "bulk" tissue from the frontal and temporal cortex [159]. Gasparoni et al. provided evidence that sorting cells improves the detection of AD-associated DNA methylation changes and reported cell-type-specific DNA methylation signatures associated with Braak stage, including loci in the MCF.2 cell-line-derived transforming sequence like (*MCF2L*), *ANK1*, microtubule-associated protein 2 (*MAP2*), volume-regulated anion channel subunit LRRC8B (*LRRC8B*), serine/threonine-protein kinase 32C (*STK32C*), and S100 calcium-binding protein B (*S100B*) genes. In the same study, a meta-analysis combining the data of significant cell sorted DMRs from neuronal and glial populations identified Braak stage—associated changes in known AD genes including *APP*, *HOXA3*, and *ADAM17*. Notably, the study also validated DNA methylation changes previously identified in AD EWAS in genes such as *HOXA3* and *ANK1*, assigning them to neurons and glia, respectively. However, it is worth noting that the experiments on FACS-isolated nuclei only utilized tissue from 15 AD and 16 control samples.

At present, the majority of EWAS of AD have utilized microarray-based technology to assess AD-associated DNA methylomic variation. Although arrays are a cost-effective method for analyzing the epigenome, especially when analyzing large sample sizes, they only cover a small proportion of potential methylation sites in the human genome, with most probes located in the CpG-rich promoter regions, covering about 95% of CpG islands. As such, disease-associated epigenetic variation in other regions in the genome might be missed. Ideally, AD EWAS would utilize sequencing-based technology to uncover AD-associated epigenetic variation; for example, whole-genome bisulfite sequencing allows the identification of DNA methylation changes at single nucleotide resolution across the entire human genome, including at non-CpG sites. However, the depth of sequencing required to accurately profile DNA methylation in a genome-wide fashion is still associated with prohibitively high costs to undertake on a sufficiently large sample size [160]. A cost-effective alternative method to analyze the DNA methylome at single-base resolution is reduced representation bisulfite sequencing (RRBS) [161,162]. This method reduces the amount of sequencing required to profile the majority of promoters and other relevant genomic regions, as it enriches the CG-rich parts of the genome (~1% of the genome) by generating small fragments containing CpG dinucleotides [161]. Looking to the future, third-generation (or long-read) sequencing technologies, currently under active development, are highly awaited, given that they can potentially allow simultaneous identification of genetic and epigenetic variations in the same sample [163]. These allow sequencing at long read lengths, as they are capable of sequencing single molecules (without the need for amplification in contrast to the previous methods) in real time.

One issue with studying epigenetic variations in diseases of inaccessible tissues, such as the brain, is that human studies are restricted to postmortem samples, which usually represent end-stage disease and so it is difficult to establish causality. Although murine models can be used for understanding disease progression, as discussed earlier, they also have their own caveats for studying a disease such as AD, for example, they do not naturally develop Aβ plaques or have overt neurodegeneration. In addition, these transgenic models are based on mutations associated with FAD and are not therefore accurate models of SAD, which has both polygenic and environmental contributions to the disease cause [5,7]. In recent years, induced pluripotent stem cells (iPSCs) have been generated from actual human patients and are becoming more widely used for studying aspects of disease that are species (human) specific. iPSCs are particularly useful for studying the impact of specific genetic variants, or combinations of genetic variants, on cellular phenotype [164]. In addition, through the clustered regularly interspersed short palindromic repeats (CRISPR)-Cas9 technology, it is possible to generate isogenic control lines where the only genetic variation is at disease-specific loci [165]. With the recent adaptations to CRISPR-Cas9 methodology, it is possible to remove or add methyl groups to specific cytosines by fusing the Cas9 protein with the enzymatic domains of TET1 or DNMT3A, respectively [166]. However, it is worth remembering that there are still drawbacks of using in vitro systems for studying complex diseases such as AD, for example, a lack of knowledge about other important cell types present and their corresponding cell–cell interactions and the normal brain microenvironment.

Finally, we are currently in the "big data" era in which large "omics" data sets with extensive clinical and demographic information have been amassed, which offers extraordinary possibilities to better understand the cause of AD. Although these studies are still in their infancy, the integration of genetic, epigenetic, transcriptomic, proteomic, and clinical data from the same individuals will certainly allow a fully comprehensive analysis of dysfunctional mechanisms and pathways involved in AD. Furthermore, by comparing similar data sets generated in mouse models with human AD data sets, a more comprehensive analysis of disease mechanisms will be achievable, by overcoming a number of caveats associated with studying human samples; for example, it will allow the mapping of loci identified in humans longitudinally in mouse models and will help map the temporospatial changes in loci back to before pathology onset. A handful of papers have pioneered this approach, exploring findings from AD mouse models in human data [116,148].

8.6 Final considerations

Our understanding of genomic regulation changes in AD has expanded considerably in the past few years; however, this only represents the tip of the iceberg and many epigenomic studies in AD are expected shortly, which will help us to better determine the processes underlying AD. Furthermore, the expected wave of "multiomics" analyses, including the integration of different levels of different data using network analyses through the joint forces of multidisciplinary teams, and the increasing improvement of high-throughput technologies will likely be the direction of the field in the coming years.

It should be noted that many other diseases besides AD are characterized by dementia. Other forms of dementia include VaD, DLB, HD, PD dementia, and FTD. The field has mainly focused on AD; however, epigenomic studies in other dementias are also needed. Moreover, many studies have mainly focused on the temporal lobe (particularly the hippocampus) and the frontal lobe (mostly the prefrontal cortex), and studies assessing other brain regions would also be important. Furthermore, if these were performed in the same individuals, it would allow the mapping of spatial epigenetic changes in the disease. As already discussed, epigenetic mechanisms are reversible, attributing to them a high therapeutic potential. Given that there are currently no therapeutic agents for AD that can reverse, or even stop, the pathologic condition, looking to the future, the possibility to specifically target epigenetic modifications to treat AD is very promising.

References

[1] Alzheimer A. Neuere Arbeiten über die Dementia senilis und die auf atheromatöser Gefässerkrankung basierenden Hirnkrankheiten. Monatsschr Psychiat Neurol 1898;3:101–15.

[2] Alzheimer A. Über eine eigenartige Erkrankung der Hirnrinde. Allgemeine Zeitschrift fur Psychiatrie und Psychisch-gerichtliche Medizin 1907;64:146–8.

[3] Blennow K, de Leon MJ, Zetterberg H. Alzheimer's disease. Lancet 2006;368(9533):387–403.

[4] Crews L, Masliah E. Molecular mechanisms of neurodegeneration in Alzheimer's disease. Hum Mol Genet 2010;19(R1):R12–20.

[5] Selkoe DJ. Alzheimer's disease. Cold Spring Harb Perspect Biol 2011;3(7).

[6] Prince M, Comas-Herrera A, Knapp M, Guerchet M, Karagiannidou M. World alzheimer report 2016. 2016. London.

[7] 2017 Alzheimer's disease facts and figures. Alzheimer's Dementia 2017;13(4):325–73.

[8] Rajan KB, Wilson RS, Weuve J, Barnes LL, Evans DA. Cognitive impairment 18 years before clinical diagnosis of Alzheimer disease dementia. Neurology 2015;85(10):898–904.

[9] O'Brien RJ, Wong PC. Amyloid precursor protein processing and Alzheimer's disease. Annu Rev Neurosci 2011;34:185–204.

[10] Wang Y, Mandelkow E. Tau in physiology and pathology. Nat Rev Neurosci 2016;17(1):5–21.

[11] Serrano-Pozo A, Frosch MP, Masliah E, Hyman BT. Neuropathological alterations in Alzheimer disease. Cold Spring Harb Perspect Med 2011;1(1):a006189.

[12] Braak H, Alafuzoff I, Arzberger T, Kretzschmar H, Del Tredici K. Staging of Alzheimer disease-associated neurofibrillary pathology using paraffin sections and immunocytochemistry. Acta Neuropathol 2006; 112(4):389–404.

[13] Braak H, Braak E. Neuropathological staging of Alzheimer-related changes. Acta Neuropathologica 1991; 82(4):239–59.

[14] Guerreiro RJ, Gustafson DR, Hardy J. The genetic architecture of Alzheimer's disease: beyond APP, PSENs and APOE. Neurobiol Aging 2012;33(3):437–56.

[15] Gatz M, Reynolds CA, Fratiglioni L, Johansson B, Mortimer JA, Berg S, et al. Role of genes and environments for explaining Alzheimer disease. Arch Gen Psychiatr 2006;63(2):168–74.

[16] Plomin R, Owen MJ, McGuffin P. The genetic basis of complex human behaviors. Science 1994;264(5166): 1733–9.

[17] Corder EH, Saunders AM, Strittmatter WJ, Schmechel DE, Gaskell PC, Small GW, et al. Gene dose of apolipoprotein E type 4 allele and the risk of Alzheimer's disease in late onset families. Science 1993; 261(5123):921–3.

[18] Strittmatter WJ, Saunders AM, Schmechel D, Pericak-Vance M, Enghild J, Salvesen GS, et al. Apolipoprotein E: high-avidity binding to beta-amyloid and increased frequency of type 4 allele in late-onset familial Alzheimer disease. Proc Natl Acad Sci U S A 1993;90(5):1977–81.

[19] Carrasquillo MM, Belbin O, Hunter TA, Ma L, Bisceglio GD, Zou F, et al. Replication of CLU, CR1, and PICALM associations with Alzheimer disease. Arch Neurol 2010;67(8):961–4.

[20] Corneveaux JJ, Myers AJ, Allen AN, Pruzin JJ, Ramirez M, Engel A, et al. Association of CR1, CLU and PICALM with Alzheimer's disease in a cohort of clinically characterized and neuropathologically verified individuals. Hum Mol Genet 2010;19(16):3295–301.

[21] Harold D, Abraham R, Hollingworth P, Sims R, Gerrish A, Hamshere ML, et al. Genome-wide association study identifies variants at CLU and PICALM associated with Alzheimer's disease. Nat Genet 2009;41(10): 1088–93.

[22] Jun G, Naj AC, Beecham GW, Wang LS, Buros J, Gallins PJ, et al. Meta-analysis confirms CR1, CLU, and PICALM as Alzheimer disease risk loci and reveals interactions with APOE genotypes. Arch Neurol 2010; 67(12):1473–84.
[23] Lambert JC, Heath S, Even G, Campion D, Sleegers K, Hiltunen M, et al. Genome-wide association study identifies variants at CLU and CR1 associated with Alzheimer's disease. Nat Genet 2009;41(10):1094–9.
[24] Zhang Q, Yu JT, Zhu QX, Zhang W, Wu ZC, Miao D, et al. Complement receptor 1 polymorphisms and risk of late-onset Alzheimer's disease. Brain Res 2010;1348:216–21.
[25] Hollingworth P, Harold D, Sims R, Gerrish A, Lambert JC, Carrasquillo MM, et al. Common variants at ABCA7, MS4A6A/MS4A4E, EPHA1, CD33 and CD2AP are associated with Alzheimer's disease. Nat Genet 2011;43(5):429–35.
[26] Lambert JC, Ibrahim-Verbaas CA, Harold D, Naj AC, Sims R, Bellenguez C, et al. Meta-analysis of 74,046 individuals identifies 11 new susceptibility loci for Alzheimer's disease. Nat Genet 2013;45(12):1452–8.
[27] Guerreiro R, Wojtas A, Bras J, Carrasquillo M, Rogaeva E, Majounie E, et al. TREM2 variants in Alzheimer's disease. N Engl J Med 2013;368(2):117–27.
[28] Sims R, van der Lee SJ, Naj AC, Bellenguez C, Badarinarayan N, Jakobsdottir J, et al. Rare coding variants in PLCG2, ABI3, and TREM2 implicate microglial-mediated innate immunity in Alzheimer's disease. Nat Genet 2017;49(9):1373–84.
[29] Cruchaga C, Karch CM, Jin SC, Benitez BA, Cai Y, Guerreiro R, et al. Rare coding variants in the phospholipase D3 gene confer risk for Alzheimer's disease. Nature 2014;505(7484):550–4.
[30] Smith AR, Mill J, Smith RG, Lunnon K. Elucidating novel dysfunctional pathways in Alzheimer's disease by integrating loci identified in genetic and epigenetic studies. Neuroepigenetics 2016;6:32–50.
[31] Jansen J, Savage K, Watanabe J, Bryois D, Williams S, Steinberg, et al. Genome-wide meta-analysis identifies new loci and functional pathways influencing Alzheimer's disease risk. Nat Genet 2019;51: 404–13. https://doi.org/10.1038/s41588-018-0311-9.
[32] Kunkle BW, Grenier-Boley B, Sims R, Bis JC, Naj AC, Boland A, et al. Genetic meta-analysis of diagnosed Alzheimer's disease identifies new risk loci and implicates Abeta. Tau, immunity and lipid processing, Nat Genet 2019;41:414–30. https://doi.org/10.1038/s41588-019-0358-2.
[33] Marioni R, Harris SE, McRae AF, Zhang Q, Hagenaars SP, Hill WD, et al. GWAS on family history of Alzheimer's disease. Transl Psychiatry 2018;8(1):99. https://doi.org/10.1038/s41398-018-0150-6.
[34] Escott-Price V, Shoai M, Pither R, Williams J, Hardy J. Polygenic score prediction captures nearly all common genetic risk for Alzheimer's disease. Neurobiol Aging 2017;49:214 e7–214 e11.
[35] Escott-Price V, Sims R, Bannister C, Harold D, Vronskaya M, Majounie E, et al. Common polygenic variation enhances risk prediction for Alzheimer's disease. Brain 2015;138(Pt 12):3673–84.
[36] Ridge PG, Mukherjee S, Crane PK, Kauwe JS. Alzheimer's disease genetics C. Alzheimer's disease: analyzing the missing heritability. PLoS One 2013;8(11):e79771.
[37] Mastroeni D, McKee A, Grover A, Rogers J, Coleman PD. Epigenetic differences in cortical neurons from a pair of monozygotic twins discordant for Alzheimer's disease. PLoS One 2009;4(8):e6617.
[38] Nee LE, Lippa CF. Alzheimer's disease in 22 twin pairs–13-year follow-up: hormonal, infectious and traumatic factors. Dement Geriatr Cogn Disord 1999;10(2):148–51.
[39] Pal S, Tyler JK. Epigenetics and aging. Sci Adv 2016;2(7):e1600584.
[40] Sen P, Shah PP, Nativio R, Berger SL. Epigenetic mechanisms of longevity and aging. Cell 2016;166(4): 822–39.
[41] Tohgi H, Utsugisawa K, Nagane Y, Yoshimura M, Genda Y, Ukitsu M. Reduction with age in methylcytosine in the promoter region -224 approximately -101 of the amyloid precursor protein gene in autopsy human cortex. Brain Res Mol Brain Res 1999;70(2):288–92.

[42] Tohgi H, Utsugisawa K, Nagane Y, Yoshimura M, Ukitsu M, Genda Y. The methylation status of cytosines in a tau gene promoter region alters with age to downregulate transcriptional activity in human cerebral cortex. Neurosci Lett 1999;275(2):89–92.
[43] Chouliaras L, Rutten BP, Kenis G, Peerbooms O, Visser PJ, Verhey F, et al. Epigenetic regulation in the pathophysiology of Alzheimer's disease. Prog Neurobiol 2010;90(4):498–510.
[44] Lunnon K, Mill J. Epigenetic studies in Alzheimer's disease: current findings, caveats, and considerations for future studies. Am J Med Genet B Neuropsychiatr Genet 2013;162B(8):789–99.
[45] Mill J. Toward an integrated genetic and epigenetic approach to Alzheimer's disease. Neurobiol Aging 2011;32(7):1188–91.
[46] Smith RG, Lunnon K. DNA modifications and alzheimer's disease. In: Delgado-Morales R, editor. Neuroepigenomics in aging and disease. Advances in experimental medicine and biology, 978. Springer; 2017. p. 303–19.
[47] Yong WS, Hsu FM, Chen PY. Profiling genome-wide DNA methylation. Epigenet Chromatin 2016;9:26.
[48] Bradley-Whitman MA, Lovell MA. Epigenetic changes in the progression of Alzheimer's disease. Mech Ageing Dev 2013;134(10):486–95.
[49] Chouliaras L, Mastroeni D, Delvaux E, Grover A, Kenis G, Hof PR, et al. Consistent decrease in global DNA methylation and hydroxymethylation in the hippocampus of Alzheimer's disease patients. Neurobiol Aging 2013;34(9):2091–9.
[50] Condliffe D, Wong A, Troakes C, Proitsi P, Patel Y, Chouliaras L, et al. Cross-region reduction in 5-hydroxymethylcytosine in Alzheimer's disease brain. Neurobiol Aging 2014;35(8):1850–4.
[51] Coppieters N, Dieriks BV, Lill C, Faull RL, Curtis MA, Dragunow M. Global changes in DNA methylation and hydroxymethylation in Alzheimer's disease human brain. Neurobiol Aging 2014;35(6):1334–44.
[52] Lashley T, Gami P, Valizadeh N, Li A, Revesz T, Balazs R. Alterations in global DNA methylation and hydroxymethylation are not detected in Alzheimer's disease. Neuropathol Appl Neurobiol 2015;41(4):497–506.
[53] Mastroeni D, Grover A, Delvaux E, Whiteside C, Coleman PD, Rogers J. Epigenetic changes in Alzheimer's disease: decrements in DNA methylation. Neurobiol Aging 2010;31(12):2025–37.
[54] West RL, Lee JM, Maroun LE. Hypomethylation of the amyloid precursor protein gene in the brain of an Alzheimer's disease patient. J Mol Neurosci 1995;6(2):141–6.
[55] Iwata A, Nagata K, Hatsuta H, Takuma H, Bundo M, Iwamoto K, et al. Altered CpG methylation in sporadic Alzheimer's disease is associated with APP and MAPT dysregulation. Hum Mol Genet 2014;23(3):648–56.
[56] Coupland KG, Kim WS, Halliday GM, Hallupp M, Dobson-Stone C, Kwok JB. Effect of PSEN1 mutations on MAPT methylation in early-onset Alzheimer's disease. Curr Alzheimer Res 2015;12(8):745–51.
[57] Rao JS, Keleshian VL, Klein S, Rapoport SI. Epigenetic modifications in frontal cortex from Alzheimer's disease and bipolar disorder patients. Transl Psychiatry 2012;2:e132.
[58] Bollati V, Galimberti D, Pergoli L, Dalla Valle E, Barretta F, Cortini F, et al. DNA methylation in repetitive elements and Alzheimer disease. Brain Behav Immun 2011;25(6):1078–83.
[59] Bakulski KM, Dolinoy DC, Sartor MA, Paulson HL, Konen JR, Lieberman AP, et al. Genome-wide DNA methylation differences between late-onset Alzheimer's disease and cognitively normal controls in human frontal cortex. J Alzheim Dis 2012;29(3):571–88.
[60] De Jager PL, Srivastava G, Lunnon K, Burgess J, Schalkwyk LC, Yu L, et al. Alzheimer's disease: early alterations in brain DNA methylation at ANK1, BIN1, RHBDF2 and other loci. Nat Neurosci 2014;17(9):1156–63.
[61] Lunnon K, Smith R, Hannon E, De Jager PL, Srivastava G, Volta M, et al. Methylomic profiling implicates cortical deregulation of ANK1 in Alzheimer's disease. Nat Neurosci 2014;17(9):1164–70.
[62] Lord J, Cruchaga C. The epigenetic landscape of Alzheimer's disease. Nat Neurosci 2014;17(9):1138–40.

[63] Yu L, Chibnik LB, Srivastava GP, Pochet N, Yang J, Xu J, et al. Association of brain DNA methylation in SORL1, ABCA7, HLA-DRB5, SLC24A4, and BIN1 with pathological diagnosis of Alzheimer disease. JAMA Neurol 2015;72(1):15–24.
[64] Smith AR, Smith RG, Condliffe D, Hannon E, Schalkwyk L, Mill J, et al. Increased DNA methylation near TREM2 is consistently seen in the superior temporal gyrus in Alzheimer's disease brain. Neurobiol Aging 2016;47:35–40.
[65] Watson CT, Roussos P, Garg P, Ho DJ, Azam N, Katsel PL, et al. Genome-wide DNA methylation profiling in the superior temporal gyrus reveals epigenetic signatures associated with Alzheimer's disease. Genome Med 2016;8(1):5.
[66] Smith RG, Hannon E, De Jager PL, Chibnik L, Lott SJ, Condliffe D, et al. Elevated DNA methylation across a 48-kb region spanning the HOXA gene cluster is associated with Alzheimer's disease neuropathology. Alzheimer's Dementia 2018;14(12):1580–8.
[67] Smith AR, Smith R, Burrage J, Troakes C, Al-Saraj S, Kalaria RN, et al. A cross-brain-regions study of ANK1 DNA methylation in different neurodegenerative diseases. Neurobiol Aging 2019:70–6.
[68] Lunnon K, Hannon E, Smith RG, Dempster E, Wong C, Burrage J, et al. Variation in 5-hydroxymethylcytosine across human cortex and cerebellum. Genome Biol 2016;17:27.
[69] Zhao J, Zhu Y, Yang J, Li L, Wu H, De Jager PL, et al. A genome-wide profiling of brain DNA hydroxymethylation in Alzheimer's disease. Alzheimer's Dementia 2017;13(6):674–88.
[70] Khare T, Pai S, Koncevicius K, Pal M, Kriukiene E, Liutkeviciute Z, et al. 5-hmC in the brain is abundant in synaptic genes and shows differences at the exon-intron boundary. Nat Struct Mol Biol 2012;19(10): 1037–43.
[71] Spiers H, Hannon E, Schalkwyk LC, Bray NJ, Mill J. 5-hydroxymethylcytosine is highly dynamic across human fetal brain development. BMC Genomics 2017;18(1):738.
[72] Devall M, Roubroeks J, Mill J, Weedon M, Lunnon K. Epigenetic regulation of mitochondrial function in neurodegenerative disease: new insights from advances in genomic technologies. Neurosci Lett 2016;625: 47–55.
[73] Blanch M, Mosquera JL, Ansoleaga B, Ferrer I, Barrachina M. Altered mitochondrial DNA methylation pattern in Alzheimer disease-related pathology and in Parkinson disease. Am J Pathol 2016;186(2):385–97.
[74] Stoccoro A, Siciliano G, Migliore L, Coppede F. Decreased methylation of the mitochondrial D-loop region in late-onset Alzheimer's disease. J Alzheim Dis 2017;59(2):559–64.
[75] Henikoff S, Smith MM. Histone variants and epigenetics. Cold Spring Harb Perspect Biol 2015;7(1): a019364.
[76] Fischer A. Targeting histone-modifications in Alzheimer's disease. What is the evidence that this is a promising therapeutic avenue? Neuropharmacology 2014;80:95–102.
[77] Zhao Y, Garcia BA. Comprehensive catalog of currently documented histone modifications. Cold Spring Harb Perspect Biol 2015;7(9):a025064.
[78] Lai WKM, Pugh BF. Understanding nucleosome dynamics and their links to gene expression and DNA replication. Nat Rev Mol Cell Biol 2017;18(9):548–62.
[79] Halder R, Hennion M, Vidal RO, Shomroni O, Rahman RU, Rajput A, et al. DNA methylation changes in plasticity genes accompany the formation and maintenance of memory. Nat Neurosci 2016;19(1):102–10.
[80] Levenson JM, O'Riordan KJ, Brown KD, Trinh MA, Molfese DL, Sweatt JD. Regulation of histone acetylation during memory formation in the hippocampus. J Biol Chem 2004;279(39):40545–59.
[81] Peleg S, Sananbenesi F, Zovoilis A, Burkhardt S, Bahari-Javan S, Agis-Balboa RC, et al. Altered histone acetylation is associated with age-dependent memory impairment in mice. Science 2010;328(5979): 753–6.
[82] Zocchi L, Sassone-Corsi P. Joining the dots: from chromatin remodeling to neuronal plasticity. Curr Opin Neurobiol 2010;20(4):432–40.

[83] Bannister AJ, Kouzarides T. Regulation of chromatin by histone modifications. Cell Res 2011;21(3):381–95.
[84] Ogawa O, Zhu X, Lee HG, Raina A, Obrenovich ME, Bowser R, et al. Ectopic localization of phosphorylated histone H3 in Alzheimer's disease: a mitotic catastrophe? Acta Neuropathologica 2003;105(5):524–8.
[85] Lithner CU, Lacor PN, Zhao WQ, Mustafiz T, Klein WL, Sweatt JD, et al. Disruption of neocortical histone H3 homeostasis by soluble Abeta: implications for Alzheimer's disease. Neurobiol Aging 2013;34(9):2081–90.
[86] Myung NH, Zhu X, Kruman II, Castellani RJ, Petersen RB, Siedlak SL, et al. Evidence of DNA damage in Alzheimer disease: phosphorylation of histone H2AX in astrocytes. Age 2008;30(4):209–15.
[87] Black JC, Van Rechem C, Whetstine JR. Histone lysine methylation dynamics: establishment, regulation, and biological impact. Mol Cell 2012;48(4):491–507.
[88] Ryu H, Barrup M, Kowall NW, McKee AC. Epigenetic modification in a monozygotic twin with Alzheimer's disease. Alzheimer's Dementia 2008;4(4).
[89] Mastroeni D, Delvaux E, Nolz J, Tan Y, Grover A, Oddo S, et al. Aberrant intracellular localization of H3k4me3 demonstrates an early epigenetic phenomenon in Alzheimer's disease. Neurobiol Aging 2015;36(12):3121–9.
[90] Anderson KW, Turko IV. Histone post-translational modifications in frontal cortex from human donors with Alzheimer's disease. Clin Proteonomics 2015;12:26.
[91] Greer EL, Shi Y. Histone methylation: a dynamic mark in health, disease and inheritance. Nat Rev Genet 2012;13(5):343–57.
[92] Buesa C, Mascaró C, Rotllant D, Griñan-Ferré C, Pallàs M, Maes T. The dual lsd1/maob inhibitor ory2001 prevents the development of the memory deficit in samp8 mice through induction of neuronal plasticity and reduction of neuroinflammation. Alzheimer's Dementia 2015;11(7).
[93] Maes T, Molinero C, Antonijoan RM, Ferrero-Cafiero JM, Martínez-Colomer J, Mascaro C, et al. First-in-Human phase I results show safety, tolerability and brain penetrance of ory-2001, an epigenetic drug targeting Lsd1 and mao-B. Alzheimer's Dementia 2017;13(7):P1573–4.
[94] Walker MP, LaFerla FM, Oddo SS, Brewer GJ. Reversible epigenetic histone modifications and Bdnf expression in neurons with aging and from a mouse model of Alzheimer's disease. Age 2013;35(3):519–31.
[95] Grinan-Ferre C, Sarroca S, Ivanova A, Puigoriol-Illamola D, Aguado F, Camins A, et al. Epigenetic mechanisms underlying cognitive impairment and Alzheimer disease hallmarks in 5XFAD mice. Aging 2016;8(4):664–84.
[96] Sharma M, Dierkes T, Sajikumar S. Epigenetic regulation by G9a/GLP complex ameliorates amyloid-beta 1-42 induced deficits in long-term plasticity and synaptic tagging/capture in hippocampal pyramidal neurons. Aging Cell 2017;16(5):1062–72.
[97] Sharma M, Raghuraman R, Sajikumar S. Epigenetics: the panacea for cognitive decline? Aging (Albany NY) 2018;10(1):1–2.
[98] Zhang K, Schrag M, Crofton A, Trivedi R, Vinters H, Kirsch W. Targeted proteomics for quantification of histone acetylation in Alzheimer's disease. Proteomics 2012;12(8):1261–8.
[99] Narayan PJ, Lill C, Faull R, Curtis MA, Dragunow M. Increased acetyl and total histone levels in post-mortem Alzheimer's disease brain. Neurobiol Dis 2015;74:281–94.
[100] Marzi S, Ribarska T, Smith AR, Hannon E, Poschmann J, Moore K, et al. A histone acetylome-wide association study of Alzheimer's disease: neuropathology-associated regulatory variation in the human entorhinal cortex. Nat Neurosci 2018;21(11):1618–27.
[101] Nativio R, Donahue G, Berson A, Lan Y, Amlie-Wolf A, Tuzer F, et al. Dysregulation of the epigenetic landscape of normal aging in Alzheimer's disease. Nat Neurosci 2018;21(4):497–505.

[102] Graff J, Rei D, Guan JS, Wang WY, Seo J, Hennig KM, et al. An epigenetic blockade of cognitive functions in the neurodegenerating brain. Nature 2012;483(7388):222–6.

[103] Odagiri S, Tanji K, Mori F, Miki Y, Kakita A, Takahashi H, et al. Brain expression level and activity of HDAC6 protein in neurodegenerative dementia. Biochem Biophys Res Commun 2013;430(1):394–9.

[104] Mehler MF, Mattick JS. Noncoding RNAs and RNA editing in brain development, functional diversification, and neurological disease. Physiol Rev 2007;87(3):799–823.

[105] Gomes AQ, Nolasco S, Soares H. Non-coding RNAs: multi-tasking molecules in the cell. Int J Mol Sci 2013;14(8):16010–39.

[106] Cech TR, Steitz JA. The noncoding RNA revolution-trashing old rules to forge new ones. Cell 2014;157(1):77–94.

[107] Bekris LM, Lutz F, Montine TJ, Yu CE, Tsuang D, Peskind ER, et al. MicroRNA in Alzheimer's disease: an exploratory study in brain, cerebrospinal fluid and plasma. Biomarkers 2013;18(5):455–66.

[108] Gstir R, Schafferer S, Scheideler M, Misslinger M, Griehl M, Daschil N, et al. Generation of a neuro-specific microarray reveals novel differentially expressed noncoding RNAs in mouse models for neurodegenerative diseases. RNA 2014;20(12):1929–43.

[109] Hu YB, Li CB, Song N, Zou Y, Chen SD, Ren RJ, et al. Diagnostic value of microRNA for alzheimer's disease: a systematic review and meta-analysis. Front Aging Neurosci 2016;8:13.

[110] Jaber V, Zhao Y, Lukiw WJ. Alterations in micro RNA-messenger RNA (miRNA-mRNA) coupled signaling networks in sporadic Alzheimer's disease (AD) hippocampal CA1. J Alzheimers Dis Parkinsonism 2017;7(2).

[111] Lau P, Bossers K, Janky R, Salta E, Frigerio CS, Barbash S, et al. Alteration of the microRNA network during the progression of Alzheimer's disease. EMBO Mol Med 2013;5(10):1613–34.

[112] Nelson PT, Wang WX. MiR-107 is reduced in Alzheimer's disease brain neocortex: validation study. J Alzheimers Dis 2010;21(1):75–9.

[113] Nunez-Iglesias J, Liu CC, Morgan TE, Finch CE, Zhou XJ. Joint genome-wide profiling of miRNA and mRNA expression in Alzheimer's disease cortex reveals altered miRNA regulation. PLoS One 2010;5(2):e8898.

[114] Patrick E, Rajagopal S, Wong HA, McCabe C, Xu J, Tang A, et al. Dissecting the role of non-coding RNAs in the accumulation of amyloid and tau neuropathologies in Alzheimer's disease. Mol Neurodegener 2017;12(1):51.

[115] Shioya M, Obayashi S, Tabunoki H, Arima K, Saito Y, Ishida T, et al. Aberrant microRNA expression in the brains of neurodegenerative diseases: miR-29a decreased in Alzheimer disease brains targets neurone navigator 3. Neuropathol Appl Neurobiol 2010;36(4):320–30.

[116] Wang LL, Min L, Guo QD, Zhang JX, Jiang HL, Shao S, et al. Profiling microRNA from brain by microarray in a transgenic mouse model of alzheimer's disease. BioMed Res Int 2017;2017. 8030369.

[117] Wang WX, Huang Q, Hu Y, Stromberg AJ, Nelson PT. Patterns of microRNA expression in normal and early Alzheimer's disease human temporal cortex: white matter versus gray matter. Acta Neuropathol 2011;121(2):193–205.

[118] Wang WX, Rajeev BW, Stromberg AJ, Ren N, Tang G, Huang Q, et al. The expression of microRNA miR-107 decreases early in Alzheimer's disease and may accelerate disease progression through regulation of beta-site amyloid precursor protein-cleaving enzyme 1. J Neurosci 2008;28(5):1213–23.

[119] Wang X, Liu P, Zhu H, Xu Y, Ma C, Dai X, et al. miR-34a, a microRNA up-regulated in a double transgenic mouse model of Alzheimer's disease, inhibits bcl2 translation. Brain Res Bull 2009;80(4–5):268–73.

[120] Cogswell JP, Ward J, Taylor IA, Waters M, Shi Y, Cannon B, et al. Identification of miRNA changes in Alzheimer's disease brain and CSF yields putative biomarkers and insights into disease pathways. J Alzheimers Dis 2008;14(1):27–41.

[121] Hebert SS, Horre K, Nicolai L, Papadopoulou AS, Mandemakers W, Silahtaroglu AN, et al. Loss of microRNA cluster miR-29a/b-1 in sporadic Alzheimer's disease correlates with increased BACE1/beta-secretase expression. Proc Natl Acad Sci U S A 2008;105(17):6415–20.
[122] Faghihi MA, Modarresi F, Khalil AM, Wood DE, Sahagan BG, Morgan TE, et al. Expression of a non-coding RNA is elevated in Alzheimer's disease and drives rapid feed-forward regulation of beta-secretase. Nat Med 2008;14(7):723–30.
[123] Massone S, Vassallo I, Fiorino G, Castelnuovo M, Barbieri F, Borghi R, et al. 17A, a novel non-coding RNA, regulates GABA B alternative splicing and signaling in response to inflammatory stimuli and in Alzheimer disease. Neurobiol Dis 2011;41(2):308–17.
[124] Ciarlo E, Massone S, Penna I, Nizzari M, Gigoni A, Dieci G, et al. An intronic ncRNA-dependent regulation of SORL1 expression affecting Abeta formation is upregulated in post-mortem Alzheimer's disease brain samples. Dis Model Mech 2013;6(2):424–33.
[125] Niu Y, Zhao X, Wu YS, Li MM, Wang XJ, Yang YG. N6-methyl-adenosine (m6A) in RNA: an old modification with a novel epigenetic function. Genom Proteom Bioinform 2013;11(1):8–17.
[126] Fransquet PD, Lacaze P, Saffery R, McNeil J, Woods R, Ryan J. Blood DNA methylation as a potential biomarker of dementia: a systematic review. Alzheimers Dement 2018;14(1):81–103.
[127] Davies MN, Volta M, Pidsley R, Lunnon K, Dixit A, Lovestone S, et al. Functional annotation of the human brain methylome identifies tissue-specific epigenetic variation across brain and blood. Genome Biol 2012; 13(6):R43.
[128] Li H, Guo Z, Guo Y, Li M, Yan H, Cheng J, et al. Common DNA methylation alterations of Alzheimer's disease and aging in peripheral whole blood. Oncotarget 2016;7(15):19089–98.
[129] Chang L, Wang Y, Ji H, Dai D, Xu X, Jiang D, et al. Elevation of peripheral BDNF promoter methylation links to the risk of Alzheimer's disease. PLoS One 2014;9(11):e110773.
[130] Ji H, Wang Y, Liu G, Chang L, Chen Z, Zhou D, et al. Elevated OPRD1 promoter methylation in Alzheimer's disease patients. PLoS One 2017;12(3):e0172335.
[131] Ferri E, Arosio B, D'Addario C, Galimberti D, Gussago C, Pucci M, et al. Gene promoter methylation and expression of Pin1 differ between patients with frontotemporal dementia and Alzheimer's disease. J Neurol Sci 2016;362:283–6.
[132] Grossi E, Stoccoro A, Tannorella P, Migliore L, Coppede F. Artificial neural networks link one-carbon metabolism to gene-promoter methylation in alzheimer's disease. J Alzheimers Dis 2016;53(4):1517–22.
[133] Arosio B, Bulbarelli A, Bastias Candia S, Lonati E, Mastronardi L, Romualdi P, et al. Pin1 contribution to Alzheimer's disease: transcriptional and epigenetic mechanisms in patients with late-onset Alzheimer's disease. Neurodegener Dis 2012;10(1–4):207–11.
[134] D'Addario C, Candia SB, Arosio B, Di Bartolomeo M, Abbate C, Case A, et al. Transcriptional and epigenetic phenomena in peripheral blood cells of monozygotic twins discordant for Alzheimer's disease, a case report. J Neurol Sci 2017;372:211–6.
[135] Yamazaki K, Yoshino Y, Mori T, Okita M, Yoshida T, Mori Y, et al. Association study and meta-analysis of polymorphisms, methylation profiles, and peripheral mRNA expression of the serotonin transporter gene in patients with Alzheimer's disease. Dement Geriatr Cognit Disord 2016;41(5–6):334–47.
[136] Yamazaki K, Yoshino Y, Mori T, Yoshida T, Ozaki Y, Sao T, et al. Gene expression and methylation analysis of ABCA7 in patients with Alzheimer's disease. J Alzheimers Dis 2017;57(1):171–81.
[137] Carboni L, Lattanzio F, Candeletti S, Porcellini E, Raschi E, Licastro F, et al. Peripheral leukocyte expression of the potential biomarker proteins Bdnf, Sirt1, and Psen1 is not regulated by promoter methylation in Alzheimer's disease patients. Neurosci Lett 2015;605:44–8.
[138] Kobayashi N, Shinagawa S, Nagata T, Shimada K, Shibata N, Ohnuma T, et al. Development of biomarkers based on DNA methylation in the NCAPH2/LMF2 promoter region for diagnosis of Alzheimer's disease and amnesic mild cognitive impairment. PLoS One 2016;11(1):e0146449.

[139] Bos I, Vos SJB, Vandenberghe R, Scheltens P, Engelborghs S, Frisoni GB, et al. European medical information framework for Alzheimer's disease (EMIF-AD): the biomarker discovery study. Alzheimer's Dementia 2017;13(7):P691−2.

[140] Vasanthakumar A, Davis JW, Idler K, Kim S, Nho K, Nudelman KNH, et al. DNA methylation dynamics in Alzheimer's disease diagnosis and progression. Alzheimer's & Dementia 2017;13(7):P297.

[141] Plagg B, Ehrlich D, Kniewallner KM, Marksteiner J, Humpel C. Increased acetylation of histone H4 at lysine 12 (H4K12) in monocytes of transgenic Alzheimer's mice and in human patients. Curr Alzheimer Res 2015;12(8):752−60.

[142] Wu HZ, Ong KL, Seeher K, Armstrong NJ, Thalamuthu A, Brodaty H, et al. Circulating microRNAs as biomarkers of alzheimer's disease: a systematic review. J Alzheimers Dis 2016;49(3):755−66.

[143] Gotz J, Ittner LM. Animal models of Alzheimer's disease and frontotemporal dementia. Nat Rev Neurosci 2008;9(7):532−44.

[144] Francis YI, Fa M, Ashraf H, Zhang H, Staniszewski A, Latchman DS, et al. Dysregulation of histone acetylation in the APP/PS1 mouse model of Alzheimer's disease. J Alzheimers Dis 2009;18(1):131−9.

[145] Marques SC, Lemos R, Ferreiro E, Martins M, de Mendonca A, Santana I, et al. Epigenetic regulation of BACE1 in Alzheimer's disease patients and in transgenic mice. Neuroscience 2012;220:256−66.

[146] Sanchez-Mut JV, Aso E, Panayotis N, Lott I, Dierssen M, Rabano A, et al. DNA methylation map of mouse and human brain identifies target genes in Alzheimer's disease. Brain 2013;136(Pt 10):3018−27.

[147] Zhang J, Hu M, Teng Z, Tang YP, Chen C. Synaptic and cognitive improvements by inhibition of 2-AG metabolism are through upregulation of microRNA-188-3p in a mouse model of Alzheimer's disease. J Neurosci 2014;34(45):14919−33.

[148] Gjoneska E, Pfenning AR, Mathys H, Quon G, Kundaje A, Tsai LH, et al. Conserved epigenomic signals in mice and humans reveal immune basis of Alzheimer's disease. Nature 2015;518(7539):365−9.

[149] Cuadrado-Tejedor M, Garcia-Barroso C, Sanchez-Arias JA, Rabal O, Perez-Gonzalez M, Mederos S, et al. A first-in-class small-molecule that acts as a dual inhibitor of HDAC and PDE5 and that rescues hippocampal synaptic impairment in Alzheimer's disease mice. Neuropsychopharmacology 2017;42(2):524−39.

[150] Govindarajan N, Agis-Balboa RC, Walter J, Sananbenesi F, Fischer A. Sodium butyrate improves memory function in an Alzheimer's disease mouse model when administered at an advanced stage of disease progression. J Alzheimers Dis 2011;26(1):187−97.

[151] Kilgore M, Miller CA, Fass DM, Hennig KM, Haggarty SJ, Sweatt JD, et al. Inhibitors of class 1 histone deacetylases reverse contextual memory deficits in a mouse model of Alzheimer's disease. Neuropsychopharmacology 2010;35(4):870−80.

[152] Ricobaraza A, Cuadrado-Tejedor M, Marco S, Perez-Otano I, Garcia-Osta A. Phenylbutyrate rescues dendritic spine loss associated with memory deficits in a mouse model of Alzheimer disease. Hippocampus 2012;22(5):1040−50.

[153] Ricobaraza A, Cuadrado-Tejedor M, Perez-Mediavilla A, Frechilla D, Del Rio J, Garcia-Osta A. Phenylbutyrate ameliorates cognitive deficit and reduces tau pathology in an Alzheimer's disease mouse model. Neuropsychopharmacology 2009;34(7):1721−32.

[154] Volmar CH, Salah-Uddin H, Janczura KJ, Halley P, Lambert G, Wodrich A, et al. M344 promotes nonamyloidogenic amyloid precursor protein processing while normalizing Alzheimer's disease genes and improving memory. Proc Natl Acad Sci U S A 2017;114(43):E9135−44.

[155] Cong L, Jia J, Qin W, Ren Y, Sun Y. Genome-wide analysis of DNA methylation in an APP/PS1 mouse model of Alzheimer's disease. Acta Neurol Belg 2014;114(3):195−206.

[156] Coupland K, Kim WS, Halliday G, Dobson-Stone C, Kwok JBJ. Mapt methylation in alzheimer's disease. Alzheimer's Dementia 2014;10(4):P317−8.

[157] Cadena-del-Castillo C, Valdes-Quezada C, Carmona-Aldana F, Arias C, Bermudez-Rattoni F, Recillas-Targa F. Age-dependent increment of hydroxymethylation in the brain cortex in the triple-transgenic mouse model of Alzheimer's disease. J Alzheimers Dis 2014;41(3):845–54.
[158] Lardenoije R, van den Hove DLA, Havermans M, van Casteren A, Le KX, Palmour R, et al. Age-related epigenetic changes in hippocampal subregions of four animal models of Alzheimer's disease. Mol Cell Neurosci 2018;86:1–15.
[159] Gasparoni G, Bultmann S, Lutsik P, Kraus TFJ, Sordon S, Vlcek J, et al. DNA methylation analysis on purified neurons and glia dissects age and Alzheimer's disease-specific changes in the human cortex. Epigenet Chromatin 2018;11(1):41.
[160] Ziller MJ, Hansen KD, Meissner A, Aryee MJ. Coverage recommendations for methylation analysis by whole-genome bisulfite sequencing. Nat Methods 2015;12(3):230–2. 1 pp. following 2.
[161] Gu H, Smith ZD, Bock C, Boyle P, Gnirke A, Meissner A. Preparation of reduced representation bisulfite sequencing libraries for genome-scale DNA methylation profiling. Nat Protoc 2011;6(4):468–81.
[162] Meissner A, Gnirke A, Bell GW, Ramsahoye B, Lander ES, Jaenisch R. Reduced representation bisulfite sequencing for comparative high-resolution DNA methylation analysis. Nucleic Acids Res 2005;33(18): 5868–77.
[163] Hirst M, Marra MA. Next generation sequencing based approaches to epigenomics. Brief Funct Genomics 2010;9(5–6):455–65.
[164] Imm J, Kerrigan TL, Jeffries A, Lunnon K. Using induced pluripotent stem cells to explore genetic and epigenetic variation associated with Alzheimer's disease. Epigenomics 2017;9(11):1455–68.
[165] Cong L, Ran FA, Cox D, Lin S, Barretto R, Habib N, et al. Multiplex genome engineering using CRISPR/Cas systems. Science 2013;339(6121):819–23.
[166] Liu XS, Wu H, Ji X, Stelzer Y, Wu X, Czauderna S, et al. Editing DNA methylation in the mammalian genome. Cell 2016;167(1):233–247 e17.

SECTION

2 Neurodevelopmental disorders

CHAPTER

9

Genetic and epigenetic influences on the phenotype of Rett syndrome

Jenny Downs[1,2], Janine M. LaSalle[3], Helen Leonard[1]

Telethon Kids Institute, The University of Western Australia, Perth, WA, Australia[1]; School of Physiotherapy and Exercise Science, Curtin University, Perth, WA, Australia[2]; Medical Microbiology and Immunology, Genome Center, MIND Institute, University of California, Davis, CA, United States[3]

Chapter Outline

9.1 Introduction **184**
9.2 The genetic cause of Rett syndrome **185**
9.3 The biology of MeCP2 .... **186**
9.3.1 Neurobiology of MeCP2 186
9.3.2 Molecular functioning of MeCP2 187
9.4 The phenotype of Rett syndrome **189**
9.4.1 Early development and regression 189
9.4.2 Diagnostic criteria 190
9.4.3 Functional impairments 190
9.4.4 Stereotypies 193
9.4.5 Comorbidities 193
9.4.6 Epidemiology 197
9.5 Evidence for epigenetic mechanisms affecting *MECP2* function and expression **198**
9.5.1 DNA methylation 199
9.5.2 Histone modifications and nucleosome and higher-order chromatin remodeling 200
9.5.3 Noncoding RNAs 201
9.5.4 RNA splicing 201
9.6 Epigenetic regulation of MeCP2 expression or phenotypes **202**
9.6.1 X chromosome inactivation 202
9.6.2 IGF1/mTOR pathway 203
9.6.3 Enriched environments 204
9.7 Inclusion of epigenetic data collection in epidemiological studies **206**
9.8 Summary **207**
References **207**

Chromatin Signaling and Neurological Disorders. https://doi.org/10.1016/B978-0-12-813796-3.00009-2

9.1 Introduction

Rett syndrome (RTT) was first described as a syndrome in 1966 in the German medical literature [1], but it was when RTT was described by Hagberg, Aicardi, Dias, and Ramos in the English literature in 1983 [2] that a period of rapid growth in its understanding began. Rett syndrome is associated with marked impairments of communication, gross motor and fine motor functioning, as well as the development of medical comorbidities such as scoliosis, epilepsy, poor growth, and autonomic dysfunction [3]. The development of database infrastructure has accelerated progress in understanding the clinical phenotype, initially with the establishment of the British Isles RTT Survey in 1990 [4]. Additional database infrastructures that continue to ascertain cases to this day were established thereafter. Based in Australia, the population-based Australian Rett Syndrome Study was established in 1993 and collects data longitudinally [5,6]; the International Rett Syndrome Phenotype Database (InterRett) was established in 2002 and collects data from families living in over 50 countries [7]; and RettBASE, the *MECP2* Variation Database, catalogs different gene variants reported by clinicians, laboratories, and in the literature [8]. In the United States, the Natural History Study was established in 2006 to collect extensive clinical data at multiple sites across the United States [9,10]. In Europe, The European Rett Syndrome Database Network (EuroRett) combines data from sources from 11 European countries [11]. The total sample from these infrastructures is substantial in size, spans the globe, and their extensive research output sets the scene for greater clarity in understanding the natural history of Rett syndrome.

RTT mainly affects females and from when the syndrome was first recognized, there was discussion that the cause was likely to be genetic. The discovery that a mutation on the *MECP2* gene was causal in 1999 [12] provided impetus for a second wave of growth in knowledge about RTT. Many individual pathogenic *MECP2* mutations have been identified with 11 of these occurring commonly [5,9]. A suite of studies has identified clear and consistent patterns between mutation type and phenotype with some mutations associated with milder clinical severity and others with greater severity. The *MECP2* gene encodes the methyl CpG binding protein 2 (MeCP2). There has been intense interest in understanding the neurobiology of MeCP2, and its role in brain function and in other nonneural cells, and the phenotype has been reversed in MeCP2 knockout mice.

However, there are additional nongenetic or environmental factors that influence phenotype, factors that directly affect the mutated *MECP2* gene and its protein production, or factors that influence phenotype through alternative avenues. Together these can be considered as epigenetic or chromatin modifications that alter phenotype without changes to DNA sequence. Epigenetics is a relatively recent field of science that is making important contributions to understanding the biological complexities surrounding monogenetic disorders such as RTT. Many epigenetic mechanisms have been identified that either promote synaptic plasticity or regulate the structure and activity of genetic function in response to various intracellular and environmental cues. The four cardinal and interdependent epigenetics mechanisms include DNA methylation; histone modifications, nucleosome, and higher order chromatin structure; noncoding RNAs; RNA editing and DNA recoding [13]. MeCP2 is itself a predicted epigenetic regulator, based on its original description of a "reader" of DNA methylation marks (see Chapter 1 for more details on readers). It is therefore important to consider that key influences can arise from both genetic and environmental cues that can modify phenotype [14,15]. For example, maternal exposure to famine around the time of conception causes decreased methylation of the gene encoding *IGF2*, accounting for later metabolic abnormalities [16]. Environmental enrichment has been found to reduce cellular and behavioral deficits in animal models of genetically caused neurological disorders including Huntington's disease, Fragile X, and Down syndrome [17]. Research on the epigenetic influences in RTT is a more recently burgeoning field.

In sum, multiple factors influence how RTT manifests, and there are important roles for both nature and nurture. Perspectives that are restricted to only phenotype, genetics, or epigenetics can only provide a cloudy vision because each portrays only a portion of the complex biological and clinical picture. Rather, combined perspectives have the capacity for greater clarity and more potential for progress along the pipeline to discover effective treatments. The main goals of this chapter are therefore threefold: (1) to describe MeCP2 structure and function, (2) to characterize the phenotype of RTT including relationships with genotype and recent findings from longitudinal datasets, and (3) to outline the known epigenetic mechanisms that influence both genetic function and phenotype in RTT. Finally, we consider how data on epigenetic influences might be collected systematically within current database infrastructures.

9.2 The genetic cause of Rett syndrome

RTT almost exclusively affects females. From the first publication describing RTT in English, it was suggested that its cause was likely to be genetic and X-linked [2]. The genetic basis of the disorder was further supported by studies of X inactivation in twins and familial examples of skewed X chromosome inactivation in asymptomatic mothers who had an affected daughter [18]. Careful exclusion mapping in a small number of rare familial cases suggested candidate areas of the X chromosome and sequencing then identified mutations in the *MECP2* gene [12]. It had been 32 years since RTT was first described by the Austrian pediatrician, Dr. Andreas Rett in 1966 [1].

Most affected individuals are female but in some of the rare cases of familial transmission, asymptomatic mothers can give birth to an affected male. Although there is variability in the severity of phenotype [19], most affected boys die during infancy because with only one X chromosome, they cannot derive benefit from wild-type *MECP2* expression in half their cells, as observed in females [19]. Furthermore, genotyping of child and parental DNA has indicated that the affected X chromosome is of paternal origin in 26 of 27 (96.3%) [20] and 84 of 88 (95.5%) sporadic Chinese cases [21], explaining the large proportion of sporadic cases because only females inherit the paternal X chromosome.

Before the discovery of the genetic cause of RTT, diagnosis depended on the use of clinical criteria but in most developed countries, genetic testing now has a fundamental role in achieving a diagnosis of RTT. Current testing uses a combined approach of mutation screening and sequencing of exons 3 and 4 for small sequence changes [22] and multiplex ligation-dependent probe amplification to screen for large deletions and duplications [23]. Screening for the much rarer exon 1 mutations [24] is also available. In some countries, direct sequencing is now being replaced with next-generation sequencing techniques such as whole-exome and whole-genome sequencing. More than 200 disease-causing *MECP2* mutations are now reported (http://mecp2.chw.edu.au) with eight of these occurring more commonly. For the common mutations, evidence is accumulating for the hypothesis that the type of genetic mutation is strongly associated with phenotype [9,25,26].

9.3 The biology of MeCP2

9.3.1 Neurobiology of MeCP2

Following the discovery of the genetic cause of RTT and in parallel with progress in understanding phenotype, there have been intensive efforts to further characterize the disorder biologically. Two *Mecp2*-null mouse models were generated in 2001 using Cre-*loxP* technology [27,28]. In each model, severe neurological symptoms were observed specifically in male *Mecp2*-null mice starting around 5–6 weeks old with death by 10–12 weeks. Consistent with clinical observations, male mice developed nervousness and trembling, abnormal gait, breathing abnormalities, and hind limb clasping [27,28]. There have also been other mouse models created with distinct phenotypes associated with *Mecp2* deletion from specific brain areas such as forebrain [29], or specific cell types such as aminergic neurons [30]. Creation of RTT mouse models has dramatically accelerated discoveries in relation to the neurobiology of MeCP2.

Mecp2-null mice exhibited gross neuroanatomical changes including microcephaly [27] and thin somatosensory and motor cortices [31] with the high density of neurons in layers II/II being maintained rather than reducing with development [31]. On closer inspection, dendrites were thinner, fewer, and less mature [27,31]. These features as well as altered calcium signaling and electrophysiological defects have been observed in induced pluripotent stem cell (iPSC) models derived from fibroblasts taken from patients with RTT, when compared to controls [32]. A neurochemical phenotype has also been identified with abnormalities in the gamma-aminobutyric acid (GABA) and glutamate pathways [33]. Deletion of *Mecp2* from GABA pathways in mouse models led to observable regression followed by the development of motor symptoms, abnormal breathing patterns, stereotypies, difficulties with learning and memory, and early death [34]. Deletion from glutamatergic pathways caused behavioral symptoms including anxiety and tremor [35]. Reduced concentrations of the aminergic metabolites of norepinephrine, dopamine, and serotonin were found in the cerebrospinal fluid of approximately 20% of 64 [30] patients with RTT compared with controls. GABA and glutamatergic neurotransmitter signaling pathways are critical to normal brain functioning, and abnormalities disrupt the necessary and finely tuned excitatory and inhibitory balance that are specific to different brain areas [33,36]. For example, reduced excitation and/or increased inhibition has been observed in the somatosensory, visual, frontal, motor, and medial prefrontal cortices [36]. On the other hand, brainstem structures such as ventrolateral medulla are more likely to demonstrate hyperexcitability [36]. Disrupted neurotransmission has major implications for health and functioning with cortical dysfunction associated with cognitive and developmental deficits and hyperactivity in brainstem regions in association with autonomic aberrations [36]. Abnormal GABA and glutamatergic pathways have specific implications for the development of epilepsy [37], which occurs in a high proportion of individuals with RTT by 15 years old [11,38–40].

Loss of MeCP2 from neuronal support cells also has a role to play. For example, MeCP2 deficiency in astrocytes has been associated with reduced blood brain-derived neurotrophic factor (BDNF) regulation, cytokine production, and dendrite development [41] and in microglia with altered glutamate synthesis [42]. In a subsequent study, preferential reexpression of MeCP2 in astrocytes resulted in some improvements in phenotype, namely better locomotion and reduced breathing abnormalities with a longer life span, as compared to MeCP2-null mice [43]. However, loss of MeCP2 function in neurons appears to be the dominant cause of RTT symptoms, consistent with the dramatic improvement in phenotype following activation of Mecp2-null allele globally [28,44] or more modest improvement when restricted to GABAnergic neurons [45]. These latter studies have illustrated that the devastating symptoms of RTT can be reversed, and there is hope that the development of therapeutic treatments in the future will be possible.

RTT is a genetically caused neurodevelopmental disability, and altered MeCP2 functioning in the central nervous system explains the neurological, behavioral, developmental, and autonomic symptoms of RTT. However, individuals with RTT experience pathologies in the peripheral body systems such as the lungs, heart, gastrointestinal tract, and bones, and evidence suggests that these consequences of MeCP2 dysfunction could be independent of central nervous system dysfunction. A peripheral *Mecp2* knockout mouse model was recently generated in which MeCP2 function was silenced in peripheral tissues but was present at near normal levels in the nervous system [46]. In these peripheral knockout mice, most neurological symptoms were absent, cardiac and autonomic function was normal but exercise capacity was reduced and there were bone abnormalities [46]. Other studies of *Mecp2* mutant mice have focused on specific body systems such as the lungs, heart, bones, and liver. In global *Mecp2*-null mouse models, lymphocytic bronchiolitis and alveolitis have been observed in the lungs [47] with reduced osteoblast functioning resulting in reduced bone volumes [48] and poorer biomechanical integrity [49] in bone tissue. Targeted deletion of *Mecp2* in the liver resulted in the fatty liver disease and dyslipidemia [50]. Together, these findings demonstrate wider effects of MeCP2 dysfunction characterized by a predominance of major defects in the central nervous system with additional and less extreme effects outside the nervous system.

9.3.2 Molecular functioning of MeCP2

RTT is a monogenic disorder and there is capacity to link downstream dysfunction with gene mutation, using the mice and cell line models discussed earlier. Genes were initially understood as discrete elements arranged in line along the DNA, providing safe storage for the protein material that is transcribed to the messenger RNA for subsequent production of a functional protein. Transcription is associated with additional complexities such as involvement with neighboring genes, transcribed regions have been found in intergenic and intronic sequences, and some transcripts have tissue-specific expression [51]. The basic structure of *MECP2* is a small gene comprising four exons from which two MeCP2 protein isoforms are expressed. Areas of importance on MeCP2 include the methyl-CpG-binding domain (MBD), a transcriptional repression domain (TRD), three AT-hooks (a DNA binding motif), and an NCOR–SMRT interaction domain within the TRD (Fig. 9.1) [52]. The protein product MeCP2 is expressed abundantly throughout the brain, and each of the protein domains has roles to play.

The MBD of MeCP2 is defined as minimal region for binding to chromatin and the different affinities to different types of DNA methylation, to be discussed in the next section. A key role of the TRD involves interactions with histone deacetylase (HDAC)-containing corepressor complexes including SIN3A [53], nuclear receptor corepressor (NCOR), and silencing mediator of retinoic acid and thyroid hormone receptor (SMRT), also known as NCOR2 [54]. The NCOR–SMRT interaction domain is located within the TRD [55]. Many of the pathogenic mutations causing RTT such as the nonsense (*) mutations p.Arg255*, p.Arg270*, and p.Arg294* are located in the TRD (Fig. 9.1). The role of the AT-hook domains in the C-terminal portion of MeCP2 is less clear, but the loss of this region is associated with chromatin differences, specifically adjustments to nucleosome positioning [56].

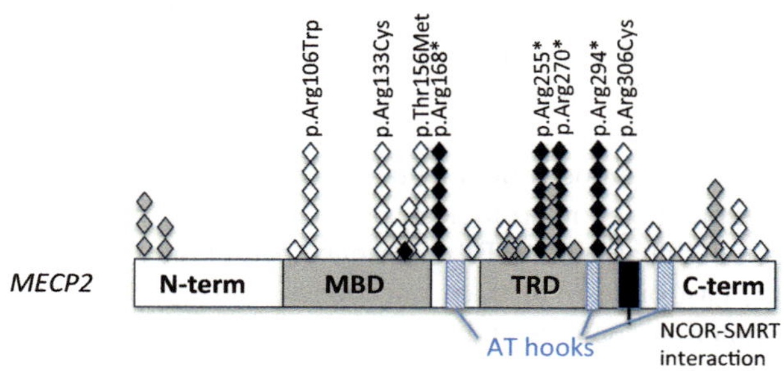

FIGURE 9.1

Map of Rett syndrome mutations in *MECP2.* The protein structure of known functional domains of MeCP2 is shown with a map of the locations and frequency of the most common mutations. The MBD, transcriptional repression domain (TRD), N-terminal, and C-terminal domains are shown. Missense mutations (white diamonds) in RTT are found most frequently in the MBD, and throughout the gene with the exception of the N-terminus. In contrast, frameshift and inframe mutations (gray diamonds) are rarely observed in the MBD, and most frequent in the N-terminal and C-terminal domains. Truncating nonsense mutations (black diamonds) are primarily observed in the TRD and interdomain between the MBD and TRD. Additional functional domains have been defined as AT-hook domains (hatched boxes) and the NCOR–SMRT interaction domain (black), and some rare RTT mutations are observed in these domains. The relative frequency of mutation positions are simplified as the number of diamonds, based on the RTT mutation database RettBASE (http://mecp2.chw.edu.au).

Complex molecular processes of MeCP2's interaction with chromatin serve to influence multiple other genes [51] through both silencing and activation [57]. MeCP2 therefore has many roles in regulating dynamic responses in cell function. Downstream effects are ubiquitous, and, in RTT, many cellular processes such as transcription and translation are disrupted. For example, MeCP2 dysfunction exposes neurons to abnormal synaptic function that differs by brain topography, including inhibition favored over excitation in cortical structures and pyramidal neurons and excitation over inhibition in the caudal brainstem [36]. These effects correspond clinically to reduced cognitive and other developmental functioning and the altered breathing patterns including breath holding and hyperventilation. *Mecp2*-null mice have demonstrated poor synaptic potentiation with reduced ability to modify synaptic responses with repeated stimulation, particularly in more severely affected mice [58]. This suggests poor plasticity and capacity to learn. *Ugcrc1* mRNA expression is elevated in MeCP2-null mice reducing the respiratory capacity of mitochondria [59], and this could account in part for poor muscle tone, growth, and developmental delay. Markers of oxidative stress are increased in the brains of *Mecp2*-null mice even before the onset of regression and can be normalized with *Mecp2* gene reactivation [60]. Similar redox imbalances have also been demonstrated in lung tissue of *Mecp2*-null mouse [47]. Demonstration of this pathology is consistent with the often subtle signs of altered development before regression including altered general movements [61], delays in the acquisition of sitting and walking [62,63], and altered quality of vocalizations [64].

9.4 The phenotype of Rett syndrome

9.4.1 Early development and regression

Before 1999, the diagnosis of RTT was based on a set of clinical criteria initially compiled in Vienna in 1985 by a group of European pediatric neurologists, headed by Professor Bengt Hagberg from Sweden [65]. A few years later, these were modified by a US group [66]. However, in both versions, the pre- and perinatal periods as well as the first 6 months of life were considered to be apparently normal before the later onset of a period of regression. At that time while the underlying probable genetic cause of RTT remained elusive, debate raged as to whether the regressive episode was a manifestation of a neurodegenerative disorder or whether it represented an arrest of development. Yet other than in the early seizure onset or Hanefeld variant form, now known to be a different disorder caused by a mutation in the *CDKL5* gene [67], the issue of early normality was never challenged. In 1993, the Australian Rett Syndrome Database, an ongoing population register of girls and women with RTT, was established [6]. In one of the first papers representing a total population of affected cases, data analysis showed that the proportions of cases with low birth weight was greater than in a population comparison group while the mean head circumference at birth was also lower after accounting for gestational age [63]. Although one in five clinicians commented on the presence of perinatal abnormalities, nearly half of the parents reported that their daughter's development or behavior was abnormal in the first 6 months of life.

Some years later and once the genetic cause of RTT had been identified, data ($n = 320$) from the Australian Rett Syndrome Database in combination with a newly established International Rett Syndrome Database were used to investigate any relationship between genotype and the presence of perinatal problems and perturbations of early development [68]. Overall, parents of nearly 60% reported some abnormalities in the first 6 months of life and ~70% in the period between 6 and 10 months. Perinatal problems were more common in those with p.Arg255*, and when a total composite score was created for abnormalities of development and mobility, the most severe profiles were seen in those with p.Arg255* and p.Arg270* mutations, the two mutations that truncate in the nuclear localizing zone of the transcription repressor domain. These findings were consistent with those of later work that demonstrated that these mutations were also associated with the greatest overall clinical severity [25] and with a UK study that found correlations between preregression and later severity scores [69].

Although these studies involved the collection of data based on parent recall of early development, other methods elsewhere were being used to investigate a similar issue. In a Scottish study, health visitors, blinded to their identity, reviewed home videos taken during the first year of life of both children who were typically developing ($n = 10$) and who were subsequently diagnosed with RTT ($n = 12$) [70]. Broad differences emerged between the groups, with the infants later diagnosed with RTT appearing less focused in their play in comparison to the typically developing group who were more interested in examining and manipulating objects. Further research using video data examined the so-called general movements (GMs) during the first 6 [71] months of life in infants subsequently diagnosed with RTT. Although a specific pattern of generalized movements was not identified, the authors concluded from their findings of abnormal GMs in both studies [71] that RTT is a disorder with subtle manifestations that appear within the first few months of life.

9.4.2 Diagnostic criteria

There have been several iterations of the diagnostic criteria for RTT with the most recent published in 2010 [72]. In addition to the four core criteria relating to loss of hand skills, loss of spoken language, gait abnormality, and stereotypic hand movements, a new mandatory criterion was introduced. This was described as "a period of regression followed by recovery or stabilization" and was required both for classical and atypical RTT. Previously, only two instead of the four core criteria were required for atypical, in contrast to the classical form. In reality, this additional criterion adds to the challenge of making a clinical diagnosis. In most developed countries, *MECP2* genetic testing can confirm diagnosis at an early stage in the disease profile, either when the child has not yet regressed or if she has, before the recovery phase has occurred. Thus, clinical diagnosis may become redundant, although the clinical criteria can still provide a guide for a clinician when assessing a patient. In contrast, the absence of available genetic tests means that the clinical criteria remain the only mechanism for diagnosis in developing countries.

If one returns to the original model of RTT depicted by Bengt Hagberg, one can envisage classical RTT in the center of a circle surrounded by a number of atypical or, in his language, "variant forms" [73]. With time, the genetic origins of these variants have become better defined either because they are associated with specific *MECP2* mutations or because that are caused by mutations in other genes. Those variants described in the 2010 clinical criteria are the Hanefeld or early seizure onset variant, the congenital or Rolando variant, and the preserved speech or Zappella variant. In 2004, the genetic etiology of the early seizure onset variant was found to be a mutation in the *CDKL5* gene [74]. This condition has now been shown to be an independent disorder from RTT, such that the clinical characteristics of many of the affected children would no longer fit with the criteria for atypical RTT, particularly because of the lack of regression [67]. Similarly, the congenital variant has now been shown to be due to mutation in the *FOXG1* gene [75]. What has been named the preserved speech or Zappella variant is a milder form of RTT, usually caused by a *MECP2* mutation and in particular the p.Arg133Cys mutation or a C-terminal deletion.

9.4.3 Functional impairments

Three of the four main criteria, including loss of hand and communication skills and impaired gait [72], relate to developmental skills. As such, the child with RTT is faced with devastating functional impairments and dependency for most tasks of daily living [76]. Strong patterns between the type of genetic mutation and skill levels in each of the core developmental domains of communication, fine and gross motor functioning, have been demonstrated [77–79].

Losses of communication and hand skills, either with a sudden or gradual onset, are usually observed during the regression period. The child may previously have babbled and used words but during regression between 12 and 18 months, these skills are lost [62,79]. The ability to use words following regression is strongly related to genotype. For example, females with the Arg133Cys mutation are more likely to use words both before and after regression, whereas those with mutations associated with greater clinical severity such as p.Arg270* or large deletions are less likely to use words [79]. Similarly, the child may have been able to grasp toys and pieces of food before regression but skills are then lost during the regression period. A study observing video data of 144 females with RTT found that nearly 60% could not pick up a small object or piece of food including nearly 30% who could not hold any object even for a short period of 2 s [77]. Hand function in RTT is also related to genotype, correlating with previously described profiles of common *MECP2* mutations in relation to overall clinical severity [77].

Impaired gross motor function is also a fundamental feature of RTT. Clinical descriptions suggest that hypotonia and weakness may be evident during the early years with dystonia (increased muscle tone with persistent abnormal postures) and bradykinesis (slowness of movement) when older [80]. During the early months, more than 90% of 293 girls learnt how to sit and the majority did so within the usual developmental timeframe [62]. The ability to walk is achieved by a smaller proportion of individuals. For example, nearly half of 293 girls learnt how to walk independently and of those occurring at a mean age of 19.6 months, illustrating delay in achieving walking in approximately half of those who did achieve walking [62]. However, gait in those who can walk is often ataxic with a wide base of support [80]. Figs. 9.2, 9.3 and 9.4 show the acquisition of sitting and walking reported by 903 families registered with either the Australian Rett Syndrome Database or the InterRett database. Fig. 9.2 confirms that approximately 90% of girls learn to sit before the age of 3 years. Fig. 9.3 confirms that approximately half will learn to walk and while this skill is usually acquired during the preschool years, walking can be achieved in a few up to 8 years. Fig. 9.4 illustrates the strong relationships between genotype and learning to walk with the majority of children with milder mutations learning to walk (e.g., C-terminal deletion, p.Arg133Cys, p.Arg294*, p.Arg306Cys) whereas the

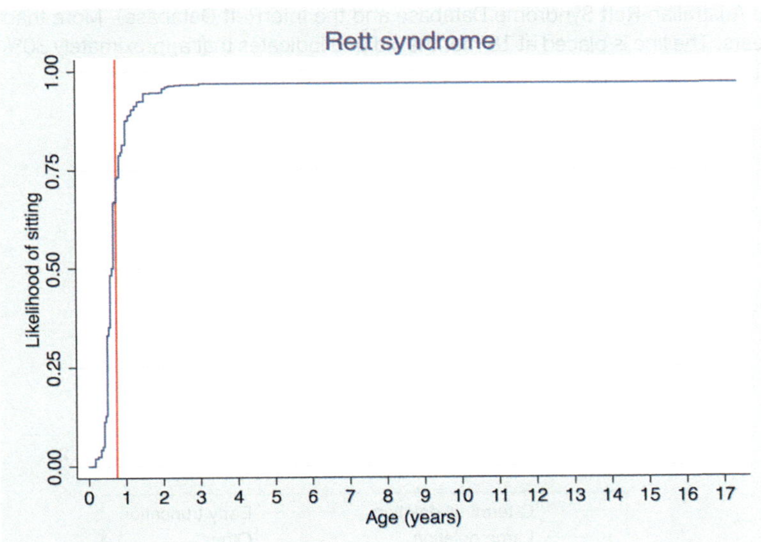

FIGURE 9.2

Kaplan–Meier time-to-event survival graph showing the time to acquisition of independent sitting ($n=903$ sourced from the Australian Rett Syndrome Database and the InterRett Database). More than 90% learn to sit before 17 years. The line is placed at 9 months old and indicates that approximately 70% learnt to sit without delay [81].

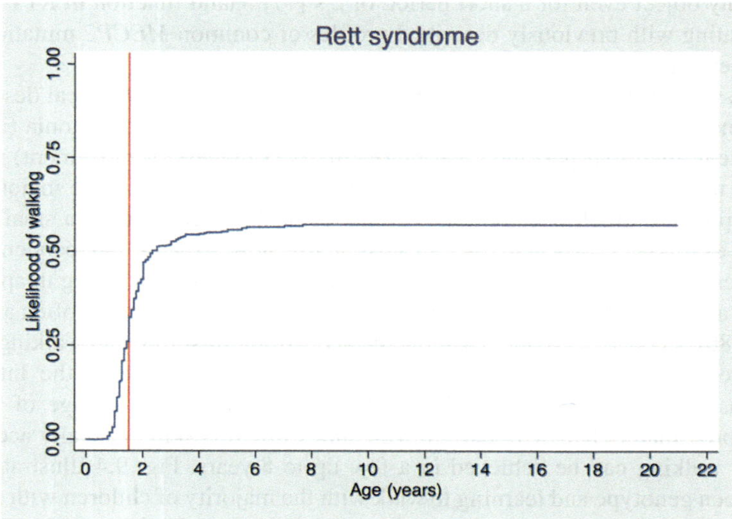

FIGURE 9.3

Kaplan–Meier time-to-event survival graph showing the time to acquisition of independent walking ($n = 903$ sourced from the Australian Rett Syndrome Database and the InterRett Database). More than 50% learn to walk before 17 years. The line is placed at 18 months old and indicates that approximately 30% learnt to walk without delay [81].

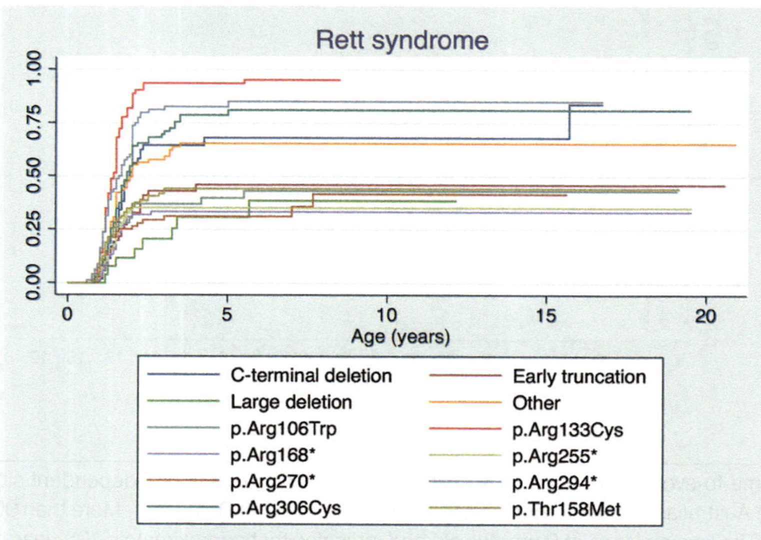

FIGURE 9.4

Kaplan–Meier time-to-event survival graph showing that individuals with a mutation associated with milder clinical severity are more likely to walk than those with a mutation associated with greater clinical severity.

minority of children with more severe mutations learn to walk (e.g., large deletion, p.Arg168*, p.Arg270*, and p.Arg255*).

The most comprehensive descriptions of gross motor function have involved assessments of video data collected from families living across Australia. Analyzing observations of 255 girls and women with RTT, motor abilities were better for those with the p.Arg133Cys, p.Arg294*, or p.Arg306Cys mutation and up to 20 points on a 45-point scale higher than those with mutations including p.Arg255*, p.Arg270*, or a large deletion after taking the effect of age into account [78]. Increasing age has been associated with greater general clinical severity [9] and taking into account the effect of mutation, decreasing gross motor skills with increasing age has also been observed in cross-sectional evaluation of gross motor functioning [78].

9.4.4 Stereotypies

Hand stereotypies are a hallmark characteristic in RTT that become apparent during the regression period. There are many recognizable patterns of movement, most of which occur in the hands, but bruxism and trunk rocking have also been classified as stereotypies [82]. Hand stereotypies frequently present as hand wringing or mouthing, clapping or tapping, or less frequently as hair pulling or sevillana [82,83]. In a study of 144 females with RTT, 60% presented with midline hand wringing, most demonstrated more than one movement pattern and with increasing age, the variety of hand stereotypy patterns decreased, and mouthing was less frequently observed [83].

9.4.5 Comorbidities

1. ***Scoliosis***—Altered muscle tone and poor motor functioning set the scene for deconditioning of the body, inadequate stretch to musculoskeletal tissues during functional activities, and thereafter the progression of deformity. During growth and development, the spine grows in synchrony with the rib cage and lungs, and both spinal and respiratory health are threatened in RTT. Scoliosis occurs in about three quarters by 15 years, particularly in association with mutations associated with phenotype and when the individual is unable to walk [5]. Spinal fusion is considered if the scoliosis progresses to greater than 45 degrees, a decision associated with considerable emotional burden for families [84]. Multiple study designs have been used to investigate whether undergoing such an extensive procedure for severe scoliosis was advantageous to health. Following spinal fusion, a Swedish case series indicated improved spinal posture and balance [85], and a qualitative study found that preoperative skills of sitting, standing, and walking were usually regained by 12 months [84]. In a cohort study, mortality was reduced in those who underwent spinal fusion ($n = 98$) compared to those who were managed conservatively ($n = 42$) with the effect particularly evident in those early onset scoliosis who also experienced a moderately reduced likelihood of later lower respiratory tract infections [86].
2. ***Epilepsy***—Epilepsy is one of the many comorbidities that occur in RTT although comparatively little has been reported on its characteristics until relatively recently. A Swedish study published in 2001 was one of the first to estimate prevalence in RTT, finding a prevalence of 95% in a small ($n = 53$) but representative series. Further to this, population-based investigations were undertaken in Australia using data from the Australian Rett Syndrome Database. In this dataset, seizures were reported in 81% of cases at a median age of 48 months, estimates being calculated

for the latter using time-to-event analysis to take into account those who had not yet had the opportunity to develop seizures [87]. In this study, those who had not learned to walk or had early developmental problems had an increased risk of seizures. Statistical modeling to identify predictive factors for onset of seizures before and after the time of diagnosis demonstrated that *MECP2* mutation negative individuals had a higher risk of seizure onset up to 4 years old. It is possible that some of the *MECP2* mutation negative cases may have been those subsequently identified with *CDKL5* mutations, based on the early seizure onset variant clinical features. Further research using the same population-based source investigated seizure frequency and its relationship with other factors, such as genotype and the use of antiepileptic drugs [40]. Seizure rates were highest in the 7–12 year age group before falling off in the older age group. They were lower in those with p.Arg294*, p.Arg255* mutations and C-terminal mutations and a little higher in those with p.Thr158Met. The individuals were being treated with multiple medications, often in combination, but in this Australian population carbamazepine, sodium valproate and lamotrigine were the most common. Since 2010, three large but not population-based studies have been published, one using the European networked database but also incorporating data from Israel [11], one from the United States [39], and one international study using the International Rett Syndrome Database [38]. In the US study, there was a lower proportion of epilepsy diagnosis with clinician confirmation [39]. The possible overdiagnosis of seizures, particularly in the non-US studies might be explained by the fact that some episodes reported and even treated as seizures may in fact be nonepileptic events related to autonomic disturbances rather than true seizures associated with electroencephalogram (EEG) seizure discharges.

3. ***Autonomic dysfunction***—Despite the body of research in animal models that has already provided the underlying rationale for a clinical trial using a serotoninergic 5HT1a agonist [88] to treat breathing irregularities in RTT, the literature on their characteristics, prevalence, and natural history in the girls and women with RTT is relatively sparse and has been limited to a handful of papers [89–91]. These breathing abnormalities usually present as either caregiver reported breath holding, affecting approximately two-thirds of girls and women, or as hyperventilation, affecting just under half [90]. Often these breathing abnormalities are also associated with abdominal bloating, which on rare occasions may lead to gastric perforation. As one caregiver described: "*It is causing pseudo bowel obstructions and she had to have a g-tube and caecostomy placed because of the air swallowing due to the breath holding. When she holds her breath, her tummy can become so distended that she looks like she is pregnant and she needs to be vented of the air throughout her stomach and intestines.*" [90].

   Onset of these breathing irregularities occurs in early childhood, but they are still observed into adulthood, where abdominal bloating, particularly, is more prevalent. The impact on the daily life of over a third of those affected can be considerable. For example, one family reported "*When she holds her breath and turns blue, it is very worrisome. She turns blue from breath holding and we usually have to shake her a bit or blow in her face to get her to start breathing again.*" When examined by genotype the impact of both breath holding and hyperventilation as well as of abdominal bloating appeared to be greater for those with an Arg294* mutation, while for the mutation p.Arg133Cys a greater impact was only seen for abdominal bloating. Interestingly, both of these mutations have been shown to have a milder overall clinical severity [25]. Genotype-specific information such as this should be helpful in targeting the most appropriate recruitment for clinical trials.

4. ***Poor growth***—Growth failure was an early-recognized feature of RTT [65]. Although head growth deceleration was the first growth abnormality to be noted, somatic growth was later shown to be affected [92]. Poor growth was initially mainly attributed to feeding difficulties associated with poor motility of the tongue, pharynx, esophagus and stomach, and frequent gastroesophageal reflux [93]. However, RTT should also be considered a pervasive disorder of growth with genetic underpinnings. This latter hypothesis was first investigated using Australian population-based data [94] and further replicated in an international dataset [95] where individuals with a C-terminal deletion were shown to have a much more favorable growth profile than those with other *MECP2* mutations. The Australian study also demonstrated that feeding difficulties were common and based on a UK classification system that nearly three quarters of those orally fed would have moderate feeding difficulties, and one in seven would have severe feeding difficulties in addition to the 20% who were already being enterally fed [94].
The Australian cohort has been followed longitudinally allowing for further characterization of the progress of feeding difficulties [96]. Subsequent analysis demonstrated that previous coughing, choking, or gagging during meals was associated with a later likelihood of subsequent gastrostomy insertion. Feeding problems, in particular choking on food, were especially common in those with a p.Arg168∗ mutation, consistent with data from the US Natural History study, where growth failure was also found to affect particularly individuals with those mutations generally considered to be severe (p.Arg270*, p.Arg255*, large deletions) or intermediate (p.Arg168* and p.Thr158Met) [97]. In guidelines developed for clinicians and families for the management of growth and feeding problems in RTT, enteral feeding is recommended when there is evidence of a failure to gain weight despite efforts to increase calorie intake, unsafe swallowing, or protracted feeding times (available at https://rett.telethonkids.org.au/resources) [98]. Currently published data from the Australian population database show that weight for age increases by almost 1SD at approximately 2 years after gastrostomy insertion [96]. Further research will investigate whether there are other benefits or disadvantages associated with this procedure. Two studies have recently shown gut microbiota alterations in RTT girls, demonstrating reduced microbial richness, changes in specific microbial taxa, including *Clostridia* and *Lactobacillus*, and corresponding changes in short-chain fatty acid profiles [99] were also noted. There are many nongenetic variables that may distinguish gut microbiota in RTT from healthy control girls, including profound differences in mobility, diet, and enteral feeding, so it remains an open question as to whether gut microbiome differences are genetic or environmental in origin.
5. ***Gall bladder disease***—In addition to growth and feeding problems, girls and women with RTT also experience other gastrointestinal problems associated with gastrointestinal motility such as constipation and gastroesophageal reflux, conditions which commonly occur in people with severe intellectual disability [100]. The gall bladder disease appears to be more common in this population than would be expected compared with the general population. Using data on individuals in the Australian [63] and international [25] databases, the incidence rate for cholelithiasis and/or cholecystectomy was 2.3 (95% confidence interval [CI]: 1.1–4.2) and 1.8 (95% CI: 1.0–3.0) per 1000 person-years for each, respectively [101]. Although it remains very important to consider gallbladder disease in the differential diagnosis of abdominal pain in RTT, one also wonders what might be the underlying pathophysiological cause contributing to this phenomenon. One possibility is an association with abnormal cholesterol metabolism that, presenting with high lipid levels, has been identified in a mouse

model of RTT [102], and has even been the motivation for a clinical trial of the use of statins in this condition (https://clinicaltrials.gov).

6. ***Sleep disturbances***—Sleep disturbances are an important feature of RTT. However, one of the first clinical studies in this area had examined the sleep/wake patterns of 20 girls with classical RTT attending the Kennedy Krieger Institute as early as 1990 [103]. The girls had less nighttime sleep, more daytime sleep, and overall more total sleep than age peers. Night waking, delayed sleep onset, and early waking were also common suggesting that children with RTT have markedly impaired sleep/wake patterns. Published in the same journal issue was research from Japan suggesting that there is an abnormality in the second epoch of circadian rhythm resulting in a disturbed sleep wake rhythm, possibly caused by dysfunction of aminergic neurons [104]. MeCP2 protein and transcript levels have also shown to oscillate with circadian rhythms [105], providing a potential mechanistic link to sleep disturbances.
   Over the last decade there has been further clinical and epidemiological research, much of it generated from the Australian and international databases. In a first Australian study restricted to the population of New South Wales ($n = 83$), many of the findings from the smaller US study were replicated with increased total sleep but lack of the age-related decrease in total and daytime sleep, although children with seizures had more and those who could walk had less daytime sleep [106]. Subsequent Australian studies utilized the Australian Database, allowing for the capture of data at different time periods. Overall, sleep problems were highly prevalent although they decreased with age, particularly night laughing (one of the more unique features of RTT), and night screaming [107,108]. There was a tendency for the prevalence of sleep problems to be highest in cases with a large *MECP2* deletion and in those with the p.Arg294* or p.Arg306Cys mutations [108], but examination of longitudinal data was unable to determine any benefit from use of medications [107]. A more recent study using international data confirmed that initiation and maintenance of sleep was most disturbed for younger children and those with a p.Arg294* mutation [109]. Unpublished data from the same dataset have again been unable to demonstrate any benefit from the use of medication but suggests benefit from the use of sleep hygiene strategies.
7. ***Bone health***—Bone health was first identified as a problem for children with RTT in the US and Australian studies, undertaken before the identification of the relationship with the MECP2 gene. The first Australian study, partly stimulated by the US findings, examined the likelihood of fracture in this population and, initially in the absence of densitometry data, assessed cortical thickness using hand X-rays [110]. It was estimated that just over 40% of girls would experience a fracture by the age of 15 years. Bone cortical thickness was much reduced compared with a normal population and in RTT was negatively affected by use of antiepileptic medication and increasing age.

This study was the impetus for subsequent research not only in Australia but also in Denmark, Israel, and the United States. Australian research further confirmed the high likelihood of fractures in this population and investigated determinants including genotype, where p.Arg270* and p.Arg168* conferred the greatest risks [111] and the relationship with medication use, where use of valproate (commonly used to treat epilepsy and bipolar disorder) was also associated with increased risk of fracture (Fig. 9.5) [112]. Bone densitometry studies were also undertaken on 97 individuals in the Australian Rett Syndrome Database, and a variety of different bone parameters were assessed. Mean values were generally low when compared with child population norms and using multiple linear

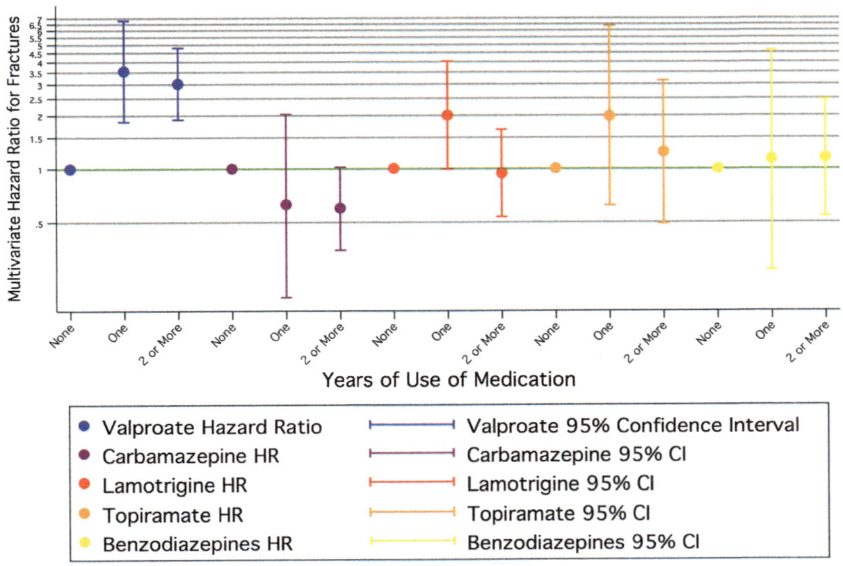

FIGURE 9.5

Risk of fracture in Rett syndrome according to the type of antiepileptic medication.

regression a strong relationship with genotype was identified with those with mutations p.Arg168* and p.Thr158Met having the lowest values.

One US study, carried out at Baylor College of Medicine demonstrated low bone mineral density (BMD) and bone mineral content (BMC) Z scores, lower in older than younger cases, in approximately half of the individuals in their study ($n = 50$) and a relationship between low BMD scores and fractures [113]. Another later study ($n = 54$), undertaken at the Kennedy Krieger Institution, found very similar results for BMD levels and nonsignificant relationships with genotype (p.Arg270* and p.Thr158Met) [114]. On the other hand, Danish researchers reported the presence of small bones associated with low bone mass [115].

It is only in recent years that more has been discovered about the genetic basis for the vulnerability to osteopenia and fractures in RTT has been identified. In 2009, it was shown that MeCP2 deficiency adversely affects bone formation in a mouse model [48]. Furthermore, recent research is now showing that it also has potentially reversible adverse effects on the biomechanical integrity of bone [49]. This is an example of how two independent lines of research can coalesce to improve knowledge and understanding of MeCP2 and RTT. It would be hoped that this may lead to the ability to raise the management of bone health in RTT to a higher level than is currently available [116].

9.4.6 Epidemiology

RTT is a rare disease predominantly affecting females, but it has always been important to find out exactly how rare it is both in terms of (a) the cumulative incidence, that is, the inverse of the

Kaplan–Meier product limit method, (b) birth prevalence, that is, the proportion of female births in which the girl will go on to develop RTT, (c) incidence of diagnosis, that is, the rate of diagnosis in a population of females followed over time, and finally (d) prevalence, that is, the proportion of girls/women who are living with the disorder. The latter of course also has to take into account the mortality from the disorder.

Epidemiological studies typically require ascertainment from an entire discrete population, as was available from the Australian Rett Syndrome Database [6]. Using the Kaplan–Meier product limit method, the cumulative incidence of diagnosis was 0.96 (95% CI: 0.78–1.19) per 10,000 females by the age of 12 years. The cumulative incidence by the age of 4 years increased from 0.2 per 10,000 in those born 1980–84 to 0.8 per 10,000 in those born between 1990 and 1993 demonstrating the increase in awareness and diagnosis of RTT in Australia at that time. Prevalence estimates for those aged 5–18 years were also estimated to allow comparability with studies elsewhere in Sweden, Scotland, Texas, and North Dakota although the age ranges included did vary by country. With a prevalence of 0.67/10,000 females, the Australian prevalence was very similar to that estimated for Sweden and Scotland but higher than estimated in Texas and North Dakota.

In 2006, the Australian estimates were revised using currently available data. The prevalence in those aged 5–18 years had increased to 0.88 per 10,000 females and the cumulative incidence to 1.09 per 10,000 till 12 years old [117]. Survival was also estimated for the first time and found to be 77.8% at 25 years, compared with 99.96% survival in the Australian female population. The main purpose of a study undertaken in 2010 was to compare survival in Rett's original Austrian cohort with that in the current Australian cohort [101]. There was a dramatic increase in survival with the probability of survival up to the age of 25 years in the Austrian cohort being 21%, compared with 71% in the Australian cohort. However, when survival was compared by birth year group in the Australian cohort there were no differences over time. The rationale for a further study undertaken in 2011 was to investigate trends in the diagnosis of RTT in Australia in relation to the availability of *MECP2* testing [118]. For this study, all four methods, a–d, were used. The study found that the median age of diagnosis fell from 4.5 years before 2000 to 3.5 years after 1999. The cumulative risk of diagnosis by 32 years at 1/8905 per 10,000 births had doubled that at 5 years old.

The final analysis in this suite of RTT epidemiological studies found that the likelihood of survival was 77.6% at 20 years, 71.5% at 25 years, and 59.8% at 37 years [119]. Using a population-based source with follow-up over 20 years, the Australian estimates were somewhat lower than the 70% survival at 45 years estimated from follow-up in the US Natural History study [10]. This could be accounted for by the fact that the Natural History study recruits a select population with better than average clinical care. Both studies investigated relationships with genotype although individual numbers were smaller for the US study because of the comparatively shorter follow-up. In the Australian study, the highest mortality risk was for the large deletion mutations followed by the p.Arg270* and this pattern was similar in the US study as well.

9.5 Evidence for epigenetic mechanisms affecting *MECP2* function and expression

MECP2's role in the pathogenesis of RTT is influenced by epigenetics in two major ways. This section will discuss the role of the MeCP2 protein functions in reading and interacting with specific epigenetic layers of the genome, while the following section will discuss how the X-linked gene *MECP2* is regulated by epigenetic mechanisms and nongenetic factors. As a highly abundant nuclear protein in the central nervous system, the protein functions of MeCP2 are diverse and complex [120,121]. Although originally named based on its binding affinity to methylated compared to unmethylated DNA, the number of ascribed functions of MeCP2 have increased to include additional chromatin and RNA-mediated functions.

9.5.1 DNA methylation

DNA methylation is an abundant epigenetic mark in the mammalian genome and serves as the first layer of epigenetic information on top of one's DNA sequence code. Methylation at cytosine prior to guanine (mCpG) is the most abundant form of methylation, as around 80% of the available CpG sites in human brain are methylated [122]. CpG dinucleotides in the mammalian genome are not randomly distributed, but instead are depleted in most regions of the genome, except at high-density clusters of CpGs, termed "CpG islands" [123]. DNA methylation can also occur much less frequently at non-CpG sites, called CpH sites, where "H" is any nucleotide other than G [124]. CpH methylation is more abundant in brain compared to most other tissues, however, and becomes most abundant around the type of synapse formation in childhood [122]. In addition to these two major classes of methylcytosine, an oxidative form called hydroxymethylcytosine (hmCpG) is also observed at higher levels in neurons compared to nonneuronal cells [125]. The transition from mCG to hmCG is mediated by Tet enzymes (Tet1-3), which is the first step in active DNA demethylation pathways that can result in the removal of mCG sites from genes that are active and accessible to Tet enzymes [126].

In addition to the emerging complexities on the different types of DNA methylation in the brain, the location of DNA methylation with regards to gene regulatory and chromatin elements is important for understanding its relationship to gene expression. Genome-wide mapping of DNA methylation by whole genome bisulfite sequencing has confirmed that active promoters and CpG island promoters are strongly depleted for DNA methylation [127]. When methylation is observed at gene promoters, there is an inverse correlation between the level of methylation and transcript levels [122,127–129]. However, outside of gene promoters, the relationship between DNA methylation is generally positively correlated with expression. CpG islands that are within genes (defined as transcription start to end sites) show higher levels of methylation for highly expressed genes. In addition, the "shores" of CpG islands, which are defined as regions 1 kb upstream or downstream of CpG islands, tend to contain the most relevant tissue-specific differences in DNA methylation levels [128,130].

MeCP2 contains a methyl-binding domain (MBD) with a higher affinity for mCpG compared to unmethylated CpG sites and is similar to other MBD protein family members in its higher affinity for methylated DNA in biochemistry experiments performed in vitro [131]. MeCP2 exhibits a lower binding affinity to mCpC, mCpT, hmCpG, and unmethylated DNA compared to mCpG, mCpA, and hmCpA sites within sequences assayed by electrophoretic mobility shift assay [124,132]. Furthermore, isothermal titration calorimetry and NMR spectroscopy analyses have demonstrated that MeCP2, but not MBD2, binds mCpA with high affinity in a strand-specific and orientation-dependent manner, a property that was sensitive to several Rett-associated MBD missense mutations [133]. These results have suggested a recently appreciated increased complexity of MeCP2 as an epigenetic reader of the DNA methylome.

Within intact brain and neurons, determination of where MeCP2 binds is achieved predominantly through variations of the method called chromatin immunoprecipitation (ChIP). For candidate gene loci where MeCP2 is predicted to bind, PCR is used to assay specific regions around the promoter. This ChIP-PCR approach has successfully identified MeCP2 binding above background levels for multiple MeCP2 target genes of interest, particularly those that respond to neuronal activity such as *Bdnf*, *Egf2*, *Arc*, and *Junb* [134–137]. In contrast, unbiased approaches using array-based (ChIP-chip) or sequencing-based (ChIP-seq) methods have given a more complete picture of the locations of MeCP2 binding in neurons and brain, revealing the association with CpG island shores of active genes [57,138]. In MeCP2 ChIP-seq analyses, because of the lower coverage per base pair and cell heterogeneity of brain or neuronal cultures, MeCP2 ChIP signal correlates highly with the negative control "input" signals, making interpretation of the wide-scale binding observed in these studies difficult [139–141]. In a more homogeneous population of olfactory neurons, MeCP2 binding was observed as distinct peaks, but correlated better with guanine-cytosine (GC) sequence content than DNA methylation [141]. Neurons exhibit the highest levels of MeCP2 and are the predominant cell-type implicated in the pathogenesis of RTT. However, other cellular players are involved in the disease severity, including astrocytes, oligodendrocytes, microglia, and macrophages, for which ChIP-seq analyses and distinct binding sites have been performed [43,142–144]. Future investigations of MeCP2 binding using sequencing approaches need to continue to focus on isolated cell types and greater depth of sequencing to more precisely define the true MeCP2 binding sites above input background and their relationship to the different types of DNA methylation. The timing of deposition of the CpH layer of DNA methylation around the time of synaptogenesis is of particular interest to understanding parallels with the regressive stage of RTT.

9.5.2 Histone modifications and nucleosome and higher-order chromatin remodeling

One of the earliest described functions of MeCP2 was its association with HDAC through the TRD [53,145]. MeCP2-deficient neurons exhibit globally elevated levels of histone core proteins H3 and H4 acetylation, consistent with a role for MeCP2 in the localization of HDAC activity [146–148]. Although increased global histone acetylation levels should be an indication of more accessible and active chromatin, paradoxically, *Mecp2*-deficient neurons are smaller, with smaller nuclear and nucleolar diameters, and more heterochromatic foci compared to wild-type neurons [149]. In addition, *Mecp2*-deficient neurons exhibit globally reduced rates of both transcription and translation [150,151], related to their reduced cellular metabolism [152]. Similarly, mice deficient in the neuronal HDAC gene Hdac3 exhibit globally reduced transcription. In addition, Mecp2 was required for Hdac3-mediated activation of neuronal activity genes such as *Bdnf*. [153] Together, these results suggest that MeCP2's role in recruiting HDACs may be important in the dynamic regulation of active genes in neurons.

A striking in vitro property of MeCP2 is its ability to compact nucleated DNA even in the absence of DNA methylation [154]. However, MeCP2 competes with linker histone H1 in chromatin binding and levels of histone H1 increase in the absence of MeCP2 [155,156]. This property could explain the increased amount of DNA observed in heterochromatic foci and overall smaller nuclei in *Mecp2*-deficient neurons [157]. A redistribution of the heterochromatic histone mark H4K20me3 in perinucleolar heterochromatin was also observed in the absence of MeCP2 in Purkinje neurons, suggesting that MeCP2 may serve to prevent higher-order chromatin aggregation in perinucleolar heterochromatin, enabling its actions as a facilitator of transcriptional activation in neurons [157].

The C-terminal domain of MeCP2 appears to be the functional domain required for higher order chromatin compaction [158]. Interestingly, different C-terminal mutations differ in their severity based on location of truncation. Explanations for the MeCP2 genotype–phenotype correlations were examined in specific mouse models of MeCP2-R270X or MeCP2-G273X mutations [56]. Three AT-hook domains in the C-terminus of MeCP2 were determined to be differentially affected by distinct C-terminal mutations, which in turn affected the localization of another heterochromatic protein ATRX to bind both MeCP2 and the cohesion complex to organize chromatin in long-range interactions [56,159]. Mutations in *ATRX* are observed in alpha thalassemia (ATR-X syndrome), a severe form of intellectual disability (discussed in Chapter 11) [160].

9.5.3 Noncoding RNAs

Noncoding RNAs have been implicated in some of MeCP2's neuronal functions. The long noncoding RNA (lncRNA) *Evf2* recruits MeCP2 and DLX to regulate transcription of the homeodomain transcription factors DLX5 and DLX6 in the developing mouse forebrain [161,162]. Through its interactions with ATRX and cohesin, MeCP2 binds to and regulates expression of two imprinted gene loci containing noncoding RNA, *H19* and *Gtl2*. [159] Interestingly, multiple miRNAs within the imprinted *Rian* transcript in the *Dlk1/Gtl2* locus were identified in an unbiased screen for miRNAs dysregulated in *Mecp2*-null mouse cortex [163]. *Rian* was enriched for MeCP2 binding in wild-type cortex, and showed altered histone acetylation peaks associated with *Mecp2* deficiency. Interestingly, two *Rian* miRNAs were predicted to degrade *Bdnf* transcripts [163]. As *Bdnf* transcript and protein levels are reduced in multiple RTT mouse models, the increased expression of *Rian* locus miRNAs may be a mechanistic explanation for the reduction of BDNF in the pathogenesis of RTT. Lastly, a screen for dysregulated lncRNAs genome-wide identified two high-confidence transcripts showing increased expression in *Mecp2*-deficient mouse brain, which localized to host coding protein gene, the gamma-aminobutyric acid receptor subunit Rho 2 (*Gabrr2*) and the Rho guanine nucleotide exchange factor encoded by *Arhgef26* [164].

9.5.4 RNA splicing

Roles for MeCP2 in RNA functions, particularly RNA splicing have also been growing in evidence and interest. Between the well-characterized MBD and TRD domains of MeCP2, lies a domain capable of binding RNA [165]. A role for MeCP2 in exon inclusion choices as part of splicing of alternative exons has been described [166]. In addition, MeCP2 has been shown to associate with a variety of RNA splicing factors, including YY-1, PRPF3, MATRIN3, SFPQ, LEDGF, and DHX9 [167]. In addition, deep RNA sequencing analyses of *Mecp2*-null versus wild-type brain has revealed hundreds of genes with altered splicing, including the glutamate receptor *Gria2* gene [168]. A mass spectrometry analysis of protein coding exons in *Mecp2*-null brain provided further independent evidence for MeCP2 functions in alternative splicing, showing enrichment for hmCpG and histone H3K4me3 marks in downregulated exons and H3K36me3 marks in upregulated exons [169]. These findings are consistent with the hypothesis that MeCP2 binding to hmCpG at promoters marked by H3K4me3 may be activating and therefore inclusive of exons at promoters, while MeCP2 binding to active gene bodies undergoing splicing and marked by mCpA and H3K36me is repressive, so loss of MeCP2 would result in exon inclusion [52]. Lastly, another proteomic analysis of MeCP2 protein binding factors and splicing alterations demonstrated a role for MeCP2 in intron retention and polymerase stalling during splicing [170]. Together, these results form strong support for a role of MeCP2 in binding to RNA and regulating RNA splicing and transcriptional progression in mouse models of RTT.

9.6 Epigenetic regulation of MeCP2 expression or phenotypes

Both MeCP2 levels and DNA methylation levels are regulated by both cell-intrinsic and cell-extrinsic factors in pre- and postnatal life [121]. Particularly important for girls with RTT is the mosaicism due to X chromosome inactivation (XCI) of cells expressing either mutant or wild type (WT) alleles. However, other nongenetic influences are implicated in the regulation of MeCP2 levels and modification of the RTT phenotypic severity in both humans and mouse models.

9.6.1 X chromosome inactivation

RTT is an X-linked dominant disorder, meaning that heterozygous females are specifically affected. Female cells are mosaic for expression of heterozygous X-linked mutations or alleles due to the random choice of X chromosome inactivation patterns that become established in each cell lineage around the time of implantation. XCI is established and maintained in subsequent daughter cells through the differential expression of oppositely oriented lncRNAs, *Xist* and *Tsix*, expressed from the inactive or active alleles, respectively [171]. The random choice of maternal versus paternal X chromosome for XCI in each cell ensures an approximately 50:50 ratio of X-linked allelic expression in females, on average. However, evidence of nonrandom XCI, also called "skewed XCI" is observed in carrier mothers of X-linked mutations as well as in women with recurrent miscarriages [172,173].

In RTT, most girls exhibit apparently random XCI patterns in blood and brain [146,174,175]. In rare cases of maternal inheritance of *MECP2* mutations, however, XCI is highly favorably skewed, explaining the asymptomatic carrier mother [176]. For RTT girls with defined mutations (i.e., p.Arg168∗, p.Thr158Met), the percentage of mutant-expressing cells correlated with clinical severity [177]. Although XCI ratios are a clearly important modifier of clinical severity, several curious exceptions to this rule perhaps could be explained by the potential for inaccuracies in this assay. For example, a few RTT girls with highly skewed XCI did not explain clinical diagnosis of classical versus preserved speech variant [178], and one asymptomatic carrier mother of a *MECP2* mutation showed an XCI pattern in blood favoring mutant expression [179]. Most puzzling was the study of *MECP2* c806delG monozygotic twins discordant for RTT severity who both showed random XCI in blood, fibroblasts, and buccal cells [180]. These discrepancies may point to other genetic and nongenetic influences that may serve to modify the clinical severity of RTT that have been underinvestigated, but will be discussed later.

Although RTT is a monogenic disorder potentially treatable by gene therapy approaches, the mosaic nature of this X-linked dominant disorder creates a very difficult problem of reexpressing *Mecp2* because wild-type expressing cells would overexpress MeCP2 and mutant-expressing cells could have a dominant-negative effect of coexpression of both wild-type and mutant *Mecp2* alleles in the same cell (Fig. 9.6). As *MECP2* duplications are a known genetic etiology of autism in males [182], overexpression of MeCP2 is expected to be detrimental to the wild-type expressing neurons.

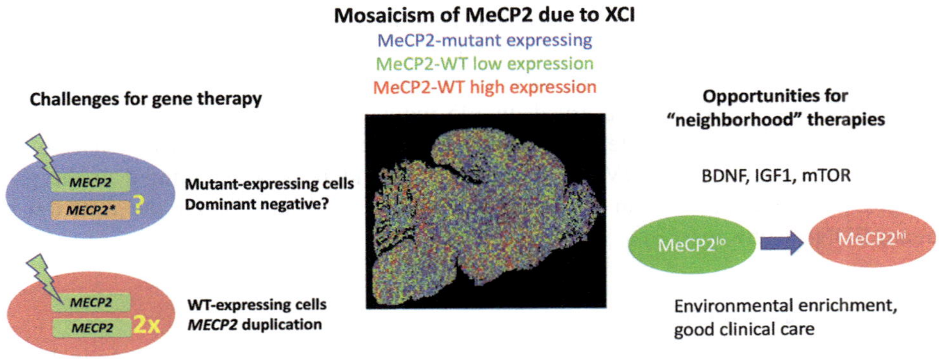

FIGURE 9.6

Consideration of Mosaicism due to XCI in Rett syndrome treatments. Center image is from a sagittal section of a female mouse model of Rett syndrome stained for MeCP2 immunofluorescence and imaged by laser scanning cytometry [181]. Cells expressing the mutant copy of *Mecp2* are colored in blue, while cell expressing the wild-type *Mecp2* are colored in either green or red based on low or high MeCP2 levels, respectively. The left panel represents the current challenges associated with gene therapy strategies for introducing MECP2 into mosaic females, because wild-type expressing cells will have overexpression and mutant-expressing cells may have a dominant-negative effect on the wild-type allele. The right panel represents opportunities for therapeutic strategies or environmental effects on improving the "bad neighborhood" that results from wild-type expressing cells with low MeCP2 levels due to noncell autonomous effects of the mutant-expressing cells in the brain.

In female mouse models of RTT, this mixture of *Mecp2* mutant and wild-type neurons and glial cells in the brain also creates both cell-autonomous and noncell-autonomous "bad neighborhood" effects in which astrocytes spread MeCP2 deficiency, affecting neuronal maturation even in wild-type expressing neurons [41,181]. These findings raise the question of whether several nongenetic factors could be investigated and utilized for "neighborhood" therapeutic improvement on the wild-type expressing cells and neuronal networks in the mosaic RTT brain.

9.6.2 IGF1/mTOR pathway

Currently, the therapeutics for RTT tend to be those that target metabolic pathways [3], fitting into the category of "neighborhood" therapies rather than gene therapies. Most of these, such as BDNF and IGF1 integrate into the central AKT/mTOR (mammalian target of rapamycin) pathways that regulate protein translation and energy homeostasis [150,152]. Additional related pathways that also feed into the central AKT/mTOR pathways that have been also attempted as RTT therapeutics in mouse models include oxidative stress [183,184] and lipid metabolism [102,185].

Mecp2-null and *Mecp2*$^{+/-}$ mouse female brain both show reduced production of BDNF [186], a protein required for normal neuronal development, synaptic signaling, and brain function [187]. A common single nucleotide polymorphism in the *BDNF* gene is substitution of a valine (Val) amino acid for methionine (Met) at codon 66. This polymorphism has been associated with poorer neurocognitive functioning in the general population and in specific groups such as those with schizophrenia [187].

In *Mecp2* knockout mice, the Val/Met polymorphism has been associated with poorer dendritic growth and formation of excitatory synapses in hypothalamic cells [188], suggesting that the val-BDNF variant contributes also to pathology in RTT. Furthermore, clinical effects of the BDNF gene on phenotype in RTT have also been observed. In 125 individuals with RTT, the presence of the met-BDNF polymorphism was associated with greater disease severity and earlier onset of seizures, compared with those with the Val/Val polymorphism, particularly when limited to the p.Arg168* mutation [189]. Treatment with an enriched environment aims to exploit the function of these downstream BDNF pathways.

9.6.3 Enriched environments

Evidence in support of both genetic and nongenetic modifiers of RTT phenotypes in humans and mouse models suggests that X chromosome inactivation skewing is not the only way to epigenetically modify RTT phenotypes [179,180,190]. Dietary changes [191–193], environmental pollutants [194], enriched environments [195,196], growth factors [197–200], genetic modifiers from copy number variants [201], and mouse strain differences [27,202] have all been shown to modify the severity of RTT phenotypes. Postnatally, *Mecp2*-deficient female mice have deficits in both the innate and learned maternal care behavior of pup retrieval [203] and neglectful behaviors are improved on a CD-1 background [202] or with CD-1 foster dams [204]. Early-life deficits in maternal care result in long-lived consequences to offspring behavior into adulthood via epigenetic mechanisms [205].

When housed in an enriched environment, *Mecp2*-null mice have demonstrated improved motor abilities and increased levels of BDNF in the brain [195,196] but for children, the evidence base is disappointingly poor. To address this gap, we recently investigated the effects of environmental enrichment on gross motor skills and blood BDNF levels in girls with RTT using an individually randomized stepped wedge design study [206]. A genetically variable group of 12 girls with a *MECP2* mutation and younger than 6 years were recruited; the baseline period was 1, 2, or 3 months as determined randomly; and thereafter, the intervention period was 6 months comprising motor learning and exercise supplemented with social, cognitive, and other sensory experiences. Gross motor assessments were conducted using the Rett Syndrome Gross Motor Scale (maximum score of 45) [78] on five occasions, two during the baseline period and three during the intervention period. Blood levels of BDNF were measured at the two baseline assessments and at the end of the intervention period. At recruitment, the mean (SD) age of the children was 3 (1.1) years ranging from 1 year 6 months to 5 years 2 months. In addition, at baseline, mean (SD) gross motor scores and blood BDNF levels were 22.7/45 (9.6) and 165.0 (28.8) ng/mL, respectively [206].

Adjusting for covariates, the enriched environment was associated with improved gross motor skills (coefficient 8.2, 95% CI: 5.1, 11.2) (Fig. 9.7) and a 321.4 ng/mL (95% CI: 272.0, 370.8) increase in blood BDNF levels (Fig. 9.8) after 6 months of treatment. Unadjusted summary data for gross motor skills scores and BDNF levels at each assessment are shown in Fig. 9.4 and 9.5. Interestingly, girls with the heterozygous polymorphism (Val/Met) had the lowest median gross motor scores and blood BDNF levels at baseline but their magnitude of increase after 6 months of treatment was similar to girls with the Val/Val or Met/Met polymorphisms. Moreover, the increase in score as a proportion of the baseline score was highest in girls with the Val/Met polymorphism (62%) compared with those with the Val/Val (29%) or Met/Met (32%) polymorphisms [206]. In sum, a 6-month period of environmental enrichment reduced the functional deficit in RTT irrespective of BDNF polymorphism. Baseline assessment

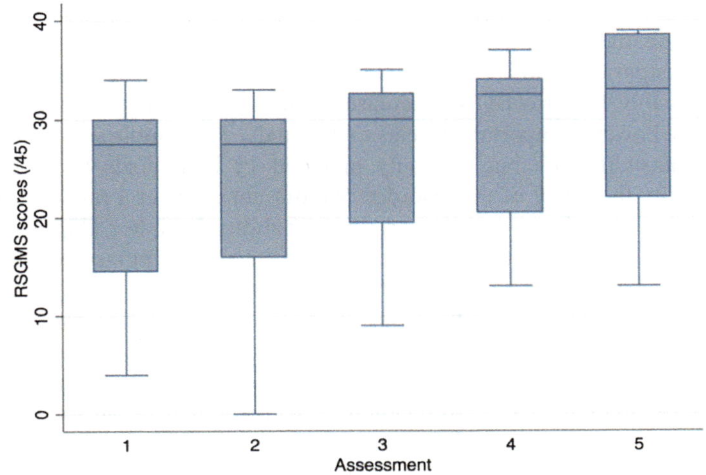

FIGURE 9.7

Box plot of Rett Syndrome Gross Motor Scale scores at each assessment. Assessments 1 and 2 show the distribution of scores at the baseline assessments, and 3, 4, and 5 show the distribution of scores at assessments in the intervention period. The boxes show, from bottom to top, the 25th percentile, median, and 75th percentile values, and the error bars indicate the lower and upper adjacent values. The upper adjacent value is defined as the largest data point $\leq$ 75th percentile + 1.5 interquartile range (IQR), and the lower adjacent value is defined as the smallest data point $\geq$ 25th percentile − 1.5 IQR.

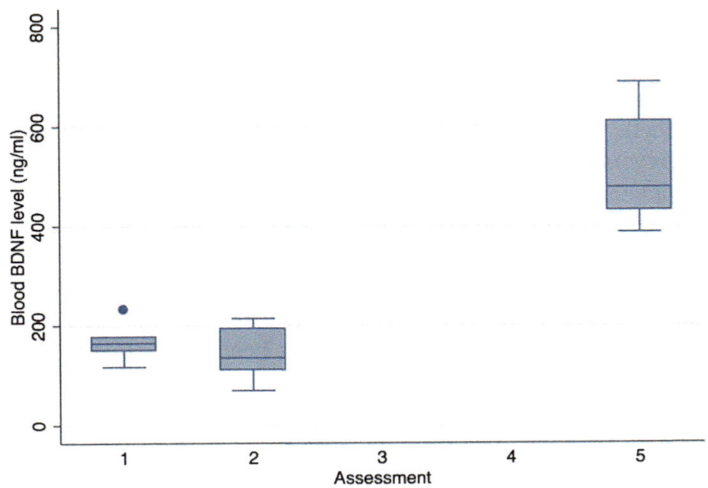

FIGURE 9.8

Box plot of blood brain-derived neurotrophic factor levels at different assessment times. Assessments 1 and 2 show the distribution of levels at the baseline assessments, and assessment 5 shows the distribution of levels after 6 months of intervention. See description of box plots in the caption of Fig. 9.6. In addition, the outside values, which are data points more extreme than the upper and lower adjacent values, are individually plotted.

indicated greater clinical severity for girls with Val/Met polymorphism consistent with previous animal [188] and clinical studies [189], but improvement was possible.

Findings are consistent with those from animal studies and illustrate that young girls with RTT can learn new skills, possibly because BDNF increased with the implementation of the environmental enrichment. There are however important caveats to this study. The sample size is the largest to date for a study of individuals with RTT but still only included 12 individuals because of difficulties in recruitment afforded by the rarity of the disorder. We did not conduct a randomized controlled trial because of sample size availability and the striking variability in motor abilities between different genotypes [9,25] that limited our capacity to randomize into two similar groups. To counter these limitations, individuals were randomized to the duration of the baseline period and then assessed under control and intervention conditions; and the assessor of gross motor skills was blinded to the order of the assessments [206]. The study provides some evidence for a causal relationship but at a lesser level of evidence than a randomized controlled trial. Replication is an important task for the future. However, it cannot be ignored that environmental enrichment will be an important adjunct in the evaluation of new drug therapies that use BDNF pathways, such as IGF1 (mecasermin) and NNZ-2566 (trofinetide), because of implications for an independent mechanism that strengthens synapses and improves functioning.

9.7 Inclusion of epigenetic data collection in epidemiological studies

The presence and type of *MECP2* mutation and features of the accompanying phenotype are fundamental to the devastating effects of RTT and data to describe these have been collected in existing database infrastructure. However, there is a growing appreciation of effects on phenotype beyond the *MECP2* gene variant, and it is now timely to build documentation of environmental and epigenetic data into these large sample repositories. Epidemiology investigates the causes and distribution of disease in populations, and this chapter has reviewed epigenetic factors that influence disease severity but which are not included in current database infrastructure. Such an expanded dataset that would comprise genetic, epigenetic, and phenotypic data could provide capacity for cohort studies where exposure to an epigenetic factor could be tested, a vital step in examining predictors of phenotype and deciphering how reversible molecular information outside the DNA sequence and the environment can be detrimental or otherwise to functioning. Put simply and as previously suggested [207], a framework for integrating genetic, environmental exposures, epigenetics, and health outcomes is a vital next step.

The declining cost of genetic testing and new investigations such as genome-wide association testing will facilitate standard collection of genetic data beyond the *MECP2* variant. Additional variables could include the presence of variants in other genes (e.g., *IRAQ1*) or the type of BDNF polymorphism. Environmental exposures could include measures of the standard of clinical care, physical activity, or markers of oxidative stress or immune functioning. The role of smart phone applications in the profiling of environmental influences should be explored. Perhaps, other 'omics, such as metabolome and microbiome analysis, could be added to further understand epigenetic implications and appropriate therapeutics. Discussion between laboratory scientists, clinicians, and researchers as to this expanded epidemiological focus is an urgent task.

9.8 Summary

The pathogenic variants in the *MECP2* gene have major consequences for the development of motor functioning, and cognitive skills and treatment to ameliorate or even cure the symptoms are desperately needed. Optimism that this will be possible is derived in part from the known monogenetic cause, the database infrastructures that have enlightened the natural history of RTT, and strong research funding support from patient organizations [3]. The preclinical and clinical findings discussed in this chapter set the scene for advances in the management of RTT, some of which may operate through pathways downstream from MeCP2 through adjustment of neurotransmitter systems, growth factor signaling, and metabolic functioning pathways [36,208]. Others could operate through epigenetic pathways influenced by metabolites or gut microbiome. Rare diseases with monogenic causes are considered increasingly amenable to new techniques for precision medicine such as use of antisense oligonucleotides. RTT is an "orphan disease" because of its low prevalence, but it has attracted extraordinary attention as evidenced by progress along the pipeline of discovery. Beyond providing information on genetics, phenotype, and epigenetics, this chapter recommends that population-level data on each of these domains combined will have an important role in accelerating the pipeline of discovery.

References

[1] Rett A. On a unusual brain atrophy syndrome in hyperammonemia in childhood. Wien Med Wochenschr 1966;116:723–6.

[2] Hagberg B, Aicardi J, Dias K, Ramos O. A progressive syndrome of autism, dementia, ataxia, and loss of purposeful hand use in girls: rett's syndrome: report of 35 cases. Ann Neurol 1983;14(4):471–9.

[3] Leonard H, Cobb S, Downs J. Clinical and biological progress over 50 years in Rett syndrome. Nat Rev Neurol 2017;13(1):37–51.

[4] Corbett J, Kerr A. Rett syndrome: from gene to gesture. J R Soc Med 1994;87:562–6.

[5] Downs J, Torode I, Wong K, Ellaway C, Elliott EJ, Christodoulou J, et al. The natural history of scoliosis in females with rett syndrome. Spine 2016;41(10):856–63.

[6] Leonard H, Bower C, English D. The prevalence and incidence of Rett syndrome in Australia. Eur Child Adolesc Psychiatry 1997;6(Suppl. 1):8–10.

[7] Louise S, Fyfe S, Bebbington A, Bahi-Buisson N, Anderson A, Pineda M, et al. InterRett, a model for international data collection in a rare genetic disorder. Res Autism Spectr Disord 2009;3(3).

[8] Krishnaraj R, Ho G, Christodoulou J. RettBASE: rett syndrome database update. Hum Mutat 2017;38(8): 922–31.

[9] Cuddapah VA, Pillai RB, Shekar KV, Lane JB, Motil KJ, Skinner SA, et al. Methyl-CpG-binding protein 2 (MECP2) mutation type is associated with disease severity in Rett syndrome. J Med Genet 2014;51(3): 152–8.

[10] Tarquinio DC, Hou W, Neul JL, Kaufmann WE, Glaze DG, Motil KJ, et al. The changing face of survival in rett syndrome and MECP2-related disorders. Pediatr Neurol 2015;53(5):402–11.

[11] Nissenkorn A, Levy-Drummer RS, Bondi O, Renieri A, Villard L, Mari F, et al. Epilepsy in Rett syndrome–lessons from the Rett networked database. Epilepsia 2015;56(4):569–76.

[12] Amir RE, Van den Veyver IB, Wan M, Tran CQ, Francke U, Zoghbi HY. Rett syndrome is caused by mutations in X-linked MECP2, encoding methyl-CpG-binding protein 2. Nat Genet 1999;23(2):185–8.

[13] Qureshi IA, Mehler MF. Understanding neurological disease mechanisms in the era of epigenetics. JAMA Neurol 2013;70(6):703–10.
[14] Champagne FA. Epigenetic legacy of parental experiences: dynamic and interactive pathways to inheritance. Dev Psychopathol 2016;28(4pt.2):1219–28.
[15] Gudsnuk K, Champagne FA. Epigenetic influence of stress and the social environment. ILAR J 2012; 53(3–4):279–88.
[16] Champagne FA. Epigenetics and developmental plasticity across species. Dev Psychobiol 2013;55(1): 33–41.
[17] Nithianantharajah J, Hannan AJ. Enriched environments, experience-dependent plasticity and disorders of the nervous system. Nat Rev Neurosci 2006;7(9):697–709.
[18] Zoghbi HY, Percy AK, Schultz RJ, Fill C. Patterns of X chromosome inactivation in the Rett syndrome. Brain Dev 1990;12(1):131–5.
[19] Villard L. MECP2 mutations in males. J Med Genet 2007;44(7):417–23.
[20] Trappe R, Laccone F, Cobilanschi J, Meins M, Huppke P, Hanefeld F, et al. MECP2 mutations in sporadic cases of Rett syndrome are almost exclusively of paternal origin. Am J Hum Genet 2001;68(5):1093–101.
[21] Zhang J, Bao X, Cao G, Jiang S, Zhu X, Lu H, et al. What does the nature of the MECP2 mutation tell us about parental origin and recurrence risk in Rett syndrome? Clin Genet 2012;82(6):526–33.
[22] Buyse IM, Fang P, Hoon KT, Amir RE, Zoghbi HY, Roa BB. Diagnostic testing for Rett syndrome by DHPLC and direct sequencing analysis of the MECP2 gene: identification of several novel mutations and polymorphisms. Am J Hum Genet 2000;67(6):1428–36.
[23] Erlandson A, Samuelsson L, Hagberg B, Kyllerman M, Vujic M, Wahlstrom J. Multiplex ligation-dependent probe amplification (MLPA) detects large deletions in the MECP2 gene of Swedish Rett syndrome patients. Genet Test 2003;7(4):329–32.
[24] Saunders CJ, Minassian BE, Chow EW, Zhao W, Vincent JB. Novel exon 1 mutations in MECP2 implicate isoform MeCP2_e1 in classical Rett syndrome. Am J Med Genet 2009;149a(5):1019–23.
[25] Bebbington A, Anderson A, Ravine D, Fyfe S, Pineda M, de Klerk N, et al. Investigating genotype-phenotype relationships in Rett syndrome using an international data set. Neurology 2008;70(11):868–75.
[26] Neul JL, Fang P, Barrish J, Lane J, Caeg EB, Smith EO, et al. Specific mutations in Methyl-CpG-binding protein 2 confer different severity in Rett syndrome. Neurology 2008;70(16):1313–21.
[27] Chen RZ, Akbarian S, Tudor M, Jaenisch R. Deficiency of methyl-CpG binding protein-2 in CNS neurons results in a Rett-like phenotype in mice. Nat Genet 2001;27(3):327–31.
[28] Guy J, Hendrich B, Holmes M, Martin JE, Bird A. A mouse Mecp2-null mutation causes neurological symptoms that mimic Rett syndrome. Nat Genet 2001;27(3):322–6.
[29] Gemelli T, Berton O, Nelson ED, Perrotti LI, Jaenisch R, Monteggia LM. Postnatal loss of methyl-CpG binding protein 2 in the forebrain is sufficient to mediate behavioral aspects of Rett syndrome in mice. Biol Psychiatry 2006;59(5):468–76.
[30] Samaco RC, Mandel-Brehm C, Chao HT, Ward CS, Fyffe-Maricich SL, Ren J, et al. Loss of MeCP2 in aminergic neurons causes cell-autonomous defects in neurotransmitter synthesis and specific behavioral abnormalities. Proc Natl Acad Sci USA 2009;106(51):21966–71.
[31] Fukuda T, Itoh M, Ichikawa T, Washiyama K, Goto Y. Delayed maturation of neuronal architecture and synaptogenesis in cerebral cortex of Mecp2-deficient mice. J Neuropathol Exp Neurol 2005;64(6):537–44.
[32] Marchetto MC, Carromeu C, Acab A, Yu D, Yeo GW, Mu Y, et al. A model for neural development and treatment of Rett syndrome using human induced pluripotent stem cells. Cell 2010;143(4):527–39.
[33] El-Khoury R, Panayotis N, Matagne V, Ghata A, Villard L, Roux JC. GABA and glutamate pathways are spatially and developmentally affected in the brain of Mecp2-deficient mice. PLoS One 2014;9(3):e92169.
[34] Chao HT, Chen H, Samaco RC, Xue M, Chahrour M, Yoo J, et al. Dysfunction in GABA signalling mediates autism-like stereotypies and Rett syndrome phenotypes. Nature 2010;468(7321):263–9.

[35] Meng X, Wang W, Lu H, He LJ, Chen W, Chao ES, et al. Manipulations of MeCP2 in glutamatergic neurons highlight their contributions to Rett and other neurological disorders. Elife 2016;5.

[36] Katz DM, Bird A, Coenraads M, Gray SJ, Menon DU, Philpot BD, et al. Rett syndrome: crossing the threshold to clinical translation. Trends Neurosci 2016;39(2):100–13.

[37] Rowley NM, Madsen KK, Schousboe A, Steve White H. Glutamate and GABA synthesis, release, transport and metabolism as targets for seizure control. Neurochem Int 2012;61(4):546–58.

[38] Bao X, Downs J, Wong K, Williams S, Leonard H. Using a large international sample to investigate epilepsy in Rett syndrome. Dev Med Child Neurol 2013;55(6):553–8.

[39] Glaze DG, Percy AK, Skinner S, Motil KJ, Neul JL, Barrish JO, et al. Epilepsy and the natural history of Rett syndrome. Neurology 2010;74(11):909–12.

[40] Jian L, Nagarajan L, de Klerk N, Ravine D, Christodoulou J, Leonard H. Seizures in Rett syndrome: an overview from a one-year calendar study. Eur J Paediatr Neurol 2007;11(5):310–7.

[41] Maezawa I, Swanberg S, Harvey D, LaSalle JM, Jin LW. Rett syndrome astrocytes are abnormal and spread MeCP2 deficiency through gap junctions. J Neurosci 2009;29(16):5051–61.

[42] Maezawa I, Jin LW. Rett syndrome microglia damage dendrites and synapses by the elevated release of glutamate. J Neurosci 2010;30(15):5346–56.

[43] Lioy DT, Garg SK, Monaghan CE, Raber J, Foust KD, Kaspar BK, et al. A role for glia in the progression of Rett's syndrome. Nature 2011;475(7357):497–500.

[44] Robinson L, Guy J, McKay L, Brockett E, Spike RC, Selfridge J, et al. Morphological and functional reversal of phenotypes in a mouse model of Rett syndrome. Brain 2012;135(Pt 9):2699–710.

[45] Ure K, Lu H, Wang W, Ito-Ishida A, Wu Z, He LJ, et al. Restoration of Mecp2 expression in GABAergic neurons is sufficient to rescue multiple disease features in a mouse model of Rett syndrome. Elife 2016;5.

[46] Ross PD, Guy J, Selfridge J, Kamal B, Bahey N, Tanner KE, et al. Exclusive expression of MeCP2 in the nervous system distinguishes between brain and peripheral Rett syndrome-like phenotypes. Hum Mol Genet 2016;25(20):4389–404.

[47] De Felice C, Rossi M, Leoncini S, Chisci G, Signorini C, Lonetti G, et al. Inflammatory lung disease in Rett syndrome. Mediat Inflamm 2014;2014:560120.

[48] O'Connor RD, Zayzafoon M, Farach-Carson MC, Schanen NC. Mecp2 deficiency decreases bone formation and reduces bone volume in a rodent model of Rett syndrome. Bone 2009;45(2):346–56.

[49] Kamal B, Russell D, Payne A, Constante D, Tanner KE, Isaksson H, et al. Biomechanical properties of bone in a mouse model of Rett syndrome. Bone 2015;71:106–14.

[50] Kyle SM, Saha PK, Brown HM, Chan LC, Justice MJ. MeCP2 co-ordinates liver lipid metabolism with the NCoR1/HDAC3 corepressor complex. Hum Mol Genet 2016;25(14):3029–41.

[51] Singh J, Saxena A, Christodoulou J, Ravine D. MECP2 genomic structure and function: insights from ENCODE. Nucleic Acids Res 2008;36(19):6035–47.

[52] Lyst MJ, Bird A. Rett syndrome: a complex disorder with simple roots. Nat Rev Genet 2015;16(5):261–75.

[53] Nan X, Ng HH, Johnson CA, Laherty CD, Turner BM, Eisenman RN, et al. Transcriptional repression by the methyl-CpG-binding protein MeCP2 involves a histone deacetylase complex. Nature 1998;393(6683):386–9.

[54] Kokura K, Kaul SC, Wadhwa R, Nomura T, Khan MM, Shinagawa T, et al. The Ski protein family is required for MeCP2-mediated transcriptional repression. J Biol Chem 2001;276(36):34115–21.

[55] Lyst MJ, Ekiert R, Ebert DH, Merusi C, Nowak J, Selfridge J, et al. Rett syndrome mutations abolish the interaction of MeCP2 with the NCoR/SMRT co-repressor. Nat Neurosci 2013;16(7):898–902.

[56] Baker SA, Chen L, Wilkins AD, Yu P, Lichtarge O, Zoghbi HY. An AT-hook domain in MeCP2 determines the clinical course of Rett syndrome and related disorders. Cell 2013;152(5):984–96.

[57] Chahrour M, Jung SY, Shaw C, Zhou X, Wong ST, Qin J, et al. MeCP2, a key contributor to neurological disease, activates and represses transcription. Science 2008;320(5880):1224–9.

[58] Weng SM, McLeod F, Bailey ME, Cobb SR. Synaptic plasticity deficits in an experimental model of rett syndrome: long-term potentiation saturation and its pharmacological reversal. Neuroscience 2011;180: 314–21.
[59] Kriaucionis S, Paterson A, Curtis J, Guy J, Macleod N, Bird A. Gene expression analysis exposes mitochondrial abnormalities in a mouse model of Rett syndrome. Mol Cell Biol 2006;26(13):5033–42.
[60] De Felice C, Della Ragione F, Signorini C, Leoncini S, Pecorelli A, Ciccoli L, et al. Oxidative brain damage in Mecp2-mutant murine models of Rett syndrome. Neurobiol Dis 2014;68:66–77.
[61] Einspieler C, Kerr AM, Prechtl HF. Abnormal general movements in girls with Rett disorder: the first four months of life. Brain Dev 2005;27(Suppl. 1):S8–s13.
[62] Fehr S, Bebbington A, Ellaway C, Rowe P, Leonard H, Downs J. Altered attainment of developmental milestones influences the age of diagnosis of rett syndrome. J Child Neurol 2011;26(8):980–7.
[63] Leonard H, Bower C. Is the girl with Rett syndrome normal at birth? Dev Med Child Neurol 1998;40(2): 115–21.
[64] Marschik PB, Pini G, Bartl-Pokorny KD, Duckworth M, Gugatschka M, Vollmann R, et al. Early speech-language development in females with Rett syndrome: focusing on the preserved speech variant. Dev Med Child Neurol 2012;54(5):451–6.
[65] Hagberg B, Goutieres F, Hanefeld F, Rett A, Wilson J. Rett syndrome: criteria for inclusion and exclusion. Brain Dev 1985;7(3):372–3.
[66] Diagnostic criteria for rett syndrome. The rett syndrome diagnostic criteria work group. Ann Neurol 1988; 23(4):425–8.
[67] Fehr S, Wilson M, Downs J, Williams S, Murgia A, Sartori S, et al. The CDKL5 disorder is an independent clinical entity associated with early-onset encephalopathy. Eur J Hum Genet 2013;21(3):266–73.
[68] Leonard H, Moore H, Carey M, Fyfe S, Hall S, Robertson L, et al. Genotype and early development in Rett syndrome: the value of international data. Brain Dev 2005;27(Suppl. 1):S59–68.
[69] Kerr AM, Prescott RJ. Predictive value of the early clinical signs in Rett disorder. Brain Dev 2005;27(Suppl. 1):S20–4.
[70] Burford B. Perturbations in the development of infants with Rett disorder and the implications for early diagnosis. Brain Dev 2005;27(Suppl. 1):S3–7.
[71] Einspieler C, Kerr AM, Prechtl HF. Is the early development of girls with Rett disorder really normal? Pediatr Res 2005;57(5 Pt 1):696–700.
[72] Neul JL, Kaufmann WE, Glaze DG, Christodoulou J, Clarke AJ, Bahi-Buisson N, et al. Rett syndrome: revised diagnostic criteria and nomenclature. Ann Neurol 2010;68(6):944–50.
[73] Hagberg BA, Skjeldal OH. Rett variants: a suggested model for inclusion criteria. Pediatr Neurol 1994; 11(1):5–11.
[74] Weaving LS, Christodoulou J, Williamson SL, Friend KL, McKenzie OL, Archer H, et al. Mutations of CDKL5 cause a severe neurodevelopmental disorder with infantile spasms and mental retardation. Am J Hum Genet 2004;75(6):1079–93.
[75] Ariani F, Hayek G, Rondinella D, Artuso R, Mencarelli MA, Spanhol-Rosseto A, et al. FOXG1 is responsible for the congenital variant of Rett syndrome. Am J Hum Genet 2008;83(1):89–93.
[76] Leonard H, Fyfe S, Leonard S, Msall M. Functional status, medical impairments, and rehabilitation resources in 84 females with Rett syndrome: a snapshot across the world from the parental perspective. Disabil Rehabil 2001;23(3–4):107–17.
[77] Downs J, Bebbington A, Jacoby P, Williams A-M, Ghosh S, Kaufmann W, et al. Level of purposeful hand function as a marker of clinical severity in Rett syndrome. Dev Med Child Neurol 2010;52(9):817–23.
[78] Downs J, Stahlhut M, Wong K, Syhler B, Bisgaard AM, Jacoby P, et al. Validating the rett syndrome gross motor scale. PLoS One 2016;11(1):e0147555.

[79] Urbanowicz A, Downs J, Girdler S, Ciccone N, Leonard H. Aspects of speech-language abilities are influenced by MECP2 mutation type in girls with Rett syndrome. Am J Med Genet 2015;167a(2):354–62.
[80] Hagberg B. Clinical manifestations and stages of Rett syndrome. Ment Retard Dev Disabil Res Rev 2002; 8(2):61–5.
[81] WHO Multicentre Growth Reference Study Group. WHO Motor Development Study: windows of achievement for six gross motor development milestones. Acta Paediatr Suppl 2006;450:86–95.
[82] Temudo T, Ramos E, Dias K, Barbot C, Vieira JP, Moreira A, et al. Movement disorders in Rett syndrome: an analysis of 60 patients with detected MECP2 mutation and correlation with mutation type. Mov Disord 2008;23(10):1384–90.
[83] Carter P, Downs J, Bebbington A, Williams S, Jacoby P, Kaufmann WE, et al. Stereotypical hand movements in 144 subjects with Rett syndrome from the population-based Australian database. Mov Disord 2010;25(3):282–8.
[84] Marr C, Leonard H, Torode I, Downs J. Spinal fusion in girls with Rett syndrome: post-operative recovery and family experiences. Child Care Health Dev 2015;41(6):1000–9.
[85] Larsson EL, Aaro S, Ahlinder P, Normelli H, Tropp H, Oberg B. Long-term follow-up of functioning after spinal surgery in patients with Rett syndrome. Eur Spine J 2009;18(4):506–11.
[86] Downs J, Torode I, Wong K, Ellaway C, Elliott EJ, Izatt MT, et al. Surgical fusion of early onset severe scoliosis increases survival in Rett syndrome: a cohort study. Dev Med Child Neurol 2016;58(6):632–8.
[87] Jian L, Nagarajan L, de Klerk N, Ravine D, Bower C, Anderson A, et al. Predictors of seizure onset in Rett syndrome. J Pediatr 2006;149(4):542–7.
[88] Abdala AP, Lioy DT, Garg SK, Knopp SJ, Paton JF, Bissonnette JM. Effect of Sarizotan, a 5-HT1a and D2-like receptor agonist, on respiration in three mouse models of Rett syndrome. Am J Respir Cell Mol Biol 2014;50(6):1031–9.
[89] Julu PO, Kerr AM, Apartopoulos F, Al-Rawas S, Engerstrom IW, Engerstrom L, et al. Characterisation of breathing and associated central autonomic dysfunction in the Rett disorder. Arch Dis Child 2001;85(1): 29–37.
[90] Mackay J, Downs J, Wong K, Heyworth J, Epstein A, Leonard H. Autonomic breathing abnormalities in Rett syndrome: caregiver perspectives in an international database study. J Neurodev Disord 2017;9:15.
[91] Weese-Mayer DE, Lieske SP, Boothby CM, Kenny AS, Bennett HL, Silvestri JM, et al. Autonomic nervous system dysregulation: breathing and heart rate perturbation during wakefulness in young girls with Rett syndrome. Pediatr Res 2006;60(4):443–9.
[92] Schultz RJ, Glaze DG, Motil KJ, Armstrong DD, del Junco DJ, Hubbard CR, et al. The pattern of growth failure in Rett syndrome. Am J Dis Child 1993;147(6):633–7.
[93] Motil KJ, Schultz RJ, Browning K, Trautwein L, Glaze DG. Oropharyngeal dysfunction and gastroesophageal dysmotility are present in girls and women with Rett syndrome. J Pediatr Gastroenterol Nutr 1999;29(1):31–7.
[94] Oddy WH, Webb KG, Baikie G, Thompson SM, Reilly S, Fyfe SD, et al. Feeding experiences and growth status in a Rett syndrome population. J Pediatr Gastroenterol Nutr 2007;45(5):582–90.
[95] Bebbington A, Percy A, Christodoulou J, Ravine D, Ho G, Jacoby P, et al. Updating the profile of C-terminal MECP2 deletions in Rett syndrome. J Med Genet 2010;47(4):242–8.
[96] Downs J, Wong K, Ravikumara M, Ellaway C, Elliott EJ, Christodoulou J, et al. Experience of gastrostomy using a quality care framework: the example of rett syndrome. Medicine (Baltim) 2014;93(28):e328.
[97] Tarquinio DC, Motil KJ, Hou W, Lee HS, Glaze DG, Skinner SA, et al. Growth failure and outcome in Rett syndrome: specific growth references. Neurology 2012;79(16):1653–61.
[98] Leonard H, Ravikumara M, Baikie G, Naseem N, Ellaway C, Percy A, et al. Assessment and management of nutrition and growth in Rett syndrome. J Pediatr Gastroenterol Nutr 2013;57(4):451–60.

[99] Borghi E, Borgo F, Severgnini M, Savini MN, Casiraghi MC, Vignoli A. Rett syndrome: a focus on gut microbiota. Int J Mol Sci 2017;18(2).

[100] Baikie G, Ravikumara M, Downs J, Naseem N, Wong K, Percy A, et al. Gastrointestinal dysmotility in Rett syndrome. J Pediatr Gastroenterol Nutr 2014;58(2):237–44.

[101] Freilinger M, Bohm M, Lanator I, Vergesslich-Rothschild K, Huber WD, Anderson A, et al. Prevalence, clinical investigation, and management of gallbladder disease in Rett syndrome. Dev Med Child Neurol 2014;56(8):756–62.

[102] Buchovecky CM, Turley SD, Brown HM, Kyle SM, McDonald JG, Liu B, et al. A suppressor screen in Mecp2 mutant mice implicates cholesterol metabolism in Rett syndrome. Nat Genet 2013;45(9): 1013–20.

[103] Piazza CC, Fisher W, Kiesewetter K, Bowman L, Moser H. Aberrant sleep patterns in children with the Rett syndrome. Brain Dev 1990;12(5):488–93.

[104] Segawa M, Nomura Y. The pathophysiology of the Rett syndrome from the standpoint of polysomnography. Brain Dev 1990;12(1):55–60.

[105] Powell WT, LaSalle JM. Epigenetic mechanisms in diurnal cycles of metabolism and neurodevelopment. Hum Mol Genet 2015;24(R1):R1–9.

[106] Ellaway C, Peat J, Leonard H, Christodoulou J. Sleep dysfunction in Rett syndrome: lack of age related decrease in sleep duration. Brain Dev 2001;23(Suppl. 1):S101–3.

[107] Wong K, Leonard H, Jacoby P, Ellaway C, Downs J. The trajectories of sleep disturbances in Rett syndrome. J Sleep Res 2015;24(2):223–33.

[108] Young D, Nagarajan L, de Klerk N, Jacoby P, Ellaway C, Leonard H. Sleep problems in Rett syndrome. Brain Dev 2007;29(10):609–16.

[109] Boban S, Wong K, Epstein A, Anderson B, Murphy N, Downs J, et al. Determinants of sleep disturbances in Rett syndrome: novel findings in relation to genotype. Am J Med Genet 2016;170(9):2292–300.

[110] Leonard H, Thomson MR, Glasson EJ, Fyfe S, Leonard S, Bower C, et al. A population-based approach to the investigation of osteopenia in Rett syndrome. Dev Med Child Neurol 1999;41(5):323–8.

[111] Downs J, Bebbington A, Woodhead H, Jacoby P, Jian L, Jefferson A, et al. Early determinants of fractures in Rett syndrome. Pediatrics 2008;121(3):540–6.

[112] Leonard H, Downs J, Jian L, Bebbington A, Jacoby P, Nagarajan L, et al. Valproate and risk of fracture in Rett syndrome. Arch Dis Child 2010;95(6):444–8.

[113] Motil KJ, Ellis KJ, Barrish JO, Caeg E, Glaze DG. Bone mineral content and bone mineral density are lower in older than in younger females with Rett syndrome. Pediatr Res 2008;64(4):435–9.

[114] Shapiro JR, Bibat G, Hiremath G, Blue ME, Hundalani S, Yablonski T, et al. Bone mass in Rett syndrome: association with clinical parameters and MECP2 mutations. Pediatr Res 2010;68(5):446–51.

[115] Roende G, Ravn K, Fuglsang K, Andersen H, Nielsen JB, Brondum-Nielsen K, et al. DXA measurements in Rett syndrome reveal small bones with low bone mass. J Bone Miner Res 2011;26(9):2280–6.

[116] Jefferson A, Leonard H, Siafarikas A, Woodhead H, Fyfe S, Ward LM, et al. Clinical guidelines for management of bone health in rett syndrome based on expert consensus and available evidence. PLoS One 2016;11(2):e0146824.

[117] Laurvick CL, de Klerk N, Bower C, Christodoulou J, Ravine D, Ellaway C, et al. Rett syndrome in Australia: a review of the epidemiology. J Pediatr 2006;148(3):347–52.

[118] Fehr S, Bebbington A, Nassar N, Downs J, Ronen GM, N DEK, et al. Trends in the diagnosis of Rett syndrome in Australia. Pediatr Res 2011;70(3):313–9.

[119] Anderson A, Wong K, Jacoby P, Downs J, Leonard H. Twenty years of surveillance in Rett syndrome: what does this tell us? Orphanet J Rare Dis 2014;9:87.

[120] Ausio J, Paz AM, Esteller M. MeCP2: the long trip from a chromatin protein to neurological disorders. Trends Mol Med 2014;20(9):487–98.

[121] Ciernia AV, LaSalle J. The landscape of DNA methylation amid a perfect storm of autism aetiologies. Nat Rev Neurosci 2016;17:411–23.
[122] Lister R, Mukamel EA, Nery JR, Urich M, Puddifoot CA, Johnson ND, et al. Global epigenomic reconfiguration during mammalian brain development. Science 2013;341(6146):1237905.
[123] Gardiner-Garden M, Frommer M. CpG islands in vertebrate genomes. J Mol Biol 1987;196(2):261–82.
[124] Guo JU, Su Y, Shin JH, Shin J, Li H, Xie B, et al. Distribution, recognition and regulation of non-CpG methylation in the adult mammalian brain. Nat Neurosci 2014;17(2):215–22.
[125] Loenarz C, Schofield CJ. Oxygenase catalyzed 5-methylcytosine hydroxylation. Chem Biol 2009;16(6): 580–3.
[126] Fritz EL, Papavasiliou FN. Cytidine deaminases: AIDing DNA demethylation? Genes Dev 2010;24(19): 2107–14.
[127] Lister R, Pelizzola M, Dowen RH, Hawkins RD, Hon G, Tonti-Filippini J, et al. Human DNA methylomes at base resolution show widespread epigenomic differences. Nature 2009;462:315–22.
[128] Kundaje A, Meuleman W, Ernst J, Bilenky M, Yen A, Heravi-Moussavi A, et al. Integrative analysis of 111 reference human epigenomes. Nature 2015;518(7539):317–30.
[129] Schroeder DI, Blair JD, Lott P, Yu HO, Hong D, Crary F, et al. The human placenta methylome. Proc Natl Acad Sci USA 2013;110(15):6037–42.
[130] Ernst J, Kheradpour P, Mikkelsen TS, Shoresh N, Ward LD, Epstein CB, et al. Mapping and analysis of chromatin state dynamics in nine human cell types. Nature 2011;473(7345):43–9.
[131] Nan X, Meehan RR, Bird A. Dissection of the methyl-CpG binding domain from the chromosomal protein MeCP2. Nucleic Acids Res 1993;21(21):4886–92.
[132] Kinde B, Gabel HW, Gilbert CS, Griffith EC, Greenberg ME. Reading the unique DNA methylation landscape of the brain: non-CpG methylation, hydroxymethylation, and MeCP2. Proc Natl Acad Sci USA 2015;112(22):6800–6.
[133] Sperlazza MJ, Bilinovich SM, Sinanan LM, Javier FR, Williams Jr DC. Structural basis of MeCP2 distribution on non-CpG methylated and hydroxymethylated DNA. J Mol Biol 2017;429(10):1581–94.
[134] Gonzales ML, Adams S, Dunaway KW, LaSalle JM. Phosphorylation of distinct sites in MeCP2 modifies cofactor associations and the dynamics of transcriptional regulation. Mol Cell Biol 2012;32(14):2894–903.
[135] Martinowich K, Hattori D, Wu H, Fouse S, He F, Hu Y, et al. DNA methylation-related chromatin remodeling in activity-dependent BDNF gene regulation. Science 2003;302(5646):890–3.
[136] Su D, Cha YM, West AE. Mutation of MeCP2 alters transcriptional regulation of select immediate-early genes. Epigenetics 2012;7(2):146–54.
[137] Swanberg SE, Nagarajan RP, Peddada S, Yasui DH, LaSalle JM. Reciprocal co-regulation of EGR2 and MECP2 is disrupted in Rett syndrome and autism. Hum Mol Genet 2009;18(3):525–34.
[138] Yasui DH, Peddada S, Bieda MC, Vallero RO, Hogart A, Nagarajan RP, et al. Integrated epigenomic analyses of neuronal MeCP2 reveal a role for long-range interaction with active genes. Proc Natl Acad Sci U S A 2007;104(49):19416–21.
[139] Chen L, Chen K, Lavery LA, Baker SA, Shaw CA, Li W, et al. MeCP2 binds to non-CG methylated DNA as neurons mature, influencing transcription and the timing of onset for Rett syndrome. Proc Natl Acad Sci USA 2015;112(17):5509–14.
[140] Gabel HW, Kinde B, Stroud H, Gilbert CS, Harmin DA, Kastan NR, et al. Disruption of DNA-methylation-dependent long gene repression in Rett syndrome. Nature 2015;522(7554):89–93.
[141] Rube HT, Lee W, Hejna M, Chen H, Yasui DH, Hess JF, et al. Sequence features accurately predict genome-wide MeCP2 binding in vivo. Nat Commun 2016;7:11025.
[142] Cronk JC, Derecki NC, Ji E, Xu Y, Lampano AE, Smirnov I, et al. Methyl-CpG binding protein 2 regulates microglia and macrophage gene expression in response to inflammatory stimuli. Immunity 2015;42(4): 679–91.

[143] Cronk JC, Derecki NC, Litvak V, Kipnis J. Unexpected cellular players in Rett syndrome pathology. Neurobiol Dis 2016;92(Pt A):64–71.
[144] Nguyen MV, Felice CA, Du F, Covey MV, Robinson JK, Mandel G, et al. Oligodendrocyte lineage cells contribute unique features to Rett syndrome neuropathology. J Neurosci 2013;33(48):18764–74.
[145] Jones PL, Veenstra GJ, Wade PA, Vermaak D, Kass SU, Landsberger N, et al. Methylated DNA and MeCP2 recruit histone deacetylase to repress transcription. Nat Genet 1998;19(2):187–91.
[146] Shahbazian M, Young J, Yuva-Paylor L, Spencer C, Antalffy B, Noebels J, et al. Mice with truncated MeCP2 recapitulate many Rett syndrome features and display hyperacetylation of histone H3. Neuron 2002;35(2):243–54.
[147] Thatcher K, LaSalle JM. Dynamic changes in Histone H3 lysine 9 acetylation localization patterns during neuronal maturation require MeCP2. Epigenetics 2006;1(1):24–31.
[148] Wan M, Zhao K, Lee SS, Francke U. MECP2 truncating mutations cause histone H4 hyperacetylation in Rett syndrome. Hum Mol Genet 2001;10(10):1085–92.
[149] Singleton MK, Gonzales ML, Leung KN, Yasui DH, Schroeder DI, Dunaway K, et al. MeCP2 is required for global heterochromatic and nucleolar changes during activity-dependent neuronal maturation. Neurobiol Dis 2011;43(1):190–200.
[150] Li Y, Wang H, Muffat J, Cheng AW, Orlando DA, Loven J, et al. Global transcriptional and translational repression in human-embryonic-stem-cell-derived Rett syndrome neurons. Cell Stem Cell 2013;13(4): 446–58.
[151] Yazdani M, Deogracias R, Guy J, Poot RA, Bird A, Barde YA. Disease modeling using embryonic stem cells: MeCP2 regulates nuclear size and RNA synthesis in neurons. Stem Cell 2012;30(10): 2128–39.
[152] Ricciardi S, Boggio EM, Grosso S, Lonetti G, Forlani G, Stefanelli G, et al. Reduced AKT/mTOR signaling and protein synthesis dysregulation in a Rett syndrome animal model. Hum Mol Genet 2011;20(6): 1182–96.
[153] Nott A, Cheng J, Gao F, Lin YT, Gjoneska E, Ko T, et al. Histone deacetylase 3 associates with MeCP2 to regulate FOXO and social behavior. Nat Neurosci 2016;19(11):1497–505.
[154] Georgel PT, Horowitz-Scherer RA, Adkins N, Woodcock CL, Wade PA, Hansen JC. Chromatin compaction by human MeCP2. Assembly of novel secondary chromatin structures in the absence of DNA methylation. J Biol Chem 2003;278(34):32181–8.
[155] Ghosh RP, Horowitz-Scherer RA, Nikitina T, Shlyakhtenko LS, Woodcock CL. MeCP2 binds cooperatively to its substrate and competes with histone H1 for chromatin binding sites. Mol Cell Biol 2010;30(19): 4656–70.
[156] Skene PJ, Illingworth RS, Webb S, Kerr AR, James KD, Turner DJ, et al. Neuronal MeCP2 is expressed at near histone-octamer levels and globally alters the chromatin state. Mol Cell 2010;37(4):457–68.
[157] Linhoff MW, Garg SK, Mandel G. A high-resolution imaging approach to investigate chromatin architecture in complex tissues. Cell 2015;163(1):246–55.
[158] Nikitina T, Shi X, Ghosh RP, Horowitz-Scherer RA, Hansen JC, Woodcock CL. Multiple modes of interaction between the methylated DNA binding protein MeCP2 and chromatin. Mol Cell Biol 2007;27(3): 864–77.
[159] Kernohan KD, Jiang Y, Tremblay DC, Bonvissuto AC, Eubanks JH, Mann MR, et al. ATRX partners with cohesin and MeCP2 and contributes to developmental silencing of imprinted genes in the brain. Dev Cell 2010;18(2):191–202.
[160] Picketts DJ, Higgs DR, Bachoo S, Blake DJ, Quarrell OW, Gibbons RJ. ATRX encodes a novel member of the SNF2 family of proteins: mutations point to a common mechanism underlying the ATR-X syndrome. Hum Mol Genet 1996;5(12):1899–907.

[161] Berghoff EG, Clark MF, Chen S, Cajigas I, Leib DE, Kohtz JD. Evf2 (Dlx6as) lncRNA regulates ultra-conserved enhancer methylation and the differential transcriptional control of adjacent genes. Development 2013;140(21):4407–16.

[162] Bond AM, Vangompel MJ, Sametsky EA, Clark MF, Savage JC, Disterhoft JF, et al. Balanced gene regulation by an embryonic brain ncRNA is critical for adult hippocampal GABA circuitry. Nat Neurosci 2009;12(8):1020–7.

[163] Wu H, Tao J, Chen PJ, Shahab A, Ge W, Hart RP, et al. Genome-wide analysis reveals methyl-CpG-binding protein 2-dependent regulation of microRNAs in a mouse model of Rett syndrome. Proc Natl Acad Sci U S A 2010;107(42):18161–6.

[164] Petazzi P, Sandoval J, Szczesna K, Jorge OC, Roa L, Sayols S, et al. Dysregulation of the long non-coding RNA transcriptome in a Rett syndrome mouse model. RNA Biol 2013;10(7):1197–203.

[165] Buschdorf JP, Stratling WH. A WW domain binding region in methyl-CpG-binding protein MeCP2: impact on Rett syndrome. J Mol Med 2004;82(2):135–43.

[166] Maunakea AK, Chepelev I, Cui K, Zhao K. Intragenic DNA methylation modulates alternative splicing by recruiting MeCP2 to promote exon recognition. Cell Research 2013;23(11):1256–69.

[167] Yasui DH, Gonzales ML, Aflatooni JO, Crary FK, Hu DJ, Gavino BJ, et al. Mice with an isoform-ablating Mecp2 exon 1 mutation recapitulate the neurologic deficits of Rett syndrome. Hum Mol Genet 2014;23(9): 2447–58.

[168] Li R, Dong Q, Yuan X, Zeng X, Gao Y, Chiao C, et al. Misregulation of alternative splicing in a mouse model of rett syndrome. PLoS Genet 2016;12(6):e1006129.

[169] Cheng TL, Chen J, Wan H, Tang B, Tian W, Liao L, et al. Regulation of mRNA splicing by MeCP2 via epigenetic modifications in the brain. Sci Rep 2017;7:42790.

[170] Wong JJ, Gao D, Nguyen TV, Kwok CT, van Geldermalsen M, Middleton R, et al. Intron retention is regulated by altered MeCP2-mediated splicing factor recruitment. Nat Commun 2017;8:15134.

[171] Chaligne R, Heard E. X-chromosome inactivation in development and cancer. FEBS Letters 2014;588(15): 2514–22.

[172] Lau AW, Brown CJ, Penaherrera M, Langlois S, Kalousek DK, Robinson WP. Skewed X-chromosome inactivation is common in fetuses or newborns associated with confined placental mosaicism. Am J Hum Genet 1997;61(6):1353–61.

[173] Sangha KK, Stephenson MD, Brown CJ, Robinson WP. Extremely skewed X-chromosome inactivation is increased in women with recurrent spontaneous abortion. Am J Hum Genet 1999;65(3):913–7.

[174] Amir RE, Zoghbi HY. Rett syndrome: methyl-CpG-binding protein 2 mutations and phenotype-genotype correlations. Am J Med Genet 2000;97(2):147–52.

[175] LaSalle JM, Goldstine J, Balmer D, Greco CM. Quantitative localization of heterogeneous methyl-CpG-binding protein 2 (MeCP2) expression phenotypes in normal and Rett syndrome brain by laser scanning cytometry. Hum Mol Genet 2001;10(17):1729–40.

[176] Augenstein K, Lane JB, Horton A, Schanen C, Percy AK. Variable phenotypic expression of a MECP2 mutation in a family. J Neurodev Disord 2009;1(4):313.

[177] Archer H, Evans J, Leonard H, Colvin L, Ravine D, Christodoulou J, et al. Correlation between clinical severity in patients with Rett syndrome with a p.R168X or p.T158M MECP2 mutation, and the direction and degree of skewing of X-chromosome inactivation. J Med Genet 2007;44(2):148–52.

[178] Bao X, Shengling J, Fuying S, Hong P, Meirong L, Wu XR. X chromosome inactivation in Rett Syndrome and its correlations with MECP2 mutations and phenotype. J Child Neurol 2008;23(1):22–5.

[179] Takahashi S, Ohinata J, Makita Y, Suzuki N, Araki A, Sasaki A, et al. Skewed X chromosome inactivation failed to explain the normal phenotype of a carrier female with MECP2 mutation resulting in Rett syndrome. Clin Genet 2008;73(3):257–61.

[180] Miyake K, Yang C, Minakuchi Y, Ohori K, Soutome M, Hirasawa T, et al. Comparison of genomic and epigenomic expression in monozygotic twins discordant for rett syndrome. PLoS One 2013;8(6):e66729.
[181] Braunschweig D, Simcox T, Samaco RC, LaSalle JM. X-Chromosome inactivation ratios affect wild-type MeCP2 expression within mosaic Rett syndrome and Mecp2-/+ mouse brain. Hum Mol Genet 2004; 13(12):1275–86.
[182] del Gaudio D, Fang P, Scaglia F, Ward PA, Craigen WJ, Glaze DG, et al. Increased MECP2 gene copy number as the result of genomic duplication in neurodevelopmentally delayed males. Genet Med 2006; 8(12):784–92.
[183] Filosa S, Pecorelli A, D'Esposito M, Valacchi G, Hajek J. Exploring the possible link between MeCP2 and oxidative stress in Rett syndrome. Free Radic Biol Med 2015;88(Pt A):81–90.
[184] Shulyakova N, Andreazza AC, Mills LR, Eubanks JH. Mitochondrial dysfunction in the pathogenesis of rett syndrome: implications for mitochondria-targeted therapies. Front Cell Neurosci 2017;11:58.
[185] Justice MJ, Buchovecky CM, Kyle SM, Djukic A. A role for metabolism in Rett syndrome pathogenesis: new clinical findings and potential treatment targets. Rare Dis 2013;1:e27265.
[186] Li W, Pozzo-Miller L. BDNF deregulation in Rett syndrome. Neuropharmacology 2014;76(Pt C):737–46.
[187] Cattaneo A, Cattane N, Begni V, Pariante CM, Riva MA. The human BDNF gene: peripheral gene expression and protein levels as biomarkers for psychiatric disorders. Transl Psychiatry 2016;6(11):e958.
[188] Xu X, Garcia J, Ewalt R, Nason S, Pozzo-Miller L. The BDNF val-66-met polymorphism affects neruonal morphology and synaptic transmission in cultured hippocampal neurons from Rett syndrome mice. Front Cell Neurosci 2017;11(203):1–12.
[189] Ben Zeev B, Bebbington A, Ho G, Leonard H, de Klerk N, Gak E, et al. The common BDNF polymorphism may be a modifier of disease severity in Rett syndrome. Neurology 2009;72(14):1242–7.
[190] Grillo E, Lo Rizzo C, Bianciardi L, Bizzarri V, Baldassarri M, Spiga O, et al. Revealing the complexity of a monogenic disease: rett syndrome exome sequencing. PLoS One 2013;8(2):e56599.
[191] Mantis JG, Fritz CL, Marsh J, Heinrichs SC, Seyfried TN. Improvement in motor and exploratory behavior in Rett syndrome mice with restricted ketogenic and standard diets. Epilepsy Behav 2009;15(2):133–41.
[192] Park MJ, Aja S, Li Q, Degano AL, Penati J, Zhuo J, et al. Anaplerotic triheptanoin diet enhances mitochondrial substrate use to remodel the metabolome and improve lifespan, motor function, and sociability in MeCP2-null mice. PLoS One 2014;9(10):e109527.
[193] Ricceri L, De Filippis B, Fuso A, Laviola G. Cholinergic hypofunction in MeCP2-308 mice: beneficial neurobehavioural effects of neonatal choline supplementation. Behav Brain Res 2011;221(2):623–9.
[194] Woods R, Vallero RO, Golub MS, Suarez JK, Ta TA, Yasui DH, et al. Long-lived epigenetic interactions between perinatal PBDE exposure and Mecp2308 mutation. Hum Mol Genet 2012;21(11):2399–411.
[195] Kondo M, Gray LJ, Pelka GJ, Christodoulou J, Tam PP, Hannan AJ. Environmental enrichment ameliorates a motor coordination deficit in a mouse model of Rett syndrome – Mecp2 gene dosage effects and BDNF expression. Eur J Neurosci 2008;27(12):3342–50.
[196] Nag N, Moriuchi J, Peitzman CGK, Ward B, Kolodny N, Berger-Sweeney J. Environmental enrichment alters locomotor behaviour and ventricular volume in Mecp2 1lox mice. Behav Brain Res 2009;196(1): 44–8.
[197] Hara M, Nishi Y, Yamashita Y, Hirata R, Takahashi S, Nagamitsu S, et al. Relation between circulating levels of GH, IGF-1, ghrelin and somatic growth in Rett syndrome. Brain Dev 2014;36(9):794–800.
[198] Khwaja OS, Ho E, Barnes KV, O'Leary HM, Pereira LM, Finkelstein Y, et al. Safety, pharmacokinetics, and preliminary assessment of efficacy of mecasermin (recombinant human IGF-1) for the treatment of Rett syndrome. Proc Natl Acad Sci USA 2014;111(12):4596–601.
[199] Schaevitz LR, Moriuchi JM, Nag N, Mellot TJ, Berger-Sweeney J. Cognitive and social functions and growth factors in a mouse model of Rett syndrome. Physiol Behav 2010;100(3):255–63.

[200] Tropea D, Giacometti E, Wilson NR, Beard C, McCurry C, Fu DD, et al. Partial reversal of Rett Syndrome-like symptoms in MeCP2 mutant mice. Proc Natl Acad Sci USA 2009;106(6):2029−34.

[201] Artuso R, Papa FT, Grillo E, Mucciolo M, Yasui DH, Dunaway KW, et al. Investigation of modifier genes within copy number variations in Rett syndrome. J Hum Genet 2011;56(7):508−15.

[202] Cobolli Gigli C, Scaramuzza L, Gandaglia A, Bellini E, Gabaglio M, Parolaro D, et al. MeCP2 related studies benefit from the use of CD1 as genetic background. PLoS One 2016;11(4):e0153473.

[203] Krishnan K, Lau BY, Ewall G, Huang ZJ, Shea SD. MECP2 regulates cortical plasticity underlying a learned behaviour in adult female mice. Nat Commun 2017;8:14077.

[204] Vogel Ciernia A, Pride MC, Durbin-Johnson B, Noronha A, Chang A, Yasui DH, et al. Early motor phenotype detection in a female mouse model of Rett syndrome is improved by cross-fostering. Hum Mol Genet 2017;26(10):1839−54.

[205] Meaney MJ, Diorio J, Francis D, Weaver S, Yau J, Chapman K, et al. Postnatal handling increases the expression of cAMP-inducible transcription factors in the rat hippocampus: the effects of thyroid hormones and serotonin. J Neurosci 2000;20(10):3926−35.

[206] Downs J, Rodger J, Chen L, Tan X, Hu N, Wong K, et al. Environmental enrichment intervention for Rett syndrome: an individually randomised stepped wedge trial. Orphanet J Rare Dis 2018;13(1):3. https://doi.org/10.1186/s13023-017-0752-8.

[207] Ladd-Acosta C, Fallin MD. The role of epigenetics in genetic and environmental epidemiology. Epigenomics 2016;8(2):271−83.

[208] Pozzo-Miller L, Pati S, Percy AK. Rett syndrome: reaching for clinical trials. Neurotherapeutics 2015;12(3):631−40.

CHAPTER

Sotos syndrome

10

Chloe Lane, Megan Freeth

Psychology Department, University of Sheffield, Sheffield, United Kingdom

Chapter Outline

10.1 Introduction 219
10.2 The genetic basis of Sotos syndrome 220
10.3 Comparing Sotos syndrome with other single-gene overgrowth syndromes 221
10.4 Neurological profile of Sotos syndrome 221
10.5 The cognitive and behavioral profile of Sotos syndrome 222
10.5.1 Cognition 222
10.5.1.1 Intellectual ability 222
10.5.1.2 Sotos syndrome cognitive profile 223
10.5.1.3 Language 224
10.5.2 Behavior 225
10.5.2.1 Attention-deficit/hyperactivity disorder 225
10.5.2.2 Anxiety 225
10.5.3 Aggression and tantrums 226
10.6 Sotos syndrome and autism spectrum disorder 226
10.7 Nuclear receptor–binding SET domain methyltransferases modify histones and affect epigenetics 228
10.8 Limitations and future research directions 229
10.9 Summary and conclusions 230
References 230

10.1 Introduction

Sotos syndrome is a congenital overgrowth disorder, initially recognized by Sotos et al. [1] who observed five patients with similar clinical features. These features were described as excessively rapid growth, acromegalic features, a nonprogressive cerebral disorder, and mental retardation. The authors considered this combination of features to be attributable to a specific syndrome, which they termed Sotos syndrome. Excessively rapid growth has been defined as advanced height, weight, and bone age; acromegalic features include a prominent forehead, high anterior hairline, prominent chin, and downslanting palpebral fissures [2]. The syndrome was described as nonprogressive, meaning that the symptoms do not significantly worsen throughout development. As macrocephaly is one of the

Chromatin Signaling and Neurological Disorders. https://doi.org/10.1016/B978-0-12-813796-3.00010-9

features of the syndrome, initial research often used the terms cerebral gigantism and Sotos syndrome interchangeably to refer to the same condition. Sotos syndrome has an estimated incidence of approximately one in 14,000 [3].

Subsequent research has confirmed these cardinal features in larger samples of individuals with Sotos syndrome. For example Cole and Hughes [4] investigated the clinical characteristics of 41 cases of Sotos syndrome. The findings confirmed overgrowth (defined as height and/or head circumference >97th percentile) with advanced bone age, macrocephaly, characteristic facial appearance, and intellectual disability as the cardinal features of the syndrome. Other health problems that are commonly experienced in children with Sotos syndrome are cardiac and genitourinary anomalies, neonatal jaundice, neonatal hypotonia, seizures, and scoliosis [3,5].

In this chapter, we outline the current knowledge of Sotos syndrome in relation to the genetic basis, including chromatin and epigenetic drivers of the condition, the neurological profile, the cognitive and behavioral profile, and the characteristics of autism spectrum disorder (ASD) in Sotos syndrome, before considering the limitations of the current knowledge and suggesting some directions for future research to further understand the genotype–phenotype relationship in Sotos syndrome.

10.2 The genetic basis of Sotos syndrome

Identification of a genetic abnormality responsible for Sotos syndrome was first established in a Japanese cohort of patients [6]. The authors identified that Sotos syndrome is caused by haploinsufficiency of the *NSD1* (nuclear receptor–binding SET domain protein 1) gene. The *NSD1* gene is involved in chromatin regulation. It encodes a SET domain–containing histone methyltransferase and is mapped on chromosome 5q35 [7]. Sotos syndrome is caused by intragenic mutations of the *NSD1* gene or 5q35 microdeletions encompassing *NSD1* and these abnormalities result in loss of function. The prevalence of NSD1 abnormalities in a cohort of 266 individuals with a clinical diagnosis of Sotos syndrome was assessed by Tatton-Brown et al. [8]. Findings from this study identified that abnormalities of the *NSD1* gene were present in 93% of individuals with a clinical diagnosis of Sotos syndrome.

Interestingly, a distinction in the prevalence of the different types of *NSD1* abnormality has been identified in individuals of differing ethnicity. In Japan, a 5q35 microdeletion encompassing the *NSD1* gene is the most common cause of Sotos syndrome [9]. However, in individuals of non-Japanese ethnicity, an intragenic mutation of the *NSD1* gene is the most common cause of Sotos syndrome, accounting for approximately 83% of clinical cases, whereas 5q35 microdeletions account for approximately 10% of cases [9]. Truncating mutations tend to be spread throughout *NSD1*, whilst clustering missense mutations tend to be found in highly conserved functional domains between exons 12 and 23 [9a]. An inversion polymorphism that predisposes to microdeletions is relatively common in Japan and it has been suggested that this may cause the reported difference in microdeletion frequency between Japan and elsewhere [8]. As Sotos syndrome is not specifically linked to the X or Y chromosomes, it affects males and females equally. In the majority of cases, the *NSD1* abnormalities that cause Sotos syndrome are de novo, meaning that they occur spontaneously [6]. However, the syndrome has an autosomal dominant inheritance pattern, meaning that a child of an individual with Sotos syndrome has a 50% chance of also having the syndrome. A small number of familial cases arising from autosomal dominant transmission have been reported in the literature [8,10].

Research has investigated potential genotype–phenotype correlations associated with the different *NSD1* abnormalities [8,11,12]. Broadly, it has been suggested that individuals with 5q35 micro-deletions encompassing the *NSD1* gene have less prominent overgrowth and more severe intellectual disability but the deletion size does not affect the clinical phenotype, when compared to individuals with an intragenic mutation of the *NSD1* gene. Additionally, it has been proposed that missense *NSD1* mutations are pathogenic only if they occur within functional domains implicated in chromatin regulation [8,12].

Genetic testing has become more widely available, with Sotos syndrome now only being diagnosed if an individual has an *NSD1* abnormality and also meets the clinical criteria for the syndrome. This means that Sotos syndrome can be diagnosed objectively and the research outlined in subsequent sections is beginning to further our understanding of the clinical, cognitive, and behavioral effects of *NSD1* abnormality associated with Sotos syndrome.

10.3 Comparing Sotos syndrome with other single-gene overgrowth syndromes

Sotos syndrome is one of several single-gene disorders associated with overgrowth and intellectual disability (OGID) and has been identified as the most prevalent OGID [13]. Other examples of OGID include Weaver syndrome [14] and Tatton–Brown Rahman syndrome (TBRS) [15]. Although the cardinal features of Sotos syndrome, Weaver syndrome, and TBRS are similar, as all these syndromes are associated with OGID, the syndromes can be differentiated by subtle differences in the phenotypes. For example, individuals with Weaver syndrome typically have a round face and almond-shaped eyes, which are not characteristic facial features associated with Sotos syndrome, as well as less prominent macrocephaly than individuals with Sotos syndrome [7,16]. Each of these OGID syndromes described are caused by an abnormality of distinct genes; Weaver syndrome is caused by mutation of the *EZH2* gene, which encodes a lysine methyltransferase [16] and TBRS is caused by mutation of the *DNMT3A* gene, which encodes a DNA methyltransferase [15].

10.4 Neurological profile of Sotos syndrome

As well as having macrocephaly, individuals with Sotos syndrome typically display distinctive neurological abnormalities. The most comprehensive study to date was conducted by Schaefer et al. [17]. Structural brain abnormalities were investigated in 40 participants with Sotos syndrome, using magnetic resonance imaging (MRI). The findings indicated that participants displayed a characteristic pattern of abnormalities. Specifically, participants displayed ventricular abnormalities (prominent trigone in 90% of cases, prominent occipital horn in 75% of cases), midline abnormalities (i.e., hypoplasia of the corpus callosum in 97.5% of cases, cavum septum pellucidum in 40% of cases, cavum vergae in 37.5% of cases), and extracerebral fluid in 70% of cases. Similar results have also been reported in a range of other smaller case series studies. Aoki et al. [18] reported neuroimaging data from two patients with Sotos syndrome. Their data indicated that shortly after birth, macrocephaly was attributed to increased volume of the cerebral parenchyma and later macrocephaly was related to retention of cerebrospinal fluid in the ventricles and subarachnoid spaces. Similar findings of ventricular abnormalities, abnormal prominence of the trigone, an occipital horn, increased extracerebral fluid levels, midline structural abnormalities thinning of the corpus callosum, cavum septum pellucidum, and cavum vergae were also observed in a case series of three individuals with Sotos syndrome by Horikoshi et al. [19]. In addition, a study by Fickie et al. [20] reported data from six individuals with Sotos syndrome displaying a range of midline structural abnormalities including cavum septum vergae and ventriculomegaly, which were observed in some individuals. Finally, similar findings were also identified in a case series of six individuals by Türkmen et al. [21]. We are not aware of any studies published to date that have investigated functional neural activity in Sotos syndrome, hence the functional impact of these structural abnormalities is currently unknown.

A number of case and cohort studies have noted that epilepsy is relatively common in Sotos syndrome. Prevalence estimates are between 15% and 50% [9,10,22]. Nicita et al. [23] reported a case series of 19 Caucasian individuals who were diagnosed with Sotos syndrome and febrile seizures (FSs) (fits that occur when an individual has a fever) or afebrile seizures (AFs) (fits that occur without a fever) in childhood, and they were followed up long term for at least 5 years after initial observation. They reported that more than half (64%) of the individuals who initially presented with FSs went on to develop AFs (epilepsy). This is a strikingly different prevalence rate to that of individuals with non-syndromic FSs who go on to develop AFs, which is estimated to be around 3–5%. In particular, temporal lobe seizures were reported in 40% of their cohort, but no clear explanation for this could be provided by MRI examination, as there were no apparent temporal lobe abnormalities. Nicita et al. [23] also found that in the vast majority of their cohort, seizures could be controlled with common anti-epileptic drugs at standard doses.

10.5 The cognitive and behavioral profile of Sotos syndrome

Cognitive and behavioral phenotyping of genetic syndromes associated with intellectual disability can provide knowledge of the strengths, difficulties and likely impact on daily living for particular syndromes. This can be very useful for families, educators and clinicians. In a systematic review [24] we found that, given the prevalence of Sotos syndrome, the cognitive and behavioral profile of this population is under-researched, with only 34 studies published up until August 2015, which have reported empirical data on any aspect of cognition or behavior in this population. The majority of the empirical literature on Sotos syndrome in relation to cognition and behavior are reports of case studies or presentations of small case series. However, since August 2015, a few studies, including some conducted in our laboratory, have reported findings from larger cohorts of individuals with the aim of outlining the cognitive and behavioral profile of Sotos syndrome using more representative samples [25–27]. Findings from all these studies are summarized in the following relevant sections.

10.5.1 Cognition

10.5.1.1 Intellectual ability

Intellectual ability, also sometimes referred to as general intelligence, is the acquired repertoire of general cognitive skills that is available to a person at a particular point in time [28] and can be measured via standardized scores on intelligence quotient (IQ) tests, also sometimes termed General Conceptual Ability (GCA) scores. The mean score for the neurotypical population is 100, and the standard deviation is 15. The International Statistical Classification of Diseases and Related Health

Problems (ICD-10) suggests the following guidelines for classification of the degree of intellectual impairment: borderline intellectual functioning, IQ of 70–84; mild intellectual disability, IQ of 50–69; moderate intellectual disability, IQ of 35–49; and severe intellectual disability, IQ of 20–34. A systematic search of the literature indicated that almost all reported cases of Sotos syndrome have a degree of intellectual disability [24]. Tatton-Brown et al. (2005) [8] found that intellectual disability was present in 97% of 266 individuals with Sotos syndrome. However, in this study intellectual ability was determined via clinical assessment in which individuals were classified as having normal intellectual ability or mild, moderate or severe intellectual disability. This classification was based on clinical observation rather than via administration of standardized cognitive assessments.

Some of our work involved assessing intellectual ability using a standardized cognitive assessment in a cohort of 52 individuals with Sotos syndrome. The findings showed that the majority of the cohort had mild intellectual disability (IQ = 50–69) or were in the borderline range (IQ = 70–84). This evidence, therefore, supports the inclusion of intellectual disability as one of the main diagnostic criteria of the syndrome [26]. However, in this study, nearly 10% of the cohort tested had average intellectual ability (IQ = 85+). This highlights the variability of intellectual ability within this population and demonstrates that some individuals with Sotos syndrome do not have intellectual disability. Consequently, milder cases of Sotos syndrome may be harder to identify and diagnose if the clinical features are less severe.

Our study also analyzed whether there were any differences in intellectual ability in relation to gender. We found that females with Sotos syndrome had significantly higher GCA scores than males with Sotos syndrome. This suggests that, on average, males with Sotos syndrome may be more likely to have a greater degree of intellectual disability than females with Sotos syndrome. No significant relationship was identified between age and GCA scores, indicating that increase or decrease in intellectual ability is not associated with age within the Sotos syndrome population. However, as our study used a cross-sectional design, it will be important for future research to utilise a longitudinal design to establish the rate and trajectory of cognitive development within this population.

10.5.1.2 Sotos syndrome cognitive profile

A cognitive profile characterizes the relative cognitive strengths and weaknesses of an individual and, in some cases, can be generalized to a specific population. Broadly, cognitive assessments can be used to determine whether individuals have an even or uneven cognitive profile by examining whether there is a significant difference between performance in tasks that assess abilities in different cognitive domains. The cognitive profile can also be assessed in further detail by examining potential differences in performance between subscales that overall comprise verbal or nonverbal reasoning ability. The identification of syndrome-specific cognitive profiles enables differentiation of the cognitive style between individuals with distinct congenital syndromes.

In our large cohort study of individuals with Sotos syndrome ($n = 52$) we assessed cognitive abilities using a standardized battery of cognitive tasks. This enabled us to establish whether individuals with Sotos syndrome display a clear and consistent profile of abilities across different cognitive domains [26]. The profile of performance on the core scales of the British Ability Scales 3 was assessed. For this analysis, we only included participants who were over 8 years of age, as all these participants completed the same battery of cognitive tests ($n = 35$). It is important to note that the focus of this approach was to establish relative cognitive strengths and weaknesses in this population. The profile analysis revealed that participants had better verbal skills than nonverbal reasoning skills.

They also displayed relative strength in visuospatial memory and relative weakness in quantitative reasoning. The finding of relative strength in visuospatial memory is a novel finding that has important implications for understanding how individuals with Sotos syndrome process and learn information. Furthermore, the finding of a relative weakness in quantitative reasoning supports a suggestion reported by Cole and Hughes [4] that individuals with Sotos syndrome display particular difficulty with numeracy. This finding indicates that individuals with Sotos syndrome may require additional support with numeracy.

In order to operationalize the cognitive profile, four specific criteria were proposed as the Sotos syndrome cognitive profile (SSCP) (see Table 10.1). The SSCP criteria sensitivity was comparable to the sensitivity of the Williams syndrome cognitive profile (WSCP) criteria, as reported by Mervis et al. [29] in which the sensitivity of the four WSCP ranged from 0.91 to 1.00 ($M = 0.95$). The WSCP was reported as Digit Recall, Naming/Definitions, or Similarities >1st percentile' (WSCP1) Pattern Construction T-score <20th percentile (WSCP2), Pattern Construction T-score < mean T-score (WSCP3), and Pattern Construction T-score < Digit Recall T-score (WSCP4). Although the WSCP reported by Mervis et al. [29] and the SSCP reported in the present study were devised on the basis of performance on the same cognitive assessment, different cognitive abilities were identified as relative strengths and weaknesses for individuals within these populations. This demonstrates that the WCSP and the SSCP are syndrome specific and can be used to differentiate between individuals with Williams syndrome and those with Sotos syndrome. Overall, the SSCP criteria provide a quantifiable and replicable characterization of the cognitive profile associated with Sotos syndrome which can be used to differentiate between individuals with and without a diagnosis of Sotos syndrome.

10.5.1.3 Language

Several studies have reported difficulties with communication and language delay in Sotos syndrome. Three case studies have reported language delays in expressive language when compared with receptive language [30–32]. However, a study by Finegan et al. [33] reported scores on standardized language tests from 27 individuals with Sotos syndrome and concluded that language ability tended to be consistent with overall intellectual ability. In our cognitive profiling study [26], we found that both expressive and receptive language tended to be relative strengths compared to nonverbal reasoning skills in participants who were 8 years old and above, whereas for younger children, expressive language tended to be more of a strength than comprehension. However, it is important to note that this

Table 10.1 Sotos syndrome cognitive profile criteria.

Criteria	Sensitivity
Verbal ability > non-verbal reasoning ability	0.94
Quantitative reasoning T-score or matrices T-score <20th percentile	0.97
Quantitative reasoning T-score < mean T-score	0.91
Recognition of designs T-score or recognition of pictures T-score > mean T-score	0.94

In total, 80% of participants met all four criteria and 97.14% met at least three of the criteria.
From Lane C, Milne E, Freeth M. The cognitive profile of Sotos syndrome. J Neuropsychol 2018. https://doi.org/10.1111/jnp.12146.

trend observed in the younger children was based on data from only 11 children. It is clear that more in-depth studies will be required to fully understand the development of language across the lifespan in Sotos syndrome.

10.5.2 Behavior

It has been suggested that children with Sotos syndrome may display more behavioral problems, compared with typically developing children [4,34]. This could be because children with Sotos syndrome are usually large for their age and are therefore often mistaken as older and more able than their actual developmental level. This assumption can lead to frustration for the child which then manifests itself in behavioral problems. A number of behavioral issues have been reported within the literature and these are summarized in the following sections.

10.5.2.1 Attention-deficit/hyperactivity disorder

Two studies have reported a high prevalence of attention-deficit/hyperactivity disorder (ADHD) in Sotos syndrome: Finegan et al. [33] found that 10 of the total 27 participants had ADHD (as measured by parental report) while Varley and Crnic [35] found that 3 of their cohort of 11 participants met diagnostic criteria for ADHD. A cohort study by Sheth et al [27] ($n = 38$) found evidence of impulsivity and overactivity in participants with Sotos syndrome, with scores being comparable to those of individuals with ASD. However, in contrast to these studies suggesting a link between Sotos syndrome and ADHD, de Boer et al. [36] found no significant difference between mean scores of their Sotos syndrome group ($n = 20$) and their control group on the 18-item Dutch ADHD list. In addition, only 4 out of the 20 participants scored in the clinical range for ADHD.

Within case studies that have measured behavioral features of individuals with Sotos syndrome, a total of 5 out of 10 individuals were reported to have a clinical diagnosis of ADHD [30,37–39]. In addition, two cases were reported of individuals who were inattentive, hyperactive, and demonstrated a lack of inhibition [30,40]. Overall, findings from these studies suggest that ADHD may be a common behavioral problem associated with Sotos syndrome; however, no systematic study in this area has yet been conducted.

10.5.2.2 Anxiety

Anxiety has been reported in individuals with Sotos syndrome; however, at present, this is an under-researched area. Sarimski [34] measured anxiety using The Children's Social Behavior Questionnaire and found that children with Sotos syndrome displayed significantly more separation anxiety and had a tendency to be more anxious in new situations when compared to a control group matched for age and cognitive ability. Furthermore, the Sotos syndrome group had higher scores for insecure/anxious behavior (as measured by the Nisonger Child Behavior Rating Form) when compared to the matched control group. In addition, Rutter and Cole [41] found that 10 of their 16 participants had some form of phobia, as described through parental report. This suggests that anxious behavior may be more prevalent within the Sotos syndrome population, compared with children of similar intellectual ability. There may also be a specific profile of anxious behavior in individuals with Sotos syndrome but this needs to be explored in future research.

10.5.3 Aggression and tantrums

Aggressive behavior and/or tantrums have been reported in a number of case studies and case series of individuals with Sotos syndrome [37,39–43]. Aggressive behavior and/or tantrums have generally been assessed via parental report or psychiatric assessment. Of these studies, five employed a case study design and only one of the case studies included a female participant [40], despite the syndrome affecting males and females equally. In a group study, Rutter and Cole [41] asked parents to describe the behavioral and emotional problems experienced by their child. Among the 16 participants, 13 were described as having tantrums in the home environment. However, it is possible that these participants may have come to medical attention as a result of behavioral issues, so this sample may not be representative of the Sotos syndrome population overall.

Aggressive behavior was also reported in a syndrome comparison study by Sheth et al. [27], finding that self-injurious behavior and destruction of property were more likely to occur in Sotos syndrome than in a control group of individuals with Down syndrome. These behaviors were found to occur in over 40% of individuals with Sotos syndrome. The prevalence of self-injury in participants with Sotos syndrome was found to be broadly comparable to that of participants with ASD and Prader–Willi syndrome.

It is important to note that, with the exception of the study by Sheth et al. [27], all the participants in the studies that have assessed aggression and tantrums have reported these behavioral issues in children. As children with Sotos syndrome are often large for their age, behavioral issues may be considered more problematic by others when the child is compared to another child of similar age and/or size. It will be important for future research to establish whether aggressive behavior and tantrums persist during adulthood in individuals with Sotos syndrome.

10.6 Sotos syndrome and autism spectrum disorder

ASD is a behaviorally defined developmental disorder associated with social communication impairment, restricted interests, and repetitive behaviors [44]. ASD is a spectrum disorder and consequently, there is significant heterogeneity and variability between individuals with ASD. It is estimated that ASD occurs in approximately 1% of the population [45,46]. Idiopathic ASD refers to individuals who have a primary diagnosis of ASD and for which the underlying cause is unknown. In contrast, syndromic ASD refers to individuals who have a diagnosis of a specific syndrome and also have a comorbid diagnosis of ASD. Neurodevelopmental disorders associated with a high prevalence of ASD can be considered as syndromic causes of ASD [47,48].

ASD symptoms have been reported in a number of congenital syndromes, including fragile X [49], Cornelia de Lange [50], and Angelman syndromes [51]. It has been suggested that approximately 10–20% of cases of ASD are caused by genetic syndromes, cytogenetics lesions, and rare *de novo* mutations [47]. Consequently, a number of aetiological genetic pathways may be implicated in ASD [47,52]. Thus investigation of the association between ASD and genetic syndromes is particularly valuable in identifying genetic mechanisms associated with ASD. Furthermore, distinct ASD phenotypes may be associated with each genetic syndrome [53]. It is therefore important to establish the profile of autistic symptoms within a syndrome, as this will facilitate understanding of both autism and genetic syndromes.

Before 2015, four studies had been published which provided data relating to Sotos syndrome and ASD. Mouridsen and Hansen [30] reported a case of a young child with Sotos syndrome who met the ICD-10 diagnostic criteria for childhood autism. Morrow et al. [54] reported a child with Sotos syndrome who, following clinical observation, was found to meet the diagnostic criteria for ASD. Additionally, Trad et al. [40] reported a case of a child with Sotos syndrome who met the *Diagnostic and Statistical Manual of Mental Disorders* (Third Edition Revised) (DSM−III−R) criteria for pervasive developmental disorder. In addition to these case studies, Zappella [55] reported a case series of 12 children with Sotos syndrome. Within this sample, it was noted that five children (42%) displayed autistic features consistent with the *DSM−III−R* criteria for autistic disorders. Although this study suggested that the incidence of ASD in Sotos syndrome is greater than in the general population, the small sample size meant that it was not possible to establish the prevalence of ASD within the Sotos syndrome population as a whole on the basis of the findings from this study.

Since 2015, three published studies have investigated the relationship between Sotos syndrome and ASD. Timonen-Soivio et al. [56] explored the relationship between ASD and Sotos syndrome in a cohort of Finnish children. Population registers were searched in order to identify the number of individuals with comorbid diagnoses of distinct congenital syndromes and ASD. The study identified a significant association between ASD and Sotos syndrome. Of the 13 children identified with Sotos syndrome, 7 (54%) had a comorbid diagnosis of ASD. Therefore, this study provides further evidence for an increased prevalence of ASD within the Sotos population, but again, the sample size was small. In addition, this study assessed the relationship between ASD and Sotos syndrome in terms of comorbid diagnoses and therefore autistic symptoms were not explicitly measured within this study. It is possible that individuals with Sotos syndrome may display behavior that would meet the diagnostic criteria for ASD but had not received a formal diagnosis.

In a cohort study of individuals with Sotos syndrome, Sheth et al. [27] reported characteristics of ASD in a sample of 38 individuals as assessed by the Social Communication Questionnaire (SCQ) [57] and the Repetitive Behavior Questionnaire (RBQ) [58]. Mean age of the participants was 17.3 years, with an age range of 6−43 years. Findings indicated that 26 of 38 participants with Sotos syndrome (68%) met clinical cutoff for ASD, as measured by the total score on the Lifetime version of the SCQ (clinical cutoff was considered as a total score $\geq$15). Data from the Sotos syndrome group were compared with data from three distinct, matched control groups: ASD, Prader−Willi syndrome, and Down syndrome. Participants with Sotos syndrome scored significantly lower than the ASD group on the repetitive behavior subscale of the SCQ but there were no significant differences between the Sotos and ASD groups on the social communication and social interaction subscales. Subsequent analyses using only the Sotos syndrome participants who scored above the clinical cutoff identified no significant differences between the Sotos syndrome and ASD groups for the three SCQ subscales. Findings from the RBQ scores indicated that the Sotos syndrome group scored significantly lower than the ASD group on the stereotyped behavior subscale but there were no significant differences between scores on the remaining subscales. Overall, the findings from this study suggest that a high proportion of individuals with Sotos syndrome display autistic characteristics of a clinical nature. Difficulties associated with repetitive behavior are less severe than observed in ASD for individuals with Sotos syndrome who do not score above the clinical cutoff, despite significant impairment in social communication and social interaction.

In 2017, we published a study [25] that complemented and extended the findings by Sheth et al. [27] in a number of important ways. Our study aimed to further understand the severity of ASD symptoms within the Sotos syndrome population by conducting a detailed profile analysis of ASD symptoms and exploring whether age and gender affect symptom severity. Behaviors associated with autism were assessed via analysis of questionnaire responses on the Social Responsiveness Scale-2 (SRS-2) provided by a close family member of 78 participants with Sotos syndrome. Within this study, 83% of the participants met clinical cutoff for ASD (scores of 60 or above). This finding suggests that the majority of individuals with Sotos syndrome display a current behavioral profile associated with the *DSM-V* criteria for ASD (social communication impairment and restricted interests and repetitive behaviors). We found no effect of gender on ASD symptom severity, indicating that there is no significant difference between the prevalence of behavioral characteristics associated with ASD in males and females with Sotos syndrome. However, it is important to note that within our sample, 16 participants had diagnoses of both Sotos syndrome and ASD, yet only 2 of these participants were female. This suggests that although males and females with Sotos syndrome appear to display a similar behavioral phenotype, there is a clear disparity between them in the diagnosis of ASD. It is perhaps the case that some clinicians miss the fact that some girls with Sotos syndrome may also qualify for a diagnosis of ASD.

In relation to the question of whether age affects ASD symptom severity in Sotos syndrome, our findings suggest that it does. Specifically, in the current sample, ASD symptoms were less severe in young children (2.5–5 years) and in adults (20+ years) than in children over the age of 5 years through to adolescence. This is an important finding because it suggests that severity of ASD symptoms may decrease as an individual transitions into adulthood.

It has been suggested that distinct profiles of ASD symptoms may be associated with different genetic syndromes [53]. The findings from the present study suggest that individuals with Sotos syndrome display trait profiles that are similar to those present in idiopathic ASD. This is supported by comparison of our Sotos syndrome data with a set of ASD data from Frazier et al. [59] on the five empirically derived subscales identified in their factor analysis of the SRS-2. This comparison demonstrated that children with Sotos syndrome appear to display behavioral characteristics that were similar in profile and severity to those identified in children with idiopathic ASD but were distinct from scores identified in the unaffected siblings of the children with ASD. Although individuals with Sotos syndrome would be considered as having syndromic ASD, the findings from our study suggest that the syndromic ASD observed in Sotos syndrome is very similar to idiopathic ASD.

10.7 Nuclear receptor–binding SET domain methyltransferases modify histones and affect epigenetics

The NSD family of proteins are involved in developmental defects such as Sotos (NSD1) and Wolf–Hirschhorn (NSD2) syndromes. Mutations of NSD1 are responsible for the genetic condition Sotos syndrome, which is characterized by growth defects and cancer predisposition. The methyltransferases NSD1–3 are capable of generating H3K36me1 and H3K36me2 [60], which are marks associated with the body of actively transcribed genes [61].

NSD1 lysine methyltransferase enzymatic activity was recognized in 2003 [62]. Interestingly, an autoregulatory loop prevents the access of H3K36 to the catalytic site of NSD1 [63]. The methylation of H3K36 is also regulated in trans by the ubiquitinylation of histone H2A (H2Aub), which inhibits NSD1 and NSD2 activity [64]. In addition, NSD proteins contain reader domains (see Chapter 1 for more information on histone mark readers) that recognize the histone H3 tail. In particular, NSD has a plant homeodomain (PHD) and a cysteine-histidine-rich domain (C5HCH) adjacent to the catalytic domain at the carboxy terminus. The PHD-C5HCH domain of NSD3 folds to bind H3K4me0K9me3, whereas NSD2 prefers to bind to unmodified H3 and NSD1 fails to associate with H3 [65], suggesting differential regulation of H3K36 methylation by NSD methyltransferases. An additional layer of regulation involves the binding of NSD2 PWWP domain with H3K36me2, which facilitates the propagation of H3K36 methylation and enhances cellular proliferation [66].

In addition to H3K36, NSD1 can methylate histones H4K44 and H1.5K168 as well as nonhistone substrates such as ATRX on K1033 [67]. Similarly, NSD3 also methylates nonhistone substrates. Specifically, NSD3 modifies epidermal growth factor receptor (EGFR) on K721, which allows EGFR to bypass epidermal growth factor activation and initiate the ERK signaling cascade independent of stimuli [68].

Regarding true epigenetic events, samples of patients with Sotos syndrome ($NSD1^{+/-}$) harbor DNA methylation (DNAme) signatures encompassing genes with neuronal differentiation functions [69]. Thus, Sotos syndrome could originate from chromatin modification defects (i.e., H3K36me, DNAme) or aberrant methylation of nonhistone proteins (e.g., ATRX, EGFR). The molecular mechanisms remain to be defined.

10.8 Limitations and future research directions

As noted by Cole and Hughes [4], a number of patients reported within the literature have come to medical attention due to developmental delay. Consequently, this may have resulted in a bias for recruitment of participants with more severe intellectual disability and/or behavioral problems in some of the studies reported in this chapter. As awareness of Sotos syndrome is fairly limited, this is a difficult issue to overcome. Any individuals who do not present with significant symptoms or who are not assessed by a clinician who is aware of the syndrome are less likely to be given a diagnosis of Sotos syndrome. Thus, until there is greater awareness of the syndrome, it will be difficult to assess cognitive and behavioral facets in a large and fully representative sample.

A fundamental methodological issue present in most of the studies included in this chapter is the limited sample size. As Sotos syndrome has a relatively low incidence, there is a limited population from which to recruit participants. It is therefore important for future research to utilize all available recruitment strategies to collect a large and representative data set. Currently, there is no published literature on longitudinal studies of Sotos syndrome; hence, knowledge of the developmental trajectory is poor. It will be important for future work to focus on a developmental approach, particularly when considering the genotype–phenotype relations between the variations in intragenic mutations and microdeletions on the NSD1 gene that result in Sotos syndrome.

Understanding of the cognitive and behavioral phenotype associated with Sotos syndrome has developed during the past couple of years. However, to date, we are not aware of any published research using electroencephalography or functional MRI to explore the relationship between brain function and behavior or cognition within the Sotos syndrome population. Future research explicitly investigating the relationship between genotype, neurology and cognitive and behavioral phenotypes within this population will improve understanding the mechanisms underlying this syndrome. Therefore it will be important for future research to use a collaborative and integrative approach to establish genotype–phenotype relationships within this population.

10.9 Summary and conclusions

Sotos syndrome is a congenital overgrowth disorder associated with intellectual disability and a range of medical issues. Approximately 70% of infants with Sotos syndrome develop neonatal jaundice and/or have difficulty with feeding. Cardiac and renal anomalies, seizures, and scoliosis occur in 15–30% of individuals with Sotos syndrome, but the type and severity of these issues vary [10]. Until the identification of the genetic abnormality associated with Sotos syndrome in 2002 [6], the syndrome was diagnosed on the basis of clinical features. However, genetic screening is now widely available and thus, opportunities to better understand the phenotypic effects of *NSD1* intragenic mutations and microdeletions are now possible.

In relation to the phenotype, research assessing cognition in Sotos syndrome has established that intellectual disability persists throughout adulthood and a cognitive profile of better verbal ability compared with nonverbal reasoning ability is characteristic of the syndrome. Other key features of the cognitive profile are that quantitative reasoning tends to be particularly poor, whereas visuospatial memory tends to be a relative strength [26]. In terms of behavior, characteristics associated with autism are the most researched behavioral issue in Sotos syndrome. Social communication impairment and difficulties with restricted interests and repetitive behaviors have been reported in 68% [27] and 83% [25] of individuals with Sotos syndrome. Other behavioral issues such as ADHD, anxiety, and aggression/tantrums have been reported in the literature, but to date, these behaviors have not been systematically investigated in large samples of individuals with Sotos syndrome so the prevalence of these behaviors is unknown [24]. In summary, syndrome-specific cognitive and behavioral profiles associated with Sotos syndrome are starting to emerge but future research is needed to fully understand the genotype–phenotype relationships within this population.

References

[1] Sotos JF, Dodge PR, Muirhead D, Crawford JD, Talbot NB. Cerebral gigantism in childhood. A syndrome of excessively rapid growth. N Engl J Med 1964;271:109–16.

[2] Dodge PR, Holmes SJ, Sotos JF. Cerebral gigantism. Dev Med Child Neurol 1983;25:248–52.

[3] Tatton-Brown K, Rahman N. Features of NSD1-positive Sotos syndrome. Clin Dysmorphol 2004;13:199–204.

[4] Cole TRP, Hughes HE. Sotos syndrome – a study of the diagnostic criteria and natural history. J Med Genet 1994;31:20–32.

[5] Opitz JM, Weaver DW, Reynolds JF. The syndromes of Sotos and weaver: reports and review. Am J Med Genet 1998;79:294–304.

[6] Kurotaki N, Imaizumi K, Harada N, Masuno M, Kondoh T, Nagai T, Ohashi H, Naritomi K, Tsukahara M, Makita Y, Sugimoto T, Sonoda T, Hasegawa T, Chinen Y, Tomita HA, Kinoshita A, Mizuguchi T, Yoshiura KI, Ohta T, Kishino T, Fukushima Y, Niikawa N, Matsumoto N. Haploinsufficiency of NSD1 causes Sotos syndrome. Nat Genet 2002;30:365–6.

[7] Tatton-Brown K, Rahman N. The NSD1 and EZH2 overgrowth genes, similarities and differences. Am J Med Genet C-Semin Med Genet 2013;163C:86–91.

[8] Tatton-Brown K, Douglas J, Coleman K, Baujat G, Cole TRP, Das S, Horn D, Hughes HE, Temple IK, Faravelli F, Waggoner D, Turkmen S, Cormier-Daire V, Irrthum A, Rahman N, Childhood Overgrowth C. Genotype-phenotype associations in Sotos syndrome: an analysis of 266 individuals with NSD1 aberrations. Am J Hum Genet 2005b;77:193–204.

[9] Tatton-Brown K, Douglas J, Coleman K, Baujat G, Chandler K, Clarke A, Collins A, Davies S, Faravelli F, Firth H, Garrett C, Hughes H, Kerr B, Liebelt J, Reardon W, Schaefer GB, Splitt M, Temple IK, Waggoner D, Weaver DD, Wilson L, Cole T, Cormier-Daire V, Irrthum A, Rahman N, Childhood Overgrowth C. Multiple mechanisms are implicated in the generation of 5q35 microdeletions in Sotos syndrome. J Med Genet 2005a; 42:307–13.

[9a] Douglas J, Hanks S, Temple IK, Davies S, Murray A, Upadhyaya M, Tomkins S, Hughes HE, Cole RT, Rahman N. NSD1 mutations are the major cause of Sotos syndrome and occur in some cases of Weaver syndrome but are rare in other overgrowth phenotypes. The American Journal of Human Genetics 2003; 72(1):132–43.

[10] Tatton-Brown K, Rahman N. Sotos syndrome. Eur J Hum Genet 2007;15:264–71.

[11] Novara F, Stanzial F, Rossi E, Benedicenti F, Inzana F, DI Gregorio E, Brusco A, Graakjaer J, Fagerberg C, Belligni E, Silengo M, Zuffardi O, Ciccone R. Defining the phenotype associated with microduplication reciprocal to Sotos syndrome microdeletion. Am J Med Genet 2014;164A:2084–90.

[12] Rio M, Clech L, Amiel J, Faivre L, Lyonnet S, LE Merrer M, Odent S, Lacombe D, Edery P, Brauner R, Raoul O, Gosset P, Prieur M, Vekemans M, Munnich A, Colleaux L, Cormier-Daire V. Spectrum of NSD1 mutations in Sotos and weaver syndromes. J Med Genet 2003;40:436–40.

[13] Tatton-Brown K, Loveday C, Yost S, Clarke M, Ramsay E, Zachariou A, Elliott A, Wylie H, Ardissone A, Rittinger O. Mutations in epigenetic regulation genes are a major cause of overgrowth with intellectual disability. Am J Hum Genet 2017;100:725–36.

[14] Weaver DD, Graham CB, Thomas I, Smith DW. A new overgrowth syndrome with accelerated skeletal maturation, unusual facies, and camptodactyly. J Pediatr 1974;84:547–52.

[15] Tatton-Brown K, Seal S, Ruark E, Harmer J, Ramsay E, del Vecchio Duarte S, Zachariou A, Hanks S, O'brien E, Aksglaede L. Mutations in the DNA methyltransferase gene DNMT3A cause an overgrowth syndrome with intellectual disability. Nat Genet 2014;46:385–8.

[16] Tatton-Brown K, Hanks S, Ruark E, Zachariou A, Duarte SDV, Ramsay E, Snape K, Murray A, Perdeaux ER, Seal S. Germline mutations in the oncogene EZH2 cause Weaver syndrome and increased human height. Oncotarget 2011;2:1127.

[17] Schaefer GB, Bodensteiner JB, Buehler BA, Lin A, Cole TRP. The neuroimaging findings in Sotos syndrome. Am J Med Genet 1997;68:462–5.

[18] Aoki N, Oikawa A, Sakai T. Serial neuroimaging studies in Sotos syndrome (cerebral gigantism syndrome). Neurol Res 1998;20(2):149–52.

[19] Horikoshi H, Kato Z, Masuno M, Asano T, Nagase T, Yamagishi Y, Kondo N. Neuroradiologic findings in Sotos syndrome. J Child Neurol 2006;21(7):614–8.

[20] Fickie MR, Lapunzina P, Gentile JK, Tolkoff-Rubin N, Kroshinsky D, Galan E, Lin AE. Adults with Sotos syndrome: review of 21 adults with molecularly confirmed NSD1 alterations, including a detailed case report of the oldest person. Am J Med Genet 2011;155(9):2105–11.

[21] Türkmen S, Şahin S, Koçer N, Peters H, Mundlos S, Tüysüz B. Neuroimaging and clinical characterization of Sotos syndrome. Genet Couns 2015;26(1):1–12.

[22] Baujat G, Cormier-Daire V. Sotos syndrome. Orphanet J Rare Dis 2007;2(1):36.

[23] Nicita F, Ruggieri M, Polizzi A, Mauceri L, Salpietro V, Briuglia S, Segni M. Seizures and epilepsy in Sotos syndrome: analysis of 19 Caucasian patients with long-term follow-up. Epilepsia 2012;53(6).

[24] Lane C, Milne E, Freeth M. Cognition and behaviour in Sotos syndrome: a systematic review. PLoS One 2016;11(2):e0149189.

[25] Lane C, Milne E, Freeth M. Characteristics of autism spectrum disorder in Sotos syndrome. J Autism Dev Disord 2017;47:135–43.

[26] Lane C, Milne E, Freeth M. The cognitive profile of Sotos syndrome. *J Neuropsychol* (in press) https://doi.org/10.1111/jnp.12146.

[27] Sheth K, Moss J, Hyland S, Stinton C, Cole T, Oliver C. The behavioral characteristics of Sotos syndrome. Am J Med Genet 2015;167:2945–56.
[28] Humphreys LG. The fleeting nature of the prediction of college academic success. J Educ Psychol 1968; 59(5):375.
[29] Mervis CB, Robinson BF, Bertrand J, Morris CA, Klein-Tasman BP, Armstrong SC. The Williams syndrome cognitive profile. Brain Cogn 2000;44(3):604–28.
[30] Mouridsen SE, Hansen MB. Neuropsychiatric aspects of Sotos syndrome. A review and two case illustrations. Eur Child Adolesc Psychiatry 2002;11:43–8.
[31] Park SH, Lee JE, Sohn YB, Ko JM. First identified Korean family with Sotos syndrome caused by a novel intragenic mutation in NSD1. Ann Clin Lab Sci 2014;44:228–31.
[32] Scarpa P, Faggioli R, Voghenzi A. Familial Sotos syndrome: longitudinal study of two additional cases. Genet Couns 1994;5:155–9.
[33] Finegan JAK, Cole TRP, Kingwell E, Smith ML, Smith M, Sitarenios G. Language and behaviour in children with Sotos syndrome. J Am Acad Child Adolesc Psychiatr 1994;33:1307–15.
[34] Sarimski K. Behavioural and emotional characteristics in children with Sotos syndrome and learning disabilities. Dev Med Child Neurol 2003;45:172–8.
[35] Varley CK, Crnic K. Emotional, behavioral, and cognitive status of children with cerebral gigantism. J Dev Behav Pediatr 1984;5:132–4.
[36] de Boer L, Roder I, Wit JM. Psychosocial, cognitive, and motor functioning in patients with suspected Sotos syndrome: a comparison between patients with and without NSD1 gene alterations. Dev Med Child Neurol 2006;48:582–8.
[37] Gajre MP, Mhatre P, Vijaykumar R. Sotos syndrome with developmental co-ordination disorder; report of a case. J Taibah Univ Sci 2014;10(2).
[38] Gosalakkal JA. Sotos syndrome (cerebral gigantism): a review of neurobehavioral, developmental and neurological manifestations. Int Pediatr J Miami Child 2004;19:147–51.
[39] Mauceri L, Sorge G, Baieli S, Rizzo R, Pavone L, Coleman M. Aggressive behavior in patients with Sotos syndrome. Pediatr Neurol 2000;22:64–7.
[40] Trad PV, Schlefer E, Hertzig M, Kernberg PF. Treatment strategies for a case of concurrent pervasive developmental disorder and cerebral gigantism. J Am Acad Child Adolesc Psychiatr 1991;30:499–506.
[41] Rutter SC, Cole TRP. Psychological characteristics of Sotos syndrome. Dev Med Child Neurol 1991;33: 898–902.
[42] Compton MT, Celentana M, Price B, Furman AC. A case of Sotos syndrome (cerebral gigantism) and Psychosis. Psychopathology 2004;37:190–3.
[43] Gomes-Silva JM, Ruviére DB, Segatto RAS, De Queiroz AM, De Freitas AC. Sotos Syndrome: a case report. Spec Care Dent 2006;26:257–62.
[44] Association AP. Diagnostic and statistical manual of mental disorders. Arlington: American Psychiatric Publishing; 2013.
[45] Baird G, Simonoff E, Pickles A, Chandler S, Loucas T, Meldrum D, Charman T. Prevalence of disorders of the autism spectrum in a population cohort of children in South Thames: the Special Needs and Autism Project (SNAP). Lancet 2006;368:210–5.
[46] Baron-Cohen S, Scott FJ, Allison C, Williams J, Bolton P, Matthews FE, Brayne C. Prevalence of autism-spectrum conditions: UK school-based population study. Br J Psychiatry 2009;194:500–9.
[47] Abrahams BS, Geschwind DH. Advances in autism genetics: on the threshold of a new neurobiology. Nat Rev Genet 2008;9:341–55.
[48] Betancur C. Etiological heterogeneity in autism spectrum disorders: more than 100 genetic and genomic disorders and still counting. Brain Res 2011;1380:42–77.

[49] Kaufmann WE, Cortell R, Kau AS, Bukelis I, Tierney E, Gray RM, Cox C, Capone GT, Stanard P. Autism spectrum disorder in fragile X syndrome: communication, social interaction, and specific behaviors. Am J Med Genet 2004;129:225–34.

[50] Moss J, Howlin P, Magiati I, Oliver C. Characteristics of autism spectrum disorder in Cornelia de Lange syndrome. J Child Psychol Psychiatry 2012;53:883–91.

[51] Peters S, Beaudet A, Madduri N, Bacino C. Autism in Angelman syndrome: implications for autism research. Clin Genet 2004;66:530–6.

[52] Zhao X, Leotta A, Kustanovich V, Lajonchere C, Geschwind DH, Law K, Law P, Qiu S, Lord C, Sebat J. A unified genetic theory for sporadic and inherited autism. Proc Natl Acad Sci Unit States Am 2007;104:12831–6.

[53] Moss J, Howlin P. Autism spectrum disorders in genetic syndromes: implications for diagnosis, intervention and understanding the wider autism spectrum disorder population. J Intellect Disabil Res 2009;53:852–73.

[54] Morrow JD, Whitman BY, Accardo PJ. Autistic disorder in Sotos syndrome: a case report. Eur J Pediatr 1990;149:567–9.

[55] Zappella M. Autistic features in children affected by cerebral gigantism. Brain Dysfunct 1990;3(5–6):241–4.

[56] Timonen-Soivio L, Vanhala R, Malm H, Hinkka-Yli-Salomäki S, Gissler M, Brown A, Sourander A. Brief report: syndromes in autistic children in a finnish birth cohort. J Autism Dev Disord 2016;46(8):1–5.

[57] Rutter M, Bailey A, Lord C. The social communication questionnaire: manual. Western Psychological Services; 2003.

[58] Moss J, Oliver C. The repetitive behaviour scale. Manual for administration and scorer interpretation. Birmingham: University of Birmingham; 2008.

[59] Frazier TW, Ratliff KR, Gruber C, Zhang Y, Law PA, Constantino JN. Confirmatory factor analytic structure and measurement invariance of quantitative autistic traits measured by the Social Responsiveness Scale-2. Autism 2014;18(1):31–44.

[60] Li Y, Trojer P, Xu CF, Cheung P, Kuo A, Drury WJ, Qiao Q, Neubert TA, Xu RM, Gozani O, Reinberg D. The target of the NSD family of histone lysine methyltransferases depends on the nature of the substrate. J Biol Chem 2009;284(49):34283–95.

[61] Barski A, Cuddapah S, Cui K, Roh TY, Schones DE, Wang Z, Wei G, Chepelev I, Zhao K. High-resolution profiling of histone methylations in the human genome. Cell 2007;129(4):823–37.

[62] Rayasam GV, Wendling O, Angrand PO, Mark M, Niederreither K, Song L, Lerouge T, Hager GL, Chambon P, Losson R. NSD1 is essential for early post-implantation development and has a catalytically active SET domain. EMBO J 2003;22(12):3153–63.

[63] Qiao Q, Li Y, Chen Z, Wang M, Reinberg D, Xu RM. The structure of NSD1 reveals an autoregulatory mechanism underlying histone H3K36 methylation. J Biol Chem 2011;286(10):8361–8.

[64] Yuan G, Ma B, Yuan W, Zhang Z, Chen P, Ding X, Feng L, Shen X, Chen S, Li G, Zhu B. Histone H2A ubiquitination inhibits the enzymatic activity of H3 lysine 36 methyltransferases. J Biol Chem 2013;288(43):30832–42.

[65] He C, Li F, Zhang J, Wu J, Shi Y. The methyltransferase NSD3 has chromatin-binding motifs, PHD5-C5HCH, that are distinct from other NSD (nuclear receptor SET domain) family members in their histone H3 recognition. J Biol Chem 2013;288(7):4692–703.

[66] Sankaran SM, Wilkinson AW, Elias JE, Gozani O. A PWWP domain of histone-lysine N-methyltransferase NSD2 binds to dimethylated Lys-36 of histone H3 and regulates NSD2 function at chromatin. J Biol Chem 2016;291(16):8465–74.

[67] Kudithipudi S, Lungu C, Rathert P, Happel N, Jeltsch A. Substrate specificity analysis and novel substrates of the protein lysine methyltransferase NSD1. Chem Biol 2014;21(2):226–37.

[68] Saloura V, Vougiouklakis T, Zewde M, Deng X, Kiyotani K, Park JH, Matsuo Y, Lingen M, Suzuki T, Dohmae N, Hamamoto R. WHSC1L1-mediated EGFR mono-methylation enhances the cytoplasmic and nuclear oncogenic activity of EGFR in head and neck cancer. Sci Rep 2017;7:40664.
[69] Choufani S, Cytrynbaum C, Chung BHY, Turinsky AL, Grafodatskaya D, Chen YA, Cohen ASA, Dupuis L, Butcher DT, Siu MT, Luk HM. NSD1 mutations generate a genome-wide DNA methylation signature. Nat Commun 2015;6:10207.

CHAPTER 11

ATRX tames repetitive DNA within heterochromatin to promote normal brain development and regulate oncogenesis

Valerie Turcotte-Cardin[1,2], Kevin G. Young[1], David J. Picketts[1,2,3]

Regenerative Medicine Program, Ottawa Hospital Research Institute, Ottawa, ON, Canada[1]; Departments of Cellular and Molecular Medicine, University of Ottawa, Ottawa, ON, Canada[2]; Departments of Biochemistry, Microbiology & Immunology, University of Ottawa, Ottawa, ON, Canada[3]

Chapter Outline

11.1 Introduction **235**

11.2 Biochemical and molecular functions of ATRX **236**

11.2.1 ATRX protein structure 236

11.2.2 ATRX is a heterochromatin interacting protein 238

11.2.3 Other critical interactions and functions of ATRX 239

11.2.4 ATRX interactions with RNA 241

11.3 Neurologic deficits and phenotypic variability in *ATRX*-associated syndromes **242**

11.4 Delineating a role for ATRX in cancer **244**

11.4.1 Cancer profiling identifies ATRX as a common mutation target 244

11.4.2 Cancers with ATRX mutations are alternative lengthening of telomere positive 245

11.4.3 Understanding the alternative lengthening of telomeres pathway 247

11.4.4 ATRX is a suppressor of the alternative lengthening of telomeres pathway 249

11.5 Conclusion **251**

List of abbreviations **251**

References **252**

11.1 Introduction

The α-thalassemia X-linked mental retardation protein, ATRX, is an essential chromatin remodeling protein. Germline mutations that affect the ATRX protein's function or expression levels are associated with several allelic syndromes whose core symptoms include severe cognitive impairment and facial dysmorphism [1]. This includes the originally described ATR-X syndrome [2], which is typified by moderate to severe intellectual disability, abnormalities in growth and genital formation, and generally mild anemia associated with abnormal hemoglobin production [3,4]. A complete loss of ATRX

Chromatin Signaling and Neurological Disorders. https://doi.org/10.1016/B978-0-12-813796-3.00011-0

function has not been observed in intellectually disabled patients, and the genetic elimination of its expression in mice causes embryonic lethality [5], indicating that human diseases associated with ATRX mutations involve partial loss-of-function, or a potential complete loss-of-function through somatic mutations within restricted cell types.

A wealth of research has demonstrated the importance of ATRX in acting as a tumor suppressor. This role arises from its involvement in regulating telomeric chromatin structure. Although there is little evidence that ATRX mutations on their own cause cancer, somatic ATRX deficiencies in combination with oncogenic mutations may be associated with >10% of all cancers [6–8]. This includes pancreatic neuroendocrine tumors (PanNETs), gliomas, neuroblastomas, and sarcomas. The high level of association of ATRX deficiency with these various cancers, along with the fact that mutations in proteins with similar or overlapping functions are associated with similar pathologic conditions, has led to intense scrutiny of the role of ATRX in cancer and whether or not it presents an opportunity for novel therapeutic intervention.

In this chapter, we will provide an overview of the molecular functions of ATRX and how the loss of one or more of these functions may generate the disease symptoms associated with ATR-X syndrome. We will then discuss how these same molecular functions help ATRX to suppress oncogenesis. The connection between this intellectual disability/autism-related protein and cancer is demonstrative of a similar link involving other nuclear proteins, including CHD2, CHD7, CHD8, ARID18, ERCC6, RAD54, and HTLF all of which have roles related to the regulation of chromatin remodeling and genome maintenance, gene transcription, and DNA repair [9]. As such, ATRX function/dysfunction provides an excellent illustration of the mechanisms underlying common pathologic conditions that affect millions of people worldwide.

11.2 Biochemical and molecular functions of ATRX

11.2.1 ATRX protein structure

ATRX is encoded by a large gene that spans ~300 kb and is located at q21.1 on the X chromosome in humans. The gene contains 40 exons and is differentially spliced to generate multiple protein-coding transcripts (~10; http://vega.archive.ensembl.org), including transcripts coding for two well-characterized protein variants. The major transcript variant is encoded on 36 exons; few descriptions of several potentially smaller ATRX variants exist. The major protein produced is 283 kDa (full-length ATRX), with the second most abundant variant having a predicted weight of 150 kDa and an apparent weight of ~200 kDa (ATRXt) [10] (Fig. 11.1). While ATRX is expressed in tissues throughout the body, the full-length transcript is particularly abundant in fetal brain and in mature skeletal muscles; the truncated ATRXt is also abundant in skeletal muscles, but not brain [10].

ATRX is a member of the SWI2/SNF2 (SWItch/sucrose nonfermenting 2) family, containing a C-terminal switch (SNF-like helicase adenosine triphosphatase (ATPase)) domain characteristic of this family [11] (Fig. 11.1). This domain, which is highly conserved within mammals, is critical to the normal functioning of ATRX [11]. A key feature of this domain is its ATPase activity, which is utilized to generate energy to remodel nucleosomes. In ATRX, the SWI/SNF chromatin remodeling capacity is unique from that of the other members of the family in that it does not randomize DNA phasing of the nucleosomes (i.e., it does not randomly reposition a nucleosome relative to a specific DNA sequence) [12]. The ability of the ATRX SWI/SNF domain to participate in triple-helix DNA displacement,

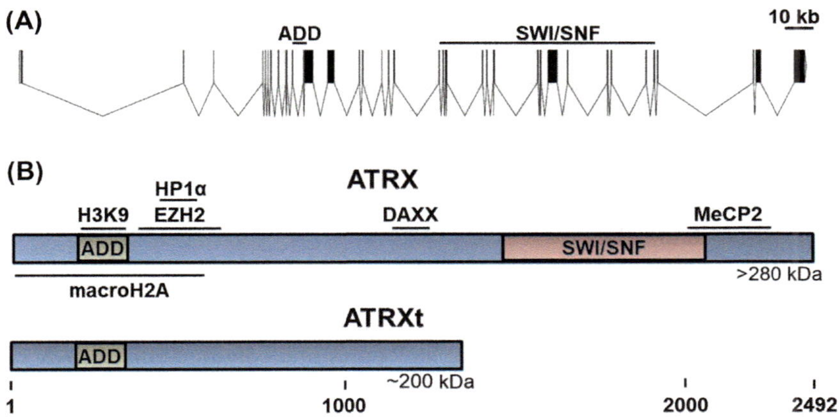

FIGURE 11.1

The human *ATRX* gene, and structures of the major proteins produced from it. (A) Located at q21.1 on the X chromosome, the *ATRX* gene spans ~300 kb. All potential exons/introns within the gene are shown, and the positions of exons encoding the ADD and SWI/SNF (SWItch/sucrose nonfermenting) domain sequences are indicated. These domain sequences are most frequently targeted by pathogenic mutations, most commonly missense. (B) The domain structure of ATRX, and structure of the truncated ATRXt are shown. Also indicated are the binding regions for many of the key ATRX-interacting proteins. Numbers at the bottom correspond to the amino acid numbering.

but not in DNA phasing of nucleosomes, is consistent with the functioning of the structurally related Rad54 protein [13], indicating potentially overlapping functions between ATRX and this well-characterized DNA repair protein.

The N-terminal ATRX, DNMT3, DNMT3L (ADD) domain is the second domain essential to normal ATRX function (Fig. 11.1). The ADD domain is composed of three subdomains. An N-terminal subdomain is structurally similar to the erythroid transcription factor GATA-1, which binds a single zinc ion through four cysteines [14]. Next to this GATA-like finger is a plant homeodomain (PHD)-type zinc finger motif that differs from many other PHD domains in that all the zinc-binding residues are cysteines. The GATA-like finger adjacent to the PHD finger is a feature shared by the DNMT3 family that prompted its nomenclature as an ADD domain [15]. The third and last subdomain is a long C-terminal α-helix that extends from the PHD finger and makes several hydrophobic connections with the N-terminal GATA-like finger bringing together the C- and N-termini of the ADD domain and forming a single globular domain [16].

A major role of the ADD domain is to bind the histone H3 tail residues. There is a pocket between the PHD domain and the GATA finger that forms an atypical binding site for H3K9me3. ADD can sense the methylation states at both H3K9 and H3K4 using its H3K9me3 recognition module [17]. Methylation at H3K4 prevents ATRX from binding to the H3K9me3 modification. To have optimal binding of ATRX to histone H3, readout of both the K4 and K9 methylation states is necessary [18].

The binding of the ADD domain to H3K9me3 was one of the first definitive indications that ATRX interacts with heterochromatin. Two important roles of H3K9me3 modified histones are in the proper localization of ATRX at pericentromeric heterochromatin and in facilitating sister chromatid cohesion during DNA synthesis. ATRX mutation in the ADD domain results in a chromosome segregation defect due to insufficient cohesion between sister chromatids, suggesting an important role in chromosome segregation [17]. A study also demonstrated that the ADD domain can recognize a second bivalent mark, namely, H3K9me3, in combination with H3S10ph within postmitotic neurons. This bivalent mark is enriched at centromeric repetitive sequences in an activity-dependent manner where it is thought to interact with ATRX to maintain silenced heterochromatin [19]. Overall, ADD provides ATRX with an ability to target the protein to specific histone modifications that are characteristic of heterochromatin.

11.2.2 ATRX is a heterochromatin interacting protein

Immunofluorescent studies with ATRX-specific antibodies first confirmed that ATRX colocalized to pericentromeric and telomeric heterochromatin throughout the cell cycle [20,21]. Other studies showed an enrichment of ATRX at the inactive X chromosome (Xi) in females [22,23]. Several mechanisms have been identified that mediate ATRX targeting to heterochromatic sites. ATRX interacts with specific histone posttranslational modifications either directly via the ADD domain described earlier or indirectly by associating with secondary interacting proteins such as HP1α or MeCP2 [24,25]. HP1α interacts with H3K9me3 through its chromodomain and recruits ATRX to these sites via an LxVxL binding site located between the ADD and ATPase domains [19]. Mutation to this motif significantly impairs ATRX localization to heterochromatin [26]. At the H3K9me3S10ph combinatorial modification, ATRX binding is dependent on the ADD domain to modulate the binding of polycomb proteins during differentiation [26,27]. This seems to be independent of HP1α because the addition of H3S10ph *in vitro* drastically impairs HP1α binding to H3K9me3, whereas no detectable effect is observed for ATRX ADD binding. Indeed, in postmitotic neurons the H3K9me3S10ph mark is found to colocalize with ATRX, but not HP1α, in the heterochromatin compartments during high cellular activity [19]. However, more work is required to fully understand the interplay of heterochromatin localization and ADD binding to H3K9me3 because it can be disrupted by H3T3ph and H3T6ph marks. Nonetheless, HP1α association with ATRX functions to stabilize the interaction of ATRX with heterochromatin.

An independent study in neurons has suggested that the localization of ATRX at heterochromatic foci is dependent on the methyl-DNA-binding protein MeCP2 [25]. MeCP2, which is highly enriched in neuronal cells, is involved in the recognition of chromatin methylation and facilitates transcriptional repression [25]. Mutations in MeCP2 are the cause of the Rett syndrome, an intellectual disability disorder that affects primarily young girls. MeCP2 targets the C-terminal helicase domain of ATRX, and in the absence of MeCP2, ATRX localization to heterochromatin loci is disrupted [21,25,28]. A subset of MeCP2 mutations found in the Rett syndrome were shown to affect the interaction between MeCP2 and ATRX [25]. As these mutations had little effect on MeCP2 targeting the methylated chromatin, the loss of interaction with ATRX indicated a mechanism underlying the deleterious nature of these mutations. Interestingly, the N-terminal fragments of ATRX localize to heterochromatic foci in an MeCP2-independent manner, whereas the C-terminal localization is dependent on the MeCP2 interaction, thereby suggesting that multiple mechanisms can recruit ATRX to heterochromatin [25]. It has also been proposed that ATRX and MeCP2 help regulate the expression of common genes within the brain [28]. The targeting of common genes for transcriptional regulation by ATRX and MeCP2 may explain some of the similarities between the ATR-X and Rett syndrome phenotypes, both of which commonly involve severe intellectual disability with signs of autism, microcephaly, dystonia, and gastrointestinal problems.

The most definitive support for an association of ATRX with heterochromatin came from chromatin immunoprecipitation sequencing (ChIPseq) experiments that showed ATRX binding sites are enriched at telomeres, centromeres, and intergenic regions with a lower frequency of binding in gene bodies [29]. Many of the intergenic sites were composed of simple tandem repeat DNA sequences particularly those with a high GC content and prone to form a secondary DNA structure known as a G-quadruplex (G4 DNA) [29]. Some of the key sequences predicted to form G4 DNA are telomeric repeats and CpG islands, although they are predicted to occur throughout the genome. The formation of such structures is presumed to present an impediment to the movement of enzymes along the DNA strand, thus affecting DNA replication and transcription, while representing a source of genetic instability [30]. ATRX is enriched at the predicted sites of G4 DNA formation and it directly interacts with G4 oligonucleotides *in vitro* [29]. The role of ATRX in preventing G-quadruplex formation will be further discussed later when we delve into the mechanisms underlying how these structures may promote some of the phenotypes observed in ATR-X syndrome and cancer.

ChIPseq also demonstrated that ATRX localized to many imprinted alleles, interstitial heterochromatin sites, and endogenous retrovirus (ERV) sequences, preferentially the long-terminal repeat sequences belonging to endogenous retroviruses family K (ERVK). This preference is determined by ATRX binding to a short sequence within the intracisternal A particle sequence of these retrotransposons [31]. In general the prevailing evidence indicates that ATRX is a heterochromatin interacting protein that functions to stabilize silenced regions of the genome.

11.2.3 Other critical interactions and functions of ATRX

Outside the two main functional domains, ATRX interactions with other proteins have also been documented and provide additional complexity to our understanding of its functions. A key interaction partner is the histone chaperone, death domain—associated protein (DAXX). ATRX and DAXX have been identified as a histone chaperone complex for the deposition of histone H3.3 in repetitive regions of the genome, including telomeres, where deposition of H3.3 with the H3.3K9me3 mark is critical for telomere maintenance [32—34]. DAXX functions as an H3.3-specific adapter and envelops an H3.3-H4 dimer by adopting an all α-helical conformation [35]. Furthermore, the promotion of H3K9 trimethylation may be facilitated by the association of DAXX and the methyltransferase SUV39H1 to generate regions enriched for nucleosomes containing H3.3K9me3 [36]. Although the presence of histone H3.3 was first identified at active genes within promoter regions, deposition in these regions is performed by the HIRA chaperone complex and not ATRX-DAXX. Aside from telomeres, ATRX/DAXX protects normally silenced repetitive sequences, including retrotransposons and telomeric sequences [36]. ATRX/DAXX loading of histone H3.3 into these regions prevents DNA secondary structure formation that stalls replication and transcription and can promote inappropriate recombination to occur at these sequences [36].

ATRX is also closely associated with promyelocytic leukemia (PML) nuclear bodies (PML-NBs) [12]. PML protein is a tumor suppressor that is predominantly found in the nucleus and bound to a network of nuclear proteins that provides structural support for organizing chromatin [37].

Specific ATRX patient mutations within the SWI/SNF domain reduce the association of ATRX with PML-NBs [38]. This indicates that this domain, along with playing a role in chromatin remodeling, is required for proper subnuclear targeting of the ATRX protein. It is thought that ATRX/DAXX localization to PML-NBs may facilitate resolution of stalled replication forks and other regions containing aberrant DNA secondary structures. Besides PML-NBs, ATRX also associates with PML within specific heterochromatin domains [39]. Thus PML is able to direct the H3.3 deposition activity carried out by ATRX/DAXX to specific heterochromatin loci.

ATRX acts as an important safeguard of chromatin integrity during cell replication. Its deficit in dividing cell populations can cause replication stress–associated defects [40–42]. ATRX is recruited to the sites of DNA damage during replication, where it associates with the DNA repair complex MRE11-RAD50-NBS1 (MRN). The MRN complex is involved in handling replication fork stalling and promotes the resumption of DNA replication following stress [43]. A deficit in ATRX activity related to replication fork stalling can result in a prolonged S phase and defects in the maintenance of telomere structure [40,41,44]. In the brain, poly(ADP-ribose) polymerase 1 (PARP1) hyperactivation during neurogenesis helps compensate for the loss of ATRX and allows for the production of late-born neurons [42]. Nonetheless, the loss of ATRX in the brain results in the generation of fewer neurons, due at least in part to an inability to produce sufficient numbers of new neurons during the rapid proliferative period that occurs during embryogenesis.

While most work has focused on loading histone H3.3 to suppress recombination and prevent replication deficits, a few studies have indicated a role for ATRX in the regulation of transcription. Several reports have shown that ATRX binds at imprinted alleles where it functions to maintain repression of the silenced alleles [28,45]. While these studies seemed to indicate a role in repression of the imprinted alleles, another study showed that loss of ATRX can interfere with active transcription [46]. At these target genes, ATRX normally binds within the gene bodies at repetitive sequences that can form G4 quadruplexes and are high in histone H3.3 occupancy. In the absence of ATRX, the RNA polymerase II (RNApolII) stalls at these sites presumably due to G4 quadruplex formation resulting in incomplete transcript formation. One target gene that accumulated RNApolII within its gene body was *Neuroligin 4*, a gene important for cognition [46].

ATRX interactions within the 3′ exons of a cluster of zinc finger genes on human chromosome 19 are reflective of the role of ATRX in maintaining normal chromatin structure and protecting against DNA damage in specific nonheterochromatin regions. Although the H3K9me3 modification is normally associated with transcriptional repression, ATRX is responsible for maintaining the presence of H3K9me3 (presumably via a methyltransferase such as SUV39H1) within this gene cluster. This is thought to protect against replicative damage and aberrant homologous recombination (HR) between different zinc finger genes thereby maintaining the region in a chromatin environment conducive to transcription [47]. The ZNF274 transcription factor helps recruit ATRX to these loci, where it further associates with a complex containing the H3K9 methyltransferase SETDB1 and the corepressor TRIM28 [47]. The 3′ ends of the zinc finger genes are unique in that they are enriched for the bivalent H3K9me3 and H3K36me3 mark. It is possible that ATRX plays a role in the maintenance of chromatin integrity at other loci containing similar atypical histone signatures.

Although much work is required to elucidate the full spectrum of ATRX functions, it is becoming clear that a key role is to prevent the formation of secondary structures that impinge on nuclear functions such as replication and transcription.

11.2.4 ATRX interactions with RNA

It is well understood that RNA transcripts can adopt many secondary structures through folding, and studies have indicated that ATRX is an RNA-interacting protein. X chromosome inactivation (XCI) in mammals is an important mechanism for dosage compensation between males and females [48]. Once the Xi has been determined, it will remain inactive for subsequent cell divisions. XCI is tightly regulated by the X-inactivation center that contains various long noncoding RNA (lncRNA) species including the sense and antisense transcripts, Xist and Tsix, respectively. Xist is negatively regulated by its antisense RNA Tsix [49] and is positively regulated by the RepA transcript, a 1.6-kb transcript comprising the same sequence as the 5′ end of the Xist RNA including the repetitive A and F motifs [50,51]. Xist can also recruit the histone methyltransferase complex polycomb repressive complex 2 (PRC2) that trimethylates histone H3 lysine 27 (H3K27me3) and represses chromatin [52]. EZH2 acts as a PRC2 subunit and catalyzes the trimethylation of H3K27 [53]. In mediating XCI in females, EZH2 forms prominent foci along the Xi [22]. A study indicated that ATRX was also involved in XCI via its interaction with EZH2 [22]. Adjacent to the ATRX ADD domain is an interaction motif for the SET domain of the chromatin-associated EZH2 protein [54]. The loss of ATRX was found to result in loss of EZH2 foci and reduction in H3K27me3 on the X chromosome. This demonstrates ATRX as a specificity determinant for PRC2 in targeting the Xi for gene repression. Interestingly, ATRX binds to RepA/Xist RNA and promotes the loading of PRC2 [22]. Furthermore, the effect of ATRX on the RepA/PRC2 association is enhanced by ATP, suggesting that ATRX remodels RepA to a more permissive conformation, similar to the function of an RNA helicase.

ATRX also interacts with another RNA motif called TERRA (telomeric repeat–containing RNA). TERRAs are lncRNA sequences that range in size from 100 bp to 100 kb and contain the typical telomeric repeat sequence UUAGGG [55]. The true function of TERRA has yet to be determined but it has been proposed to regulate recombination between telomeric ends, as it is a major contributor to telomere structure [56]. Interestingly, it was demonstrated that the location of TERRA on chromatin correlates with higher ATRX density. When TERRA and ATRX share binding sites, they had opposite effects. TERRA enhances and ATRX represses gene expression at both telomeric and nontelomeric sites. Within telomeres, TERRA transcripts regulate ATRX localization by competing with ATRX–DNA interactions because TERRA–ATRX forms a more stable complex than ATRX–DNA [56]. Whether or not ATRX and TERRA compete for telomeric DNA binding at the sites of G4 DNA formation remains to be demonstrated, but multiple evidence suggests that this is likely [57]. Thus TERRA is essential to modulate transcription along with ATRX at nontelomeric sites as well as to serve as an important regulator of telomere structure.

Over two-thirds of the RNA transcripts produced by human cells do not code for proteins and are classified as lncRNAs [58]. The affinity of ATRX for RNA and the growing recognition of the role of lncRNAs in chromatin remodeling [59] suggests that ATRX may have a broader role in the regulation of lncRNA function. In yeast, four different ATP-dependent chromatin-remodeling complexes were demonstrated to repress over 250 antisense lncRNAs to maintain the normal expression of corresponding sense transcripts [60]. This suggests that ATP-dependent chromatin remodelers may have a general role as lncRNA repressors [59]. The maintenance of transcriptional repression within heterochromatin has also been demonstrated to be dependent on the interaction of SUV39H1, a DAXX interaction partner, with chromatin-associated RNA [61]. Further exploration of ATRX's ability to bind different lncRNA transcripts should help broaden our understanding of the role of chromatin remodeling proteins in contributing to lncRNA functions.

11.3 Neurologic deficits and phenotypic variability in *ATRX*-associated syndromes

The relatively high expression of ATRX in fetal brain suggests an important role for this protein in neurodevelopmental processes. It is no surprise, then, that a major phenotype associated with ATR-X syndrome is intellectual disability, and the disability associated with the syndrome is not progressive. This suggests that a key role for ATRX is in ensuring normal brain development and that it may play a less important role in ensuring normal brain function following maturation. The molecular functions of ATRX, as discussed previously, may be most critical to ensure that cycling cells survive critical periods of rapid proliferation during development.

ATRX is mutated in several allelic syndromic disorders. These include the Carpenter–Waziri syndrome, the Holmes–Gang syndrome, the Chudley–Lowry syndrome, and X-linked mental retardation-arch fingerprints-hypotonia syndrome [62]. The main characteristics of these include facial dysmorphism and psychomotor disability [1]. Mutations in one of the two main ATRX functional domains, ADD and SWI/SNF, are responsible for the majority of these cases [62,63] (Fig. 11.2). As already stated, all known germline *ATRX* mutations result in impaired ATRX function or a reduction of *ATRX* expression, but not a complete loss of ATRX protein.

Along with several other core symptoms of ATR-X syndrome, intellectual disability is highly variable. The location of the specific *ATRX* mutation is one key factor that contributes to this variability [64]. For instance, some patients diagnosed with either ATR-X syndrome or the Chudley–Lowry syndrome are affected by a nonsense mutation (R37X) within exon 2 that results in the inefficient use of an internal methionine (M40) to produce a slightly N-terminally truncated protein [65–67]. While there is conflicting evidence whether this mutation also results in altered splicing to facilitate the production of this functional variant, the ATRX protein is produced in a reduced amount, resulting in a milder intellectual disability phenotype. Mice generated with an *Atrx* mutation that mimics this patient mutation via the complete deletion of exon 2 exhibit altered neuronal signaling affecting glutamate-responsive

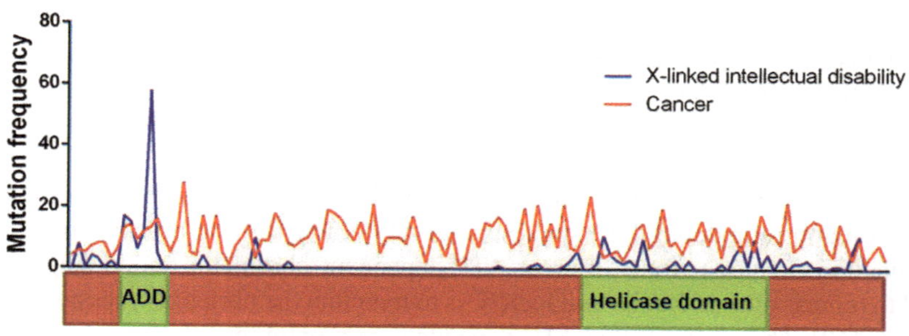

FIGURE 11.2

Mutation frequency in the *ATRX* gene. The graph depicts the frequency of mutations across the *ATRX* gene in X-linked intellectual disability (*blue*) and cancer (*red*). Each data point represents the mutations within a bin window of 20 amino acids. A schematic diagram of the *ATRX* gene to the same scale is shown below the graph.

neurons [68,69]. This contributes to the generation of a mild behavioral phenotype that includes impaired memory and cognition.

A review of magnetic resonance imaging/computed tomography (MRI/CT) assessments performed on 27 Japanese patients possessing *ATRX* mutations in various regions of the gene has also demonstrated variability in brain abnormalities correlating in part with the position of the mutation [70]. Mutations affecting ATRX expression levels were observed to potentially associate more with myelination defects, as were mutations affecting the SWI/SNF domain. Mutations that affected the ADD domain were more frequently associated with gray matter abnormalities and may contribute to more severe psychomotor deficiencies. A correlation was found between the general severity of the ATRX-syndrome and the localization of the mutation. Indeed, patients with mutations in the PHD-like domain have a more severe phenotype and permanent psychomotor impairment, whereas patients with mutations in the helicase domain have a delayed motor development but eventually become normal [71].

In addition to the location of the *ATRX* mutation, the broader genetic background of the patient can greatly affect the resulting characteristics of the disorder. Although neurologic deficits are no doubt altered by the genetic background of individual patients, the effect of genetic background on the resultant phenotype is best exemplified by the variation in the reduction of α-globin synthesis of patients with ATR-X syndrome, which results in the production of an abnormal hemoglobin, called HbH hemoglobin (generated by the tetramerization of four β-globin monomers and lacking any α-globin content). Patients with identical *ATRX* mutations have been demonstrated to exhibit highly variable levels of α-thalassemia associated with varying levels of HbH hemoglobin production. For instance, one study demonstrated how distantly related cousins with an identical *ATRX* mutation either had or lacked α-thalassemia [4]. This was thought to depend on altered abilities to regulate the splicing of the mutant *ATRX* transcript, allowing for the cousin lacking the disease phenotype to produce a more normal transcript. However, this can now be attributed to ATRX binding at an upstream variable number tandem repeat (VNTR) sequence (see later discussion). In addition, 15 different patients who each had an identical mutation in the ADD domain coding region presented with highly variable levels of HbH inclusions within their red blood cells [72]. This strongly demonstrated the importance of other genetic factors in determining the final phenotype.

The *HBA2/HBA1* (α-globin) locus is located in a G-rich region of the genome that replicates in the early S-phase. It is also in an "open" chromatin environment where there is a high gene density. On the other hand, *HBB* (β-globin) is replicated later in the S-phase, located in a GC-poor region that is also a region of low gene density, and is associated with a "closed" chromatin environment [73]. ChIPseq was used to determine that ATRX binding sites coincide with CpG islands and that these sites adjacent to the *HBA2/HBA1* locus consist of VNTRs [29]. The repetitive G-rich DNA associated with the *HBA2/HBA1* locus makes it uniquely prone to forming G-quadruplex structures, in contrast to the *HBB* locus. The variability in the α-thalassemia severity in ATR-X syndrome coincides with the length of the upstream VNTRs at the *HBA2/HBA1* locus. Indeed, a longer VNTR region is associated with the increased formation of HbH inclusions in red blood cells. Therefore the VNTR polymorphisms surrounding the *HBA1/HBA2* locus contribute to the severity of the HbH-associated anemia.

Another possible explanation for the reduction of α-globin levels in ATR-X syndrome involves the histone variant macroH2A. MacroH2A is normally associated with transcriptionally silent chromatin. Also, an mH2A1 domain, a spliced variant of macroH2A, was found associated with the α-globin cluster. ATRX acts as a negative regulator of macroH2A incorporation into chromatin [74]. Thus the loss of ATRX increases the incorporation of macroH2A into α-globin clusters, leading to the repression of the *HBA2/HBA1* locus. More studies are required to determine whether macroH2A incorporation and VNTR length cooperate to silence the α-globin gene in the absence of ATRX or represent distinct silencing mechanisms.

Along with altered α-globin synthesis, ATRX has been demonstrated to regulate the expression of several other genes that are in close proximity to VNTR sequences targeted by ATRX [29]. These include *NME4*, *RASA3*, *SLC7A5*, and *GAS8*. ATRX also facilitates the transcription of *Neuroligin 4* in the brain, possibly by helping resolve G-quadruplexes that occur within this gene [46]. Thus along with facilitating cell survival through its roles in promoting DNA replication and maintaining normal heterochromatin structure during proliferative periods of early brain development, ATRX may help ensure normal brain function and development by regulating the expression of specific neuronal proteins.

11.4 Delineating a role for ATRX in cancer

11.4.1 Cancer profiling identifies ATRX as a common mutation target

Next-generation sequencing techniques are being used by many groups to sequence tumor DNA and further identify common changes that help delineate differences and catalog tumors with phenotypes. They are also clinically being used to personalize treatments for the improvement of disease management and prognostics in patients with cancer. Although every tumor is different from one another, with a unique mutation profile, ATRX has been identified as a commonly mutated gene in PanNETs, gliomas, neuroblastoma [75], and sarcomas [76] (Fig. 11.2). Surprisingly, there is no clinical evidence to suggest that patients with ATR-X syndrome are subject to premature development of malignancy or to have an increased risk of cancer [77].

Gliomas are the most common type of primary brain tumors in children and adults. They are classified into four groups according to the cell of origin: astrocytic gliomas, oligodendral tumors, mixed oligoastrocytomas, and ependymal tumors [78]. Gliomas can be further classified based on the tumor's histologic profile, ranging from grade I to IV based on the World Health Organization's scoring system. Typically, grade I, pilocytic astrocytoma, and grade II, diffuse astrocytomas, are referred as low-grade gliomas, whereas grade III, anaplastic glioma, and grade IV, glioblastoma (GBM) multiforme and GBM, are considered higher grade gliomas [78].

Studies have provided new markers to further characterize specific subtypes of gliomas [75]. Mutations in the isocitrate dehydrogenase (IDH) 1 and 2 genes have been found in 80% of grade II diffuse anaplastic gliomas and secondary GBM. There are two additional mutations that characterize the grade II and grade III gliomas that are tightly associated with IDH1/2 mutation. The first mutation resides in the *TP53* gene, which encodes the p53 protein. Mutations in this tumor suppressor gene are detected in 60% of diffuse and anaplastic gliomas. Also, 86% of pediatric GBM samples that harbor mutations in ATRX/DAXX and/or H3.3 had a mutation in *TP53* [8]. The second type of mutation is 1p19q codeletions that are found in 80% of oligodendrocytomas [75], but rarely correlates with *ATRX* mutation [79]. Interestingly, mixed oligoastrocytomas can harbor either of these hallmark mutations in combination with *IDH1/2* mutations. Pediatric gliomas are most commonly low grade and they are rarely associated with *IDH1/2* mutations.

Research has identified a recurrent driver mutation in both the histone H3.3 variant and *ATRX* mutations that occurred in 36% and 33% of pediatric GBMs, respectively. In adults, truncating and missense *ATRX* mutations were found in 47.8% of adult glioma samples, with 37% of these mutations resulting in a truncated protein and 11.4% in missense mutations that are mainly located within the helicase domain [75]. However, *ATRX* mutations are only found in tumors of astrocytic lineage; no mutation has been found in oligodendrogliomas. Interestingly, all grade II *ATRX*-altered gliomas and 92% of grade III gliomas of the astrocytic lineage also contain *IDH* mutations. Thus *ATRX* alterations are strongly associated with mutations in either *IDH1* or *IDH2* gene, and gliomas that display a decrease in *ATRX* mRNA expression are associated with an *IDH1* mutation. It was proposed that *ATRX* alteration from grade II to grade IV astrocytomas might be a potential marker of progression, as its prevalence increases with grades.

Another type of cancer that is affected by *ATRX* mutations is the PanNETs, which are the second most common malignancy of the pancreas, affecting 2.2 per 1,000,000 individuals per year, with a male predominance [80]. The cell type that is thought to give rise to PanNETs is the pluripotent cell found in the exocrine pancreas. Because of the nature of the affected cells, the tumors can produce a variety of hormones including insulin, glucagon, somatostatin, and vasoactive intestinal peptide. In some cases, patients can display systematic symptoms related to the overproduction of hormones, in which case the PanNET is referred to as functional. When the excessive secretion of hormones does not induce symptoms, clinicians refer to it as nonfunctional. Typically, PanNETs are sporadic but they can arise commonly in multiple endocrine neoplasia type 1 (MEN1) syndrome. Other syndromes that can less frequently be associated with PanNETs are von Hippel—Lindau disease, neurofibromatosis type 1, and tuberous sclerosis complex (TSC) [81].

The genetic exploration of PanNETs has identified several genes of interest by comparing mutation frequency within these tumors. *MEN1* and *ATRX/DAXX* are the most commonly altered genes in functional PanNETs, whereas *MEN1* and allelic loss of chromosome 11q alterations are found in nonfunctioning PanNETs [82,83]. PanNETs can have mutations in either *MEN1* or *ATRX/DAXX*, but one tumor was also found to harbor both mutations. Members of the mTOR (mechanistic target of rapamycin) pathway have also been identified in the functional form of these tumors. *PTEN* was found mutated in 7.3% of PanNETs, *TSC2* in 8.8%, and *PIK3CA* in 1.4%. Unlike the gliomas, these tumors rarely have alterations in *TP53* [84]. Sporadic well-differentiated PanNETs have shown that 44% of the time, *DAXX* or *ATRX* is mutated and, typically, insertion/deletion or nonsense mutations are observed in these genes [84]. Patients with *DAXX/ATRX* and/or *MEN1* mutations have a prolonged survival relative to those with other mutations. Indeed, they can survive at least 10 more years, whereas 60% of patients with PanNET without *DAXX/ATRX* mutations die within 5 years [84]. *ATRX* mutations have also been associated with a telomerase-independent pathway, called alternative lengthening of telomeres (ALT), to prevent the shortening of the telomeres [7]. An in vitro study showed that 90% of human cancer cell lines that have *ATRX* mutations maintain their telomeres by ALT, suggesting that ATRX is an important mediator of this pathway.

11.4.2 Cancers with ATRX mutations are alternative lengthening of telomere positive

ALT is a cell process used by certain types of tumors relying on recombination-mediated elongation to maintain telomere length. Activation of the ALT pathway is typically found in PanNETs [6], gliomas [85], neuroblastomas [86], and sarcomas [76]. Telomere-specific fluorescence in situ hybridization (FISH) was performed on a pool of 41 sporadic nonfunctional PanNETs [6]. Interestingly, 61% of tumors displayed large telomeres and an ultrabright FISH signal, consistent with an ALT phenotype.

Furthermore, PanNETs with *ATRX* and *DAXX* mutations are also ALT positive in this sample set. A small portion of PanNETs with no detectable point mutation, insertion or deletion of *ATRX* and *DAXX*, were found to be ALT positive. Upon further investigation, these tumors nonetheless have lost nuclear expression of either *ATRX* or *DAXX*, whereas tumors with no ALT phenotype have a robust expression of both proteins [6].

Interestingly, loss of nuclear expression of *ATRX* and/or *DAXX* and the presence of ALT only occur in patients with PanNET tumors measuring ≥3 cm and with *MEN1* mutation. This suggests that these events are happening later in the PanNET tumorigenesis, after the neoplasm has grown beyond the size of microadenomas (0.2 cm). This finding has prompted the development of a model describing the possible steps for the formation of metastasis in PanNETs [87]. Basically, *DAXX/ATRX* mutations occur during uncontrolled cell division, which then induces chromosomal instability (CIN) and triggers ALT to maintain telomere length, leading to the heterogeneity of telomere sizes. CIN is allowing the affected cells to acquire chromosomal aberrations and a large spectrum of mutations that will confer selective advantages for cell proliferation and the establishment of dominance in a subpopulation of cells. This suggests that *DAXX/ATRX* loss in combination with ALT activation is a cause of CIN, which characterized a specific subset of PanNETs. In nonfunctional PanNETs, the presence of CIN correlates with larger tumor size, a high level of chromosomal aberrations, and metastases, whereas in functional PanNETs, particularly insulinomas, it correlates with poor outcome. Thus CIN is proposed as a diagnostic tool for the development of metastatic disease [88].

Interestingly, loss of *ATRX* nuclear expression in pediatric GBM multiforme, adult GBM, oligodendrogliomas, and medulloblastomas also correlates with an ALT phenotype [6]. The presence of ALT phenotype in grade IV astrocytomas (GBM multiforme) correlates with a positive prognostic outcome. Indeed, median survival of patients with ALT-positive tumors is 542 days, whereas for patients with nonpositive tumors, it is 247 days [89]. ALT-positive tumors were also found to increase survival rate of lower grade astrocytomas [90]. The presence of telomerase did not alter or change the survival of patients with this type of tumor. Loss of *ATRX* expression in low-grade astrocytomas does not correlate with an ALT phenotype [85]. However, there is an association of *ATRX* with ALT positivity in high-grade pediatric and adult astrocytomas (i.e., anaplastic astrocytoma and GBM).

Neuroblastoma is a cancer of the sympathetic nervous system and it is considered the most common extracranial solid tumor among children [91]. More than 90% patients with neuroblastoma are diagnosed within the first 5 years of life often due to their association with life-threatening symptoms. Neuroblastomas that harbor an ALT phenotype are associated with a resistance to chemotherapy without rapid growth and unfavorable outcomes [92]. Also, alterations in either *ATRX* or *DAXX* were found in all the ALT-positive neuroblastoma tumors [86]. In some rare cases, mutations in *ATRX* or *DAXX* are linked to a higher risk of late recurrence.

Sarcomas can be ALT positive, more frequently in tumors of mesenchymal origin than in those derived from the epithelium [76]. The ALT phenotype is detected in 35% of osteosarcomas and 35% of soft tissue sarcomas (STSs) [90]. The largest group of STS malignancy is the liposarcoma, which accounts for 20% of all tumors of mesenchymal origin [93]. Telomere maintenance mechanisms (TMMs) in liposarcoma rely on the ALT pathway 24% of the time, essentially double the overall incidence of the ALT phenotype [90]. There is a positive correlation between the utilization of the ALT pathway as a TMM and the tumor grade in STS [94]. Indeed, approximately 70% of grade I liposarcoma do not display any TMM, whereas in grades II and III, 70% of the tumors have either ALT or telomerase or both. Therefore TMM could be a potential marker for malignancy in tumors of mesenchymal origin and could be ultimately a target for chemotherapeutic interventions [76]. Thus sarcomas with nonspecific complex karyotypes (i.e., osteosarcoma) more frequently use ALT as their TMM than sarcomas with simple karyotypes with specific translocation [95].

As previously mentioned, both TMMs have very distinct characteristics and one of them is the difference in CIN [76]. A study, using FISH on the osteosarcoma cell line U-2 OS, showed that ALT-positive cells displayed a wide variety of telomere lengths, unusual configurations of telomere repeats and loci, and dicentric marker chromosomes [96]. Comparison of two osteosarcoma cell lines with telomerase activity showed an increase in translocation, deletion, and complex chromosomal rearrangements. Furthermore, ALT-positive cells had a higher frequency of DNA bridges, a direct measure of chromosomal aberrations, likely due to free chromosomal ends [96]. These free ends allow end-to-end chromosomal fusion leading to breakage-fusion-bridge cycles and increase in gross chromosomal rearrangements. Taken together, these findings demonstrate that mesenchymal tumors using ALT have a more complex chromosomal background and that their mutations are nonspecific and unpredictable. It has been suggested that ALT may be less efficient at maintaining telomeres because it can create a series of events that lead to CIN [76]. Another possibility is that ALT induces and/or requires these genetic alterations to properly function.

The involvement of *ATRX* in sarcoma is similar to other types of cancer. In a study that examined 573 sarcomas, 58 of them had a complete loss of ATRX nuclear expression and were classified as high-grade tumors [97]. In another set of samples, almost all (41 out of 42) of the sarcomas with a complete loss of ATRX were also positive for ALT.

Sarcomas have a different outcome than gliomas when it comes to ALT-positive tumors. Indeed, 60% of patients with ALT-positive osteosarcoma have a 5-year survival period, whereas 90% of patients with both nonpositive ALT tumors and negative telomerase activity have a lower mortality rate (5-year survival) [98]. Similarly, a study examining liposarcoma tumors and patient survival found that patients with ALT phenotype have highly aggressive tumors and a higher mortality rate, whereas patients with telomerase-positive tumors were aggressive but showed differences in survival [94]. Overall, both TMMs have different influences on tumorigenesis and patient survival and the effect highly depends on the specific type of malignancy observed [76].

Taken together, these findings showed that *ATRX* mutations are crucial events for the development of the ALT pathway during tumorigenesis. The appearance of the ALT phenotype is a late event that correlates with more advanced and higher grade tumors. The presence of ALT is associated with a positive prognosis in most cancers and could be a potential therapeutic target. Also, *ATRX* maintenance of genome integrity is critical both in the development and prevention of malignancy in oncogenic cells; however, the underlying mechanisms are still unclear and will require further work.

11.4.3 Understanding the alternative lengthening of telomeres pathway

The association between the ALT pathway and tumorigenesis is well established; however, the mechanism of this pathway is still not well understood. It would seem odd that such a pathway would develop purely to facilitate tumor formation, as this would hardly lead to a competitive advantage for the species making use of such a pathway. Indeed, although it was first identified in immortalized and tumorigenic cells, ALT may routinely be used in normal somatic cells [99]. In early development, a specialized role for ALT in early cleavage embryos, which lack telomerase function, has been demonstrated [100]. The prevalence of ALT in cancers indicates that a subset of cancers, particularly those involving CIN, have readily adopted ALT as a useful mechanism to maintain telomeres and escape senescence.

One crucial characteristic of cancer cells is their immortality. Cancer cells have a high proliferation index, which means that without any TMM, DNA damage response (DDR) factors would induce replicative senescence or apoptosis [101]. As the extremities of chromosomes, telomeres are crucial for the protection of DNA during replication. At the termini of the telomere DNA sequence are TTAGGG tandem repeats. In yeast and protozoans, the TTAGGG repeat is present for hundreds of base pairs, whereas in vertebrae, the pattern is repeated for several thousand base pairs. Following the hexanucleotide sequence, the 3′ strand protrudes beyond the 5′ strand to form an overhang enriched in guanosine residues called the G-tail. The G-tail is folded on itself to form a t-loop to prevent the DNA repair machinery from recognizing this overhang as damaged DNA and thus preventing it from exonuclease degradation [102].

During chromosome replication, DNA polymerase binds to the DNA primer and elongates the DNA chains at the 3′ end until the end of the replication fork, which results in a completely copied leading strand and a shorter lagging strand. To prevent the length reduction of the lagging strand, telomeric repeat sequences are added via a protein—RNA complex enzyme called telomerase. This enzyme is only active in embryonic and adult stem cells [103]. These repeats protect the chromosome from degradation following each cell cycle.

There are two TMMs that a cancer cell can use to become immortal and it is dependent on the origin of the tumor. Reactivation of telomerase is used in approximately 85% of human cancers [103], whereas ALT, a telomerase-independent mechanism, is found in 10%—15% of tumors [104]. The ALT pathway depends on DNA repair and HR to lengthen the telomeres [105] (Fig. 11.3).

ALT-positive cells have four specific characteristic features that are consistent with hyperactive HR [7]. The first one is the large heterogeneous telomere length [106]. In telomerase-negative cells, telomeres can vary from 3 to 50 kb in length, whereas in telomerase-positive cells, they are consistently around 10 kb. The second characteristic is the presence of several classes of extrachromosomal telomeric repeats (ECTRs) [107]. ECTRs are single- and double-stranded circular DNA molecules that contain telomeric repeats. Whether ECTRs contribute to the formation or are a byproduct of ALT is still unknown, but single-stranded telomeric DNA circles (C-circles) are associated with ALT activity and can be used as a marker to identify ALT-positive cells [108]. The third feature is an increase in telomere sister chromatid exchange [109]. Lastly, ALT-positive cells contain an increase in PML-NBs that are associated with telomeres (APBs). PML-NBs are implicated in several cellular functions including induction of apoptosis, cellular senescence, inhibition of proliferation, and maintenance of genomic stability [37]. They are primarily composed of the PML protein, SP100, and the small ubiquitin-related modifier, while hundreds of other proteins can also associate with them, including ATRX and DAXX [110]. Some proteins such as DAXX and the Bloom syndrome protein are not tightly attached to the nuclear structure and can rapidly move in and out of PML-NBs [111]. APBs include all the components of the PML-NBs and the telomere-associated shelterin proteins TRF1, TRF2, POT1, or RAP1 and the factors that are involved in the DDR [101]. It is also thought that proteins involved in HR (e.g., the MRN complex) are also localized in APBs [101]. The exact mechanism by which the ALT pathway is triggered is still unclear, but it is thought to be due to a series of deregulation events [101].

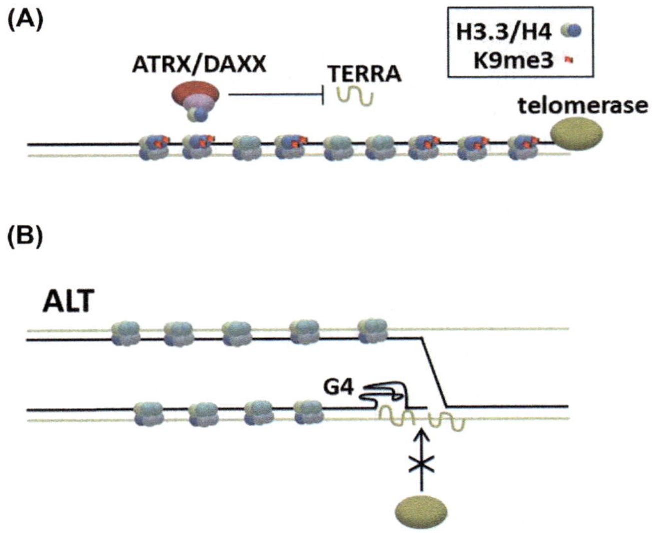

FIGURE 11.3

Loss of ATRX function at telomeres leads to loss of H3.3K9me3, derepression of telomeric repeat–containing RNA (TERRA), and initiation of alternative lengthening of telomeres (ALT). (A) ATRX/DAXX deposits H3.3/H4 dimers into telomeric nucleosomes, which are modified to contain K9me3 on the H3.3. ATRX also competes with TERRA to inhibit it. Lengthening of telomeres with ATRX present is performed by telomerase. (B) In the absence of ATRX function, G-quadruplex (G4) DNA structures can form within telomeres, and TERRA binding is increased. These events may inhibit telomerase activity and promote the use of homologous recombination–driven ALT.

11.4.4 ATRX is a suppressor of the alternative lengthening of telomeres pathway

The ALT pathway is associated with a perturbation of chromatin at the telomeric region. Although the activation mechanism of the ALT pathway is still unclear, studies have provided new clues to understand how it might work. Human cultured cells have a fairly low frequency of conversion to ALT [7]. It typically requires viral intervention and many months of culture. This low frequency is suggestive that mutations and/or epigenetic alterations are required to activate this pathway. One interesting study found that codepletion of the HIRA H3.3 histone chaperone complex subunits ASF1a and ASF1b induce all the features of the ALT phenotype [112]. This study suggested that ALT may be a consequence of poor histone management.

As previously mentioned, ATRX helps in the deposition of the histone variant H3.3 into telomeric and pericentromeric chromatin. Furthermore, studies have identified inactivating mutations of the ATRX/DAXX/H3.3 complex in ALT-positive cells [6,76,85,86]. Therefore ATRX was thought to be a potential candidate for the regulation of the ALT pathway. Genetic changes were monitored in several ALT-positive cell lines [7]. Large deletions in the *ATRX* gene, ranging from 4 to 26 exons in size, were identified. Some of these mutations included translocations, promotor or splicing mutations, and epigenetic modifications. Therefore loss of *ATRX* expression might be an initiating event for the establishment of the ALT pathway, and ATRX is involved in the repression of ALT in human cells (Fig. 11.3). However, *ATRX/DAXX* depletion is not sufficient by itself to trigger the activation of the ALT pathway [7]. It is possible that depletion of ATRX/DAXX is compensated by other H3.3 chaperones (e.g., HIRA, ASF1a, ASF1b) to prevent ALT. In this regard, depletion of ASF1a and 1b leads to the induction of all ALT hallmarks in human primary and cancer cell lines, suggesting it may be the primary factor critical for H3.3 management and ALT formation.

The role of ATRX in the ALT pathway has been further studied using the human osteosarcoma U-2 OS cell line. The characteristics of this cell line include telomerase negative activity and the lack of expression of the *ATRX* gene [113]. Exons 2–19 of the *ATRX* gene are missing in these cells, leading to the inactivation of the protein and, ultimately, its nuclear depletion [6]. A pivotal study examined the effect of *ATRX* expression in U-2 OS cells on the hallmarks of ALT, including telomere length, APBs, and ECTRs [113]. Reexpression of *ATRX* led to a progressive reduction in C-circles and APBs and the subsequent removal of *ATRX* led to their augmentation. Furthermore, telomeres in ALT-positive cells contain several DNA damage foci, also called telomere dysfunction–induced foci (TIFs). Following reintroduction of *ATRX*, a significant reduction in TIFs and a progressive decrease in telomere length were observed [113]. Taken together, these results suggest that the loss of *ATRX* expression is critical for the maintenance of the ALT phenotype. Moreover, reinsertion of the gene prevents the functioning of the ALT pathway suggesting that restoring *ATRX* expression could potentially alleviate the growth of these cancers.

How exactly ATRX mediates the suppression of the ALT pathway is an important question remaining to be answered. The initial studies focused on known functions of ATRX and its interactors. There appears to be a clear involvement of the ATRX/DAXX/H3.3 complex [113]. In U-2 OS cells, ATRX alters the levels of telomeric H3.3 and this alteration is prevented by the removal of DAXX. Furthermore, in the absence of DAXX, ATRX reexpression fails to reduce C-circle and APB production. Therefore ATRX-mediated suppression of the ALT pathway is dependent on DAXX. The roles of ATRX in suppressing telomeric G-quadruplex structures through H3.3 deposition [29] and in suppressing TERRA binding at telomeres [56] also suggest possible mechanisms for the ATRX suppression of ALT (Fig. 11.3).

Other studies are suggesting a role for the MRN complex, which physically interacts with ATRX, in the ALT phenotype. The MRN complex is required for homology-directed repair (HDR) and restart of stalled replication forks. Studies performed in ALT-positive cells showed that depletion of any component of the MRN complex, or sequestration of the MRN complex, suppressed APB formation and telomeric recombination, demonstrating the implication of the complex as a driver of the ALT phenotype [114,115]. Similarly, depletion of the APB-associated proteins MUS81 or FANCD2 also results in suppressed telomeric recombination [116,117]. Furthermore, the reintroduction of ATRX in ALT cells induces the relocalization of MRE11 to a nontelomeric region and away from PML-NBs [113]. The association of the MRN complex with PML is required for ALT activity [118]. Similarly, sequestration of the MRN complex away from APBs by the overexpression of SP100 also leads to the suppression of the ALT phenotype [114]. Therefore ATRX might have a comparable effect to the overexpression of SP100, where reexpression of ATRX in ALT-positive cells reduces replication fork stalling, which could lead to a more limited substrate for HDR and, ultimately, reduction or suppression of the ALT pathway [113]. Thus evidence indicates that ATRX suppresses the ALT pathway by actively sequestering the MRN complex away from telomeres and APBs.

In ALT-positive cell lines, genomic alterations are numerous and complex and are thought to be a result of a telomere dysfunction prior to their immortalization [7]. Genome instability can be studied by examining lagging chromosomes, which are single kinetochore-positive chromosomes that are located between the segregated chromosomes during anaphase. Several of these cell lines have an increase in micronucleation frequency, which is a consequence of lagging chromosomes and an indicator of genome instability. Interestingly, depletion of ATRX by short hairpin RNA in HeLa cells induces the formation of micronuclei [7].

These findings clearly show that alteration of ATRX is a crucial event for the initiation and the maintenance of the ALT pathway. Also, ATRX-mediated suppression of the ALT pathway is dependent on DAXX and could be due to the sequestration of the MRN complex away from telomeres. ATRX is a crucial player in tumorigenesis and therefore could be a potential therapeutic target in preventing the formation of higher grade tumors.

11.5 Conclusion

The *ATRX* gene was first discovered as the cause of the ATR-X syndrome, and since then, numerous studies have been conducted to determine its role in chromosome structure maintenance in repetitive regions of the genome, such as telomeres. ATRX is required for the deposition of histone H3.3 within repetitive regions in heterochromatin to facilitate DNA replication and prevent aberrant gene transcription. Mutation of ATRX causes the ATR-X syndrome that is characterized by a highly variable range of learning and memory deficits as well as some psychomotor impairment. Additional studies are required to associate specific cellular defects caused by ATRX mutations and possibly extrapolate these defects into specific phenotypic features of the syndrome. Furthermore, loss of ATRX expression strongly correlates with an ALT phenotype in certain tumors. Even though there is some evidence that could lead to the underlying mechanisms by which ATRX suppresses the ALT pathway, the clear cascade of events has yet to be determined. Mutation of the *ATRX* gene and activation of the ALT pathway are exclusively found together in higher grade tumors. Therefore ATRX could be a potential target to block the activation of the ALT phenotype and prevent subsequent tumor development.

List of abbreviations

ALT	Alternative lengthening of telomeres
APB	Promyelocytic leukemia nuclear body associated with telomere
BLM	The Bloom syndrome protein
ChIPseq	Chromatin immunoprecipitation sequencing
CIN	Chromosomal instability
CNS	Central nervous system
DAXX	Death domain–associated protein
DDR	DNA damage response
ECTR	Extrachromosomal telomeric repeat
ERV	Endogenous retrovirus
ERVK	Endogenous retroviruses family K
ESC	Embryonic stem cell
FISH	Fluorescence in situ hybridization

GBM	Glioblastoma
HDR	Homology directed repair
HP1α	Heterochromatin protein 1α
HR	Homologous recombination
IAP	Intracisternal A particle
IDH	Isocitrate dehydrogenase
lncRNA	Long noncoding RNA
LTR	Long-terminal repeat
MEN1	Multiple endocrine neoplasia type 1
MRN	Complex containing MRE11, RAD50, and NBS1
NF1	Neurofibromatosis type 1
NGS	Next-generation sequencing
PanNET	Pancreatic neuroendocrine tumor
Parp-1	Poly(ADP-ribose) polymerase 1
PHD	Plant homeodomain
PML	Promyelocytic leukemia
PML-NB	Promyelocytic leukemia nuclear body
PRC2	Polycomb repressive complex 2
shRNA	Short hairpin RNA
STS	Soft tissue sarcoma
SUMO	Small ubiquitin-related modifier
TERRA	Telomeric repeat–containing RNA
TIF	Telomere dysfunction–induced foci
TMM	Telomere maintenance mechanism
TSC	Tuberous sclerosis complex
T-SCE	Telomere sister chromatid exchange
VHL	The von Hippel–Lindau disease
VIP	Vasoactive intestinal peptide
VNTR	Variable number tandem repeat
XCI	X chromosome inactivation
Xi	Inactive X chromosome
WHO	World Health Organization
WT	Wild type

References

[1] Gibbons RJ. alpha-Thalassemia, mental retardation, and myelodysplastic syndrome. Cold Spring Harb Perspect Med 2012;2(10).

[2] Weatherall DJ, et al. Hemoglobin H disease and mental retardation: a new syndrome or a remarkable coincidence? N Engl J Med 1981;305(11):607–12.

[3] Gibbons RJ, et al. A newly defined X linked mental retardation syndrome associated with alpha thalassaemia. J Med Genet 1991;28(11):729–33.

[4] Villard L, et al. Splicing mutation in the ATR-X gene can lead to a dysmorphic mental retardation phenotype without alpha-thalassemia. Am J Hum Genet 1996;58(3):499–505.

[5] Garrick D, et al. Loss of Atrx affects trophoblast development and the pattern of X-inactivation in extra-embryonic tissues. PLoS Genet 2006;2(4):e58.

[6] Heaphy CM, et al. Altered telomeres in tumors with ATRX and DAXX mutations. Science 2011; 333(6041):425.
[7] Lovejoy CA, et al. Loss of ATRX, genome instability, and an altered DNA damage response are hallmarks of the alternative lengthening of telomeres pathway. PLoS Genet 2012;8(7):e1002772.
[8] Schwartzentruber J, et al. Driver mutations in histone H3.3 and chromatin remodelling genes in paediatric glioblastoma. Nature 2012;482(7384):226–31.
[9] Crawley JN, Heyer WD, LaSalle JM. Autism and cancer share risk genes, pathways, and drug targets. Trends Genet 2016;32(3):139–46.
[10] Garrick D, et al. A conserved truncated isoform of the ATR-X syndrome protein lacking the SWI/SNF-homology domain. Gene 2004;326:23–34.
[11] Picketts DJ, et al. ATRX encodes a novel member of the SNF2 family of proteins: mutations point to a common mechanism underlying the ATR-X syndrome. Hum Mol Genet 1996;5(12):1899–907.
[12] Xue Y, et al. The ATRX syndrome protein forms a chromatin-remodeling complex with Daxx and localizes in promyelocytic leukemia nuclear bodies. Proc Natl Acad Sci USA 2003;100(19):10635–40.
[13] Jaskelioff M, et al. Rad54p is a chromatin remodeling enzyme required for heteroduplex DNA joint formation with chromatin. J Biol Chem 2003;278(11):9212–8.
[14] Omichinski JG, et al. NMR structure of a specific DNA complex of Zn-containing DNA binding domain of GATA-1. Science 1993;261(5120):438–46.
[15] Xie S, et al. Cloning, expression and chromosome locations of the human DNMT3 gene family. Gene 1999; 236(1):87–95.
[16] Argentaro A, et al. Structural consequences of disease-causing mutations in the ATRX-DNMT3-DNMT3L (ADD) domain of the chromatin-associated protein ATRX. Proc Natl Acad Sci USA 2007;104(29): 11939–44.
[17] Iwase S, et al. ATRX ADD domain links an atypical histone methylation recognition mechanism to human mental-retardation syndrome. Nat Struct Mol Biol 2011;18(7):769–76.
[18] Dhayalan A, et al. The ATRX-ADD domain binds to H3 tail peptides and reads the combined methylation state of K4 and K9. Hum Mol Genet 2011;20(11):2195–203.
[19] Noh KM, et al. ATRX tolerates activity-dependent histone H3 methyl/phos switching to maintain repetitive element silencing in neurons. Proc Natl Acad Sci USA 2015;112(22):6820–7.
[20] Berube NG, Smeenk CA, Picketts DJ. Cell cycle-dependent phosphorylation of the ATRX protein correlates with changes in nuclear matrix and chromatin association. Hum Mol Genet 2000;9(4):539–47.
[21] McDowell TL, et al. Localization of a putative transcriptional regulator (ATRX) at pericentromeric heterochromatin and the short arms of acrocentric chromosomes. Proc Natl Acad Sci USA 1999;96(24): 13983–8.
[22] Sarma K, et al. ATRX directs binding of PRC2 to Xist RNA and polycomb targets. Cell 2014;159(4): 869–83.
[23] Baumann C, De La Fuente R. ATRX marks the inactive X chromosome (Xi) in somatic cells and during imprinted X chromosome inactivation in trophoblast stem cells. Chromosoma 2009;118(2):209–22.
[24] Kourmouli N, et al. Epigenetic regulation of mammalian pericentric heterochromatin in vivo by HP1. Biochem Biophys Res Commun 2005;337(3):901–7.
[25] Nan X, et al. Interaction between chromatin proteins MECP2 and ATRX is disrupted by mutations that cause inherited mental retardation. Proc Natl Acad Sci USA 2007;104(8):2709–14.
[26] Eustermann S, et al. Combinatorial readout of histone H3 modifications specifies localization of ATRX to heterochromatin. Nat Struct Mol Biol 2011;18(7):777–82.
[27] Sabbattini P, et al. An H3K9/S10 methyl-phospho switch modulates Polycomb and Pol II binding at repressed genes during differentiation. Mol Biol Cell 2014;25(6):904–15.

[28] Kernohan KD, et al. ATRX partners with cohesin and MeCP2 and contributes to developmental silencing of imprinted genes in the brain. Dev Cell 2010;18(2):191–202.
[29] Law MJ, et al. ATR-X syndrome protein targets tandem repeats and influences allele-specific expression in a size-dependent manner. Cell 2010;143(3):367–78.
[30] Valton AL, Prioleau MN. G-quadruplexes in DNA replication: a problem or a necessity? Trends Genet 2016;32(11):697–706.
[31] Sadic D, et al. Atrx promotes heterochromatin formation at retrotransposons. EMBO Rep 2015;16(7): 836–50.
[32] Voon HP, Wong LH. New players in heterochromatin silencing: histone variant H3.3 and the ATRX/DAXX chaperone. Nucleic Acids Res 2016;44(4):1496–501.
[33] Tang J, et al. A novel transcription regulatory complex containing death domain-associated protein and the ATR-X syndrome protein. J Biol Chem 2004;279(19):20369–77.
[34] Udugama M, et al. Histone variant H3.3 provides the heterochromatic H3 lysine 9 tri-methylation mark at telomeres. Nucleic Acids Res 2015;43(21):10227–37.
[35] Elsasser SJ, et al. DAXX envelops a histone H3.3-H4 dimer for H3.3-specific recognition. Nature 2012; 491(7425):560–5.
[36] He Q, et al. The daxx/atrx complex protects tandem repetitive elements during DNA hypomethylation by promoting H3K9 trimethylation. Cell Stem Cell 2015;17(3):273–86.
[37] Bernardi R, Pandolfi PP. Structure, dynamics and functions of promyelocytic leukaemia nuclear bodies. Nat Rev Mol Cell Biol 2007;8(12):1006–16.
[38] Berube NG, et al. Patient mutations alter ATRX targeting to PML nuclear bodies. Eur J Hum Genet 2008; 16(2):192–201.
[39] Delbarre E, et al. PML protein organizes heterochromatin domains where it regulates histone H3.3 deposition by ATRX/DAXX. Genome Res 2017;27(6):913–21.
[40] Leung JW, et al. Alpha thalassemia/mental retardation syndrome X-linked gene product ATRX is required for proper replication restart and cellular resistance to replication stress. J Biol Chem 2013;288(9): 6342–50.
[41] Clynes D, et al. ATRX dysfunction induces replication defects in primary mouse cells. PLoS One 2014; 9(3):e92915.
[42] Huh MS, et al. Stalled replication forks within heterochromatin require ATRX for protection. Cell Death Dis 2016;7:e2220.
[43] Tittel-Elmer M, et al. The MRX complex stabilizes the replisome independently of the S phase checkpoint during replication stress. EMBO J 2009;28(8):1142–56.
[44] Huh MS, et al. Compromised genomic integrity impedes muscle growth after Atrx inactivation. J Clin Invest 2012;122(12):4412–23.
[45] Voon HP, et al. ATRX plays a key role in maintaining silencing at interstitial heterochromatic loci and imprinted genes. Cell Rep 2015;11(3):405–18.
[46] Levy MA, et al. ATRX promotes gene expression by facilitating transcriptional elongation through guanine-rich coding regions. Hum Mol Genet 2015;24(7):1824–35.
[47] Valle-Garcia D, et al. ATRX binds to atypical chromatin domains at the 3' exons of zinc finger genes to preserve H3K9me3 enrichment. Epigenetics 2016;11(6):398–414.
[48] Huynh KD, Lee JT. X-chromosome inactivation: a hypothesis linking ontogeny and phylogeny. Nat Rev Genet 2005;6(5):410–8.
[49] Lee JT, Davidow LS, Warshawsky D. Tsix, a gene antisense to Xist at the X-inactivation centre. Nat Genet 1999;21(4):400–4.
[50] Hoki Y, et al. A proximal conserved repeat in the Xist gene is essential as a genomic element for X-inactivation in mouse. Development 2009;136(1):139–46.

[51] Zhao J, et al. Polycomb proteins targeted by a short repeat RNA to the mouse X chromosome. Science 2008;322(5902):750–6.
[52] Simon JA, Kingston RE. Occupying chromatin: polycomb mechanisms for getting to genomic targets, stopping transcriptional traffic, and staying put. Mol Cell 2013;49(5):808–24.
[53] Kim KH, Roberts CW. Targeting EZH2 in cancer. Nat Med 2016;22(2):128–34.
[54] Cardoso C, et al. Specific interaction between the XNP/ATR-X gene product and the SET domain of the human EZH2 protein. Hum Mol Genet 1998;7(4):679–84.
[55] Azzalin CM, Lingner J. Telomere functions grounding on TERRA firma. Trends Cell Biol 2015;25(1):29–36.
[56] Chu HP, et al. TERRA RNA antagonizes ATRX and protects telomeres. Cell 2017;170(1):86–101. e16.
[57] Rippe K, Luke B. TERRA and the state of the telomere. Nat Struct Mol Biol 2015;22(11):853–8.
[58] Derrien T, et al. The GENCODE v7 catalog of human long noncoding RNAs: analysis of their gene structure, evolution, and expression. Genome Res 2012;22(9):1775–89.
[59] Han P, Chang CP. Long non-coding RNA and chromatin remodeling. RNA Biol 2015;12(10):1094–8.
[60] Alcid EA, Tsukiyama T. ATP-dependent chromatin remodeling shapes the long noncoding RNA landscape. Genes Dev 2014;28(21):2348–60.
[61] Johnson WL, et al. RNA-dependent stabilization of SUV39H1 at constitutive heterochromatin. Elife 2017;6.
[62] Gibbons RJ, et al. Mutations in the chromatin-associated protein ATRX. Hum Mutat 2008;29(6):796–802.
[63] Giardine B, et al. Systematic documentation and analysis of human genetic variation in hemoglobinopathies using the microattribution approach. Nat Genet 2011;43(4):295–301.
[64] Young KGaP, David J. Role of ATRX chromatin remodelling factor in α-thalassaemia X-linked mental retardation. In: eLS. John Wiley & Sons; 2016.
[65] Abidi FE, et al. Mutation in the 5' alternatively spliced region of the XNP/ATR-X gene causes Chudley-Lowry syndrome. Eur J Hum Genet 2005;13(2):176–83.
[66] Guerrini R, et al. A nonsense mutation of the ATRX gene causing mild mental retardation and epilepsy. Ann Neurol 2000;47(1):117–21.
[67] Howard MT, et al. Attenuation of an amino-terminal premature stop codon mutation in the ATRX gene by an alternative mode of translational initiation. J Med Genet 2004;41(12):951–6.
[68] Nogami T, et al. Reduced expression of the ATRX gene, a chromatin-remodeling factor, causes hippocampal dysfunction in mice. Hippocampus 2011;21(6):678–87.
[69] Shioda N, et al. Aberrant calcium/calmodulin-dependent protein kinase II (CaMKII) activity is associated with abnormal dendritic spine morphology in the ATRX mutant mouse brain. J Neurosci 2011;31(1):346–58.
[70] Wada T, et al. Neuroradiologic features in X-linked alpha-thalassemia/mental retardation syndrome. AJNR Am J Neuroradiol 2013;34(10):2034–8.
[71] Badens C, et al. Mutations in PHD-like domain of the ATRX gene correlate with severe psychomotor impairment and severe urogenital abnormalities in patients with ATRX syndrome. Clin Genet 2006;70(1):57–62.
[72] Gibbons RJ, et al. Mutations in transcriptional regulator ATRX establish the functional significance of a PHD-like domain. Nat Genet 1997;17(2):146–8.
[73] Craddock CF, et al. Contrasting effects of alpha and beta globin regulatory elements on chromatin structure may be related to their different chromosomal environments. EMBO J 1995;14(8):1718–26.
[74] Ratnakumar K, et al. ATRX-mediated chromatin association of histone variant macroH2A1 regulates alpha-globin expression. Genes Dev 2012;26(5):433–8.
[75] Liu XY, et al. Frequent ATRX mutations and loss of expression in adult diffuse astrocytic tumors carrying IDH1/IDH2 and TP53 mutations. Acta Neuropathol 2012;124(5):615–25.

[76] Johnson JE, Broccoli D. Telomere maintenance in sarcomas. Curr Opin Oncol 2007;19(4):377–82.
[77] Gibbons RJ, et al. Mutations in a putative global transcriptional regulator cause X-linked mental retardation with alpha-thalassemia (ATR-X syndrome). Cell 1995;80(6):837–45.
[78] Louis DN, et al. The 2007 WHO classification of tumours of the central nervous system. Acta Neuropathol 2007;114(2):97–109.
[79] Jiao Y, et al. Frequent ATRX, CIC, FUBP1 and IDH1 mutations refine the classification of malignant gliomas. Oncotarget 2012;3(7):709–22.
[80] Halfdanarson TR, et al. Pancreatic neuroendocrine tumors (PNETs): incidence, prognosis and recent trend toward improved survival. Ann Oncol 2008;19(10):1727–33.
[81] Capurso G, et al. Molecular pathology and genetics of pancreatic endocrine tumours. J Mol Endocrinol 2012;49(1):R37–50.
[82] Moore PS, et al. Role of disease-causing genes in sporadic pancreatic endocrine tumors: MEN1 and VHL. Genes Chromosomes Cancer 2001;32(2):177–81.
[83] Perren A, et al. Multiple endocrine neoplasia type 1 (MEN1): loss of one MEN1 allele in tumors and monohormonal endocrine cell clusters but not in islet hyperplasia of the pancreas. J Clin Endocrinol Metab 2007;92(3):1118–28.
[84] Jiao Y, et al. DAXX/ATRX, MEN1, and mTOR pathway genes are frequently altered in pancreatic neuroendocrine tumors. Science 2011;331(6021):1199–203.
[85] Abedalthagafi M, et al. The alternative lengthening of telomere phenotype is significantly associated with loss of ATRX expression in high-grade pediatric and adult astrocytomas: a multi-institutional study of 214 astrocytomas. Mod Pathol 2013;26(11):1425–32.
[86] Kurihara S, et al. Clinical features of ATRX or DAXX mutated neuroblastoma. J Pediatr Surg 2014;49(12): 1835–8.
[87] Marinoni I, et al. Loss of DAXX and ATRX are associated with chromosome instability and reduced survival of patients with pancreatic neuroendocrine tumors. Gastroenterology 2014;146(2):453–460 e5.
[88] Jonkers YM, et al. Chromosomal instability predicts metastatic disease in patients with insulinomas. Endocr Relat Cancer 2005;12(2):435–47.
[89] Hakin-Smith V, et al. Alternative lengthening of telomeres and survival in patients with glioblastoma multiforme. Lancet 2003;361(9360):836–8.
[90] Henson JD, et al. A robust assay for alternative lengthening of telomeres in tumors shows the significance of alternative lengthening of telomeres in sarcomas and astrocytomas. Clin Canc Res 2005;11(1):217–25.
[91] Cheung NK, et al. Association of age at diagnosis and genetic mutations in patients with neuroblastoma. J Am Med Assoc 2012;307(10):1062–71.
[92] Onitake Y, et al. Telomere biology in neuroblastoma: telomere binding proteins and alternative strengthening of telomeres. J Pediatr Surg 2009;44(12):2258–66.
[93] Sandberg AA. Updates on the cytogenetics and molecular genetics of bone and soft tissue tumors: liposarcoma. Cancer Genet Cytogenet 2004;155(1):1–24.
[94] Costa A, et al. Telomere maintenance mechanisms in liposarcomas: association with histologic subtypes and disease progression. Cancer Res 2006;66(17):8918–24.
[95] Montgomery E, et al. Telomere lengths of translocation-associated and nontranslocation-associated sarcomas differ dramatically. Am J Pathol 2004;164(5):1523–9.
[96] Scheel C, et al. Alternative lengthening of telomeres is associated with chromosomal instability in osteosarcomas. Oncogene 2001;20(29):3835–44.
[97] Koelsche C, et al. Differential nuclear ATRX expression in sarcomas. Histopathology 2016;68(5):738–45.
[98] Ulaner GA, et al. Divergent patterns of telomere maintenance mechanisms among human sarcomas: sharply contrasting prevalence of the alternative lengthening of telomeres mechanism in Ewing's sarcomas and osteosarcomas. Genes Chromosomes Cancer 2004;41(2):155–62.

[99] Neumann AA, et al. Alternative lengthening of telomeres in normal mammalian somatic cells. Genes Dev 2013;27(1):18–23.
[100] Liu L, et al. Telomere lengthening early in development. Nat Cell Biol 2007;9(12):1436–41.
[101] Chung I, et al. PML body meets telomere: the beginning of an ALTernate ending? Nucleus 2012;3(3): 263–75.
[102] Nezu T, et al. Telomere G-tail length is a promising biomarker related to white matter lesions and endothelial dysfunction in patients with cardiovascular risk: a cross-sectional study. EBioMedicine 2015; 2(8):960–7.
[103] Shay JW, Bacchetti S. A survey of telomerase activity in human cancer. Eur J Cancer 1997;33(5):787–91.
[104] Gilson E, Geli V. How telomeres are replicated. Nat Rev Mol Cell Biol 2007;8(10):825–38.
[105] Henson JD, Reddel RR. Assaying and investigating alternative lengthening of telomeres activity in human cells and cancers. FEBS Lett 2010;584(17):3800–11.
[106] Henson JD, et al. Alternative lengthening of telomeres in mammalian cells. Oncogene 2002;21(4): 598–610.
[107] Cesare AJ, Griffith JD. Telomeric DNA in ALT cells is characterized by free telomeric circles and heterogeneous t-loops. Mol Cell Biol 2004;24(22):9948–57.
[108] Henson JD, et al. DNA C-circles are specific and quantifiable markers of alternative-lengthening-of-telomeres activity. Nat Biotechnol 2009;27(12):1181–5.
[109] Londono-Vallejo JA, et al. Alternative lengthening of telomeres is characterized by high rates of telomeric exchange. Cancer Res 2004;64(7):2324–7.
[110] Yeager TR, et al. Telomerase-negative immortalized human cells contain a novel type of promyelocytic leukemia (PML) body. Cancer Res 1999;59(17):4175–9.
[111] Weidtkamp-Peters S, et al. Dynamics of component exchange at PML nuclear bodies. J Cell Sci 2008; 121(Pt 16):2731–43.
[112] O'Sullivan RJ, et al. Rapid induction of alternative lengthening of telomeres by depletion of the histone chaperone ASF1. Nat Struct Mol Biol 2014;21(2):167–74.
[113] Clynes D, et al. Suppression of the alternative lengthening of telomere pathway by the chromatin remodelling factor ATRX. Nat Commun 2015;6:7538.
[114] Jiang WQ, et al. Suppression of alternative lengthening of telomeres by Sp100-mediated sequestration of the MRE11/RAD50/NBS1 complex. Mol Cell Biol 2005;25(7):2708–21.
[115] Zhong ZH, et al. Disruption of telomere maintenance by depletion of the MRE11/RAD50/NBS1 complex in cells that use alternative lengthening of telomeres. J Biol Chem 2007;282(40):29314–22.
[116] Fan Q, et al. A role for monoubiquitinated FANCD2 at telomeres in ALT cells. Nucleic Acids Res 2009; 37(6):1740–54.
[117] Zeng S, et al. Telomere recombination requires the MUS81 endonuclease. Nat Cell Biol 2009;11(5): 616–23.
[118] Dunham MA, et al. Telomere maintenance by recombination in human cells. Nat Genet 2000;26(4): 447–50.

SECTION

Neuropsychiatric disorders

3

CHAPTER

12 Epigenetic dysregulation in the fragile X-related disorders

Karen Usdin, Daman Kumari

Section on Gene Structure and Disease, Laboratory of Cell and Molecular Biology, National Institute of Diabetes, Digestive and Kidney Diseases, National Institutes of Health, Bethesda, MD, United States

Chapter Outline

12.1 Introduction 261
12.2 Clinical features of the FXDs 262
12.2.1 Fragile X syndrome 262
12.2.2 Fragile X-associated tremor/ataxia syndrome 263
12.2.3 Fragile X-associated primary ovarian insufficiency 263
12.3 Genetics of the FXDs 264
12.4 The pathological basis of FXTAS 265
12.5 The pathological basis of FXPOI 267
12.6 The pathological basis of FXS 268
12.7 Epigenetic abnormalities associated with the FXDs 268
12.8 Resolving the repeat paradox 270
12.9 Prospects and challenges for epigenetic therapies for the FXDs 271
12.10 Concluding remarks 273
Grant Sponsor 273
References 273

12.1 Introduction

Gene expression in mammals is regulated at a number of different levels. Information encoded in the primary DNA sequence is important for specifying the binding of important transcriptional activators and repressors. However, access to this primary sequence represents another important level of regulation. In eukaryotes, this access to the DNA is regulated via a complex network of epigenetic changes both to DNA and to the unstructured N-terminal tails of histones that project out of the nucleosome around which DNA is wrapped [1]. These tails are the substrates for a number of posttranslational modifications (PTMs) including acetylation, methylation, phosphorylation, and ubiquitination of specific lysines, arginines, serines, threonines, and tyrosines. The PTM profile of any

Chromatin Signaling and Neurological Disorders. https://doi.org/10.1016/B978-0-12-813796-3.00012-2

region of the genome is determined by the antagonistic effects of the chromatin writers that deposit these marks and the chromatin erasers that remove them. These PTMs act as docking sites for chromatin readers that modulate the extent of chromatin compaction via the recruitment of chromatin remodelers [2]. In addition to histone modifications, the DNA itself is subject to its own, albeit more limited, modifications. The most prevalent of these modifications in mammals is cytosine methylation that occurs primarily at CpG residues. There is cross talk between the DNA and histone modifications to regulate accessibility to the transcription machinery and thus both are key to the proper regulation of gene expression [1].

Dysregulation of these epigenetic processes results in aberrant gene expression that is responsible for a growing list of human disorders that includes the fragile X-related disorders (FXDs). The FXDs include a neurodegenerative condition, fragile X-associated tremor/ataxia syndrome (FXTAS; OMIM 300623), a form of female infertility, fragile X-associated primary ovarian insufficiency (FXPOI; OMIM 300624), and the neurodevelopmental disorder, fragile X syndrome (FXS; OMIM 300624), the leading heritable cause of intellectual disability and most common monogenic cause of autism. As will be discussed in more detail later in this chapter, the mutation responsible for these disorders has two different direct epigenetic effects on the affected gene that significantly impacts local gene expression. These effects are responsible for FXS and are considered to exacerbate disease pathology in the case of FXTAS and perhaps FXPOI. The mutation also has an indirect global epigenetic effect in individuals with FXS.

12.2 Clinical features of the FXDs

12.2.1 Fragile X syndrome

FXS, the best known of the FXDs, was first described many decades ago by Martin and Bell [3]. It was later named for the folate-sensitive fragile site that was subsequently found to be a characteristic cytogenetic, and early diagnostic, feature seen in affected individuals [4,5]. This fragile site is located on the long arm of the X chromosome where the mutation responsible for FXS is now known to be located. As the gene is X-linked, females have a higher likelihood of inheriting an affected allele but because of X chromosome inactivation, they are generally less severely affected.

The clinical presentation of FXS can be quite variable. Features often include moderate to severe cognitive impairment, with almost all males having intelligence quotients scores of 70 or less [7]. A number of behavioral abnormalities are seen in this patient population including attention-deficit hyperactivity disorder, aggressive behavior, impulsivity, self-injury, insomnia, depression, and social anxiety. Unusual speech patterns and language delay are common [8]. Half of the males and 20% of the females meet the Diagnostic and Statistical Manual of Mental Disorders, Fifth Edition criteria for autism spectrum disorder (ASD) [9]. ASD symptoms seen in individuals with FXS include impaired social interaction, repetitive behaviors such as hand flapping, perseverative speech, and gaze avoidance. About 14% of boys and 6% of girls have seizures that begin between 4 and 10 years of age, the most common form being complex partial seizures [10]. These are easily controlled and most often resolve during childhood. Recurrent otitis media and gastrointestinal problems are seen more frequently than in typically developing children [11], and there is a high incidence of ocular disorders including refractive errors, nystagmus, and strabismus [12]. Males, in particular, can have a physical presentation that includes macroorchidism, hyperflexible joints, large ears, and a long face.

12.2.2 Fragile X-associated tremor/ataxia syndrome

Unlike FXS, which is primarily a neurodevelopmental disorder, FXTAS is an adult-onset neurodegenerative disorder characterized by cerebellar ataxia, intention tremor, parkinsonism, executive function deficits, and sometimes dementia [13]. Peripheral neuropathy, autonomic dysfunction as well as disinhibition, anxiety, irritability, agitation, apathy, and depression are frequently seen [14–16]. Radiological features of FXTAS include T2 weighted magnetic resonance imaging signal hyperintensities in the middle cerebellar peduncles, subcortical white matter, and splenium of the corpus callosum. Cerebellar, cerebral, and brainstem atrophy is also seen [17,18], and there is significant cerebellar Purkinje cell loss [19]. Neuropathological findings include intranuclear neuronal inclusions in multiple brain regions [19]. The inclusions contain a number of chaperone proteins including HSP27, HSP70, and αβ-crystallin and a number of proteins involved in the DNA damage response [20,21]. However, the inclusions do not contain polyglutamine proteins, α-synuclein or tau proteins [19,20]. The inclusions also contain lamin A/C [20] and fibroblasts from individuals with FXTAS show an altered distribution of lamin A/C and an altered nuclear shape [22].

In the brains of a mouse model of FXTAS, resting cytoplasmic calcium is chronically elevated, as are the levels of activated ATM, Bax, and Bcl-2 [23] proteins. Mitochondrial dysfunction is also seen in the brains of mouse models and the fibroblasts and brain tissue of human premutation (PM) carriers [23–26].

12.2.3 Fragile X-associated primary ovarian insufficiency

FXPOI is the most common single-gene cause of primary ovarian insufficiency (POI), a form of ovarian dysfunction resulting from the failure of the ovary to respond normally to appropriate gonadotropin stimulation by the hypothalamus and pituitary. FXPOI is characterized by elevated serum gonadotropin levels, reduced estrogen levels, and reduced levels of antimüllerian hormone, consistent with a diminished pool of small growing ovarian follicles [27,28]. The net effect is irregular menses, fertility problems [29], and poor response to ovarian stimulation in younger women [30]. A clinical diagnosis of FXPOI is based on cessation of menses for at least 4 months before the age of 40. However, ~33% of affected women reach menopause before the age of 30 and ~4.5% do so before the age of 18 [31]. The earlier than normal age at menopause also puts these women at increased risk of cardiovascular disease, cognitive decline, and osteoporosis. FXPOI is responsible for ~3% of idiopathic cases of infertility and ~11% of familial cases [32,33]. The penetrance of FXPOI is ~20%. However, a shift toward younger age at menopause is seen for all carriers of the affected gene, and occult FXPOI involving varying degrees of ovarian dysfunction is seen in many women [34,35].

The ovaries of affected women are similar to that of unaffected women in terms of general histopathology [36]. Inclusions of the sort seen in neurons and astrocytes of individuals with FXTAS are seen in the ovarian stroma of women with FXPOI, but not in ovarian follicles [36,37]. In mouse models of FXPOI, the ovaries have fewer granulosa cells and the remaining granulosa cells show evidence of mitochondrial dysfunction [26,38,39].

12.3 Genetics of the FXDs

All three disorders result from a larger than normal CGG-repeat tract in exon 1 of the X-linked gene, *FMR1* (OMIM *309550). The repeat is located just 5′ of the start of the coding sequence for the *FMR1* gene product, FMRP. Normal individuals have fewer than 45 repeats, most frequently 29 or 30. Alleles with 45–54 repeats are referred to as intermediate alleles or gray-zone alleles. The pathology associated with these alleles is still the subject of debate [40–47]. Alleles with 55–200 repeats are known for historical reasons as premutation (PM) alleles, as they were once considered to be asymptomatic alleles that were only at risk of generating larger symptomatic full mutation (FM) alleles. However, it is now known that PM alleles confer risk of FXTAS and FXPOI. PM allele prevalence is 1 in 130–250 females and 1 in 250–810 males [48]. However, for reasons that are not well understood, only 20% of female PM carriers develop overt FXPOI, while 40% of male PM carriers and 16% of female PM carriers develop FXTAS. The severity of FXTAS symptoms is directly related to the repeat number [49,50]. In contrast, there is a nonlinear relationship between repeat number and incidence of FXPOI such that FXPOI risk peaks at ~100 repeats, after which point it begins to decline [46,51–55].

FXS results from inheritance of FM alleles. The prevalence of FM alleles in the United States is 1 in ~5000 males and 1 in ~2500–8000 females [6]. Historically, FM alleles are considered to have >200 repeats. However, recent data suggest that the minimum repeat number associated with FXS may be closer to 400 repeats [56,57].

The different pathologies seen in PM and FM carriers are considered to arise from the paradoxical effects of the repeats on gene expression. As illustrated in Fig. 12.1, PM carriers, for reasons that are still not clear, show an increase in the rate of transcription initiation with the result that the PM transcript is present at 2–8 times the level that is seen in normal individuals [58,59]. This RNA is

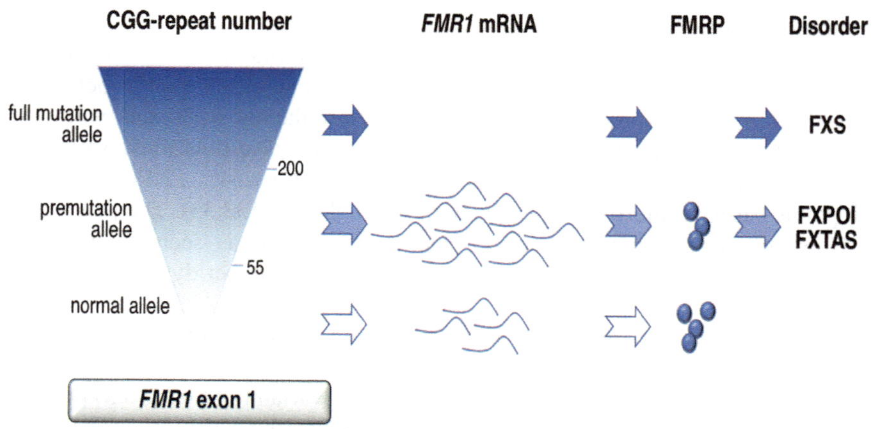

FIGURE 12.1 Genetic and Molecular Correlates in the FXDs.

Diagrammatic representation of the relationship between repeat number, transcription, translation, and disease pathology. The repeat numbers are shown as a spectrum rather than as distinct classes to reflect the overlap in disease symptoms resulting from heterogeneity both in repeat number and methylation mosaicism that blurs the distinctions between PM and FM alleles.

considered to have deleterious consequences that will be discussed further later in this chapter. In contrast, FM alleles undergo an incompletely understood process of epigenetic silencing [60–66]. The net effect is that FM carriers make little or no FMRP. Such a loss of function mutation generated by repeat-mediated gene silencing is consistent with the fact that inactivating point mutations in FMRP phenocopy the expansion mutation [67–69].

Despite gene silencing, most FM carriers make some *FMR1* mRNA [70], and those rare individuals that escape gene silencing can make large amounts of this mRNA [71–76]. These carriers of unmethylated FM alleles (UFMs) can exhibit the symptoms of both FXTAS and FXS [71,77]. Furthermore, many PM carriers show reduced levels of the *FMR1* gene product either because they show some DNA methylation [78] or because of the inefficient translation of transcripts with large numbers of CGG repeats [79,80]. This, together with the mosaicism for repeat number that is frequently seen [81], can produce a continuum of symptoms, some resulting from the toxic effects of the RNA and some from the FMRP deficiency. This has led to the suggestion that fragile X spectrum disorder (FXSD) may be a better name for these disorders [82].

12.4 The pathological basis of FXTAS

As the CGG-repeat tract responsible for FXTAS and FXPOI is located outside the FMRP open reading frame and FMRP levels in most PM carriers are only moderately reduced relative to unaffected individuals, current thinking is that pathology arises from some deleterious effect of the transcript containing large numbers of CGG repeats [83]. This deleterious effect would then be exacerbated by hyperexpression of the PM allele [83]. An RNA gain-of-function model has been proposed for FXTAS based on what is known about the pathogenic mechanism in myotonic dystrophy type I and 2. In these disorders, the CUG and CCUG repeats bind important proteins including those of the Muscleblind-like family [84]. This binding effectively reduces the availability of these proteins to carry out their normal role in mRNA processing. In the FXTAS version of this model, pathology arises because the repeat-containing transcript sequesters vital CGG-RNA-binding proteins (as illustrated in Fig. 12.2). *FMR1* mRNA is found in intranuclear inclusions and some of the proteins in these inclusions are considered to bind to the CGG repeat itself, including the DiGeorge Syndrome Critical Region Gene 8 (DGCR8) protein involved in miRNA biogenesis [85], the multifunctional RNA processing protein heterogeneous nuclear ribonucleoprotein (hnRNP) A2/B1 [86], and the purine-rich binding protein α (Purα), an RNA- and DNA-binding protein implicated in RNA transport and translation [87].

A role for sequestration of DGCR8 in FXTAS pathology is suggested by the altered miRNA profiles seen in blood of individuals with FXTAS and in Drosophila overexpressing CGG-RNA [85,93,94]. There is evidence to suggest that hnRNP A2/B1-sequestration by CGG-RNA can be deleterious in different ways. For example, FXTAS neurons show reduced delivery of hnRNP A2/B1-target mRNAs [95]. The protein also binds retrotransposons in Drosophila and recruits the heterochromatin protein HP1 to silence them. Overexpression of CGG-RNA in flies is associated with activation of these retrotransposons [96]. Suppression of retrotransposon activation reduces signs of neurodegeneration, suggesting that transposon activation contributes in part to CGG-RNA-induced pathology at least in flies [96]. Whether it acts the same way in humans remains to be seen. Overexpression of human hnRNP A2/B1 or its Drosophila orthologs reduces the rough eye phenotype

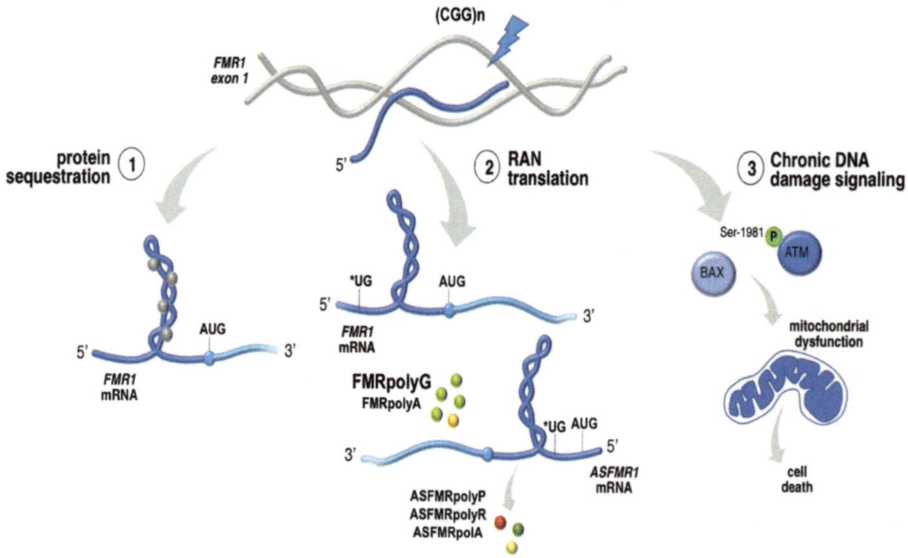

FIGURE 12.2 Current Models for PM Pathology.

1. Protein sequestration model: the hairpins formed by the CGG-RNA in the PM transcript are proposed to bind and sequester proteins such as DCRG8, Purα, and hnRNP A2/B [85–88]. This results in pathology because the CGG-RNA-bound proteins are unable to carry out their normal function, thus effectively generating a loss-of-function mutation. 2. RAN translation model: Repeat-mediated initiation of translation at near AUG codons results in the production of proteins with homopolymeric amino acid tracts. At least some of resultant proteins are thought to have deleterious properties either related to the homopolymeric tract or to the increased stability that they confer on the protein [89–91]. 3. Chronic DNA damage model: Single-stranded DNA resulting from the formation of a cotranscriptional R-loop may make the *FMR1* locus prone to DNA damage [92]. This in turn may lead to increased phosphorylation of ATM on serine 1981 and increased BAX expression [23], which may account for the increased mitochondrial damage that is observed in mouse and human PM carriers [23–26,38,39]. Cell death may then result from the elevated mitochondrial damage and other consequences of chronic activation of the DNA damage response.

associated with overexpression of CGG-RNA in flies [86,97]. Purα is found associated with the cytoplasmic inclusions seen in Drosophila overexpressing 90 CGG repeats but whether it is present in the inclusions in mammals is still the subject of some debate. In Drosophila, Purα binds Rm62, the ortholog of the mammalian P68/DDX5 helicase that is involved in nuclear transport [98]. Consistent with a role for Purα in neurodegeneration is the observation that flies overexpressing CGG-RNA show increased nuclear retention of the Rm62 mRNA targets, while overexpression of Rm62 rescues CGG-RNA toxicity.

A second model for PM pathology, also illustrated in Fig. 12.2, is based on the observation that tandem repeats can promote initiation of translation at non-AUG codons, a phenomenon known as repeat-associated non-AUG (RAN) translation [99]. In principle, this can result in translation of the repeats in all three open reading frames from the sense transcript and translation in all three open reading frames on any antisense transcript as illustrated in Fig. 12.2. In FXTAS, the sense transcript would generate FMRpolyR, a protein containing a polyarginine tract that would be in frame with FMRP, FMRpolyG, containing a tract of glycines and a 42 amino acid C-terminal extension, and FMRpolyA, which contains a tract of alanines [89]. The antisense transcript encodes a putative polyproline protein, ASFMRpolyP, via initiation at an AUG start codon. In addition, RAN translation has the potential to generate a polyarginine containing peptide, ASFMRpolyR, and a polyalanine containing peptide, ASFMRpolyA [100]. FMRPolyG, ASFMRpolyP, ASFMRpolyR, and ASFMR-polyA are present in the intranuclear inclusions seen in FXTAS brain tissue, while FMRpolyR and FMRpolyA have not been detected [89,90,100].

Evidence in support of the RAN translation model for FXTAS comes from experiments where a stop codon placed between the repeat and the near-cognate initiation codons upstream of the repeat eliminated the retinal degeneration seen in the Drosophila model [89]. Consistent with this, mice containing exon 1 of the human *FMR1* gene have intranuclear neuronal inclusions and exhibit locomotor abnormalities, Purkinje cell loss, and early death, while mice lacking the region 5′ of the repeat do not [90]. How the RAN proteins cause toxicity is still not fully understood. Overexpression of the 42 amino acid C-terminal of the FMRpolyG protein is sufficient to cause toxicity and nuclear lamina abnormalities. The C-terminus binds proteins that include Lap2β, a lamin-associated protein and overexpression of this protein is able to reduce FMRpolyG toxicity in neuronal cells [90]. Other potentially pathological roles for FMRpolyG include impairment of the ubiquitin proteasome system [101] that may account for the accumulation of proteasomal components including ubiquitin, heat shock proteins, and αβ-crystallin in FXTAS inclusions [20].

It has also been suggested that persistent DNA damage associated with the repeats triggers cell death [92]. *FMR1* alleles form R-loops, triple-stranded structures in which the transcript forms an RNA:DNA hybrid with the template strand for transcription leaving the nontemplate strand free [102–104]. R-loops are known to be vulnerable to DNA breakage [105]. Consistent with this idea, γH2AX, a histone variant associated with DNA double-strand breaks (DSBs) is found in nuclear inclusions [21]. In addition, as illustrated in the right hand side of Fig. 12.2, activation of ATM, the DSB-activated damage checkpoint kinase, and elevated levels of BAX are seen in both cell and mouse models of FXTAS [23].

12.5 The pathological basis of FXPOI

Comparatively little work has been done to specifically examine the basis of pathology in the ovary, but it is reasonable to assume that the mechanisms invoked to explain FXTAS could equally be relevant to FXPOI. However, it should be noted that despite evidence of ovarian follicle abnormalities in mouse models of FXPOI [26,38,39,106], nuclear inclusions have not been seen in either the oocyte or granulosa cells in an FXPOI mouse model or in women with FXPOI [36,37]. They are only seen in the ovarian stroma [36,37], although the stroma is not visibly abnormal in either mice or women with the PM. It is possible that pathology in follicles precedes the appearance of these inclusions or that the inclusions are protective rather than pathogenic. It is also possible that, despite their microscopically normal appearance, the stroma is not functioning properly.

Another perplexing observation is that, unlike the linear relationship between repeat number and pathology risk that is seen in FXTAS, the risk of FXPOI increases until about 100 repeats, after which point the risk seems to decline. One possible explanation for this observation is that RNA toxicity, whatever its molecular basis, is less apparent in the ovaries of larger PM allele carriers because the repeat has expanded into the FM range and become silenced. An alternative explanation is suggested by the observation that in a mouse model of FXPOI, the activity of mammalian target of rapamycin (mTOR) is downregulated [39]. Inhibition of mTOR activity is associated with compromised granulosa cell proliferation and reduced follicle growth [107] reminiscent of what is seen in mouse models of FXPOI [26,38,39]. Activation of mTOR also reduces neurodegeneration in flies expressing 90 CGG repeats supporting the role of mTOR activity in FXTAS pathology as well [108]. In contrast, the loss of FMRP is associated with an increase in mTOR activity [109–111]. It may be that as the repeat number rises, inhibition of mTOR activity increases as a result of the deleterious effects of the CGG-RNA. However, when the repeat number increases above 100 in oocytes, FMRP levels drop because of the difficulty in translation of the mRNA [79,80]. This may be exacerbated in the oocyte by expansion into the FM range. As a result, mTOR activity begins to increase and offset the negative effect of the CGG-RNA on mTOR activity. We speculate that the linear relationship between repeat number and disease severity in FXTAS is related to a less severe effect of the repeats on FMRP translation in brain or the failure of neuronal PM alleles to expand into the FM range. As a result, the CGG-RNA-associated inhibition of mTOR activity is not offset by the effect of decreased FMRP on mTOR activation. The net effect being that pathology in the brain continues to rise as the repeat number increases.

12.6 The pathological basis of FXS

The *FMR1* gene product, FMRP, is missing or dramatically reduced in individuals with FXS. FMRP is a multifunctional RNA-binding protein that is highly expressed in brain and gonads. This protein is known to negatively regulate the translation of a subset of brain mRNAs in response to synaptic activation. In particular, FMRP acts as a brake on protein production in response to mGluR5 receptor activation [112]. There is also a decrease in GABA(A) receptor expression [113–115]. In addition to an effect on translation, FMRP plays a role in stabilizing mRNAs such as *PSD-95* [116] and regulates the transport of a variety of transcripts important for synaptogenesis and plasticity [117,118]. It is also recruited to chromatin in response to replication stress and DNA damage during gametogenesis [119]. FMRP overexpression is associated with markers of aggressive breast cancer, with increased risk of lung metastases and with melanoma invasiveness [120,121]. This may be related to FMRP's binding to *E-cadherin* and *Vimentin* mRNAs that are involved in the epithelial-to-mesenchyme transition and invasion [120]. Other FMRP targets include transcriptional activators such as BRD4 [122] and, ironically given that the proximal cause of FXS is *FMR1* gene silencing, *Fmr1* KO mice show globally elevated levels of some of the histone modifications characteristic of active genes. Suppressing these modifications reduces some FXS pathology in a mouse model [122]. Thus, the epigenetic silencing of *FMR1* has a secondary epigenetic consequence that impacts disease pathology.

12.7 Epigenetic abnormalities associated with the FXDs

Relatively little is known about the epigenetic landscape of PM alleles beyond the fact that such alleles are associated with increased levels of H3 and H4 acetylation [91]. This hyperacetylation can be reversed by treatment with histone acetyltransferase (HAT) inhibitors. Furthermore, HAT inhibitors or overexpression of histone deacetylases 3, 6 or 11 suppresses neurodegeneration in a *Drosophila* model consistent with a possible contribution of *FMR1* hyperexpression to disease pathology [91].

In differentiated cells such as fibroblasts and lymphoblastoid cells from individuals with FXS, FM alleles are transcriptionally silent and have extensive DNA methylation of the promoter and exon 1. They show lower levels of marks of active chromatin than normal alleles and are enriched for an array of different marks of repressed chromatin [63–66]. It has been suggested that silencing occurs relatively late in embryonic development after weeks of differentiation [123]. However, most FXS embryonic stem cells (ESCs) show some DNA methylation [56,124], suggesting that silencing may actually occur much earlier than this. A minority of FM alleles do not become silenced [71–76,123]. These UFM alleles generally have <400 repeats and recent work suggests that, rather than these individuals having a genetic background that enabled them to escape silencing, 400 repeats may be a more robust threshold for the definition of FM alleles than the current threshold of 200 [56,57].

Silenced alleles in patient cells can be partially reactivated by treatment with the DNA methyltransferase inhibitor, 5-aza-2′-deoxycytidine (AZA) [64,125]. Splitomicin-mediated inhibition of SIRT1, a Class III histone deacetylase that deacetylates both H3K9 and H4K16, or the overexpression of a dominant negative version of SIRT1, also reactivates silenced alleles [126]. This reactivation is seen despite the fact that DNA methylation levels remain unchanged after splitomicin treatment. AZA treatment also increases H4K16 acetylation but not acetylation of H3K9 [126]. These data suggest that H3K9 deacetylation is independent of, or precedes, DNA methylation. H4K16 deacetylation, on the other hand, likely occurs downstream of DNA methylation. However, histone deacetylase inhibitors that target Class I and Class II histone deacetylases, such as trichostatin A, do not reactivate FM alleles in the absence of AZA [64,127]. This suggests that SIRT1 may be the major histone deacetylase (HDAC) responsible for H4K16 deacetylation at the *FMR1* locus.

Repressive histone methylation marks associated with the silenced *FMR1* gene include those typical of constitutive heterochromatin, with H4K20me3 and H3K9me3 marks concentrated in the vicinity of the repeat, and those more typically associated with facultative heterochromatin, H3K27me3 and H3K9me2, that are more broadly distributed across the locus [65]. The enrichment of H4K20me3 and H3K9me3 in the vicinity of the repeat suggests that the repeat is the nucleation site for heterochromatinization. Interestingly, AZA treatment results in an increase in the levels of H3K27me3 on exon 1 of FM alleles. These levels rise as the *FMR1* transcript levels increase, and decrease as the *FMR1* transcript levels drop after AZA is withdrawn and the *FMR1* gene returns to its silenced state [128]. This increase in H3K27 trimethylation is reduced if the *FMR1* transcript is depleted by treatment with *FMR1* siRNA suggesting that the transcript is required for deposition of this repressive histone mark [128]. H3K27 trimethylation is carried out by the Polycomb Repressive Group 2 (PRC2) complex. Inhibition of EZH2, the catalytic component of PRC2, significantly delays the resilencing of the FX allele that occurs after AZA withdrawal, consistent with the idea that PRC2 and the *FMR1* transcript are integral to the silencing process [104]. A role for the *FMR1* transcript in FM gene silencing would be consistent with reports of the silencing described in neurons differentiated from FXS ESCs [123]. EZH2 inhibitors have no effect on patient cells that have not been pretreated with AZA. DNA methylation is a very stable epigenetic mark that is clonally propagated even if the initial events that trigger methylation are blocked or removed. This would mask any effect of inhibitors of steps in the silencing pathway that precede DNA methylation. If true, it would suggest that PRC2 recruitment also precedes DNA methylation in the silencing process.

Recently, Argonaute-1 (AGO1), a member of the argonaute family of small RNA-binding proteins, was also reported to be associated with the 5′ end of silenced alleles in FXS lymphoblastoid cells [129]. Furthermore, when FXS lymphoblasts were electro-fused with mouse ESCs depleted of DICER, an RNAse III enzyme, the silenced *FMR1* allele was reactivated [129]. AGO1 and DICER are important components of the RNA interference (RNAi) machinery that is involved in transcriptional and posttranscriptional silencing [130]. DICER processes double-stranded RNA (dsRNA) into 21–24 nucleotide (nt) fragments. In the transcriptional silencing pathway in model organisms, these short interfering RNAs (siRNAs) are incorporated into RNA-induced transcriptional silencing (RITS) complexes that include AGO1. The siRNAs target the RITS complexes to their genomic templates via interaction with nascent transcripts. This results in subsequent H3K9 methylation and gene silencing. A two to three fold increase in the levels of a ~21 nucleotide CGG repeat containing RNA has been reported for FX embryonic stem cells [129]. It was suggested that these small RNAs represent the key elements for an RNAi-based mechanism for silencing of FX FM alleles [129].

A PRC2-based mechanism of silencing and an RNA interference-based process are not mutually exclusive. For example, both PRC2 [131] and AGO1 [132] have been implicated in the silencing of D4Z4 repeats in human cells. However, no effect of DICER depletion on gene silencing has been seen in other studies [123]. Whether these differences are related to differences in experimental methodology or they reflect differences in the model systems used is unclear. More work is thus needed to properly understand the full complement of repressive chromatin modifiers that are associated with the silencing of FM alleles.

12.8 Resolving the repeat paradox

Any model for epigenetic regulation of *FMR1* in the FXDs has to account for the paradoxical effects of the repeats on gene expression, specifically how 55–200 repeats cause *FMR1* hyperexpression while >200 repeats result in *FMR1* gene silencing. As mentioned earlier, R-loops are known to form at the 5′ end of normal and PM alleles, as well as FM alleles that have been reactivated with AZA [102–104]. Similar R-loops have been implicated in heterochromatin formation in Friedreich ataxia, a disorder resulting from the presence of long GAA tracts in the first intron of the *Frataxin* gene [102]. R-loops are an appealing way of explaining the dependence of FM silencing on the *FMR1* transcript [104,123,128]. It is also possible that an R-loop accounts for hyperexpression of PM alleles.

R-loops can have paradoxical effects. They are a characteristic feature of active, unmethylated CpG-island promoters [133] as well as sites of transcription termination [134,135], gene repression, and heterochromatin formation [136–138]. Why some R-loops are associated with gene activation and others with gene repression is unknown. One possibility is that this difference is related to the stability and structure of the R-loop. At the *FMR1* locus, the R-loops associated with PM alleles would be more stable than those formed on normal alleles and less stable than those formed on FM alleles. By analogy with events considered to be involved at active CpG promoters, the R-loop may facilitate PM hyperexpression by decreasing nucleosome occupancy and inhibiting *de novo* DNA methylation by DNA methyltransferase DNMT3B [133,139]. Promoter proximal R-loops are known to recruit the Tip60-p400 histone acetyltransferase complex [140]. The single-stranded DNA in the R-loop may also recruit single-stranded binding proteins including members of the AID/APOBEC family of cytosine deaminases proposed to cause active DNA demethylation [137,141,142], and histone methyltransferases, that are responsible for the deposition of H3K4me3 [143], a histone mark associated with active transcription. It has also been suggested that promoter-associated transcripts compete with the promoter for binding to the maintenance DNA methyltransferase, DNMT1, thereby contributing to reduced levels of DNA methylation [144]. A similar competition between mRNA and DNA for PRC2 binding has also been suggested [145]. The RNA would be acting as a decoy as long as it was in the R-loop and after the R-loop was resolved, the repressive epigenetic modifiers would be swept away along with the transcript [144].

In contrast, the more stable R-loops formed by FM alleles might have properties that are more likely to lead to gene silencing. Longer CGG-repeat tracts are more likely to form intrastrand DNA structures than shorter ones [146–149]. As a result, the free DNA strand in the R-loop formed on FM alleles would have less single-stranded character [103]. Thus, the R-loop formed on active FM alleles may not bind some gene activators as well as smaller R-loops. In addition, the residence time in the vicinity of the *FMR1* promoter of mRNA-bound repressive epigenetic complexes, such as PRC2, would be increased. This would increase the likelihood that the deposition of H3K27me3 and other repressive histone modifications would occur. Longer R-loops are likely to be associated with increased antisense transcription [136]. This could allow a critical amount of dsRNA to be produced for entry into an RNAi-based silencing pathway. The longer repeat tracts in transcripts produced from FM alleles would also produce longer, more stable RNA hairpins than smaller alleles [150,151]. As these hairpins are DICER substrates [150], they may also contribute to the pool of dsRNA produced from the *FMR1* locus. Some combination of these processes could result in the creation of a suitable binding platform for the recruitment of other chromatin modifiers, such as the histone deacetylases, H3K9 methyltransferases, and de novo DNA methyltransferases necessary for downstream events in the gene-silencing process.

However, it should be noted that while a number of laboratories have detected R-loops on normal, PM and active FM alleles in dermal fibroblasts and lymphoblastoid cells using DNA:RNA immunoprecipitation assays [102–104], no promoter-associated transcript was seen on normal or PM alleles in human ESCs using a chromatin isolation by RNA purification (ChIRP) assay [123]. Even the transcript associated with active FM promoters was limited to a narrow developmental window ~45 days after the initiation of neuronal differentiation in these cell lines [123]. It is possible that the apparent absence of a promoter-associated transcript in the ESC studies reflects the relative differences in R-loop stability with only the most stable or long-lived R-loops being detected with the ChIRP assay [123].

12.9 Prospects and challenges for epigenetic therapies for the FXDs

Normalizing the epigenetic dysregulation seen in the FXDs has the potential to ameliorate disease symptoms. Although this approach has its problems, the difficulty of compensating for the loss of FMRP given its many crucial cellular roles, make gene reactivation strategies worth consideration. The modulation of key chromatin modifiers using small molecule activators or inhibitors is the simplest approach. For example, HAT inhibitors reduce *FMR1* hyperexpression in cells from FXTAS patients and extend the lifespan of Drosophila-expressing long CGG-repeat tracts [91]. The use of small molecules that target chromatin modifiers globally might be expected to have deleterious consequences due to epigenetic effects elsewhere in the genome. Nonetheless, DNA methyltransferase inhibitors and histone deacetylase inhibitors are in use or have shown efficacy in preclinical studies and/or clinical trials for a variety of other disorders resulting from aberrant epigenetic modifications. These disorders include myelodysplastic syndrome and two other diseases resulting from repeat-mediated epigenetic changes, Huntington disease (HD) and Friedreich ataxia (FRDA) [152–156]. For example, a SIRT1 inhibitor, selisistat, has been shown to be well tolerated in a short-term study in HD [157]. However, whether a similar pharmacological approach, with potentially global effects on gene expression, would ever be acceptable for long-term use in the treatment of nonfatal illnesses is unclear. Furthermore, given that the loss of FMRP is associated with hyperexpression of many chromatin activators [122], nonspecific treatment with drugs that upregulate such activators or inhibition of enzymes that offset their effects may be particularly problematic in FXS.

Ideally, a more targeted gene-specific approach would be desirable for the treatment of the FXDs. This could involve small molecules or modified oligonucleotides that interact specifically with the *FMR1* transcript or locus. For example, NSC311153 (9-hydroxy-5,11-dimethyl-2-(2-(piperidin-1-yl) ethyl)-6H-pyrido[4,3-b]carbazol-2-ium, aka Compound 1a), was identified as a CGG-binding ligand that is able to disrupt protein binding to CGG-RNA [158]. It is able to prevent silencing in cells in which DNA methylation has not yet taken place and to delay the resilencing of *FMR1* after AZA treatment [104,123], perhaps by interfering with PRC2 recruitment to the *FMR1* locus [104]. Thus, small molecules, such as 1a, that block recruitment of chromatin repressors in a sequence-specific manner, have the potential to be therapeutically useful.

The successful outcome of the Phase III trial of Nusinersen, an antisense oligonucleotide for the treatment of spinal muscular atrophy, and the preclinical and early Phase I/IIa results for antisense oligonucleotides in the treatment of HD [159–162], suggest that oligonucleotide-based approaches may be clinically useful even in disorders of the central nervous system. Our current understanding of the repeat-mediated epigenetic events suggest a number of places where oligonucleotide-based approaches may be fruitful. For example, reducing R-loop formation may help normalize *FMR1* transcription in PM carriers and by analogy with FRDA [102], prevent or reduce gene silencing in FM carriers. It may be possible to reduce R-loop formation using modified oligonucleotides that compete with the template strand for binding to the transcript. Such an approach using locked nucleic acids complementary to the transcript has been used to increase gene expression in FRDA cells [163]. Any approach aimed at reducing PM transcription would have to take into consideration the fact that although the *FMR1* gene in PM carriers is hyperexpressed, FMRP levels are not elevated. However, as FMRP levels are very variable in the normal population, this may not be a serious issue. Approaches aimed at gene reactivation would also need to overcome a number of additional obstacles. First, data suggest that once silencing had occurred, prior treatment with a DNA methylation inhibitor will be required for reactivation [104,128]; second, *FMR1* transcripts with large CGG-repeat numbers are often poorly translated [79,80]; and third, expression of high levels of *FMR1* mRNA with large repeat numbers is deleterious [89,90,164,165]. Although deleterious symptoms are only seen relatively late in life, a strategy that combines gene reactivation with improved FMRP translation and reduced RAN translation may be the best approach.

The recent discovery that a number of FMRP targets are chromatin activators [122] creates an exciting new avenue for epigenetic treatments for FXS symptoms. JQ1, an inhibitor of the bromo- and extraterminal domain proteins including the FMRP target BRD4, reversed some of the transcriptional changes seen in *Fmr1* KO mice, reduced the abnormal spine density in hippocampal neurons and improved some of the behavioral abnormalities seen in these animals [122]. However, this work also demonstrated that tight control of transcription is essential in neurons. Encouragingly, a combination of low doses of JQ1 and an inhibitor of Casein Kinase 2 (Ck2), a protein that phosphorylates BRD4, successfully reversed some phenotypes without any observable side effects [122]. However, as interventions that are effective in KO mice have not, thus far, translated well to improvements in the clinic [166–169], and as BRD4 is just one of many chromatin modifiers that are misregulated in FXS [122], it remains to be seen whether this approach will be effective in the treatment of FXS.

12.10 Concluding remarks

The FXDs provide a number of examples of ways in which large tandem-repeat tracts can have epigenetic consequences for human health. However, many questions related to the exact mechanisms involved remain unresolved, including the timing of gene silencing, the role of AGO1 and DICER in this process, and the contribution of R-loops to hyperexpression of PM alleles and/or the silencing of FM alleles. Progress in our understanding of the molecular basis of these epigenetic changes, along with work in other epigenetic disorders, suggests that the reversal of these changes may ultimately be clinically useful. However, much work is still necessary to assess whether such epigenetic approaches can be refined to the point that off-target effects are minimized while still maintaining clinical efficacy.

Grant Sponsor

Intramural program of the NIDDK, NIH (DK057808).

Conflict of Interest

The authors have no conflict of interest.

References

[1] Allis CD, Jenuwein T. The molecular hallmarks of epigenetic control. Nat Rev Genet 2016;17(8):487–500.
[2] Allshire RC, Madhani HD. Ten principles of heterochromatin formation and function. Nat Rev Mol Cell Biol 2018;19(4):229–44.
[3] Martin JP, Bell J. A pedigree of mental defect showing sex-linkage. J Neurol Psychiatr 1943;6(3–4):154–7.
[4] Lubs HA. A marker X chromosome. Am J Hum Genet 1969;21(3):231–44.
[5] Proops R, Webb T. The 'fragile' X chromosome in the Martin-Bell-Renpenning syndrome and in males with other forms of familial mental retardation. J Med Genet 1981;18(5):366–73.
[6] Coffee B, Keith K, Albizua I, Malone T, Mowrey J, Sherman SL, Warren ST. Incidence of fragile X syndrome by newborn screening for methylated FMR1 DNA. Am J Hum Genet 2009;85(4):503–14.
[7] Hessl D, Nguyen DV, Green C, Chavez A, Tassone F, Hagerman RJ, Senturk D, Schneider A, Lightbody A, Reiss AL, Hall S. A solution to limitations of cognitive testing in children with intellectual disabilities: the case of fragile X syndrome. J Neurodev Disord 2009;1(1):33–45.
[8] Abbeduto L, Brady N, Kover ST. Language development and fragile X syndrome: profiles, syndrome-specificity, and within-syndrome differences. Ment Retard Dev Disabil Res Rev 2007;13(1):36–46.

[9] Kaufmann WE, Kidd SA, Andrews HF, Budimirovic DB, Esler A, Haas-Givler B, Stackhouse T, Riley C, Peacock G, Sherman SL, Brown WT, Berry-Kravis E. Autism spectrum disorder in fragile X syndrome: cooccurring conditions and current treatment. Pediatrics 2017;139(Suppl. 3):S194–206.
[10] Berry-Kravis E, Raspa M, Loggin-Hester L, Bishop E, Holiday D, Bailey DB. Seizures in fragile X syndrome: characteristics and comorbid diagnoses. Am J Intellect Dev Disabil 2010;115(6):461–72.
[11] Hagerman RJ, Altshul-Stark D, McBogg P. Recurrent otitis media in the fragile X syndrome. Am J Dis Child 1987;141(2):184–7.
[12] Kidd SA, Lachiewicz A, Barbouth D, Blitz RK, Delahunty C, McBrien D, Visootsak J, Berry-Kravis E. Fragile X syndrome: a review of associated medical problems. Pediatrics 2014;134(5):995–1005.
[13] Hagerman P. Fragile X-associated tremor/ataxia syndrome (FXTAS): pathology and mechanisms. Acta Neuropathol 2013;126(1):1–19.
[14] Berry-Kravis E, Abrams L, Coffey SM, Hall DA, Greco C, Gane LW, Grigsby J, Bourgeois JA, Finucane B, Jacquemont S, Brunberg JA, Zhang L, Lin J, Tassone F, Hagerman PJ, Hagerman RJ, Leehey MA. Fragile X-associated tremor/ataxia syndrome: clinical features, genetics, and testing guidelines. Mov Disord 2007;22(14):2018–30. quiz 2140.
[15] Bacalman S, Farzin F, Bourgeois JA, Cogswell J, Goodlin-Jones BL, Gane LW, Grigsby J, Leehey MA, Tassone F, Hagerman RJ. Psychiatric phenotype of the fragile X-associated tremor/ataxia syndrome (FXTAS) in males: newly described fronto-subcortical dementia. J Clin Psychiatr 2006;67(1):87–94.
[16] Bourgeois JA, Seritan AL, Casillas EM, Hessl D, Schneider A, Yang Y, Kaur I, Cogswell JB, Nguyen DV, Hagerman RJ. Lifetime prevalence of mood and anxiety disorders in fragile X premutation carriers. J Clin Psychiatr 2011;72(2):175–82.
[17] Jacquemont S, Hagerman RJ, Leehey M, Grigsby J, Zhang L, Brunberg JA, Greco C, Des Portes V, Jardini T, Levine R, Berry-Kravis E, Brown WT, Schaeffer S, Kissel J, Tassone F, Hagerman PJ. Fragile X premutation tremor/ataxia syndrome: molecular, clinical, and neuroimaging correlates. Am J Hum Genet 2003;72(4):869–78.
[18] Cohen S, Masyn K, Adams J, Hessl D, Rivera S, Tassone F, Brunberg J, DeCarli C, Zhang L, Cogswell J, Loesch D, Leehey M, Grigsby J, Hagerman PJ, Hagerman R. Molecular and imaging correlates of the fragile X-associated tremor/ataxia syndrome. Neurology 2006;67(8):1426–31.
[19] Greco CM, Hagerman RJ, Tassone F, Chudley AE, Del Bigio MR, Jacquemont S, Leehey M, Hagerman PJ. Neuronal intranuclear inclusions in a new cerebellar tremor/ataxia syndrome among fragile X carriers. Brain 2002;125(Pt 8):1760–71.
[20] Iwahashi CK, Yasui DH, An HJ, Greco CM, Tassone F, Nannen K, Babineau B, Lebrilla CB, Hagerman RJ, Hagerman PJ. Protein composition of the intranuclear inclusions of FXTAS. Brain 2006;129(Pt 1):256–71.
[21] Hoem G, Raske CR, Garcia-Arocena D, Tassone F, Sanchez E, Ludwig AL, Iwahashi CK, Kumar M, Yang JE, Hagerman PJ. CGG-repeat length threshold for FMR1 RNA pathogenesis in a cellular model for FXTAS. Hum Mol Genet 2011;20(11):2161–70.
[22] Garcia-Arocena D, Yang JE, Brouwer JR, Tassone F, Iwahashi C, Berry-Kravis EM, Goetz CG, Sumis AM, Zhou L, Nguyen DV, Campos L, Howell E, Ludwig A, Greco C, Willemsen R, Hagerman RJ, Hagerman PJ. Fibroblast phenotype in male carriers of FMR1 premutation alleles. Hum Mol Genet 2010;19(2):299–312.
[23] Robin G, Lopez JR, Espinal GM, Hulsizer S, Hagerman PJ, Pessah IN. Calcium dysregulation and Cdk5-ATM pathway involved in a mouse model of fragile X-associated tremor/ataxia syndrome. Hum Mol Genet 2017;26(14):2649–66.
[24] Ross-Inta C, Omanska-Klusek A, Wong S, Barrow C, Garcia-Arocena D, Iwahashi C, Berry-Kravis E, Hagerman RJ, Hagerman PJ, Giulivi C. Evidence of mitochondrial dysfunction in fragile X-associated tremor/ataxia syndrome. Biochem J 2010;429(3):545–52.

[25] Hukema RK, Buijsen RA, Raske C, Severijnen LA, Nieuwenhuizen-Bakker I, Minneboo M, Maas A, de Crom R, Kros JM, Hagerman PJ, Berman RF, Willemsen R. Induced expression of expanded CGG RNA causes mitochondrial dysfunction in vivo. Cell Cycle 2014;13(16):2600–8.

[26] Conca Dioguardi C, Uslu B, Haynes M, Kurus M, Gul M, Miao DQ, De Santis L, Ferrari M, Bellone S, Santin A, Giulivi C, Hoffman G, Usdin K, Johnson J. Granulosa cell and oocyte mitochondrial abnormalities in a mouse model of fragile X primary ovarian insufficiency. Mol Hum Reprod 2016;22(6):384–96.

[27] Rohr J, Allen EG, Charen K, Giles J, He W, Dominguez C, Sherman SL. Anti-Mullerian hormone indicates early ovarian decline in fragile X mental retardation (FMR1) premutation carriers: a preliminary study. Hum Reprod 2008;23(5):1220–5.

[28] Welt CK, Smith PC, Taylor AE. Evidence of early ovarian aging in fragile X premutation carriers. J Clin Endocrinol Metab 2004;89(9):4569–74.

[29] Wheeler AC, Raspa M, Green A, Bishop E, Bann C, Edwards A, Bailey Jr DB. Health and reproductive experiences of women with an FMR1 premutation with and without fragile X premature ovarian insufficiency. Front Genet 2014;5:300.

[30] Bibi G, Malcov M, Yuval Y, Reches A, Ben-Yosef D, Almog B, Amit A, Azem F. The effect of CGG repeat number on ovarian response among fragile X premutation carriers undergoing preimplantation genetic diagnosis. Fertil Steril 2010;94(3):869–74.

[31] De Caro JJ, Dominguez C, Sherman SL. Reproductive health of adolescent girls who carry the FMR1 premutation: expected phenotype based on current knowledge of fragile x-associated primary ovarian insufficiency. Ann N Y Acad Sci 2008;1135:99–111.

[32] Crawford DC, Acuna JM, Sherman SL. FMR1 and the fragile X syndrome: human genome epidemiology review. Genet Med 2001;3(5):359–71.

[33] Cronister A, Teicher J, Rohlfs EM, Donnenfeld A, Hallam S. Prevalence and instability of fragile X alleles: implications for offering fragile X prenatal diagnosis, Obstet. Gynecology 2008;111(3):596–601.

[34] Sullivan SD, Welt C, Sherman S. FMR1 and the continuum of primary ovarian insufficiency. Semin Reprod Med 2011;29(4):299–307.

[35] Allen EG, Grus WE, Narayan S, Espinel W, Sherman SL. Approaches to identify genetic variants that influence the risk for onset of fragile X-associated primary ovarian insufficiency (FXPOI): a preliminary study. Front Genet 2014;5:260.

[36] Rinn JL, Chang HY. Genome regulation by long noncoding RNAs. Annu Rev Biochem 2012;81:145–66.

[37] Buijsen RA, Visser JA, Kramer P, Severijnen EA, Gearing M, Charlet-Berguerand N, Sherman SL, Berman RF, Willemsen R, Hukema RK. Presence of inclusions positive for polyglycine containing protein, FMRpolyG, indicates that repeat-associated non-AUG translation plays a role in fragile X-associated primary ovarian insufficiency. Hum Reprod 2016;31(1):158–68.

[38] Hoffman GE, Le W, Entezam A, Otsuka N, Tong Z-B, Nelson L, Flaws JA, McDonald JH, Jafar S, Usdin K. Ovarian abnormalities in a mouse model of Fragile X-associated primary ovarian insufficiency. J Histochem Cytochem 2012;60(6):439–56.

[39] Lu C, Lin L, Tan H, Wu H, Sherman SL, Gao F, Jin P, Chen D. Fragile X premutation RNA is sufficient to cause primary ovarian insufficiency in mice. Hum Mol Genet 2012;21(23):5039–47.

[40] Karimov CB, Moragianni VA, Cronister A, Srouji S, Petrozza J, Racowsky C, Ginsburg E, Thornton KL, Welt CK. Increased frequency of occult fragile X-associated primary ovarian insufficiency in infertile women with evidence of impaired ovarian function. Hum Reprod 2011;26(8):2077–83.

[41] Madrigal I, Xuncla M, Tejada MI, Martinez F, Fernandez-Carvajal I, Perez-Jurado LA, Rodriguez-Revenga L, Mila M. Intermediate FMR1 alleles and cognitive and/or behavioural phenotypes. Eur J Hum Genet 2011;19(8):921–3.

[42] Voorhuis M, Onland-Moret NC, Janse F, Ploos van Amstel HK, Goverde AJ, Lambalk CB, Laven JS, van der Schouw YT, Broekmans FJ, Fauser BC, C. Dutch Primary Ovarian Insufficiency. The significance of

fragile X mental retardation gene 1 CGG repeat sizes in the normal and intermediate range in women with primary ovarian insufficiency. Hum Reprod 2014;29(7):1585–93.
[43] Debrey SM, Leehey MA, Klepitskaya O, Filley CM, Shah RC, Kluger B, Berry-Kravis E, Spector E, Tassone F, Hall DA. Clinical phenotype of adult fragile X gray zone allele carriers: a case series. Cerebellum 2016;15(5):623–31.
[44] Hall D, Tassone F, Klepitskaya O, Leehey M. Fragile X-associated tremor ataxia syndrome in FMR1 gray zone allele carriers. Mov Disord 2012;27(2):296–300.
[45] Loesch D, Hagerman R. Unstable mutations in the FMR1 gene and the phenotypes. Adv Exp Med Biol 2012;769:78–114.
[46] Spath MA, Feuth TB, Smits AP, Yntema HG, Braat DD, Thomas CM, van Kessel AG, Sherman SL, Allen EG. Predictors and risk model development for menopausal age in fragile X premutation carriers. Genet Med 2011;13(7):643–50.
[47] Bennett CE, Conway GS, Macpherson JN, Jacobs PA, Murray A. Intermediate sized CGG repeats are not a common cause of idiopathic premature ovarian failure. Hum Reprod 2010;25(5):1335–8.
[48] Tassone F, Iong KP, Tong TH, Lo J, Gane LW, Berry-Kravis E, Nguyen D, Mu LY, Laffin J, Bailey DB, Hagerman RJ. FMR1 CGG allele size and prevalence ascertained through newborn screening in the United States. Genome Med 2012;4(12):100.
[49] Jacquemont S, Leehey MA, Hagerman RJ, Beckett LA, Hagerman PJ. Size bias of fragile X premutation alleles in late-onset movement disorders. J Med Genet 2006;43(10):804–9.
[50] Leehey MA, Berry-Kravis E, Goetz CG, Zhang L, Hall DA, Li L, Rice CD, Lara R, Cogswell J, Reynolds A, Gane L, Jacquemont S, Tassone F, Grigsby J, Hagerman RJ, Hagerman PJ. FMR1 CGG repeat length predicts motor dysfunction in premutation carriers. Neurology 2008;70(16 Pt 2):1397–402.
[51] Ennis S, Ward D, Murray A. Nonlinear association between CGG repeat number and age of menopause in FMR1 premutation carriers. Eur J Hum Genet 2006;14(2):253–5.
[52] Allen EG, Sullivan AK, Marcus M, Small C, Dominguez C, Epstein MP, Charen K, He W, Taylor KC, Sherman SL. Examination of reproductive aging milestones among women who carry the FMR1 premutation. Hum Reprod 2007;22(8):2142–52.
[53] Mailick MR, Hong J, Greenberg J, Smith L, Sherman S. Curvilinear association of CGG repeats and age at menopause in women with FMR1 premutation expansions. Am J Med Genet B Neuropsychiatr Genet 2014; 165B(8):705–11.
[54] Tejada MI, Garcia-Alegria E, Bilbao A, Martinez-Bouzas C, Beristain E, Poch M, Ramos-Arroyo MA, Lopez B, Fernandez Carvajal I, Ribate MP, Ramos F. Analysis of the molecular parameters that could predict the risk of manifesting premature ovarian failure in female premutation carriers of fragile X syndrome. Menopause 2008;15(5):945–9.
[55] Sullivan AK, Marcus M, Epstein MP, Allen EG, Anido AE, Paquin JJ, Yadav-Shah M, Sherman SL. Association of FMR1 repeat size with ovarian dysfunction. Hum Reprod 2005;20(2):402–12.
[56] Zhou Y, Kumari D, Sciascia N, Usdin K. CGG-repeat dynamics and FMR1 gene silencing in fragile X syndrome stem cells and stem cell-derived neurons. Mol Autism 2016;7:42.
[57] Brykczynska U, Pecho-Vrieseling E, Thiemeyer A, Klein J, Fruh I, Doll T, Manneville C, Fuchs S, Iazeolla M, Beibel M, Roma G, Naumann U, Kelley N, Oakeley EJ, Mueller M, Gomez-Mancilla B, Buhler M, Tabolacci E, Chiurazzi P, Neri G, Bouwmeester T, Di Giorgio FP, Fodor BD. CGG repeat-induced FMR1 silencing depends on the expansion size in human iPSCs and neurons carrying unmethylated full mutations. Stem Cell Reports 2016;7(6):1059–71.
[58] Tassone F, Hagerman RJ, Taylor AK, Gane LW, Godfrey TE, Hagerman PJ. Elevated levels of FMR1 mRNA in carrier males: a new mechanism of involvement in the fragile-X syndrome. Am J Hum Genet 2000;66(1):6–15.
[59] Tassone F, Beilina A, Carosi C, Albertosi S, Bagni C, Li L, Glover K, Bentley D, Hagerman PJ. Elevated FMR1 mRNA in premutation carriers is due to increased transcription. RNA 2007;13(4):555–62.

[60] Pieretti M, Zhang FP, Fu YH, Warren ST, Oostra BA, Caskey CT, Nelson DL. Absence of expression of the FMR-1 gene in fragile X syndrome. Cell 1991;66(4):817–22.

[61] Sutcliffe JS, Nelson DL, Zhang F, Pieretti M, Caskey CT, Saxe D, Warren ST. DNA methylation represses FMR-1 transcription in fragile X syndrome. Hum Mol Genet 1992;1(6):397–400.

[62] Oberle I, Rousseau F, Heitz D, Kretz C, Devys D, Hanauer A, Boue J, Bertheas MF, Mandel JL. Instability of a 550-base pair DNA segment and abnormal methylation in fragile X syndrome. Science 1991; 252(5009):1097–102.

[63] Coffee B, Zhang F, Ceman S, Warren ST, Reines D. Histone modifications depict an aberrantly heterochromatinized FMR1 gene in fragile x syndrome. Am J Hum Genet 2002;71(4):923–32.

[64] Coffee B, Zhang F, Warren ST, Reines D. Acetylated histones are associated with FMR1 in normal but not fragile X-syndrome cells. Nat Genet 1999;22(1):98–101.

[65] Kumari D, Usdin K. The distribution of repressive histone modifications on silenced FMR1 alleles provides clues to the mechanism of gene silencing in fragile X syndrome. Hum Mol Genet 2010;19(23):4634–42.

[66] Pietrobono R, Tabolacci E, Zalfa F, Zito I, Terracciano A, Moscato U, Bagni C, Oostra B, Chiurazzi P, Neri G. Molecular dissection of the events leading to inactivation of the FMR1 gene. Hum Mol Genet 2005; 14(2):267–77.

[67] Handt M, Epplen A, Hoffjan S, Mese K, Epplen JT, Dekomien G. Point mutation frequency in the FMR1 gene as revealed by fragile X syndrome screening. Mol Cell Probes 2014;28(5–6):279–83.

[68] Gronskov K, Brondum-Nielsen K, Dedic A, Hjalgrim H. A nonsense mutation in FMR1 causing fragile X syndrome. Eur J Hum Genet 2011;19(4):489–91.

[69] De Boulle K, Verkerk AJ, Reyniers E, Vits L, Hendrickx J, Van Roy B, Van den Bos F, de Graaff E, Oostra BA, Willems PJ. A point mutation in the FMR-1 gene associated with fragile X mental retardation. Nat Genet 1993;3(1):31–5.

[70] Tassone F, Hagerman RJ, Taylor AK, Hagerman PJ. A majority of fragile X males with methylated, full mutation alleles have significant levels of FMR1 messenger RNA. J Med Genet 2001;38(7):453–6.

[71] Basuta K, Schneider A, Gane L, Polussa J, Woodruff B, Pretto D, Hagerman R, Tassone F. High functioning male with fragile X syndrome and fragile X-associated tremor/ataxia syndrome. Am J Med Genet A 2015; 167A(9):2154–61.

[72] Hagerman RJ, Hull CE, Safanda JF, Carpenter I, Staley LW, O'Connor RA, Seydel C, Mazzocco MM, Snow K, Thibodeau SN, et al. High functioning fragile X males: demonstration of an unmethylated fully expanded FMR-1 mutation associated with protein expression. Am J Med Genet 1994;51(4):298–308.

[73] Wohrle D, Salat U, Glaser D, Mucke J, Meisel-Stosiek M, Schindler D, Vogel W, Steinbach P. Unusual mutations in high functioning fragile X males: apparent instability of expanded unmethylated CGG repeats. J Med Genet 1998;35(2):103–11.

[74] Tabolacci E, Moscato U, Zalfa F, Bagni C, Chiurazzi P, Neri G. Epigenetic analysis reveals a euchromatic configuration in the FMR1 unmethylated full mutations. Eur J Hum Genet 2008;16(12):1487–98.

[75] Loesch DZ, Sherwell S, Kinsella G, Tassone F, Taylor A, Amor D, Sung S, Evans A. Fragile X-associated tremor/ataxia phenotype in a male carrier of unmethylated full mutation in the FMR1 gene. Clin Genet 2012;82(1):88–92.

[76] Santa Maria L, Pugin A, Alliende MA, Aliaga S, Curotto B, Aravena T, Tang HT, Mendoza-Morales G, Hagerman R, Tassone F. FXTAS in an unmethylated mosaic male with fragile X syndrome from Chile. Clin Genet 2014;86(4):378–82.

[77] Basuta K, Narcisa V, Chavez A, Kumar M, Gane L, Hagerman R, Tassone F. Clinical phenotypes of a juvenile sibling pair carrying the fragile X premutation. Am J Med Genet A 2011;155A(3):519–25.

[78] Pretto DI, Mendoza-Morales G, Lo J, Cao R, Hadd A, Latham GJ, Durbin-Johnson B, Hagerman R, Tassone F. CGG allele size somatic mosaicism and methylation in FMR1 premutation alleles. J Med Genet 2014;51(5):309–18.

[79] Feng Y, Zhang F, Lokey LK, Chastain JL, Lakkis L, Eberhart D, Warren ST. Translational suppression by trinucleotide repeat expansion at FMR1. Science 1995;268(5211):731–4.
[80] Primerano B, Tassone F, Hagerman RJ, Hagerman P, Amaldi F, Bagni C. Reduced FMR1 mRNA translation efficiency in fragile X patients with premutations. RNA 2002;8(12):1482–8.
[81] Nolin SL, Glicksman A, Houck Jr GE, Brown WT, Dobkin CS. Mosaicism in fragile X affected males. Am J Med Genet 1994;51(4):509–12.
[82] Lozano R, Rosero CA, Hagerman RJ. Fragile X spectrum disorders. Intractable Rare Dis Res 2014;3(4): 134–46.
[83] Hagerman PJ, Hagerman RJ. Fragile X-associated tremor/ataxia syndrome (FXTAS). Ment Retard Dev Disabil Res Rev 2004;10(1):25–30.
[84] Mankodi A, Urbinati CR, Yuan QP, Moxley RT, Sansone V, Krym M, Henderson D, Schalling M, Swanson MS, Thornton CA. Muscleblind localizes to nuclear foci of aberrant RNA in myotonic dystrophy types 1 and 2. Hum Mol Genet 2001;10(19):2165–70.
[85] Sellier C, Freyermuth F, Tabet R, Tran T, He F, Ruffenach F, Alunni V, Moine H, Thibault C, Page A, Tassone F, Willemsen R, Disney MD, Hagerman PJ, Todd PK, Charlet-Berguerand N. Sequestration of DROSHA and DGCR8 by expanded CGG RNA repeats alters microRNA processing in fragile X-associated tremor/ataxia syndrome. Cell Rep 2013;3(3):869–80.
[86] Sofola OA, Jin P, Qin Y, Duan R, Liu H, de Haro M, Nelson DL, Botas J. RNA-binding proteins hnRNP A2/B1 and CUGBP1 suppress fragile X CGG premutation repeat-induced neurodegeneration in a Drosophila model of FXTAS. Neuron 2007;55(4):565–71.
[87] Jin P, Duan R, Qurashi A, Qin Y, Tian D, Rosser TC, Liu H, Feng Y, Warren ST. Pur alpha binds to rCGG repeats and modulates repeat-mediated neurodegeneration in a Drosophila model of fragile X tremor/ataxia syndrome. Neuron 2007;55(4):556–64.
[88] Sellier C, Rau F, Liu Y, Tassone F, Hukema RK, Gattoni R, Schneider A, Richard S, Willemsen R, Elliott DJ, Hagerman PJ, Charlet-Berguerand N. Sam68 sequestration and partial loss of function are associated with splicing alterations in FXTAS patients. EMBO J 2010;29(7):1248–61.
[89] Todd PK, Oh SY, Krans A, He F, Sellier C, Frazer M, Renoux AJ, Chen KC, Scaglione KM, Basrur V, Elenitoba-Johnson K, Vonsattel JP, Louis ED, Sutton MA, Taylor JP, Mills RE, Charlet-Berguerand N, Paulson HL. CGG repeat-associated translation mediates neurodegeneration in fragile X tremor ataxia syndrome. Neuron 2013;78(3):440–55.
[90] Sellier C, Buijsen RA, He F, Natla S, Jung L, Tropel P, Gaucherot A, Jacobs H, Meziane H, Vincent A, Champy MF, Sorg T, Pavlovic G, Wattenhofer-Donze M, Birling MC, Oulad-Abdelghani M, Eberling P, Ruffenach F, Joint M, Anheim M, Martinez-Cerdeno V, Tassone F, Willemsen R, Hukema RK, Viville S, Martinat C, Todd PK, Charlet-Berguerand N. Translation of expanded CGG repeats into FMRpolyG is pathogenic and may contribute to fragile X tremor ataxia syndrome. Neuron 2017;93(2):331–47.
[91] Todd PK, Oh SY, Krans A, Pandey UB, Di Prospero NA, Min KT, Taylor JP, Paulson HL. Histone deacetylases suppress CGG repeat-induced neurodegeneration via transcriptional silencing in models of fragile X tremor ataxia syndrome. PLoS Genet 2010;6(12):e1001240.
[92] Hagerman RJ, Hagerman P. Fragile X-associated tremor/ataxia syndrome - features, mechanisms and management. Nat Rev Neurol 2016;12(7):403–12.
[93] Alvarez-Mora MI, Rodriguez-Revenga L, Madrigal I, Torres-Silva F, Mateu-Huertas E, Lizano E, Friedlander MR, Marti E, Estivill X, Mila M. MicroRNA expression profiling in blood from fragile X-associated tremor/ataxia syndrome patients. Genes Brain Behav 2013;12(6):595–603.
[94] Tan H, Poidevin M, Li H, Chen D, Jin P. MicroRNA-277 modulates the neurodegeneration caused by Fragile X premutation rCGG repeats. PLoS Genet 2012;8(5):e1002681.
[95] Muslimov IA, Patel MV, Rose A, Tiedge H. Spatial code recognition in neuronal RNA targeting: role of RNA-hnRNP A2 interactions. J Cell Biol 2011;194(3):441–57.

[96] Tan H, Qurashi A, Poidevin M, Nelson DL, Li H, Jin P. Retrotransposon activation contributes to fragile X premutation rCGG-mediated neurodegeneration. Hum Mol Genet 2012;21(1):57–65.

[97] He F, Krans A, Freibaum BD, Taylor JP, Todd PK. TDP-43 suppresses CGG repeat-induced neurotoxicity through interactions with HnRNP A2/B1. Hum Mol Genet 2014;23(19):5036–51.

[98] Qurashi A, Li W, Zhou JY, Peng J, Jin P. Nuclear accumulation of stress response mRNAs contributes to the neurodegeneration caused by Fragile X premutation rCGG repeats. PLoS Genet 2011;7(6):e1002102.

[99] Cleary JD, Ranum LP. Repeat-associated non-ATG (RAN) translation in neurological disease. Hum Mol Genet 2013;22(R1):R45–51.

[100] Krans A, Kearse MG, Todd PK. Repeat-associated non-AUG translation from antisense CCG repeats in fragile X tremor/ataxia syndrome. Ann Neurol 2016;80(6):871–81.

[101] Oh SY, He F, Krans A, Frazer M, Taylor JP, Paulson HL, Todd PK. RAN translation at CGG repeats induces ubiquitin proteasome system impairment in models of fragile X-associated tremor ataxia syndrome. Hum Mol Genet 2015;24(15):4317–26.

[102] Groh M, Lufino MM, Wade-Martins R, Gromak N. R-loops associated with triplet repeat expansions promote gene silencing in Friedreich ataxia and fragile X syndrome. PLoS Genet 2014;10(5):e1004318.

[103] Loomis EW, Sanz LA, Chedin F, Hagerman PJ. Transcription-associated R-loop formation across the human FMR1 CGG-repeat region. PLoS Genet 2014;10(4):e1004294.

[104] Kumari D, Usdin K. Sustained expression of FMR1 mRNA from reactivated fragile X syndrome alleles after treatment with small molecules that prevent trimethylation of H3K27. Hum Mol Genet 2016;25(17): 3689–98.

[105] Aguilera A, Garcia-Muse T. R loops: from transcription byproducts to threats to genome stability. Mol Cell 2012;46(2):115–24.

[106] Entezam A, Biacsi R, Orrison B, Saha T, Hoffman GE, Grabczyk E, Nussbaum RL, Usdin K. Regional FMRP deficits and large repeat expansions into the full mutation range in a new Fragile X premutation mouse model. Gene 2007;395(1–2):125–34.

[107] Yu J, Yaba A, Kasiman C, Thomson T, Johnson J. mTOR controls ovarian follicle growth by regulating granulosa cell proliferation. PLoS One 2011;6(7):e21415.

[108] Lin Y, Tang C, He H, Duan R. Activation of mTOR ameliorates fragile X premutation rCGG repeat-mediated neurodegeneration. PLoS One 2013;8(4):e62572.

[109] Sharma A, Hoeffer CA, Takayasu Y, Miyawaki T, McBride SM, Klann E, Zukin RS. Dysregulation of mTOR signaling in fragile X syndrome. J Neurosci 2010;30(2):694–702.

[110] Hoeffer CA, Sanchez E, Hagerman RJ, Mu Y, Nguyen DV, Wong H, Whelan AM, Zukin RS, Klann E, Tassone F. Altered mTOR signaling and enhanced CYFIP2 expression levels in subjects with fragile X syndrome. Genes Brain Behav 2012;11(3):332–41.

[111] Kumari D, Bhattacharya A, Nadel J, Moulton K, Zeak NM, Glicksman A, Dobkin C, Brick DJ, Schwartz PH, Smith CB, Klann E, Usdin K. Identification of fragile X syndrome specific molecular markers in human fibroblasts: a useful model to test the efficacy of therapeutic drugs. Hum Mutat 2014;35(12): 1485–94.

[112] Bear MF, Huber KM, Warren ST. The mGluR theory of fragile X mental retardation. Trends Neurosci 2004; 27(7):370–7.

[113] El Idrissi A, Ding XH, Scalia J, Trenkner E, Brown WT, Dobkin C. Decreased GABA(A) receptor expression in the seizure-prone fragile X mouse. Neurosci Lett 2005;377(3):141–6.

[114] Gantois I, Vandesompele J, Speleman F, Reyniers E, D'Hooge R, Severijnen LA, Willemsen R, Tassone F, Kooy RF. Expression profiling suggests underexpression of the GABA(A) receptor subunit delta in the fragile X knockout mouse model. Neurobiol Dis 2006;21(2):346–57.

[115] D'Hulst C, De Geest N, Reeve SP, Van Dam D, De Deyn PP, Hassan BA, Kooy RF. Decreased expression of the GABAA receptor in fragile X syndrome. Brain Res 2006;1121(1):238–45.

[116] Zalfa F, Eleuteri B, Dickson KS, Mercaldo V, De Rubeis S, di Penta A, Tabolacci E, Chiurazzi P, Neri G, Grant SG, Bagni C. A new function for the fragile X mental retardation protein in regulation of PSD-95 mRNA stability. Nat Neurosci 2007;10(5):578−87.

[117] Dictenberg JB, Swanger SA, Antar LN, Singer RH, Bassell GJ. A direct role for FMRP in activity-dependent dendritic mRNA transport links filopodial-spine morphogenesis to fragile X syndrome. Dev Cell 2008;14(6):926−39.

[118] Wang H, Dictenberg JB, Ku L, Li W, Bassell GJ, Feng Y. Dynamic association of the fragile X mental retardation protein as a messenger ribonucleoprotein between microtubules and polyribosomes. Mol Biol Cell 2008;19(1):105−14.

[119] Alpatov R, Lesch BJ, Nakamoto-Kinoshita M, Blanco A, Chen S, Stutzer A, Armache KJ, Simon MD, Xu C, Ali M, Murn J, Prisic S, Kutateladze TG, Vakoc CR, Min J, Kingston RE, Fischle W, Warren ST, Page DC, Shi Y. A chromatin-dependent role of the fragile X mental retardation protein FMRP in the DNA damage response. Cell 2014;157(4):869−81.

[120] Luca R, Averna M, Zalfa F, Vecchi M, Bianchi F, La Fata G, Del Nonno F, Nardacci R, Bianchi M, Nuciforo P, Munck S, Parrella P, Moura R, Signori E, Alston R, Kuchnio A, Farace MG, Fazio VM, Piacentini M, De Strooper B, Achsel T, Neri G, Neven P, Evans DG, Carmeliet P, Mazzone M, Bagni C. The fragile X protein binds mRNAs involved in cancer progression and modulates metastasis formation. EMBO Mol Med 2013;5(10):1523−36.

[121] Zalfa F, Panasiti V, Carotti S, Zingariello M, Perrone G, Sancillo L, Pacini L, Luciani F, Roberti V, D'Amico S, Coppola R, Abate SO, Rana RA, De Luca A, Fiers M, Melocchi V, Bianchi F, Farace MG, Achsel T, Marine JC, Morini S, Bagni C. The fragile X mental retardation protein regulates tumor invasiveness-related pathways in melanoma cells. Cell Death Dis 2017;8(11):e3169.

[122] Korb E, Herre M, Zucker-Scharff I, Gresack J, Allis CD, Darnell RB. Excess translation of epigenetic regulators contributes to fragile X syndrome and is alleviated by Brd4 inhibition. Cell 2017;170(6): 1209−1223.e20.

[123] Colak D, Zaninovic N, Cohen MS, Rosenwaks Z, Yang WY, Gerhardt J, Disney MD, Jaffrey SR. Promoter-bound trinucleotide repeat mRNA drives epigenetic silencing in fragile X syndrome. Science 2014; 343(6174):1002−5.

[124] Avitzour M, Mor-Shaked H, Yanovsky-Dagan S, Aharoni S, Altarescu G, Renbaum P, Eldar-Geva T, Schonberger O, Levy-Lahad E, Epsztejn-Litman S, Eiges R. FMR1 epigenetic silencing commonly occurs in undifferentiated fragile X-affected embryonic stem cells. Stem Cell Reports 2014;3(5):699−706.

[125] Chiurazzi P, Pomponi MG, Willemsen R, Oostra BA, Neri G. In vitro reactivation of the FMR1 gene involved in fragile X syndrome. Hum Mol Genet 1998;7(1):109−13.

[126] Biacsi R, Kumari D, Usdin K. SIRT1 inhibition alleviates gene silencing in Fragile X mental retardation syndrome. PLoS Genet 2008;4(3):e1000017.

[127] Chiurazzi P, Pomponi MG, Pietrobono R, Bakker CE, Neri G, Oostra BA. Synergistic effect of histone hyperacetylation and DNA demethylation in the reactivation of the FMR1 gene. Hum Mol Genet 1999; 8(12):2317−23.

[128] Kumari D, Usdin K. Polycomb group complexes are recruited to reactivated FMR1 alleles in Fragile X syndrome in response to FMR1 transcription. Hum Mol Genet 2014;23(24):6575−83.

[129] Hecht M, Tabib A, Kahan T, Orlanski S, Gropp M, Tabach Y, Yanuka O, Benvenisty N, Keshet I, Cedar H. Epigenetic mechanism of FMR1 inactivation in Fragile X syndrome. Int J Dev Biol 2017;61(3−4-5):285−92.

[130] Moazed D. Small RNAs in transcriptional gene silencing and genome defence. Nature 2009;457(7228): 413−20.

[131] Cabianca DS, Casa V, Bodega B, Xynos A, Ginelli E, Tanaka Y, Gabellini D. A long ncRNA links copy number variation to a polycomb/trithorax epigenetic switch in FSHD muscular dystrophy. Cell 2012; 149(4):819−31.

[132] Lim JW, Snider L, Yao Z, Tawil R, Van Der Maarel SM, Rigo F, Bennett CF, Filippova GN, Tapscott SJ. DICER/AGO-dependent epigenetic silencing of D4Z4 repeats enhanced by exogenous siRNA suggests mechanisms and therapies for FSHD. Hum Mol Genet 2015;24(17):4817–28.

[133] Ginno PA, Lott PL, Christensen HC, Korf I, Chedin F. R-loop formation is a distinctive characteristic of unmethylated human CpG island promoters. Mol Cell 2012;45(6):814–25.

[134] Tous C, Aguilera A. Impairment of transcription elongation by R-loops in vitro. Biochem Biophys Res Commun 2007;360(2):428–32.

[135] Belotserkovskii BP, Soo Shin JH, Hanawalt PC. Strong transcription blockage mediated by R-loop formation within a G-rich homopurine-homopyrimidine sequence localized in the vicinity of the promoter. Nucleic Acids Res 2017;45(11):6589–99.

[136] Skourti-Stathaki K, Kamieniarz-Gdula K, Proudfoot NJ. R-loops induce repressive chromatin marks over mammalian gene terminators. Nature 2014;516(7531):436–9.

[137] Castellano-Pozo M, Santos-Pereira JM, Rondon AG, Barroso S, Andujar E, Perez-Alegre M, Garcia-Muse T, Aguilera A. R loops are linked to histone H3 S10 phosphorylation and chromatin condensation. Mol Cell 2013;52(4):583–90.

[138] Sun Q, Csorba T, Skourti-Stathaki K, Proudfoot NJ, Dean C. R-loop stabilization represses antisense transcription at the Arabidopsis FLC locus. Science 2013;340(6132):619–21.

[139] Boque-Sastre R, Soler M, Oliveira-Mateos C, Portela A, Moutinho C, Sayols S, Villanueva A, Esteller M, Guil S. Head-to-head antisense transcription and R-loop formation promotes transcriptional activation. Proc Natl Acad Sci USA 2015;112(18):5785–90.

[140] Chen PB, Chen HV, Acharya D, Rando OJ, Fazzio TG. R loops regulate promoter-proximal chromatin architecture and cellular differentiation. Nat Struct Mol Biol 2015;22(12):999–1007.

[141] Popp C, Dean W, Feng S, Cokus SJ, Andrews S, Pellegrini M, Jacobsen SE, Reik W. Genome-wide erasure of DNA methylation in mouse primordial germ cells is affected by AID deficiency. Nature 2010;463(7284): 1101–5.

[142] Franchini DM, Chan CF, Morgan H, Incorvaia E, Rangam G, Dean W, Santos F, Reik W, Petersen-Mahrt SK. Processive DNA demethylation via DNA deaminase-induced lesion resolution. PLoS One 2014;9(7):e97754.

[143] Krajewski WA, Nakamura T, Mazo A, Canaani E. A motif within SET-domain proteins binds single-stranded nucleic acids and transcribed and supercoiled DNAs and can interfere with assembly of nucleosomes. Mol Cell Biol 2005;25(5):1891–9.

[144] Di Ruscio A, Ebralidze AK, Benoukraf T, Amabile G, Goff LA, Terragni J, Figueroa ME, De Figueiredo Pontes LL, Alberich-Jorda M, Zhang P, Wu M, D'Alo F, Melnick A, Leone G, Ebralidze KK, Pradhan S, Rinn JL, Tenen DG. DNMT1-interacting RNAs block gene-specific DNA methylation. Nature 2013; 503(7476):371–6.

[145] Davidovich C, Zheng L, Goodrich KJ, Cech TR. Promiscuous RNA binding by polycomb repressive complex 2. Nat Struct Mol Biol 2013;20(11):1250–7.

[146] Nadel Y, Weisman-Shomer P, Fry M. The fragile X syndrome single strand d(CGG)n nucleotide repeats readily fold back to form unimolecular hairpin structures. J Biol Chem 1995;270(48):28970–7.

[147] Fry M, Loeb LA. The fragile X syndrome d(CGG)n nucleotide repeats form a stable tetrahelical structure. Proc Natl Acad Sci USA 1994;91(11):4950–4.

[148] Usdin K, Woodford KJ. CGG repeats associated with DNA instability and chromosome fragility form structures that block DNA synthesis in vitro. Nucleic Acids Res 1995;23(20):4202–9.

[149] Mitas M, Yu A, Dill J, Haworth IS. The trinucleotide repeat sequence d(CGG)15 forms a heat-stable hairpin containing Gsyn. Ganti base pairs. Biochemistry 1995;34(39):12803–11.

[150] Handa V, Saha T, Usdin K. The fragile X syndrome repeats form RNA hairpins that do not activate the interferon-inducible protein kinase, PKR, but are cut by Dicer. Nucleic Acids Res 2003;31(21):6243–8.

[151] Zumwalt M, Ludwig A, Hagerman PJ, Dieckmann T. Secondary structure and dynamics of the r(CGG) repeat in the mRNA of the fragile X mental retardation 1 (FMR1) gene. RNA Biol 2007;4(2):93–100.

[152] Smith MR, Syed A, Lukacsovich T, Purcell J, Barbaro BA, Worthge SA, Wei SR, Pollio G, Magnoni L, Scali C, Massai L, Franceschini D, Camarri M, Gianfriddo M, Diodato E, Thomas R, Gokce O, Tabrizi SJ, Caricasole A, Landwehrmeyer B, Menalled L, Murphy C, Ramboz S, Luthi-Carter R, Westerberg G, Marsh JL. A potent and selective Sirtuin 1 inhibitor alleviates pathology in multiple animal and cell models of Huntington's disease. Hum Mol Genet 2014;23(11):2995–3007.

[153] Herman D, Jenssen K, Burnett R, Soragni E, Perlman SL, Gottesfeld JM. Histone deacetylase inhibitors reverse gene silencing in Friedreich's ataxia. Nat Chem Biol 2006;2(10):551–8.

[154] Rai M, Soragni E, Chou CJ, Barnes G, Jones S, Rusche JR, Gottesfeld JM, Pandolfo M. Two new pimelic diphenylamide HDAC inhibitors induce sustained frataxin upregulation in cells from Friedreich's ataxia patients and in a mouse model. PLoS One 2010;5(1):e8825.

[155] Rai M, Soragni E, Jenssen K, Burnett R, Herman D, Coppola G, Geschwind DH, Gottesfeld JM, Pandolfo M. HDAC inhibitors correct frataxin deficiency in a Friedreich ataxia mouse model. PLoS One 2008;3(4):e1958.

[156] Saba HI, Wijermans PW. Decitabine in myelodysplastic syndromes. Semin Hematol 2005;42(3 Suppl. 2): S23–31.

[157] Sussmuth SD, Haider S, Landwehrmeyer GB, Farmer R, Frost C, Tripepi G, Andersen CA, Di Bacco M, Lamanna C, Diodato E, Massai L, Diamanti D, Mori E, Magnoni L, Dreyhaupt J, Schiefele K, Craufurd D, Saft C, Rudzinska M, Ryglewicz D, Orth M, Brzozy S, Baran A, Pollio G, Andre R, Tabrizi SJ, Darpo B, Westerberg G, Consortium P. An exploratory double-blind, randomized clinical trial with selisistat, a SirT1 inhibitor, in patients with Huntington's disease. Br J Clin Pharmacol 2015;79(3):465–76.

[158] Disney MD, Liu B, Yang WY, Sellier C, Tran T, Charlet-Berguerand N, Childs-Disney JL. A small molecule that targets r(CGG)(exp) and improves defects in fragile X-associated tremor ataxia syndrome. ACS Chem Biol 2012;7(10):1711–8.

[159] Finkel RS, Mercuri E, Darras BT, Connolly AM, Kuntz NL, Kirschner J, Chiriboga CA, Saito K, Servais L, Tizzano E, Topaloglu H, Tulinius M, Montes J, Glanzman AM, Bishop K, Zhong ZJ, Gheuens S, Bennett CF, Schneider E, Farwell W, De Vivo DC, Group ES. Nusinersen versus sham control in infantile-onset spinal muscular atrophy. N Engl J Med 2017;377(18):1723–32.

[160] Wild EJ, Tabrizi SJ. Therapies targeting DNA and RNA in Huntington's disease. Lancet Neurol 2017; 16(10):837–47.

[161] Datson NA, Gonzalez-Barriga A, Kourkouta E, Weij R, van de Giessen J, Mulders S, Kontkanen O, Heikkinen T, Lehtimaki K, van Deutekom JC. The expanded CAG repeat in the huntingtin gene as target for therapeutic RNA modulation throughout the HD mouse brain. PLoS One 2017;12(2):e0171127.

[162] Rue L, Banez-Coronel M, Creus-Muncunill J, Giralt A, Alcala-Vida R, Mentxaka G, Kagerbauer B, Zomeno-Abellan MT, Aranda Z, Venturi V, Perez-Navarro E, Estivill X, Marti E. Targeting CAG repeat RNAs reduces Huntington's disease phenotype independently of huntingtin levels. J Clin Invest 2016; 126(11):4319–30.

[163] Li L, Matsui M, Corey DR. Activating frataxin expression by repeat-targeted nucleic acids. Nat Commun 2016;7:10606.

[164] Handa V, Goldwater D, Stiles D, Cam M, Poy G, Kumari D, Usdin K. Long CGG-repeat tracts are toxic to human cells: implications for carriers of Fragile X premutation alleles. FEBS Lett 2005;579(12):2702–8.

[165] Hashem V, Galloway JN, Mori M, Willemsen R, Oostra BA, Paylor R, Nelson DL. Ectopic expression of CGG containing mRNA is neurotoxic in mammals. Hum Mol Genet 2009;18(13):2443–51.

[166] Berry-Kravis E, Hagerman R, Visootsak J, Budimirovic D, Kaufmann WE, Cherubini M, Zarevics P, Walton-Bowen K, Wang P, Bear MF, Carpenter RL. Arbaclofen in fragile X syndrome: results of phase 3 trials. J Neurodev Disord 2017;9:3.

[167] Youssef EA, Berry-Kravis E, Czech C, Hagerman RJ, Hessl D, Wong CY, Rabbia M, Deptula D, John A, Kinch R, Drewitt P, Lindemann L, Marcinowski M, Langland R, Horn C, Fontoura P, Santarelli L, Quiroz JA, FragXis Study G. Effect of the mGluR5-NAM basimglurant on behavior in adolescents and adults with fragile X syndrome in a randomized, double-blind, placebo-controlled trial: FragXis phase 2 results. Neuropsychopharmacology 2018;43(3):503–12.
[168] Ligsay A, Van Dijck A, Nguyen DV, Lozano R, Chen Y, Bickel ES, Hessl D, Schneider A, Angkustsiri K, Tassone F, Ceulemans B, Kooy RF, Hagerman RJ. A randomized double-blind, placebo-controlled trial of ganaxolone in children and adolescents with fragile X syndrome. J Neurodev Disord 2017;9(1):26.
[169] Berry-Kravis E, Des Portes V, Hagerman R, Jacquemont S, Charles P, Visootsak J, Brinkman M, Rerat K, Koumaras B, Zhu L, Barth GM, Jaecklin T, Apostol G, von Raison F. Mavoglurant in fragile X syndrome: results of two randomized, double-blind, placebo-controlled trials. Sci Transl Med 2016;8(321):321ra5.

CHAPTER

The epigenetics of autism

13

Aicha Massrali[1], Varun Warrier[1], Arkoprovo Paul[1], Dwaipayan Adhya[1], Deepak P. Srivastava[2,3], Mark Kotter[4], Simon Baron-Cohen[1]

Autism Research Centre, Department of Psychiatry, University of Cambridge, Cambridge, United Kingdom[1]; Department of Basic and Clinical Neuroscience, Maurice Wohl Clinical Neuroscience Institute, Institute of Psychiatry, Psychology and Neuroscience, King's College London, London, United Kingdom[2]; MRC Centre for Neurodevelopmental Disorders, King's College London, London, United Kingdom[3]; Department of Clinical Neurosciences, Ann McLaren Laboratory of Regenerative Medicine, University of Cambridge, Cambridge, United Kingdom[4]

Chapter Outline

13.1 Autism **285**
13.1.1 Heritability and genetics 286
13.2 Epigenetics of autism **287**
13.2.1 DNA methylation and hydroxymethylation in autism 288
13.2.1.1 Candidate gene methylation studies in humans 289
13.2.1.2 Methylome-wide association studies 290
13.2.2 Histone modifications 292
13.2.2.1 Histone methylation and acetylation 292
13.2.2.2 Chromatin modifying and remodeling complexes 293
13.2.3 Risk factors affecting the epigenetics of autism 294
13.2.3.1 Hormones 295
13.3 Discussion **296**
Acknowledgments **297**
References **297**

13.1 Autism

The term autism refers to a group of neurodevelopmental conditions characterized by difficulties in social interactions alongside stereotyped behavior and unusually narrow interests [1]. In addition, autistic individuals usually have other comorbid conditions including intellectual disability, epilepsy, sleep difficulties, and attention deficits (attention-deficit/hyperactivity disorder [ADHD]/attention deficit disorder) [1]. Males are diagnosed three times more with autism than females, reflecting biological sex differences in the condition or an underdiagnosis of females, or a combination of both. However, this sex difference in autism reduces with intellectual disability [2]. In the

Chromatin Signaling and Neurological Disorders. https://doi.org/10.1016/B978-0-12-813796-3.00013-4

fifth edition of the *Diagnostic and Statistical Manual of Mental Disorders* (*DSM-V*), the Asperger syndrome and pervasive developmental disorder not otherwise specified (PDD-NOS) fall under the umbrella of autism spectrum disorders (ASDs) [3]. The use of the term autism spectrum conditions (ASCs) instead of ASDs is to avoid the pejorative connotation of the word "disorder," given that, although some characteristics of autism are disabling, others (such as excellent attention to and memory for detail) may even confer talent in some autistic individuals [1]. The prevalence of diagnosed autism has been steadily increasing since the term was coined (70 years ago), with a substantial rise in the past two decades. The worldwide prevalence of autism is currently 1–2% [4–7]. This increase in autism prevalence can be attributable to changes in the diagnostic criteria and greater public awareness, rather than an absolute increase in the condition [5]. Nonetheless, an increase in a risk factor could also be partially involved [1]. Epidemiologic studies have reported a link between some environmental factors (e.g., maternal health during pregnancy, prenatal exposure to chemicals, toxins) that put individuals at higher risk of developing ASCs. However, none of the studies prove a causal link with autism [8].

13.1.1 Heritability and genetics

It is clear that a considerable proportion of the risk for autism is genetic. In one of the earliest studies investigating the heritability of autism, Folstein and Rutter [9] studied a sample of 21 twin pairs and reported a monozygotic twin concordance of 36% and a dizygotic twin concordance of 0% for autism. Subsequently, several studies in much larger cohorts have established relatively high twin heritability for autism. A meta-analysis of twin studies [10] reported a heritability of 64–91%. While these studies focus on clinically defined autism, studies have also investigated the heritability of autistic traits (i.e., subthreshold autistic traits that are thought to be representative of the liability model of autism). Ronald and Hoekstra [11], in a meta-analysis, reported a heritability of between 60% and 90% for autistic traits. Heritability can also be investigated using familial recurrence. A study investigating insurance claims from over a third of the entire US population identified a heritability of approximately 92% for autism, which had the highest heritability of the 149 diseases and conditions investigated [12]. Similarly, familial recurrences in Sweden and Finland have identified a sibling recurrence risk of ~10% [13,14] and heritability estimates of 83% [15].

Evidence of molecular genetics provides further support for the considerable heritability of autism. Several studies have investigated the additive single nucleotide polymorphism (SNP) heritability (i.e., the proportion of variance explained by all SNPs investigated) using multiple models. Cumulatively, SNPs investigated are thought to contribute to between 15% and 50% of the total risk for autism, with a higher contribution in multiplex autism families [16–18]. Although, *en masse*, these SNPs contribute to a substantial proportion of the heritability, the per-SNP effect is very small, suggesting a highly polygenic architecture of autism, similar to that of other complex conditions.

Several genome-wide association studies (GWASs) have sought to identify significant SNPs associated with autism at a *P*-value threshold $<5 \times 10^{-8}$. In a GWAS of autism, five independent loci were identified [19]. Candidate genes identified from this analysis includes *KMT2E*, which encodes the MLL5 tumor suppressor, and *MACROD2*, which encodes a macrodomain deacetylase that removes ADP-ribose from mono-ADP-ribosylated proteins [19]. Early studies used a case-pseudocontrol design in which they tested differences in allele frequencies between transmitted and nontransmitted alleles in simplex trio families, which is statistically underpowered [20]. Newer studies have included a case-control design and have increased sample sizes by meta-analyzing genetic data from multiple different cohorts [19,21].

GWASs have also allowed for the dissection of directional pleiotropy between autism and other related traits by using genetic correlations. Multiple studies have identified that there is a modest positive genetic correlation between the measures of intelligence, including educational attainment and cognitive aptitude, and the risk for autism [22,23]. This is in contrast to studies on rare genetic variation, which identify an enrichment for *de novo* protein-truncating mutations in autistic individuals with below average IQ [24,25]. Genetic correlations have also identified a positive genetic correlation between autism and other psychiatric conditions such as depression, schizophrenia, and ADHD, suggesting shared underlying liabilities for autism and other psychiatric conditions [19]. Studies from our laboratory have shown that social traits related to autism, such as empathy and social relationship satisfaction, are also genetically correlated with autism: common alleles that are associated with the increased risk for autism are also associated with decreased social relationship satisfaction and empathetic abilities [26,27]. In contrast, alleles that are associated with the increased risk for autism are also associated with increased systemizing, in support of the hypersystemizing theory of autism [28].

In addition to common genetic variations, other classes of genetic variations contribute to the risk for autism. Notably, although rare, protein-truncating *de novo* mutations are enriched in the probands of simplex autism families, in autistic individuals with comorbid intellectual disability, and in female autistic probands [29]. Although the individual effect sizes are large, together, *de novo* variants are thought to account for a smaller proportion of the risk for autism. For example, rare genetic variants, both *de novo* and inherited, are thought to account for the risk in 10–30% of autistic individuals [30]. As of December 2017, as many as 78 genes have been identified as enriched for rare variation in autism (http://spark-sf.s3.amazonaws.com/SPARK_gene_list.pdf). Many of these genes encode chromatin modifiers (e.g., ASH1L, EHMT1, EP300, NSD1, SUV420H1) and transcription regulators (e.g., CHD8, MECP2, MED13L).

13.2 Epigenetics of autism

Epigenetic modifications function to maintain the architecture and the appropriate segregation of chromosomes during mitosis, to regulate repetitive elements and endogenous retroelements, and to regulate gene expression [34]. There are two main molecular epigenetic mechanisms involved in gene expression: covalent modification of DNA through methylation and chromatin structure regulation via histone posttranslational modification (PTM) [35]. These modifications arise from heritable modifications, environmental factors, and replication errors [31]. While the genetic replication error rate is 1:1000000, the usual error rate for replicating epigenetic elements is closer to 1:1000 [32]. There are also spontaneous changes that are thought to occur to the epigenome, known as epigenetic drift. This may begin to explain the stochastic nature of the epigenomes and the divergence from their identical starting conditions [33].

There are three main molecular epigenetic mechanisms involved in the regulation of gene expression: covalent modification of DNA through methylation, chromatin modifiers, and ATP-dependent chromatin remodelers [35].

13.2.1 DNA methylation and hydroxymethylation in autism

The majority of studies that have investigated the epigenetic architecture of autism have studied DNA methylation, as it acts at the interface between genetic and environmental factors and may hold an answer to autism's complex etiology [36]. Previous studies have identified perinatal environmental exposures to various environmental factors and their link to impaired methylation in children with autism [37–40]. DNA methylation is the covalent binding of a methyl group (-CH_3) to a cytosine C5 side chain. This chemical modification is catalyzed by DNA methyltransferases (DNMTs), a class of enzymes that transfer a methyl group specifically to cytosines that are usually followed by a guanine, forming the CpG dinucleotide sequences, or followed by any other base, forming a CpH. More than two-thirds of the CpG dinucleotides in the genome are methylated, whereas the remaining occur in CpG islands (small clusters). The methylated CpG dinucleotides are found mainly in intra- and intergenic sequences within DNA repeats, whereas the unmethylated CpGs occur in regions that actively regulate gene transcription [35]. Although Wyatt and Cohen reported the presence of 5-hydroxymethylcytosine (5-hmC) in the early 1950s, their findings were not recognized until 2009 when the dynamic role of 5-hmC in DNA demethylation was demonstrated [41].

The novel ten–eleven translocation (TET) protein family acts as an "eraser" and is responsible for the conversion of 5-methylcytosine (5-mC) to 5-hmC followed by oxidization to produce 5-formylcytosine (5-fC) and 5-carboxylcytosine (5-caC) in the demethylation process [42]. Furthermore, 5-mC and 5-hmC can be directly converted to thymine and 5-hydroxymethyluracil, respectively. This is mediated by activation-induced cytidine deaminase and apolipoprotein B mRNA editing enzyme, a catalytic polypeptide-like (APOBEC) family. Thymine DNA glycosylase then recognizes and removes the mismatched thymine, 5-hmU, 5-fC, or 5-caC from the genome and replaces them with unmodified cytosine to initiate base excision repair [43].

Several lines of evidence suggest an important role for epigenetic regulation in the risk for autism. One of the earliest genes implicated in autism is *MECP2*, which encodes the methyl-CpG-binding protein causing the Rett syndrome [44] (see also Chapter 9), and *de novo* frameshift mutations in the transcriptional activator CREB-binding protein that cause the Rubinstein–Taybi syndrome (RSTS). Missense mutations in the Tenascin C (TNC) gene, insulin-like growth factor–binding protein, and acid-labile subunit (IGFALS) gene have been reported in whole-exome sequencing of a patient with RSTS presenting with an autism phenotype [45]. Duplications of *MECP2* cause MECP2 duplication syndrome, which also has similarities to autism [46]. Canonically, MECP2 is thought to bind to methylated DNA to condense chromatin, repressing transcription. However, studies suggest that MECP2 might also act as an enhancer, indicating a more complex role for the protein during neurodevelopment [47].

Studies have associated TET and DNMT mutations with intellectual disability and autism [29]. The Angelman syndrome and the Prader–Willi syndrome are closely related to the autism spectrum. They result from the loss of maternal and paternal alleles, respectively, of 15q11.2–q13.3. In some rare cases, the imprinting control region is unmethylated in the Angelman syndrome and hemimethylated in the Prader–Willi syndrome. So although this locus is deleted in both syndromes, it is duplicated in autism [48,49].

13.2.1.1 Candidate gene methylation studies in humans

Akin to candidate gene association studies, a few studies have investigated methylation modifications in candidate genes in autism. One such gene is *SHANK3*, which encodes a synaptic scaffolding protein, involved in the formation and maturation of dendritic spines. Mutations in this gene have been associated with autism [50]. Zhu and colleagues [51] conducted a DNA methylation profiling of five CpG islands (CGI-1 to CGI-5) in the *SHANK3* gene in postmortem brain tissues from 54 cases of ASCs and 43 controls. This study was the first to report an altered methylation pattern. Specifically, a significant increase in overall DNA methylation of three intragenic CGIs (CGI-2, CGI-3, and CGI-4) of *SHANK3* in ASC brain tissues was observed. Notably, 15% of ASC brain samples revealed an increase in DNA methylation specifically in CpG-2 and CpG-4 of *SHANK3*. Moreover, there was an altered splicing pattern of *SHANK3* isoforms, with concurrent increase in methylation of specific CGIs in ASC brain tissues. The findings suggest that *SHANK3* has an important role in causing ASCs and may serve as a valuable biomarker to dissect the contributions of genetic and environmental components.

Oxytocin receptor (*OXTR*), which encodes a G-protein-coupled receptor that binds to the neurotransmitter peptide hormone oxytocin, is another important gene implicated in ASC etiology [52]. Gregory and colleagues [53] performed a high-resolution genome-wide microarray and comparative genomic hybridization on 119 probands from multiplex ASC families to identify copy number variants in the *OXTR* gene. The results from this study showed a genomic deletion in the proband and his mother, although the affected sibling did not harbor a deletion but displayed aberrant gene silencing in the region of *OXTR* promoter by DNA methylation, with the highest increase in methylation at a single CpG being 37%. They also carried out a DNA methylation analysis on peripheral blood and temporal cortex from ASC cases and matched controls from independent data sets. The methylation status of several CGIs in *OXTR* was found to be significantly increased by 23% in DNA extracted from peripheral blood in 20 sex-matched ASC cases when compared with controls. Further analysis on DNA methylation of temporal cortex brain tissues in independent data sets from 10 ASC cases and sex-matched controls revealed up to 41% higher methylation at specific CGIs in males with ASCs. This study provided the first evidence supporting the role of epigenetic regulation of OXTR in causing ASCs.

Glutamate decarboxylase (*GAD*), which encodes an enzyme that catalyzes the conversion of the excitatory neurotransmitter glutamate to inhibitory neurotransmitter γ-aminobutyric acid (GABA) and thereby plays a pivotal role in maintaining the excitatory–inhibitory balance, is considered to be another relevant gene in causing ASCs [54,55].

MECP2 is an important epigenetic regulator in human brain development. As a member of the methyl-binding domain family of proteins, it is highly enriched in the central nervous system, especially in GABAergic interneurons. MECP2 binds to methylated CpGs and can act as either a gene activator or a repressor depending on the PTM or methylation status (5-mC or 5-hmC) of the associated target gene. The binding affinity of MECP2 to 5-mC or 5-hmC is an important factor to account for when considering its differential regulatory action. As glutamic acid decarboxylase 67 (GAD1), glutamate decarboxylase 65 (GAD2), and reelin (RELN) have been previously reported to be downregulated in the postmortem brains of ASC cases when compared with controls [56,57], a study aimed to investigate the binding of MECP2 and its role in regulating the expression of those genes. The authors reported a significant increase in MECP2 binding specifically to the promoter regions of *GAD1* and *RELN* in the brains of patients with ASCs. They used methyl DNA immunoprecipitation (MeDIP) and hydroxymethyl DNA immunoprecipitation (hMeDIP) to measure the total 5-mC and 5-hmC levels, respectively. They reported an enrichment of 5-hmC levels relative to the levels of 5-mC at the regulatory regions of the investigated genes. This is associated with an increased MECP2 binding at the *GAD1* and *RELN* promoters. These results suggest that increased levels of 5-hmC at regulatory gene domains improve MECP2 binding and downregulate the expression of the target genes in ASCs [58].

13.2.1.2 Methylome-wide association studies

In contrast to candidate gene methylation studies, advances in methylation bead arrays have allowed for the investigation of methylation patterns across the genome (called methylome-wide association studies, or, in some instances, epigenome-wide association study (EWAS), but the term epigenome is not restricted to methylation patterns). Methylome-wide association studies are usually conducted following two different approaches: those based on sequencing followed by bisulfite treatment and those based on the affinity for methylcytosine [59]. However, sequencing-based protocols are expensive, time consuming, and require expertise [60,61]. For these reasons, new-generation arrays have been developed to interrogate a wide range of CpG sites throughout the genome [62,63].

One of the challenges of methylation studies is tissue source. Methylation patterns vary widely across tissues; however, some of these signatures are preserved across tissues. For autism, investigating methylation patterns in neural tissue will provide the most accurate mechanistic insights into the biology of the condition. A few studies have investigated differential methylation in autistic brains compared with nonautistic brains using postmortem tissue. Ginsberg and colleagues [64] compared differentially methylated gene networks in DNA isolated from postmortem cerebellar cortex and occipital cortex of nine cases with idiopathic ASC and nine age-matched typically developing controls using HM27 arrays. No significant ASC-specific differentially methylated probes (DMPs) were discovered at a false discovery rate (FDR) of 0.05 but downregulation of genes of mitochondrial oxidative phosphorylation and protein translation was identified. Ladd-Acosta and colleagues [65] investigated methylation profiles in ASC brain samples from the prefrontal cortex (6 cases and 5 controls), temporal cortex (6 cases and 10 controls), and cerebellum (7 cases and 6 controls) using the Illumina 450 K array and were able to identify four significant autism-associated differentially methylated regions (DMRs).

Using the same type of Illumina 450 K array, Nardone and colleagues [66] identified 5329 differentially methylated CpG sites in the prefrontal cortex (12 cases and 12 controls) and 10,745 significant CpGs in the anterior cingulate gyrus (11 cases and 11 controls) between the control and autistic postmortem brain tissues. To understand the effect of methylation on gene expression in autistic brains, they compared the DMRs from this study with transcriptomic data from autistic postmortem brains published by Voineagu and colleagues [67]. Interestingly, the hypomethylated regions overlapped with genes showing higher expression, and the genes with altered CpG methylation and altered expression were enriched for immune-related functions [66,67]. Extending this work, Nardone and colleagues investigated methylation signatures specifically in neurons in postmortem autistic and nonautistic brains. They measured methylation using the Illumina 450 K array in fluorescence-activated cell-sorted neuronal nuclei. Although they did not identify any significant CpGs that differed in methylation levels in autistic and control brains, they identified 58 DMRs [68].

To identify sets of methylated regions that are correlated in their methylation, the authors conducted weighted gene coexpression network analyses and identified eight modules that were significantly correlated with case-control status. Two of these modules were significantly enriched for synaptic processes and one for immune-related genes mirroring gene-set enrichment in transcriptomic studies [67,69]. The two modules enriched for synaptic processes were also enriched for candidate genes implicated in autism and for neurodevelopment-specific methylation signatures.

In parallel to studies on the postmortem brain, a few studies have also investigated methylation in peripheral tissues in autism. Although epigenetic signatures differ across tissues, several studies have demonstrated the use of studying methylation in peripheral tissues. The first methylome-wide study of autism investigated methylation patterns in twins discordant and concordant for autism. Nguyen and colleagues performed global methylation analysis on the lymphoblastoid cell lines derived from three pairs of discordant monozygotic twins for autism and siblings from two of these pairs. At a liberal FDR threshold of 10%, the authors identified 73 differentially methylated CpG islands. This study revealed many candidate genes that were differentially methylated between discordant monozygotic twins as well as between both the twins and unaffected siblings [70]. Genes reported included *BCL-2* and retinoic acid–related orphan receptor alpha (*RORA*). Their methylation status was further confirmed by bisulfite sequencing and methylation-specific polymerase chain reaction. Immunohistochemistry analysis showed decreased expression of both genes in the cerebellum and frontal cortex in the autistic brain when compared with controls.

A similar approach was taken by Wong and colleagues [71], in which they investigated methylation patterns in 100 twins (50 pairs) for both case-control status and quantitative traits related to autism and measured them using the Childhood Autism Symptom Test (CAST). Methylation was measured from blood samples. The authors identified differences in regional methylation in twin pairs discordant for autism. They further examined methylation differences in quantitative traits associated with autism and identified several CpGs that were associated with scores on the CAST.

The largest methylation study [72] of autism to date investigated methylation patterns using neonatal dried blood spots in 1263 individuals in the MINERvA cohort from Denmark, of which approximately half of the participants had an autism diagnosis. Methylation was investigated using the Infinium HumanMethylation450k array. Regression analyses were conducted using polygenic risk score for autism from the iPSYCH autism study as the predictor variable and methylation as the dependent variable. The study identified two epigenome-wide significant CpGs (cg02771117: $P = 3.14 \times 10^{-8}$ and cg27411982: $P = 8.38 \times 10^{-8}$), with several additional CpGs with $P < 1 \times 10^{-5}$. The two significant probes flank a region implicated by GWASs for autism, suggesting a contribution of both genetic and nongenetic effects to methylation in autism. To identify if case status is associated with methylation, the authors meta-analyzed the data from the MINERvA cohort and two additional cohorts (methylation from blood samples), but they did not identify any significant signals in the total sample of 1425 ASD cases and 1492 controls. Further investigating the genetic contribution to methylation in autism, the authors conducted bayesian colocalization using methylation quantitative trait locus (QTL) data from blood. They identified 91 loci that contribute to both autism risk and DNA methylation in autism.

To further understand the role of genetic variation in autism, Andrews and colleagues [73] created methylation QTL maps across multiple tissues including fetal brain, blood, and cord blood. The study identified relatively high concordance in methylation QTLs across blood and cord blood, but to a lesser extent across fetal brain and the two peripheral tissues. GWAS loci associated with autism at various P-value thresholds were enriched for methylation QTLs in the fetal brain and to a lesser extent in blood and cord blood. To understand the biological processes underlying both the polygenic and methylation architecture in autism, the authors conducted gene-set enrichments of methylation QTL targets of the autism-associated SNPs. Several of the processes involved were related to immune regulation.

A few studies have investigated methylation in other tissues. Feinberg et al. [74] tested the relationship paternal sperm DNA methylation with autism risk in offspring. They examined genome-wide DNA methylation in paternal semen from an enriched-risk cohort of fathers of autistic children using CHARM 3.0 array and Illumina Infinium HumanMethylation450K BeadChip array. They also compared it with the methylation profiles previously obtained from postmortem brain in the study from Ladd-Acosta and colleagues [65]. With overlapping CHARM data with 450K data, 193 DMRs were identified. Furthermore, 18 of 75 (24%) 450K array probes showed consistent differences in the cerebellums of autistic individuals when compared with controls. This indicated the role of epigenetic differences in the paternal sperm as a potential risk factor to autism risk in the progeny [74].

While the BeadChip technology focuses on investigating CpG regions, other studies took a more global approach to study methylation not only at CpG regions but also at CpH sites throughout the genome. A study used reduced representation bisulfite sequencing to analyze methylation status in 63 postmortem cortical brain samples (Brodmann area 19) from 29 autism-affected and 34 control individuals. Although no significant DMRs were found in this study, a marked enrichment of CpH methylation in autistic brains was reported [75].

13.2.2 Histone modifications

13.2.2.1 Histone methylation and acetylation

Histone PTMs can be residue and site specific and they are far more diverse than DNA methylation. They include mono- (me1), di- (me2), and tri- (me3) methylation; acetylation and crotonylation; polyADP-ribosylation; and small protein (ubiquitin, SUMO) modification of specific lysine residues, as well as arginine (R) methylation and citrullination, serine (S) phosphorylation, tyrosine (Y) hydroxylation, and several others.

These modifications are coregulated and occur as a group of histone PTMs affecting chromatin structure and function. For example, histone H3 lysine 4 (H3K4) methylation and acetylation marks are found in many active promoters. The trimethylated forms of H3K9, H3K27, and H4K20 are found to cluster at the same loci in the genome and are known to be repressive PTMs. These are specific examples, but generally, acetylation of histones marks active regions of the genome, while histone methylation marks either active or inactive chromatin region loci depending on the state and site of methylation.

To date, most EWASs in autism have focused on DNA methylation, whereas very little is known about other chromatin modifications such as histone acetylation. Autism has been previously linked to deleterious mutations in genes involved in the regulation of chromatin structure and function, such as histone deacetylase HDAC4, histone H3 lysine 9 (H3K9) methyltransferase EHMT1/KMT1D/GLP, and H3K4 demethylase JARIDIC/KDM5C/SMCX, as reported in Balemans et al. [76] and Adegbola et al. [77] respectively. H3K4me3, the trimethylated form of H3K4, is enriched at the transcriptional start site of genes and has a major role in neuronal health because of its role in chromatin remodeling complexes of genes involved in differentiation, growth, and plasticity required for hippocampal learning and memory [78]. Shulha and colleagues [79] reported the first neuron-specific study of H3K4me3 in autism. They used prefrontal cortex neurons (sorted and separated from postmortem tissue) from 16 subjects with autism and 16 control subjects ranging in age from 6 months to 70 years to identify autism-associated H3K4me3 changes using anti-H3K4me3-ChIP followed by deep sequencing. Comparing the H3K4me3 epigenomes in the two groups and across a wide range of age revealed that the generalized disruption of developmentally regulated chromatin remodeling in infant prefrontal cortex neurons is only found in the normal cohort and not autism. They also identified an excess (503 loci) and loss (208 loci) of H3K4me3 marks in autistic subjects. Most of these loci represent transcription start sites and overlap with previously annotated autism-risk loci. This reflects a shared genetic and epigenetic risk architecture in autism. To clearly identify and assess the acclaimed role of H3K4me3 abnormalities in autism pathophysiology, additional research with larger cohorts is required.

Sun and colleagues [80] conducted a genome-wide histone acetylation study on 257 postmortem samples (including three brain regions: prefrontal cortex, temporal cortex, and cerebellum) from syndromic and idiopathic ASCs and age-matched controls by using chromatin immunoprecipitation sequencing (ChIP-seq). Genome-wide acetylome profile from this study revealed epigenetic variation in genes involved in synaptic transmission, chemokinesis, ion transport, epilepsy, behavioral abnormality, and immunity.

13.2.2.2 Chromatin modifying and remodeling complexes

Methods such as large-scale exome and whole-genome sequencing have identified the role of chromatin modifiers in causing autism. Modification of histones can occur via two main mechanisms. The first is to directly alter the overall structure of chromatin and the second is to bind to effector molecules such as chromatin-modifying enzymes, or chromatin-remodeling complexes. These complexes affect transcription initiation and/or elongation. Many of the genes identified through whole exome sequencing in autism belong to the family of such complexes. For example, the chromodomain-coding genes *CHD2*, *CHD7*, and *CHD8* are all more frequently mutated in autistic individuals than in nonautistic individuals [29]. All three genes encode ATP-dependent helicases that are involved in chromatin remodeling. These proteins alter the placement of the nucleosome, thereby making the DNA more accessible to transcription factors. One of the most frequently mutated genes in autism is *CHD8* (chromodomain helicase DNA–binding 8). Mutations in *CHD8* are thought to contribute to risk in as much as 0.5% of autistic individuals [81]. Autistic individuals with *CHD8* mutations often have distinct phenotypes including macrocephaly and gastrointestinal difficulties [81]. CHD8 is thought to repress Wnt/β-catenin target genes, and CHD8 targets are significantly enriched for autism-risk genes [82].

The CHD family of chromatin remodelers are not the only genes implicated in autism. *ARID1B* is another gene that is frequently mutated in autism and is a component of the ATP-dependent human SWI/SNF (or BAF) chromatin-remodeling complex [83,84]. Two other genes, *BCL11A* and *ADNPç* identified to be frequently mutated in autism, encode proteins that directly interact with members of the SWI/SNF complex [85]. Mutations in *ADNP* contribute to consistent features in autistic individuals, including attention deficit/hyperactivity disorder and facial dysmorphism such as a prominent forehead and broad nasal bridge [85]. Individuals with intellectual disability also frequently have mutations in other genes in the SWI/SNF complex, highlighting the central role of this complex in typical neural and cognitive development [86].

Chromatin modifiers have also been identified in the list of 78 genes that are enriched for protein-truncating mutations. ASH1L is a histone-lysine N-methyltransferase and contains an SET domain that methylates histones, thereby altering the accessibility of the chromatin-bound DNA. Other examples include the (WW domain-containing adaptor with coiled-coil) also known as WAC; it contains a WW domain protein module found in a wide range of signaling proteins. WAC acts as a linker between histone H2B monoubiquitylation and RNA transcriptional machinery and regulates transcription. Another gene prioritized in a study is *KMT2C*, which is also a methyltransferase that methylates lysine 4 of histone H3 [84].

Statistical evaluation of genes identified using exome sequencing supports the role of chromatin modification in autism. Modeling based on gene coexpression patterns in the human midfetal cortex identified a significant enrichment for genes involved in transcriptional and chromatin regulation [84]. However, this is not unique to autism, or to rare genetic variants. For instance, one study integrating genetic GWAS data for three adult psychiatric conditions (schizophrenia, depression, and bipolar disorder) also identified an enrichment for histone methylation pathways [87]. Another study investigating case-control and trio-based exome sequencing data in schizophrenia identified an enrichment in loss-of-function mutations in *SETD1A* (*KMT2F*), which encodes for a H3K4 histone-lysine N-methyltransferase [88]. In the recent GWAS of autism, there was a nominal enrichment for histone H3 lysine 4 monomethylation marks (H3K4me1), which are histone marks on nucleosomes near enhancers [19,21]. Interestingly, *KMT2C* (*MLL3*), a gene prioritized from exome sequencing and from the recent GWAS of autism, is responsible for the deposition of the H3K4me1 marks, suggesting that rare and common variants may converge on similar pathways. In addition, *KMT2E* (*MLL5*), another SET domain-containing protein but catalytically inactive, was also identified in these GWAS studies.

13.2.3 Risk factors affecting the epigenetics of autism

The GxE (gene × environment) concept represents a combination of genetic susceptibility (G) and environmental factors (E) in understanding complex conditions such as autism. A model has been proposed in which the environmental factors change the gene expression via DNA and histone modifications rather than changes in the DNA sequence [89]. Some environmental factors are known to affect gene expression through epigenetic mechanisms. These changes sometimes persist and can even be transferred to the next cellular generation or furthermore to the next organism generation. This is especially the case during critical periods of embryo formation and development and the fetus is at risk for epigenetic and genomic imprinting disorders such as autism.

Although exposure to environmental factors can be harmful, they cannot cause autism on their own, but they are rather considered as risk factors that in combination with genetic susceptibility increase the risk of developing autism. Many factors have been studied and are divided into prenatal, natal, and postnatal risk factors.

Evidence from epidemiologic and genetic studies showed that increase in parental age (both the father and mother) increases the risk of the offspring developing autism [90]. Studies have yielded conflicting results in determining whether this risk is caused by maternal age, paternal age, or both. Parner and colleagues [91] found that both maternal and paternal ages were associated with a greater risk of ASD in the offspring. The mechanism through which maternal or paternal age impact this risk is different. Whole-genome sequencing studies showed that older fathers have a higher rates of *de novo* mutations. Shelton et al. [92] showed that advancing maternal age increases the risk for autism in the offspring regardless of the paternal age, while increased risk associated with advanced paternal age is observed among younger mothers. This suggests that maternal age plays a more major role than paternal age as a risk factor for later causing autism [92]. Although maternal age is of major importance, younger age of mothers is also critical, as studies reported that mothers younger than 20 years pose a greater risk of growth retardation of fetus and preterm birth, which are both associated with increased risk of autism [93,94].

Maternal well-being is of crucial importance during pregnancy, both physical and mental health. Bleeding and metabolic syndromes such as diabetes, hypertension, and obesity significantly increase the risk of having a child with autism. This can be explained by the resulting hypoxia in utero that alters normal fetal brain development [95]. Viral infections also pose a risk on the fetus due to the abnormal maternal immune activation and elevated cytokines that interfere with brain development epigenetically [93].

Maternal medication use during pregnancy is associated with 46% increased risk for the fetus to develop autism. Placental crossing of different kinds of drugs increases risks for the fetus to develop autism later. For example, antiepileptic drugs, valproic acid, and acetaminophen cause oxidative stress and affect gene expression resulting in abnormal brain development and delays, deficient motor activities, and deficient social behaviors. Other studies showed that analgesic/antipyretic drugs induce apoptosis and necrosis, which are observed in autistic brains [96]. Other drugs have also been reported (see Karimi et al. [93] for review).

Similar to maternal physical well-being, mental health impact is of major importance especially during pregnancy. Mothers who experience depression, anxiety, and mental stress during pregnancy can have indirect effects on genes involved in neurobiological metabolism and in the physiology of the developing fetus through various epigenetic mechanisms. Longer periods of stress and aggression in pregnant mothers exposes the fetus to elevated steroid hormone levels, which have epigenetic effects on fetal brain development [93].

13.2.3.1 Hormones

Steroid hormones play an important role in epigenetic fetal programming on early brain development [97]. In 2015, Baron-Cohen and colleagues [98] showed that levels of amniotic fluid steroid hormones (delta-4 pathway) and cortisol are elevated in those who were later diagnosed with an ASD. These findings were important to elucidate the importance of prenatal environment in the early development of autism. Sex differences are known to exert influences on brain function and neural development. After the bipotential gonad is differentiated by chromosomal sex, steroid hormones are produced and through them the brain acquires its sexual phenotype [99]. Chromosomal sex directs the differentiation of the bipotential gonad, and it is the steroid hormone production by the gonad that then determines the phenotype of the brain in regard to relative masculinization or feminization [100]. Steroid hormones, also known as gonadal hormones, are androgens, estrogens, and progestagens. They are produced in the gonads, the cortex of the suprarenal gland, the placental tissue, and some peripheral tissues. Their secretion is mediated via the hypothalamic-pituitary-gonadal axis. The hypothalamic neurons synthesize and release the gonadotropin-releasing hormone. This stimulates the anterior pituitary to secrete luteinizing hormone (LH) and follicle-stimulating hormone (FSH). Together, the LH and FSH regulate the level of sex hormones derived from cholesterol found in the male testes or the female ovaries. Cholesterol is converted to the sex steroid hormones via a cascade of enzymatic events, mediated by P450 enzymes [101].

During the perinatal window, the human brain is sensitive to the organization effects of sex steroid hormones. In the male brain, testosterone levels have three distinct surges, the first one is mid-gestational and two postnatal peaks [102]. Hence, the exposure of neural tissue to testosterone and/or estradiol (aromatized form of testosterone) during gestation is necessary for both defeminization and masculinization of the developing male brain [100,102,103]. Estradiol-mediated masculinization of the brain induced male sexual behavior in the adult, while defeminization is the process by which the female specific traits are suppressed. This restricts the release of LH and FSH from the anterior pituitary and thus reduces female sexual behavior in the adult [104].

The preoptic area (POA) is perhaps the most reliable sexually dimorphic brain area. It contains the medial preoptic nucleus, which is critical for control of male sexual behavior. It also includes the anteroventral periventricular nucleus (AVPV), which is important for gonadotropin secretion and LH surge in females during ovulation. In the neonatal rodent brain, estradiol protects the sexually dimorphic nucleus of the POA from apoptosis, while it induces apoptosis in the AVPV to defeminize the brain. In addition to those opposing actions to modulate cell death and survival, estradiol also mediates sexually dimorphic synaptogenesis in the MPN. It triggers a twofold increase in putative excitatory synapses and higher density of dendritic spine synapses when compared with females [103,104]. To determine the role of methylation in the estradiol-mediated masculinization of the POA, Nugent and colleagues [105] measured the Dnmt enzymatic activity and expression in male, female, and estradiol-treated female rat pups. They then inhibited Dnmt activity and performed whole-genome bisulfite sequencing. They reported that the Dnmt inhibition mimics steroids and results in masculinized neuronal markers and male sexual behavior. Also, females showed higher levels of methylation than males. The permanent perinatal organizational effects of steroids on the developing brain is long standing, but what this study aimed to investigate is the involvement of DNA methylation in this process and their results show that the feminization of the brain is maintained through methylation and repression of masculinization [105]. This sheds light on how anomalies of prenatal steroidal exposure might be an important risk factor for brain development in autism.

13.3 Discussion

In this chapter, we have discussed the epigenetic mechanisms, such as DNA methylation and histone modifications, implicated in causing ASCs. However, it is worth mentioning that epigenetic variation can be causal or can arise as a consequence of a neurodevelopmental condition. Examples of epigenetic alterations resulting from a disease state could include alterations of expression in immune cells in autoimmune disorders, somatic mutation–induced epigenetic changes in cancer, etc. [60]. Nevertheless, it is not straightforward to disentangle this from epigenetic mechanisms that could play a causal role in a disease state or mental health condition. Therefore it is crucial to consider whether the changes arise before the symptoms of the condition become apparent. Three key factors in elucidating this could be based on whether the variations are inherited and present in all tissues, emerge stochastically during development in utero and are restricted to specific tissues, or arise postnatally because of environmental or lifestyle-related risk factors [60]. Tissue-specific epigenetic variations pose a challenge to compare and replicate findings on ASC-specific biomarkers from independent studies because of the heterogeneity of biological samples (e.g., postmortem brains, blood, saliva).

Another limitation in all these studies is the low number of cases and controls, which is in most cases less than 100. Another concern is the age of the participants, as ASC-discordant pairs are likely to grow more epigenetically discordant over time due to stochastic or environmental risk factors or progression of the condition and may obscure the causal pathway. The differences in the method of autism diagnosis may also act as a potential confounding factor mainly because DMPs/DMRs are set specific for each diagnostic tool. Diversity in the methods of analyses may also introduce difficulties in comparison across multiple studies. For example, in the majority of the EWAS reports discussed previously, some studies focus on probe-based methods (HM27, HM450), whereas other studies use region-based techniques (MIRA, MeDIP). An ideal way for future studies would be to overcome the analysis biases and confounds by using a minimal effect-size cutoff and multiple comparison adjustment. Besides, the use of independent methods in the same study cohort and replication in independent population may resolve the problem of dealing with confounds. Another effective strategy could be standardization of differentially epigenetically regulated probes and regions with sufficient sample size.

The role of environmental and lifestyle-related risk factors on epigenetic programming, especially during fetal development in ASCs, are some of the areas that have gained scientific attention, and in the future, studies may lead toward integrating information from several independent platforms ranging from genomic, epigenomic, transcriptomic, and proteomic levels to thoroughly elucidating the role of epigenetic biomarkers and gene—environment interactions in ASC diagnostics and therapeutics.

Acknowledgments

The authors were supported by the Autism Research Trust, the Templeton World Charitable Foundation (TWCF), the Medical Research Council UK, the Brain and Behaviour Research Foundation, and the Cambridge Commonwealth Trusts during the period of this work.

References

[1] Lai M-C, Lombardo MV, Baron-Cohen S. Autism. Lancet September 2013. https://doi.org/10.1016/S0140-6736(13)61539-1.

[2] Halladay AK, Bishop SL, Constantino JN, et al. Sex and gender differences in autism spectrum disorder: summarizing evidence gaps and identifying emerging areas of priority. Mol Autism 2015;6(1):36. https://doi.org/10.1186/s13229-015-0019-y.

[3] Schaafsma SM, Pfaff DW. Etiologies underlying sex differences in autism spectrum disorders. Front Neuroendocrinol 2014;35:255—71.

[4] Russell G, Rodgers LR, Ukoumunne OC, et al. Prevalence of parent-reported ASD and ADHD in the UK: findings from the millennium cohort study. J Autism Dev Disord 2014;44(1):31—40. https://doi.org/10.1007/s10803-013-1849-0.

[5] Leventhal BL, Koh Y, Laska E, et al. Prevalence of autism spectrum disorders in a total population sample. Am J Psychiatry 2011;168(1):904—12. https://doi.org/10.1176/appi.ajp.2011.10101532.

[6] Saemundsen E, Magnússon P, Georgsdóttir I, et al. Prevalence of autism spectrum disorders in an Icelandic birth cohort. BMJ Open 2013;3(6):1—7. https://doi.org/10.1136/bmjopen-2013-002748.

[7] Wan Y, Hu Q, Li T, et al. Prevalence of ASD among children in China. Shanghai Arch Psychiatry 2013; 25(2):70–80.

[8] Parellada M, Penzol MJ, Pina L, et al. The neurobiology of autism spectrum disorders. Eur Psychiatry 2014; 29(1):11–9. https://doi.org/10.1016/j.eurpsy.2013.02.005.

[9] Folstein SE, Rutter M. Infantile autism: a genetic study of 21 twin pairs. JCPP (J Child Psychol Psychiatry) 1977;18(4):297–321. https://doi.org/10.1111/j.1469-7610.1977.tb00443.x.

[10] Tick B, Bolton PF, Happé F, et al. Heritability of autism spectrum disorders: a meta-analysis of twin studies. J Child Psychol Psychiatry Allied Discip 2016;57(5):585–95. https://doi.org/10.1111/jcpp.12499.

[11] Ronald A, Hoekstra RA. Autism spectrum disorders and autistic traits: a decade of new twin studies. Am J Med Genet Part B Neuropsychiatr Genet 2011;156(3):255–74. https://doi.org/10.1002/ajmg.b.31159.

[12] Wang K, Gaitsch H, Poon H, et al. Classification of common human diseases derived from shared genetic and environmental determinants. Nat Genet 2017;49(9):1319–25. https://doi.org/10.1038/ng.3931.

[13] Jokiranta-Olkoniemi E, Cheslack-Postava K, Sucksdorff D, et al. Risk of psychiatric and neuro developmental disorders among siblings of probands with autism spectrum disorders. JAMA Psychiatry 2016; 73(6):622. https://doi.org/10.1001/jamapsychiatry.2016.0495.

[14] Sandin S, Lichtenstein P, Kuja-Halkola R, et al. The familial risk of autism. J Am Med Assoc 2014;311(17): 1770–7. https://doi.org/10.1001/jama.2014.4144.

[15] Sandin S, Lichtenstein P, Kuja-Halkola R, et al. The heritability of autism spectrum disorder. J Am Med Assoc 2017;318(12):1182. https://doi.org/10.1001/jama.2017.12141.

[16] Klei LL, Sanders SJ, Murtha MT, et al. Common genetic variants, acting additively, are a major source of risk for autism. Mol Autism 2012;3(1):9. https://doi.org/10.1186/2040-2392-3-9.

[17] Gaugler T, Klei LL, Sanders SJ, et al. Most genetic risk for autism resides with common variation. Nat Genet 2014;46(8):881–5. https://doi.org/10.1038/ng.3039.

[18] Cross-Disorder Group of the Psychiatric Genomics Consortium. Identification of risk loci with shared effects on five major psychiatric disorders: a genome-wide analysis. Lancet 2013;381(9875):1371–9. https://doi.org/10.1016/S0140-6736(12)62129-1.

[19] Grove J, Ripke S, Als TD, et al. Common risk variants identified in autism spectrum disorder. bioRxiv November 2017:224774. https://doi.org/10.1101/224774.

[20] Peyrot WJ, Boomsma DI, Penninx BWJH, et al. Disease and polygenic architecture: avoid trio design and appropriately account for unscreened control subjects for common disease. Am J Hum Genet 2016;98(2): 382–91. https://doi.org/10.1016/j.ajhg.2015.12.017.

[21] The Autism Spectrum Disorders Working Group of The Psychiatric Genomics Consortium. Meta-analysis of GWAS of over 16,000 individuals with autism spectrum disorder highlights a novel locus at 10q24.32 and a significant overlap with schizophrenia. Mol Autism 2017;8(1):21. https://doi.org/10.1186/s13229-017-0137-9.

[22] Bulik-Sullivan BK, Finucane HK, Anttila V, et al. An atlas of genetic correlations across human diseases and traits. Nat Genet 2015;47(11):1236–41. https://doi.org/10.1038/ng.3406.

[23] Sniekers S, Stringer S, Watanabe K, et al. Genome-wide association meta-analysis of 78,308 individuals identifies new loci and genes influencing human intelligence. Nat Genet May 2017. https://doi.org/10.1038/ng.3869.

[24] Samocha KE, Robinson EB, Sanders SJ, et al. A framework for the interpretation of de novo mutation in human disease. Nat Genet 2014;46(9):944–50. https://doi.org/10.1038/ng.3050.

[25] Robinson EB, Samocha KE, Kosmicki JA, et al. Autism spectrum disorder severity reflects the average contribution of de novo and familial influences. Proc Natl Acad Sci Unit States Am 2014;111(42):15161–5. https://doi.org/10.1073/pnas.1409204111.

[26] Warrier V, the 23andMe Research team, Bourgeron T, Baron-Cohen S. Genome-wide association study of social relationship satisfaction: significant loci and correlations with psychiatric conditions. bioRxiv October 2017. https://doi.org/10.1101/196071.
[27] Warrier V, Toro R, Chakrabarti B, et al. Genome-wide analyses of self-reported empathy: correlations with autism, schizophrenia, and anorexia nervosa. bioRxiv November 2017. https://doi.org/10.1101/050682. 50682.
[28] Baron-Cohen S. The hyper-systemizing, assortative mating theory of autism. Prog Neuro-Psychopharmacol Biol Psychiatry 2006;30(5):865–72. https://doi.org/10.1016/j.pnpbp.2006.01.010.
[29] Sanders SJ, He X, Willsey AJ, et al. Insights into Autism Spectrum Disorder genomic architecture and biology from 71 risk loci. Neuron 2015;87(6):1215–33. https://doi.org/10.1016/j.neuron.2015.09.016.
[30] Vorstman JAS, Parr JR, Moreno-De-Luca D, et al. Autism genetics: opportunities and challenges for clinical translation. Nat Rev Genet 2017;18(6):362–76. https://doi.org/10.1038/nrg.2017.4.
[31] Mbadiwe T, Millis RM. Epigenetics and autism. Autism Res Treat 2013;2013. https://doi.org/10.1155/2013/826156. 826156.
[32] Riggs AD, Xiong Z, Wang L, et al. Methylation dynamics, epigenetic fidelity and X chromosome structure. Novartis Found Symp 1998;214:214–32.
[33] Houston I, Peter CJ, Mitchell A, et al. Epigenetics in the human brain. Neuropsychopharmacology 2013; 38(1):183. https://doi.org/10.1038/npp.2012.78.
[34] Adalsteinsson BT, Ferguson-Smith AC. Epigenetic control of the genome-lessons from genomic imprinting. Genes 2014;5(3):635–55. https://doi.org/10.3390/genes5030635.
[35] Sweatt JD, Meaney MJ, Nestler EJ, et al. Epigenetic regulation in the nervous system. Elsevier; 2013. https://doi.org/10.1016/B978-0-12-391494-1.00012-4.
[36] Lasalle JM. Epigenomic strategies at the interface of genetic and environmental risk factors for autism. J Hum Genet 2013;58(7):396–401. https://doi.org/10.1038/jhg.2013.49.
[37] James SJ, Cutler P, Melnyk S, et al. Metabolic biomarkers of increased oxidative stress and impaired methylation capacity in children with autism. Am J Clin Nutr 2004;80(1):1611–7.
[38] Rusiecki JA, Baccarelli A, Bollati V, et al. Global DNA hypomethylation is associated with high sserum-persistent organic pollutants in Greenlandic Inuit. Environ Health Perspect 2008;116(11):1547–52. https://doi.org/10.1289/ehp.11338.
[39] Dolinoy DC, Weidman JR, Jirtle RL. Epigenetic gene regulation : linking early developmental environment to adult disease. Reprod Toxicol 2007;23(1):297–307. https://doi.org/10.1016/j.reprotox.2006.08.012.
[40] Jirtle RL, Skinner MK. Environmental epigenomics and disease susceptibility. Nat Genet 2007;8(1): 253–62. https://doi.org/10.1038/nrg2045.
[41] Kriaucionis S, Heintz N. The nuclear DNA base 5-hydroxymethylcytosine is present in purkinje neurons and the brain. Science 2009;324(5929):929–30.
[42] Bachman M, Uribe-Lewis S, Yang X, et al. 5-Hydroxymethylcytosine is a predominantly stable DNA modification. Nat Chem 2014;6(12):1049–55. https://doi.org/10.1038/nchem.2064.
[43] Cheng Y, Bernstein A, Chen D, et al. 5-Hydroxymethylcytosine: a new player in brain disorders? Exp Neurol 2015;268:3–9. https://doi.org/10.1016/j.expneurol.2014.05.008.
[44] LaSalle JM, Yasui DH. Evolving role of MeCP2 in Rett syndrome and autism. Epigenomics 2009;1(1): 119–30. https://doi.org/10.2217/epi.09.13.
[45] Yoo HJ, Kim K, Kim IH, et al. Whole exome sequencing for a patient with Rubinstein-Taybi Syndrome reveals *de Novo* variants besides an overt *CREBBP* mutation. Int J Mol Sci 2015;16(3):5697–713. https://doi.org/10.3390/ijms16035697.
[46] Ramocki MB, Peters SU, Tavyev YJ, et al. Autism and other neuropsychiatric symptoms are prevalent in individuals with *MeCP* duplication syndrome. Ann Neurol 2009;66(6):771–82. https://doi.org/10.1002/ana.21715.

[47] Chahrour M, Jung SY, Shaw C, et al. MeCP2, a key contributor to neurological disease, activates and represses transcription. Science 2008;320(5880):1224–9. https://doi.org/10.1126/science.1153252.

[48] LaSalle JM, Reiter LT, Chamberlain SJ. Epigenetic regulation of UBE3A and roles in human neurodevelopmental disorders. Epigenomics 2015;7(7):1213–28. https://doi.org/10.2217/epi.15.70.Epigenetic.

[49] Leung KN, Chamberlain S, Lalande M, et al. Neuronal chromatin dynamics of imprinting in development and disease. J Cell Biochem 2011;112(2):365–73. https://doi.org/10.1002/jcb.22958.Neuronal.

[50] Uchino S, Waga C. SHANK3 as an autism spectrum disorder-associated gene. Brain Dev 2013;35(2):106–10. https://doi.org/10.1016/j.braindev.2012.05.013.

[51] Zhu L, Wang X, Li XL, et al. Epigenetic dysregulation of SHANK3 in brain tissues from individuals with autism spectrum disorders. Hum Mol Genet 2014;23(6):1563–78. https://doi.org/10.1093/hmg/ddt547.

[52] LoParo D, Waldman ID. The oxytocin receptor gene (OXTR) is associated with autism spectrum disorder: a meta analysis. Mol Psychiatr 2015;20(5):640–6. https://doi.org/10.1038/mp.2014.77.

[53] Gregory SG, Connelly JJ, Towers AJ, et al. Genomic and epigenetic evidence for oxytocin receptor deficiency in autism. BMC Med 2009;7:62. https://doi.org/10.1186/1741-7015-7-62.

[54] Yip J, Soghomonian JJ, Blatt GJ. Decreased GAD67 mRNA levels in cerebellar Purkinje cells in autism: pathophysiological implications. Acta Neuropathol 2007;113:559–68. https://doi.org/10.1007/s00401-006-0176-3.

[55] Yip J, Soghomonian JJ, Blatt GJ. Increased GAD67 mRNA expression in cerebellar interneurons in autism: implications for Purkinje cell dysfunction. J Neurosci Res 2008;86:525–30. https://doi.org/10.1002/jnr.21520.

[56] Fatemi SH, Halt AR, Stary JM, Kanodia R, Schulz SC, Realmuto GR. Glutamic acid decarboxylase 65 and 67 kDa proteins are reduced in autistic parietal and cerebellar cortices. Biol Psychiatry 2002;52:805–10.

[57] Fatemi SH, Stary JM, Halt AR, et al. Dysregulation of Reelin and Bcl-2 proteins in autistic cerebellum. J Autism Dev Disord 2001;31:529–35.

[58] Zhubi A, Chen Y, Dong E, et al. Increased binding of MeCP2 to the GAD1 and RELN promoters may be mediated by an enrichment of 5-hmC in autism spectrum disorder (ASD) cerebellum. Transl Psychiatry 2014;4(1):e349. https://doi.org/10.1038/tp.2013.123.

[59] Bernstein BE, Meissner A, Lander ES. The mammalian epigenome. Cell 2007;128:669–81.

[60] Rakyan VK, Down TA, Balding DJ, Beck S. Epigenome-wide association studies for common human diseases. Nat Rev Genet 2011;12:529–41.

[61] Olkhov-Mitsel E, Bapat B. Strategies for discovery and validation of methylated and hydroxymethylated DNA biomarkers. Cancer Med 2012;1(2):237–60. https://doi.org/10.1002/cam4.22.

[62] Stewart SK, Morris TJ, Guilhamon P, et al. OxBS-450K: a method for analysing hydroxymethylation using 450K BeadChips. Methods 2015;72(C):9–15. https://doi.org/10.1016/j.ymeth.2014.08.009.

[63] Dedeurwaerder S, Defrance M, Bizet M, et al. A comprehensive overview of Infinium Human-Methylation450 data processing. Briefings Bioinf 2013;15(6):929–41. https://doi.org/10.1093/bib/bbt054.

[64] Ginsberg MR, Rubin RA, Falcone T, et al. Brain transcriptional and epigenetic associations with autism. PLoS One 2012;7:e44736. https://doi.org/10.1371/journal.pone.0044736.

[65] Ladd-Acosta C, Hansen KD, Briem E, et al. Common DNA methylation alterations in multiple brain regions in autism. Mol Psychiatr 2014;19(8):862–71. https://doi.org/10.1038/mp.2013.114.

[66] Nardone S, Sams DS, Reuveni E, et al. DNA methylation analysis of the autistic brain reveals multiple dysregulated biological pathways. Transl Psychiatry 2014;4(9):1–9. https://doi.org/10.1038/tp.2014.70.

[67] Voineagu I, Wang X, Johnston P, et al. Transcriptomic analysis of autistic brain reveals convergent molecular pathology. Nature 2011;474(7351):380–4. https://doi.org/10.1038/nature10110.

[68] Nardone S, Sams DS, Zito A, et al. Dysregulation of cortical neuron DNA methylation profile in autism spectrum disorder. Cerebr Cortex December 1, 2017;27(12):5739–54. https://doi.org/10.1093/cercor/bhx250.

[69] Parikshak NN, Swarup V, Belgard TG, et al. Genome-wide changes in lncRNA, splicing, and regional gene expression patterns in autism. Nature 2016;540(7633):423–7. https://doi.org/10.1038/nature20612.

[70] Nguyen A, Rauch TA, Pfeifer GP, et al. Global methylation profiling of lymphoblastoid cell lines reveals epigenetic contributions to autism spectrum disorders and a novel autism candidate gene, RORA, whose protein product is reduced in autistic brain. FASEB J August 2010;24(8):3036–51. https://doi.org/10.1096/fj.10-154484.

[71] Wong CCY, Meaburn EL, Ronald A, et al. Methylomic analysis of monozygotic twins discordant for autism spectrum disorder and related behavioural traits. Mol Psychiatr 2014;19(4):495–503. https://doi.org/10.1038/mp.2013.41.

[72] Hannon E, Schendel D, Ladd-Acosta C, et al. Elevated polygenic burden for autism is associated with differential DNA methylation at birth. bioRxiv November 2017. https://doi.org/10.1101/225193.

[73] Andrews SV, Ellis SE, Bakulski KM, et al. Cross-tissue integration of genetic and epigenetic data offers insight into autism spectrum disorder. Nat Commun 2017;8 (1):1011. https://doi.org/10.1038/s41467-017-00868-y.

[74] Feinberg JI, Bakulski KM, Jaffe AE, et al. Paternal sperm DNA methylation associated with early signs of autism risk in an autism-enriched cohort. Int J Epidemiol 2015;44(4):1199–210. https://doi.org/10.1093/ije/dyv028.

[75] Ellis SE, Gupta S, Moes A, et al. Exaggerated CpH methylation in the autism-affected brain. bioRxiv 2016: 69120. https://doi.org/10.1101/069120.

[76] Balemans MC, Huibers MM, Eikelenboom NW, et al. Reduced exploration, increased anxiety, and altered social behavior: autistic-like features of euchromatin histone methyltransferase 1 heterozygous knockout mice. Behav Brain Res 2010;208(1):47–55.

[77] Adegbola A, Gao H, Sommer S, et al. A novel mutation in JARID1C/SMCX in a patient with autism spectrum disorder (ASD). Am J Med Genet 2008;146A(4):505–11.

[78] Shilatifard A. Molecular implementation and physiological roles for histone H3 lysine 4 (H3K4) methylation. Curr Opin Cell Biol 2008;20(3):341–8.

[79] Shulha HP, Cheung I, Whittle C, et al. Epigenetic signatures of autism trimethylated H3K4 landscapes in prefrontal neurons. Arch Gen Psychiatr 2012;69(3):314–24. https://doi.org/10.1001/archgenpsychiatry.2011.151.

[80] Sun W, Poschmann J, et al. Histone acetylome-wide association study of autism spectrum disorder. Cell 2016;167(Issue 5):1385–1397.e11.

[81] Bernier R, Golzio C, Xiong B, et al. Disruptive CHD8 mutations define a subtype of autism early in development. Cell 2014;158(2):263–76. https://doi.org/10.1016/j.cell.2014.06.017.

[82] Cotney J, Muhle RA, Sanders SJ, et al. The autism-associated chromatin modifier CHD8 regulates other autism risk genes during human neurodevelopment. Nat Commun 2015;6:6404. https://doi.org/10.1038/ncomms7404.

[83] Hamdan FF, Srour M, Capo-Chichi J-M, et al. De novo mutations in moderate or ssevere intellectual disabilityCooper GM, editor. PLoS Genet 2014;10(10):e1004772. https://doi.org/10.1371/journal.pgen.1004772.

[84] De Rubeis S, He X, Goldberg AP, et al. Synaptic, transcriptional and chromatin genes disrupted in autism. Nature 2014;515(7526):209–15. https://doi.org/10.1038/nature13772.

[85] Helsmoortel C, Vulto-van Silfhout AT, Coe BP, et al. A SWI/SNF-related autism syndrome caused by de novo mutations in ADNP. Nat Genet 2014;46(4):380–4. https://doi.org/10.1038/ng.2899.

[86] Ronan JL, Wu W, Crabtree GR. From neural development to cognition: unexpected roles for chromatin. Nat Rev Genet 2013;14(5):347–59. https://doi.org/10.1038/nrg3413.

[87] O'Dushlaine C, Rossin L, Lee PH, et al. Psychiatric genome-wide association study analyses implicate neuronal, immune and histone pathways. Nat Neurosci 2015;18(2):199–209. https://doi.org/10.1038/nn.3922.
[88] Singh T, Kurki MI, Curtis D, et al. Rare loss-of-function variants in SETD1A are associated with schizophrenia and developmental disorders. Nat Neurosci 2016;19(4):571–7. https://doi.org/10.1038/nn.4267.
[89] Kubota T, Mochizuki K. Epigenetic effect of environmental factors on autism spectrum disorders. Int J Environ Res Publ Health 2016;13(5):504. https://doi.org/10.3390/ijerph13050504.
[90] Durkin MS, Maenner MJ, Newschaffer CJ, et al. Advanced parental age and the risk of autism spectrum disorder. Am J Epidemiol 2008;168(11):1268–76. https://doi.org/10.1093/aje/kwn250.
[91] Parner ET, Baron-Cohen S, Lauristen B, et al. Parental age and autism spectrum disorders. Ann Epidemiol 2012;22(3):143–50. https://doi.org/10.1016/j.annepidem.2011.12.006.
[92] Shelton JF, Tancredi DJ, Hertz-Picciotto I. Independent and dependent contributions of advanced maternal and paternal ages to autism risk. Autism Res 2010;3(1):30–9. https://doi.org/10.1002/aur.116.
[93] Karimi P, Kamali E, Mousavi SM, et al. Environmental factors influencing the risk of autism. J Res Med Sci 2017;22:27. https://doi.org/10.4103/1735-1995.200272.
[94] Larsson HJ, Eaton WW, et al. Risk factors for autism: perinatal factors, parental psychiatric history, and ssocioeconomic status. Am J Epidemiol May 15 , 2005;161(Issue 10):916–25.
[95] Krakowiak P, Walker CK, Bremer AA, et al. Maternal metabolic conditions and risk for autism and other neurodevelopmental disorders. Pediatrics 2012;129:e1121–8.
[96] Bauer AZ, Kriebel D. Prenatal and perinatal analgesic exposure and autism: an ecological link. Environ Health 2013;12:41. https://doi.org/10.1186/1476-069X-12-41.
[97] Sarachana T, Hu VW. Differential recruitment of coregulators to the RORA promoter adds another layer of complexity to gene (dys) regulation by sex hormones in autism. Mol Autism 2013;4(1):39. https://doi.org/10.1186/2040-2392-4-39.
[98] Baron-Cohen S, Auyeung B, Nørgaard-Pedersen B, et al. Elevated fetal steroidogenic activity in autism. Mol Psychiatr 2015;20(3):369–76. https://doi.org/10.1038/mp.2014.48.
[99] McCarthy MM, Auger AP, Bale TL, et al. The epigenetics of sex differences in the brain. J Neurosci 2009; 29(41):12815–23. https://doi.org/10.1523/JNEUROSCI.3331-09.2009.
[100] Ghahramani N. The effects of perinatal testosterone exposure on the DNA methylome of the mouse brain are late-emerging. Biol Sex Differ 2014;5(8):1–18.
[101] Lenz B, Müller CP, Stoessel C, et al. Sex hormone activity in alcohol addiction: integrating organizational and activational effects. Prog Neurobiol 2012;96(1):136–63. https://doi.org/10.1016/j.pneurobio.2011.11.001.
[102] McCarthy MM. Estradiol and the developing brain. Physiol Rev 2008;88(1):91–124. https://doi.org/10.1152/physrev.00010.2007.
[103] Arnold AP, Chen X. What does the "four core genotypes" mouse model tell us about sex differences in the brain and other tissues? Front Neuroendocrinol 2009;30(1):1–9. https://doi.org/10.1016/j.yfrne.2008.11.001. What].
[104] Nugent BM, Schwarz JM, McCarthy MM. Hormonally mediated epigenetic changes to steroid receptors in the developing brain: implications for sexual differentiation. Horm Behav 2011;59(3):338–44. https://doi.org/10.1016/j.yhbeh.2010.08.009.
[105] Nugent BM, Wright CL, Shetty AC, et al. Brain feminization requires active repression of masculinization via DNA methylation. Nat Neurosci 2015;18(5):690–7. https://doi.org/10.1038/ja.2013.113.

CHAPTER

Chromatin modification and remodeling in schizophrenia

14

Jubao Duan[1,2]

Center for Psychiatric Genetics, NorthShore University HealthSystem, Evanston, IL, United States[1]; *Department of Psychiatry and Behavioral Neurosciences, The University of Chicago, Chicago, IL, United States*[2]

Chapter Outline

14.1 Introduction **304**
14.2 SZ GWAS implicate gene expression and chromatin regulation as a possible causal molecular mechanism **305**
14.3 SZ and DNA methylation **306**
14.3.1 Aberrant DNA methylation in SZ 307
14.3.2 Genetic control of DNA methylation and its relevance to SZ 308
14.4 SZ and histone modifications **309**
14.4.1 Histone acetylation 310
14.4.2 Histone methylation 311
14.5 SZ and 2D chromatin structure **312**
14.6 SZ and higher-order chromatin structure **313**
14.7 SZ genetic risk variants affect chromatin remodeling gene pathway **315**
14.7.1 Analysis of common SZ variants implicates the dysregulated chromatin-signaling pathway 315
14.7.2 Rare SZ coding risk variants and chromatin remodeling 316
14.7.3 SZ-associated CNVs and abnormal chromatin organization 317
14.8 hiPSC model combined with CRISPR editing for studying SZ-relevant chromatin function **318**
14.8.1 hiPSC-derived neurons as a cellular model for neurodevelopmental disorder 318
14.8.2 CRISPR-based approaches for genome/epigenome perturbation 319
14.8.3 CRISPR-based 2D and 3D chromatin perturbation relevant to SZ in hiPSC models 320
14.9 Therapeutic drugs that target chromatin structure and activity in SZ **321**
14.10 Conclusion and perspectives **322**
Acknowledgments **323**
References **323**

Chromatin Signaling and Neurological Disorders. https://doi.org/10.1016/B978-0-12-813796-3.00014-6

14.1 Introduction

Schizophrenia (SZ) is a devastating mental disorder that afflicts ~1% of the population worldwide. The core features of SZ include both positive symptoms (delusions, hallucinations, and disordered thinking) and negative symptoms (social withdrawal, lack of motivation, and poverty of speech), as well as some cognitive impairments [1]. SZ patients also often experience anxiety, depression, or substance-abuse disorders [2]. Males are more often affected and experience more severe symptoms. Schizophrenia, such as other psychiatric disorders, is a syndrome with substantial heterogeneity [1]. Unfortunately, there are no diagnostic tests or biomarkers for SZ. The exact pathophysiology of SZ remains unclear; multiple lines of evidence suggest SZ as a neurodevelopmental disorder with impaired frontal cortical development [3–8], possibly involving dysfunction of multiple major neurotransmitter systems (i.e. dopaminergic-DA, glutamatergic, and GABAergic). Although medications can improve SZ symptoms, most available antipsychotic drugs are still based on the blockade of the dopamine D2 receptor (DRD2), a mechanism discovered over 50 years ago [9]. A better understanding of SZ etiology and disease biology will likely allow identification of more effective therapeutic targets.

The recent success of schizophrenia genome-wide association studies (GWAS) has given the field a greater understanding of SZ etiology. SZ GWAS support the polygenic nature of SZ, meaning that many genes are involved, but each contributing only small effect [10]. SZ GWAS have identified over 100 significant susceptibility loci with common variants associated with the disease [10–15], providing unprecedented opportunity to understand the newly revealed disease biology and identify novel drug targets. However, these genetic findings only explain a portion of population risk to SZ, implying the importance of nongenetic factors. This is consistent with the observations in SZ twin studies, where there is only ~50% concordance rate for developing SZ. The high discordance rate for SZ twins suggests a critical role of environmental factors in disease etiology. Possible environmental factors, such as stress events, viral infections, or nutritional deficits during the pregnancy may act independently, or through interaction with genetic factors to cause lasting changes of disease-relevant gene expression.

As a major epigenetic mechanism, chromatin states strongly influence gene transcription, cellular differentiation, and function. The mammalian genome is organized in the form of chromatin, with nucleosomes as the fundamental building block, composed of ~146 bp of DNA and octamers of histone proteins [16] (see Chapter 1). In some genomic regions, the nucleosomes are packed very loosely (i.e., euchromatin or open chromatin) and are easily accessible to protein regulatory factors, while other regions are packed very compactly (i.e., heterochromatin). Open chromatin is associated with active gene transcription, while closed chromatin is often associated with transcriptional repression. Chromatin structure, in 2D (the linear nucleosome units) or 3D (the high order of chromatin with physical interactions) form, is dynamic and exerts profound influence over gene expression and fundamental cellular processes. Both DNA methylation and histone biochemical modification determine the chromatin structure and impact gene expression [17]. We and others have recently reported how regulatory sequences associated with SZ can alter chromatin structures, thereby changing downstream molecular and cellular phenotypes [18,19]. This chapter covers the current status of research on the link between SZ and DNA methylation, histone modifications, and 2D/3D chromatin structures. There is also discussion of how emerging cutting-edge technologies, such as patient-specific induced pluripotent stem cells (iPSCs) and clustered regularly interspaced short palindromic repeats (CRISPR)/CRISPR-associated (Cas)9-mediated genome/epigenome editing, may further understand disease biology

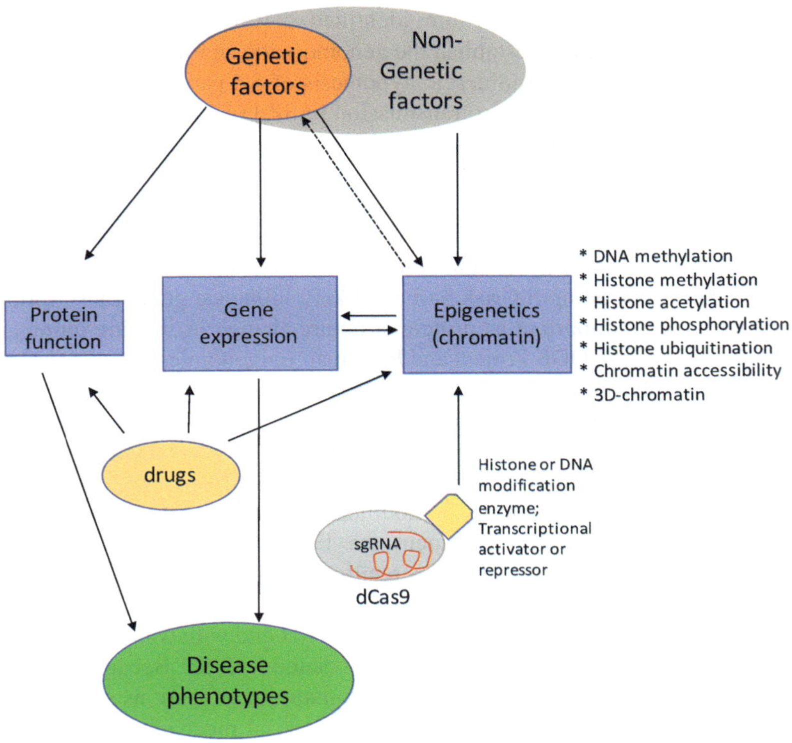

FIGURE 14.1

A diagram that shows the information flow from genetics/nongenetics risk factors of schizophrenia to disease phenotypes. Very few genetic risk variants (common polymorphisms) affect protein function and most others (noncoding risk variants) influence gene transcription through acting on epigenome (chromatin level). Listed are major chromatin modifications discussed in this chapter. Both drugs and CRISPR/dCas9-mediated epigenome editing can perturb chromatin signaling at epigenetic level.

related to chromatin signaling and aid in development of novel drug targets. The central aim of the chapter is to describe the information flow at a complex system level of how the SZ genetic and epigenetic factors (chromatin signaling) interact with each other to influence intermediate phenotype (i.e., gene expression) and disease phenotypes (Fig. 14.1).

14.2 SZ GWAS implicate gene expression and chromatin regulation as a possible causal molecular mechanism

GWAS is a systematic approach to interrogate common genetic variations in the entire genome for genetic association with a disease in a well-powered sample. Recent SZ GWAS have identified a plethora of risk loci [10–15]. With 36,989 SZ cases and 113,075 controls, the landmark study carried out by the Psychiatric Genomics Consortium identified >100 risk loci, spanning hundreds of genes [15]. These discoveries not only establish the genome-wide significant association to the *DRD2* locus [15], central to the classical dopaminergic hypothesis of SZ pathogenesis, but also support the hypothesis of abnormal glutamatergic neurotransmission in SZ [14,15]. Despite the exciting findings from SZ GWAS, a substantial proportion of genetic risk remains unexplained, suggesting the possible influence of epigenetic factors.

SZ GWAS findings also strongly imply the importance of chromatin modification features in helping to identify the putatively functional disease risk variants. Most SZ-associated variants, as for other complex disorders, are located in the noncoding part of the genome, for which the functionality is difficult to predict. These noncoding SZ-risk variants likely influence gene transcription. Expression variation is expected to be as important as protein structure change in shaping human-specific brain function [20,21]. The recent ENCODE Project and Roadmap Epigenomics Program [22–24] provide a rich empirical resource of chromatin state marks and transcription factor-binding sites (TFBS) in 349 cell and tissue samples for bioinformatic annotation of functional noncoding sequences [25]. These genome-wide chromatin marks can help to predict promoters, enhancers, insulators, and TFBS. Indeed, similar to other common diseases, SZ-associated variants are enriched in open chromatin [15,18,24,26,27]. An effort to profile chromatin modifications in human brains of psychiatric patients and controls (PsychENCODE) may complement the ENCODE-annotated chromatin marks, by providing more disease-relevant information, thus facilitating the functional interpretation of SZ-risk variants of regulatory potential.

The link between chromatin modifications and SZ does not end with using chromatin modification marks to help infer the functional noncoding variants associated with SZ. Because temporal and spatial gene expression in mammals is tightly regulated by chromatin modifications and remodeling, the SZ GWAS finding of gene expression alteration as a possible causal mechanism also implies a possible link between chromatin signaling and SZ pathogenesis. In fact, gene network analyses of both common risk variants implicated by SZ GWAS and rare risk variants identified by large-scale sequencing efforts have strongly nominated chromatin remodeling and modification as one of the most relevant pathogenic pathways for SZ (see later).

14.3 SZ and DNA methylation

Methylation is a major type of DNA modification in mammals. This chemical process is mainly catalyzed by three DNA methyltransferases (DNMTs), DNMT1, 3a, and 3b, occurring at the 5-carbon position of the cytosine (5 mC). Although 5 mC exists mostly in the CpG dinucleotide context, about 70%–80% of CpGs are methylated in mammalian genomes. The methylation of DNA can be enzymatically reversed by ten-eleven translocation (TET) family enzymes (TET 1, 2, and 3) via oxidizing 5-methylcytosine. Methylated DNA regions are often associated with repressed transcription, likely mediated by weakening transcription factor (TF)-binding and/or altering chromatin state. The dynamics of DNA methylation and demethylation profoundly influence many biological processes, including transcriptional regulation, genomic stability, chromatin structure modulation, and development [28]. As one of the major epigenetic modifications, the role of DNA methylation has been extensively studied in peripheral blood and postmortem brain tissues for neuropsychiatric disorders, including SZ.

14.3.1 Aberrant DNA methylation in SZ

Earlier studies of aberrant DNA methylation in SZ have been focused on some specific genes of pathophysiological relevance to SZ, and more recently on a genome-wide scale (see review [29]). Studies of aberrant DNA methylation in SZ have been historically focused on genes of dopaminergic, GABAergic, and serotonergic systems. For dopaminergic genes, the catechol-O-methyltransferase (*COMT*) gene, which encodes the enzyme to metabolize dopamine, has been the major focus. The hypomethylation of the *COMT* promoter region has been reported in postmortem brain samples of SZ [29], and in saliva DNA of SZ and bipolar patients [30]. However, this observation does not seem to reconcile with the hyperdopaminergic activity in SZ brains; the hypomethylation of *COMT* in SZ is expected to lead to a higher level of COMT and, consequently, greater dopamine degradation. For GABAergic genes, in general, multiple lines of evidence suggest that the increased expression of DNMT genes in SZ postmortem brains has led to the hypermethylation of GABAergic genes [29]. For instance, the promoter hypermethylation of reelin (*RELN*) and glutamic acid decarboxylase (*GAD1*) have been reported in SZ cortex, accompanied by the reduced expression of RELN and GAD1, and increased expression of DNMT1 [31]. Besides *RELN* and *GAD1*, a subset of GAD1-regulated genes (*MSX1*, *CCND2*, and *DAXX* a histone chaperone) have also been shown to be differentially methylated in the hippocampus of SZ patients [32]. With regard to serotonergic genes, promoter hypermethylation of both serotonin receptor type-1 (*HTR1A*) and type-2 (*HTR2A*) has been reported in blood and postmortem brains of SZ patients, consistent with the reduced expression of both genes in SZ [29].

Although these focused studies seem to support the role of aberrant methylation in SZ, they are often limited by a biased selection of small sets of genes and by small cohorts. With recent technological development of epigenomic profiling, systematic studies of genome-wide methylation in blood or postmortem brain tissues of SZ cases and controls are gaining momentum. In a recent genome-wide DNA methylation analysis of postmortem human brain tissues of 24 SZ patients and 24 controls, methylation of 4641 probes corresponding to 2929 unique genes was found to be associated with SZ case/control status. Some of the genes were previously implicated in SZ [33] (i.e., *NOS1*, *AKT1*, *DTNBP1*, *DNMT1*, *PPP3CC*, and *SOX10*) and showed aberrant methylation in SZ. Overall, the sample size of these genome-wide methylation studies is relatively small, which may partially explain why these studies replicated very few of the previous individual candidate gene studies, e.g., aberrant methylation of some GABAergic genes and DNMT1 in SZ (see review [29]).

The largest genome-wide methylation study of SZ consisted of 759 SZ cases and 738 controls, in a relatively homogenous Swedish population [34]. This study used a methyl-CpG-binding domain protein to enrich the methylated genomic fraction, followed by next-generation DNA sequencing. Despite having a large sample, only 25 sites showed significant differential methylation after Bonferroni correction and 139 differentially methylated sites were identified at a more relaxed statistical cut-off (false discovery rate of 0.01). The most significant locus contains *FAM63B*, a gene that is part of the microRNA-regulated networks related to neuronal differentiation and dopaminergic gene expression [34]. It is noteworthy that *RELN*, one of the most frequently studied candidate genes for methylation [29], was found to show differential methylation associated with SZ. However, most other hits are related to immune rather than brain function, which is not unexpected given that peripheral bloods were used for methylation profiling. Peripheral bloods are a type of sample at convenience and easy to collect in large scale but may not be so relevant to disease biology of SZ.

With more relevant tissues, a more recent genome-wide methylation study in dorsolateral prefrontal cortex (DLPFC) of a relatively large SZ case/control sample (191 SZ patients and 335 controls) presented more biologically meaningful supporting evidence for aberrant DNA methylation in SZ [35]. The Illumina HumanMethylation450 (450k) microarray was used for methylation profiling. Although the initial analysis of all 191 adult SZ patients and the 240 nonpsychiatric controls older than 16 did not yield any significant finding, reanalyzing a subset of the sample (108 patients and 136 controls), after excluding possible technical artifacts such as experimental batch effects that might have masked the biological differences, identified 2104 CpG sites that showed significant differential methylation. These SZ-associated differentially methylated CpGs were in or near genes significantly enriched for embryo development, cell fate commitment, and nervous system differentiation, suggesting possible abnormal methylation in neurodevelopmental aspects of SZ. However, almost all the differences of methylation involved hypomethylation in SZ patients, which is inconsistent with previous findings that exclusively showed hypermethylation of SZ candidate genes in SZ patients [29]. Furthermore, very few overlapping loci were found compared to previous genome-wide methylation studies in SZ postmortem brains [29,36]. These inconsistencies may be due to small sample sizes in previous studies and, more likely, the cellular heterogeneity and other cryptic effects of nondisease-related factors inherent in postmortem brain studies. For instance, the SZ patients in this large postmortem methylation study are older and heavy smokers, and their brain tissues have a lower pH and longer postmortem interval that may affect the fidelity of the chromatin structure, and the majority were on antipsychotic medications at the time of death (64.0%) [35]. In summary, although both candidate gene studies and genome-wide profiles of methylation in SZ have suggested a possible role of aberrant DNA methylation in the disease etiology, the evidence remains inconclusive.

14.3.2 Genetic control of DNA methylation and its relevance to SZ

The differences in DNA methylation between SZ patients and controls may be explained by both nongenetic factors, and an individual's genetic makeup. Indeed, DNA methylation, as a quantitative trait, has been shown to be under tight genetic control. There are abundant genetic variants in the human genome that are associated with DNA methylation, which are called methylation quantitative trait loci (meQTLs). Identifying meQTLs that are also associated with SZ may provide a direct mechanistic link between a disease risk variant and disease biology. Some of the recent meQTL mapping studies in a large number of samples (blood or postmortem brain tissues) are highlighted here.

The largest meQTL mapping study was performed in human blood cells of 697 normal control subjects, which is part of the SZ case/control sample that was used for identifying aberrant methylation in SZ [35]. Next-generation sequencing was used to assay DNA methylation at approximately 4.5 million loci, each comprising about 2.9 CpGs. Methylation measures at each locus were tested for association with about 4.5 million SNPs to screen for meQTLs. This study found that genetic influence on the human blood methylome may be common, with ~15% of methylation sites having meQTLs. Most meQTLs are local, with the associated SNP adjacent to the methylation site. Although there are 393 local meQTLs that overlap with disease GWAS risk loci in general, no specific SZ-relevant meQTLs were found, this may be due to the use of blood cells.

In contrast, two recent human postmortem brain meQTL mapping studies reported potential relevance of meQTLs to SZ [37,38]. Jaffe and colleagues tested association between ~7.4 million common SNPs (minor allele frequency, MAF > 5%) and methylation measurements of ~478,000 CpG sites in 258 postmortem brains (age > 13), and identified 4.1 million significant meQTLs at false discovery rate (FDR) < 1%. About 25% of GWAS-suggestive loci manifest as significant meQTLs. For SZ, 59.6% of genome-wide significant loci had a risk or proxy SNP that was a meQTL. Consistently, Hannon and colleagues identified abundant meQTLs in developing brains and their enrichment in SZ-risk loci [38]. With 166 human fetal brain samples, spanning 56–166 days postconception, they examined association between 430,000 common SNPs and 315,000 DNA methylation sites, and identified >16,000 fetal brain meQTLs, most of which are *cis*-acting (i.e., local) and showed substantial overlap with genetic variants that were also associated with gene expression (i.e., eQTLs) in the brain. There was a highly significant enrichment (~fourfold) of genome-wide significant SZ-risk variants among fetal brain meQTLs. Using imputed genotypes for 5 million common variants, Hannon and colleagues identified an additional 256,000 meQTLs, enabling the identification of 1067 SNPs that were associated with both DNA methylation and SZ.

Altogether, the enrichment of brain meQTLs for SZ GWAS risk variants and the colocalization of variants associated with both DNA methylation and SZ support the possible pathophysiological relevance of DNA methylation to SZ. However, meQTLs, like any disease GWAS risk variants, are association-based, in other words an associated variant may not be the causal one. As a result, the interpretation of the mechanistic link between DNA methylation and SZ-risk variants remains a challenge, and may need integrative analyses with other dimension of functional genomics data (e.g., transcriptomic profiles and histone modifications).

14.4 SZ and histone modifications

Histone modifications determine local chromatin structure and function. Together with DNA methylation, histone modification plays important roles in epigenetic regulation of gene transcription. Histone modifications mainly include histone lysine acetylation, arginine and lysine methylation, phosphorylation, and ubiquitination, of which histone acetylation and methylation are commonly studied. Histones can be modified by enzymes that add (writers) or remove (erasers) various covalent modifications, while the functional consequences of these histone modifications are mediated in part through proteins (readers) that bind to specific modified residues and influence the chromatin states and gene transcription (see Chapter 1 for more details). Histone acetylation and methylation mostly occur on N-terminal lysine (K) residues of histones H3 and H4, modifications that are catalyzed by specific enzymes [39]. Histone acetylation generally promotes an open chromatin state and gene transcription, while histone methylation has more diverse functions, depending on the positions of the methylated K residue and the number of the added methyl group (see subsections later). Both histone methylation and histone acetylation can be biochemically reversed, providing a molecular mechanism that enables the switch between active and repressive chromatin and transcriptional states. The diverse histone modifications constitute a complex and dynamic process that modulates cellular growth, differentiation, and function. Abnormal histone modifications have been reported to play important roles in the etiology of neurodevelopmental disorders, including SZ [40].

14.4.1 Histone acetylation

Histone H3 is primarily acetylated at lysine (K) residues positions K9, K14, K18, K23, K27, and K56. Histone acetylation and deacetylation are modulated by histone acetyltransferases (HATs) and histone deacetylases (HDACs), respectively. Histone acetylation by HATs is often associated with a more open chromatin and active transcription, while histone deacetylation by HDACs often leads to a more condensed chromatin and repressive transcription. Acetylation of histone H3 at lysine 27 (H3K27ac) together with H3K64ac and H3K122ac, and an H4 tail acetylation, H4K16ac, are enriched at active enhancers [41]. The role of histone acetylation and deacetylation in SZ has been interrogated in both peripheral blood cells and in postmortem brains.

Earlier studies of histone acetylation in SZ were performed in blood cells, aiming to identify possible biomarkers that can be easily implemented for clinical diagnosis. An initial study with very small number of SZ and bipolar patients ($n = 14$) were treated with a clinically proven drug, valproic acid (VA), which is also a potent inhibitor of HDACs [42]. The acetylated histones H3 and H4 in SZ lymphocytes were found to be irresponsive to VA treatment [42]. The baseline levels of H3 K9/K14 acetylation in SZ blood cells were also found to be lower (measured by Western blot) than in controls [43]. Consistent with the result from blood cells, later study of postmortem brain histone H3 K9/K14 was found to be hypoacetylated (measured by chromatin immunoprecipitation-qPCR) at the promoter regions of several SZ candidate genes, including *GAD1*, 5-hydroxytryptamine (serotonin) receptor 2C, and the myelin-related genes in SZ [44]. These studies, although very few, suggest that the reduced H3 K9/K14 acetylation might be associated with SZ.

The association of histone acetylation changes with SZ in the brain has not been systematically examined. However, genome-wide histone acetylation study on autism (ASD) may provide some insights on SZ, give that both are neurodevelopmental disorders that partially share molecular neuropathology [45]. A H3K27ac (active enhancers) chromatin immunoprecipitation sequencing (ChIP-seq) on 257 postmortem brain samples from ASD patients and matched controls identified a common acetylome signature for >5000 *cis*-regulatory elements in the prefrontal and temporal cortex [46]. By further correlating histone acetylation with genotypes, the authors discovered >2000 brain histone acetylation quantitative trait loci (haQTLs), some of which overlap with putative causal variants for SZ and ASD (e.g., rs4765905 in *CACNA1* and rs8054791 in *GRIN2A*) [46]. It is noteworthy that ~77% of the binding sites of TCF4, a transcription factor encoded by a leading SZ GWAS risk gene, overlap with the H3K27ac histone modification [47].

In addition to possible aberrant histone acetylation in SZ, abnormal expression for enzymes that modulate histone acetylation or deacetylation, mainly HDACs, has also been implicated in SZ. In a study with a large well-characterized human postmortem brain sample ($n = 700$), the transcripts of HDAC2, but not HDAC1, in SZ-relevant adult DLPFC were decreased by 34% in SZ compared to controls [48]. Although this interesting observation may be confounded by use of antipsychotic drugs or other environmental factors, it was noted that neither smoking nor therapeutic drugs impacted HDAC2 levels [48]. However, the observed reduction of HDAC2 expression in SZ postmortem brains was not replicated by some other studies. In one study, HDAC1 levels were found to be increased in blood samples of SZ patients who had encountered early-life stress (ELS), compared to patients without ELS [49]. In another study, the hippocampus of SZ-like mice showed H3K9 deacetylation that was regulated by an increase in both HDAC1 and HDAC3 [50]. Therefore, it remains unclear which types of HDACs are relevant to SZ, and how they may impair the SZ phenotypes. More systematic study of HDACs in SZ is needed, which may implicate other HDACs, such as HDAC9 (see later section), in SZ pathogenesis.

14.4.2 Histone methylation

Histone methylation mostly occurs at lysine (K) residues positions of H3 or H4, resulting in various types of histone marks including H3K4, H3K9, H3K27, H3K36, H3K79, and H4K20. Histone methylation can cause transcriptional activation or repression, depending on the position of the lysine and the number of the methyl groups added to the lysine residue. Specifically, lysine residues can be mono- (me1), di- (me2), or trimethylated (me3) (see review [39]). Histone methylation marks in the genome can be mapped by ChIP-seq. Different types of regulatory DNA sequences (e.g., promoter and enhancer) are often marked by specific histone methylation states or a combination of histone modifications. Generally, H3K4me3 marks promoters, H3K4me1 marks enhancers, while H3K4me1/H3K27ac marks active enhancers, and H3K4me1/H3K27me3 marks inactive enhancers. In addition, H3K4me1 and H3K4me3 represent transcriptional active states, while H3K9me3 and H3K27me3 are repressive chromatin marks. Histone methylation and demethylation are modulated by different enzymes with opposite functions: lysine methyltransferases (KMTs) to add the methyl group and histone lysine demethylases (KDMs) to remove it. There are more than 60 KMTs in human cells [39,50a]. Some KMTs (e.g., SET1, MLL, and SMYD3) are specifically responsible for methylation of H3K4, while others (e.g., G9a, EZH2, and SETDB1) catalyze methylation of histone H3K9 or H3K27. Different KDMs demethylate different types of histone methylation (see review [39]). The role of histone methylation, mostly at H3K4 and H3K9 sites, in SZ has been investigated in both peripheral blood cells and postmortem brain tissues.

Although the relevance of blood histone methylation to SZ pathogenesis may not be so clear, increased H3K9me2 has been repeatedly reported (by Western blot) in SZ lymphocytes [51,52]. Concordantly, the levels of three histone methyltransferase enzymes (i.e., G9a, GLP, and SETDB1) responsible for most H3K9 methylation modifications were also found to be elevated in SZ [51]. In human brains, Akbarian and colleagues first reported the association of histone modifications with SZ by showing high levels of histone H3R17 methylation in the prefrontal cortex [53]. The increased H3K9me2 level in SZ was replicated in cortical brains and peripheral lymphocytes of SZ [54]. For other histone methylation, decreased levels of H3K4me3 and elevated levels of H3K27me3 (by Western blot) were reported in SZ postmortem prefrontal cortex as well [51]. Although there has been no report on genome-wide H3K4 methylation in SZ brain tissue, one recent H3K4me3 profiling in immature neurons derived from olfactory epithelium cells of four SZ subjects and four matched controls showed SZ-associated H3K4 methylation changes for hundreds of loci [55]. A total of 72 genes, some of which were of relevance for oxidative stress, metabolism, and cellular signaling (e.g., synaptic), were affected by altered H3K4me3 in conjunction with altered expression [55]. At the individual gene level, decreased H3K4me3, predominantly in females, at *GAD1* locus was also observed in SZ postmortem prefrontal cortex [56]. It would be interesting to investigate the relevance of H3K4 and H3K27 readers to SZ pathogenesis in the future.

Overall, histone modification and its relevance to SZ is an emerging research field. Compared to histone acetylation and methylation, other histone modifications are understudied. Furthermore, it remains to be established whether genome-wide histone modifications are associated with SZ in human brain samples and/or hiPSC-derived neuronal samples. In the absence of such comprehensive disease-relevant histone modification datasets, the ENCODE project mapped histone modification marks (e.g., H3K4me1/H3K27ac) from the very limited number of postmortem brain samples have been used to show the enrichment of SZ GWAS risk [57], which suggests that SZ genetic factors likely affect disease phenotypes through acting on histone modifications.

14.5 SZ and 2D chromatin structure

Because of chromatin modifications and remodeling, some parts of the mammalian genome are open with loosely organized euchromatin and some other parts are more condensed with tightly packed heterochromatin. Accessible, or open chromatin, is associated with active transcription. Open chromatin is correlated with epigenomic histone modifications associated with active enhancers and promoters (e.g., H3K4me1 and H3K4me3) [27,58–61]. More than 60% of methylation-eQTLs [62] are also within open chromatin. A major determinant factor of transcription is chromatin accessibility [22,27]. Open chromatin overlies >97% of *cis*-regulatory sequences [22,27,62]. Open chromatin is thus a versatile index of regulatory sequence elements, and a powerful assay for screening *cis*-regulatory variants. Traditionally, open chromatin is assayed by a sequencing-based DNaseI hypersensitive site (DHS) assay (DNase-seq) [25]. More recently, open chromatin profiles are mapped by a much simpler method, Assay for Transposase-Accessible Chromatin by sequencing (ATAC-seq) [63], which requires a very small number of cells and is thus very suitable for studying human brain cells or cultured neurons. Open chromatin has been recently studied for its relevance to SZ in both human postmortem brains and neurons derived from hiPSC.

Fullard and colleagues first applied ATAC-seq to map OCR profiles in sorted neuronal and nonneuronal nuclei isolated from frozen postmortem human brain [64]. They found that most OCRs are differentially accessible between neurons and nonneuronal cells, and showed enrichment for known promoters and enhancers [64]. They further found enrichment for SZ-risk genes in both neuronal and nonneuronal OCRs, with neuronal OCRs showing higher enrichment (2-fold vs. 1.6-fold). This supports the possible involvement of both neuronal cells and nonneuronal cells in SZ pathogenesis. Because one of the major functional characteristics of open chromatin is its accessibility to TF binding, Fullard et al. further identified the TF-binding footprints based on the OCR profiles. They found that the neuronal TF-binding footprints showed much higher enrichment of SZ GWAS risk variants compared to TF-binding footprints of nonneuronal cells [64], suggesting some of the SZ GWAS causal variants likely act through altering chromatin accessibility to neuronal TFs. However, although this result is consistent with previously reported enrichment of SZ GWAS risk variants in open chromatin (mapped by DNAse-seq) of human postmortem brain [15], it is inconsistent with the other study using brain DNAse-seq profiles [65].

The inconsistent results of these enrichment analyses of postmortem brain OCRs for SZ-risk variants may be partially due to the well-known limitations of using postmortem brain tissues, such as the data quality can be easily confounded by tissue/cell heterogeneity and environmental factors [66]. Furthermore, postmortem brain data often do not capture changes at early neurodevelopmental stages [67]. In a recent study, we have mapped global OCRs by ATAC-seq in differentiating neurons from hiPSCs [18], a cellular model for studying neurodevelopmental disorders such as SZ. We found that the OCRs are highly dynamic and that neuronal OCRs are enriched for GWAS-implicated SZ-risk variants. Of the > 3500 SZ-associated GWAS variants, ~100 can be prioritized to be putatively functional ones based on their physical colocalization with neuronal OCRs that flank TF-binding sites [18]. At a leading SZ-risk locus spanning *MIR137*, we narrowed down common GWAS risk variants to a single putatively functional SNP rs1198588 in a neuronal OCR. To demonstrate the functional effects of the prioritized SZ-risk variants in OCR, we further carried out CRISPR/Cas9 editing of an SZ patient line containing *MIR137* risk variants located in OCRs, and we showed that the prioritized putative functional noncoding SZ-risk variants are associated with lower *MIR137* expression, altered OCR dynamics, and accelerated neuronal maturation [18]. Our study suggests that noncoding disease SZ variants in OCRs may impact neurodevelopment, and that neuronal OCRs can help to prioritize functional noncoding variants for neurodevelopmental disorders such as SZ.

Therefore, as an index of regulatory sequences, open chromatin may synergize with the genetic effects of SZ-risk variants on gene transcription and cellular phenotypes. Such chromatin-mediated functional effects of disease variants are expected to influence the downstream neuronal activity and cognitive phenotypes that are relevant to SZ. However, until recently, it has remained an unexplored question whether, and to what extent, neuronal activity may trigger changes in chromatin accessibility [68]. In a newly published study, Su and colleagues compared open chromatin landscapes of adult mouse dentate granule neurons in vivo before and after synchronous neuronal activation. They found genome-wide changes after 1 h of activation, some of which remained stable for at least 24 h [68]. Neuronal activity-altered OCRs are enriched for binding sites of AP1-complex components, including c-Fos, which was further shown to be critical for initiating neuronal activity-induced chromatin opening [68]. This study suggests a transcriptional mechanism by which transient neuronal activation leads to dynamic changes of chromatin accessibility. It would be important to investigate whether the observed neuronal activity-dependent chromatin remodeling is relevant to SZ, and how it may interact with SZ genetic factors to influence disease phenotypes.

14.6 SZ and higher-order chromatin structure

The mammalian genome is organized in a three-dimensional (3D) structure, with physically interacting chromatin domains that coordinate cellular gene expression programs [69]. Some parts of the genome are compartmentalized into "topologically associated domains" (TADs) typically encompassing several hundreds of kilobases (kb) of genomic segments, bound together with regulatory protein factors, such as cohesin and the multifunctional CCCTC-binding transcription factor (CTCF) [69–71]. TADs thus control expression of a cluster of TAD-associated genes that may be far apart from each other, serving as a master regulator of gene expression. 3D chromatin structure plays an important role in defining the functional units that mediate the effect of a *cis*-regulatory DNA sequence element on expression of its distal genes. Such 3D chromatin structures between regulatory sequences and their target genes are often dynamic and tissue or cell type-specific. Given the emerging role of chromatin remodeling in neurodevelopment and cognition [40], understanding the temporal and spatial changes of 3D chromatin structure in developing human brains is expected to inform the molecular mechanisms of neurodevelopmental disorders such as SZ.

To explore how changes of 3D chromatin structure in the developing human brain can help to understand the molecular basis of neuropsychiatric disorders, Won and colleagues carried out a genome-wide chromatin conformation capture by high-throughput sequencing approach (i.e., Hi–C) in developing human cerebral cortex, during the peak of neurogenesis and migration from the cortical plate (CP) and the germinal zone (GZ) [72]. These two brain regions contain postmitotic neurons and mitotically active neural progenitor cells (NPCs), respectively. The 3D structure changes between CP and GZ were found to correlate with changes in chromatin accessibility and other chromatin modification marks (e.g., histone modification patterns), as well as gene expression. Won and colleagues further intersected the SZ GWAS risk SNPs with their interacting genes defined by Hi–C chromatin structure profiles [72]. There are about 500 such SZ-risk SNP-interacting genes that are not adjacent to the index SZ GWAS risk SNPs, or in linkage disequilibrium (LD) with them, suggesting the importance of 3D chromatin structure in mediating long-range gene expression. These putative SZ candidate genes are enriched for genes related to postsynaptic density, neuronal differentiation, and chromatin remodeling. Notably, such joint analysis of 3D chromatin structure and SZ credible GWAS risk SNPs informed possible pathogenic mechanisms that are consistent with some of the classical pathogenic hypotheses for SZ. For instance, SZ GWAS risk SNPs interact with the promoter of *DRD2*, a target of most classical antipsychotic drugs [9], and with enhancers/promoters of *GRIA1* and *GRIN2A*, supporting the important role of abnormal glutamatergic neurotransmission in SZ pathophysiology [4,5,14,15,73]. Credible SZ-risk SNPs also interact with promoters of several acetylcholine receptors (*CHRM2*, *CHRM4*, *CHRNA2*, *CHRNA3*, *CHRNA5*, and *CHRNB4*), further supporting the possible role of abnormal cholinergic neurotransmission in SZ [74,75]. These results suggest that 3D chromatin structural changes, specifically TADs in the developing brain, are relevant to SZ, potentially by mediating the regulatory effects of genetic risk factors.

However, it remains largely unexplored how 3D chromatin structures mechanistically function in the brain. For the first time, Jiang and colleagues reported the essential role of the SET domain bifurcated 1 (SETDB1; also known as ESET or KMT1E), a histone H3K9 methyltransferase, in the neuronal maintenance of very large TADs (superTADs) that modulate gene expression at the SZ-relevant protocadherin (cPcdh) gene cluster [19]. In the Setdb1-deficient mouse brain, although the global high-order chromatin structures were unchanged, there were about 110 long-range (>200 kb) loop contacts that were affected in the mutant neurons. The altered TADs involve the protocadherin (cPcdh) locus that harbors 58 genes encoding cell adhesion molecules, linearly arranged as three gene clusters (*Pcdh-α*, *Pcdh-β*, and *Pcdh-γ*), which encode the Pcdha, Pcdhb, and Pcdhg that regulate neuronal connectivity, respectively) [19]. Excess amounts of CTCF occupancy were found in Setdb1-deficient neuronal genomes, converting the local chromatin to an "open" state that is associated with reduced DNA methylation levels. Setdb1 exerts unique transcriptional control across the cPcdh domain, specifically in mature neurons. The deletion of Setdb1 substantially and preferentially upregulates transcripts at the *cPcdh* locus. In correlation with the gene expression changes of *cPcdh* locus, Setdb1-deficient neurons morphologically show ~50% increased spine density, with increased size. Jiang and colleagues further showed the possible relevance of this SETDB1-controlled cPcdh TAD domain to SZ [19]. A leading SZ GWAS risk variant at this locus (rs111896713; a small insertion–deletion) is colocalized with the Setdb1 chromatin interaction peak, which was "replaced" by de novo CTCF peak after Setdb1 ablation. DNA sequences close to the SETDB1 peak flanking the SZ-risk variant were associated with the reduced expression of *PCDHGB6* (but not *PCDHGA3*), suggesting that SETDB1-mediated long-range 3D chromatin interaction domain has a repressive effect on protocadherin gene expression [19]. However, whether this specific SZ-risk variant or any other proxy SNPs in LD with it can affect the SETDB1-mediated 3D chromatin structure and protocadherin gene expression remains to be examined.

Repressive epigenome modifications associated with SETDB1 have been previously implicated in SZ [51] and show sex-dependent characteristics [54], which is consistent with the well-known sex difference of SZ prevalence. Additionally, the functional effect of SETDB1-associated TAD on the *cPCDH* locus could have even broader implications for other neuropsychiatric diseases for the following reasons: (1) microdeletions of SETDB1 are associated with neurodevelopmental delay [76], (2) CpG hypermethylation of the CTCF-binding sites within the *PCDH* gene cluster is associated with

Down syndrome (trisomy 21) [77], and (3) *cPcdh* DNA promoter methylation has been linked to depression and anxiety [78,79]. Besides affecting cPcdh gene clusters, neuronal deletion or overexpression of Setdb1 H3K9-methyltransferase (and H3K4-methyltransferase Kmt2a/Mll1) also leads to higher-order chromatin changes at the distal *Grin2b* locus, and impairs working memory [80]. Genetic perturbation (naturally occurring SNPs or genetically engineered genomic deletions) of such 3D chromatin interaction confers liability for cognitive performance, showing decreased expression of GRIN2B [80]. The 3D chromatin architecture of other putative SZ-risk loci, such as the *GAD1* (GABAergic system) locus in neurons, has also been suggested to be compromised by the disease process [81]. Recently, we have also demonstrated that a rare SZ-associated SNP 1:g.98515539A>T and a common SZ GWAS risk variant rs1198588 at a leading SZ GWAS locus that flanks a noncoding microRNA, MIR137, influence the chromatin accessibility of the promoter region of the adjacent *MIR137* [18,82].

Therefore, changes of high-order chromatin structure may play an important role in modulating temporal and spatial expression of genes in SZ-relevant pathways, thereby conferring disease risk. A comprehensive map of 3D chromatin structures in human brains and, ideally, in different subtypes of neuronal cells, may further shed light on molecular causal mechanisms of SZ.

14.7 SZ genetic risk variants affect chromatin remodeling gene pathway

Although chromatin structure and function are epigenetic characteristics, they may also be influenced by genetic factors. For instance, a disease causal variant may disrupt a DNA sequence that undergoes DNA methylation or binding to chromatin remodeler/TF, thereby affecting chromatin structure (2D or 3D) and gene transcription. Alternatively, a disease variant may specifically alter the protein sequence/function or expression of a gene that is critical for DNA methylation or histone modifications. For SZ, the possible effects of disease variants on chromatin structure and function are supported by the above-described enrichments of SZ GWAS risk variants or colocalization in different epigenetic marks (DNA methylation or histone modifications). Both common SZ-risk variants with subtle effects and rare risk variants with relatively high penetrance, including copy number variants (CNVs), implicate the involvement of chromatin remodeling pathway in SZ pathogenesis.

14.7.1 Analysis of common SZ variants implicates the dysregulated chromatin-signaling pathway

SZ GWAS have identified over 100 loci, spanning hundreds of genes, each contributing very small effect. This polygenic nature of SZ suggests that these potential risk genes likely function in gene networks. Identifying core gene networks is important to understanding the disease biology underlying these genetic associations. Analyzing GWAS data of 60,000 participants on three major psychiatric disorders (SZ, major depression, and bipolar disorder) from the Psychiatric Genomics Consortium shows that the most enriched gene pathways are related to histone methylation processes [83]. The epigenetic nature of the implicated pathways is consistent with the contribution of the well-replicated environmental risks for SZ occurring at critical periods early in development, such as in the Dutch Hunger Winter and Chinese famine studies [84]. The early developmental stage is a time when the epigenome is particularly labile and is correlated with rapid cell replication, differentiation, and tissue specification. The pathway analysis of GWAS on SZ and other major psychiatric disorders thus suggests that dysregulation of the genes in histone methylation and chromatin organization may be a common etiological mechanism of SZ [83].

Gene network analyses which integrate SZ GWAS data and human brain and/or neuronal transcriptomic data also implicate the possible pathogenic role of chromatin modifications. In our recent effort to identify genes that may serve as master regulators of expression of a large body of genes in SZ [85], we have reanalyzed the CommonMind transcriptomic dataset of 307 SZ patients and 245 controls [86]. Among the top five putative master regulators of SZ gene networks [85], both TCF4 and HDAC9 may function through affecting histone modifications. TCF4 encodes a TF, and is one of the top hits in SZ GWAS, and ~77% of the TCF4-binding sites overlap with the H3K27ac marks in the genome [47]. HDAC9 is a histone deacetylase that potentiates hippocampal-dependent memory and synaptic plasticity, leading to different neuropsychiatric conditions [87,88]. Furthermore, hemizygous deletion of HDAC9 has been found in some SZ patients [89]. Both SZ gene networks and their master regulators thus suggest the important role of histone acetylation in SZ pathogenesis. The importance of histone acetylation in SZ was further supported by a family-based genetic association study of SZ, where the association with one maker rs14251 (*HDAC3*) was replicated ($P = .04$) and remained significant in the whole sample ($P = .004$), and a significant three-locus interaction model was detected involving rs17265596 (*HDAC9*), rs7290710 (*HDAC10*), and rs7634112 (*HDAC11*) [90]. These results further support the possible involvement of HDAC genes in the etiology of SZ [90].

Genetic association studies of SZ also strongly support the involvement of the chromatin remodeling gene network. Genetic variations in the bromodomain containing 1 (*BRD1*) gene located on 22q13.33 have repeatedly been associated with both SZ and bipolar disorder [10,91,92]. The *BRD1* locus approached genome-wide significance ($P = 3.31 \times 10^{-7}$) in Psychiatric Genomics Consortium SZ GWAS (PGC2) [15], and reached genome-wide significance by using an Empirical Bayes statistical approach [93]. Interestingly, BRD1 was recently found to physically interact with chromatin remodeling proteins (e.g., PBRM1) as well as histone modifiers (e.g., ING4, ING5, MEAF6, and MYST2) along with histone H3 [94]. This strongly implies the role of BRD1 and its interaction gene network that are essential for chromatin remodeling. Altogether, some genes implicated by SZ GWAS may thus act as epigenetic regulatory hubs in certain core SZ gene networks that impair some neurodevelopmental aspects of SZ.

14.7.2 Rare SZ coding risk variants and chromatin remodeling

Recent whole exome sequencing (WES) of SZ samples has identified many rare coding risk variants, including de novo mutations (DNMs) and inherited mutations. The hope of these WES projects is to identify some rare SZ-associated variants with much higher penetrance than common variants implicated by SZ GWAS. However, none of the individual rare coding variants shows genome-wide significant association with SZ. This promotes the pathway analyses for genes that carry deleterious and likely pathogenic variants, aiming to understand the underlying disease biology. Notably, consistent with pathway analysis of common SZ GWAS risk variants [83], the pathway analyses of genes enriched for rare SZ-risk variants strongly support the etiological role of chromatin remodeling and histone modifications in SZ.

In an earlier WES study of 57 trios with sporadic or familial SZ to identify DNMs [95], a significant enrichment of nonsense DNMs was found in sporadic trios. Genes with DNMs overlapped with genes implicated in autism (e.g., *AUTS2*, *CHD8*, and *MECP2*) and intellectual disability (e.g., *HUWE1* and *TRAPPC9*), supporting a shared genetic etiology between these disorders. Because CHD8, MECP2, and HUWE1 functionally converge on epigenetic regulation (DNA methylation and chromatin structure) of transcription, this observation supports chromatin modification as a risk mechanism for SZ [95]. Collectively, nonsense DNMs were significantly enriched among a set of 419 genes characterized by domains highly specific to chromatin modification. More importantly, across all mutational categories, except for silent mutations, there was a significant overrepresentation of genes involved in chromatin organization [95].

In a later WES of a relatively large SZ case/control sample (2045 schizophrenic patients and 2045 controls) [96], weighted burden pathway analysis identified a statistically significant excess of pathways with more rare and putatively functional variants in cases than in controls. Consistent with the results from analyzing SZ DNMs [95], histone modification is among the top-ranking gene pathways [96]. Most recently, by performing a meta-analysis of WES from 4133 SZ cases and 9274 controls, de novo mutations in 1077 family trios, and CNVs from 6882 cases and 11,255 controls, Singh and colleagues showed that several gene sets related to chromatin modification and organization are enriched for rare coding variants conferring risk for SZ (at FDR $< 1\%$) [97]. These include a set of 519 genes under gene ontology (GO) term (biological process) "chromatin modification" (GO: 0016568). Interestingly, these also include a gene set of about 3000 genes whose promoters are potentially targeted by autism-associated CHD8 (as demonstrated by ChIP-seq in human NSCs and human brain tissue) [98], suggesting potentially shared gene pathways between SZ and autism. These results further support the evidence that chromatin modification pathways are important in the etiology of schizophrenia.

Although no single WES study or meta-analysis identified a genome-wide significant rare SZ-risk variant, the deleterious mutations in *SETD1A* (also known as KMT2F) are collectively associated with SZ with a genome-wide significance [99,100]. Multiple extremely rare loss of function (LoF) mutations, de novo or inherited, were found only in SZ patients, and other neurodevelopmental abnormalities. Because SETD1A is a H3K4 trimethyltransferase, the SZ-associated heterozygous LoF mutations may reduce the enzymatic activity of the C-terminal SET domain, or disrupt a canonical splice acceptor site and cause exon 16 skipping. Given that H3K4me3 plays a predominant role in recruiting gene transcription machinery, a dysfunctional SETD1A is expected to cause haploinsufficiency of genes essential for many biological processes. Interestingly, a recent study found that SETD1A-mediated increase of H3K4me3 levels and β-catenin expression may be important for regulating neural progenitor cell proliferation and neurogenesis [101]. However, SETD1A seemed to be constitutively expressed in hiPSCs and during neuronal differentiation (Duan, unpublished). The brain-specific roles of SETD1A and its relevance to SZ and other neurodevelopmental disorders remain to be illustrated. Given the relatively high penetrance of LoF mutations in *SETD1A*, identifying the specific targets of SETD1A in neuronal cells and the brain-specific roles of SETD1A may help establish the causal link between chromatin signaling and the pathogenesis of SZ.

14.7.3 SZ-associated CNVs and abnormal chromatin organization

A recent genome-wide approach has also implicated multiple rare and large recurrent CNVs of larger effect on increasing risk to SZ [73,102–104]. These CNVs are often >100 kb genomic segments of duplications or deletions, spanning many genes. Genome-wide significance evidence was obtained for eight loci, including 1q21.1, 2p16.3 (*NRXN1*), 3q29, 7q11.2, 15q13.3, distal 16p11.2, proximal 16p11.2, and 22q11.2. There was a global enrichment of CNV burden in SZ cases (odds ratio (OR) = 1.11; SZ cohort of 21,094 cases and 20,227 controls) [104]. Although with relatively strong penetrance, these CNVs are all characterized by extensive phenotypic variability. For example, 22q11.2 deletion, the SZ CNV with highest penetrance, leads to 22q11.2 syndrome (22q11DS), with variable phenotypic expression such as congenital heart, craniofacial, and neurobehavioral abnormalities, of which only ~30% of the carriers develop into SZ. One explanation for the resultant phenotypic heterogeneity is that these CNVs are large genomic regions that often span many genes. However, given the known effect of chromosomal microdeletion and duplication events on specific 3D chromatin TADs [105,106], it is conceivable that some of the SZ-associated CNVs may confer disease risk through impairing local or distal chromatin architecture, and subsequently altering expression of a cluster of genes within or outside the CNV region.

The 22q11.2 deletion is a good example that a CNV that may function through affecting chromatin modifications and remodeling. Comparing genome-wide chromatin interaction profiles between 22q11DS and unaffected individuals showed that some long-range or trans- (i.e., in other chromosomes) chromatin interactions anchored with the deletion region are present in normal B lymphocyte cell lines, but absent in deletion-carrying cell lines [107], suggesting that 22q11.2 deletion region affects expression of genes outside the CNV region at long range. Consistent with this observation, in our recent study of the transcriptomic signature of SZ-associated rare CNVs in lymphoblastoid cell lines, we identified abundant gene expression changes (~500 genes; FDR < 5%) outside of the 22q11.2 deletion region. Interestingly, these 22q11 deletion-altered genes are highly enriched with GO terms for chromatin modification and organization [108]. The broad transcriptomic effects of 22q11 deletion may be attributed to the CNV-induced changes of 3D chromatin TADs [105,106]. Alternatively, it may be also due to some specific genes within the 22q11.2 region (e.g., the histone chaperone HIRA). It is noteworthy that HIRA was found to interact with SETD1A, a H3K4 trimethyltransferase that has LoF mutations strongly associated with SZ [99,100], to increase H3K4me3 levels, and to activate β-catenin expression, thereby regulating neural progenitor cell proliferation and neurogenesis [101]. These observations suggest that 22q11 deletion, and other large SZ-associated CNVs, can have widespread effects on chromatin organization, which may contribute to the high penetrance of the CNV as well as the inherent phenotypic variability.

14.8 hiPSC model combined with CRISPR editing for studying SZ-relevant chromatin function

Most knowledge about chromatin signaling in SZ was gained from using human postmortem brains, peripheral blood cells, or rodent animals. Each experimental model has its own pros and cons in faithfully capturing the epigenome changes associated with disease pathogenesis. Neuronal cells derived from hiPSCs have emerged as a promising cellular model for neuropsychiatric disorders [109,110], offering an excellent alternative to the use of human postmortem brains. In particular, when combined with genome editing technology, hiPSC-derived neurons represent a powerful cellular model that can be genetically or epigenetically manipulated to understand the cell type and developmental stage-specific chromatin regulation and disease biology [111].

14.8.1 hiPSC-derived neurons as a cellular model for neurodevelopmental disorder

Reprogramming human somatic cells into hiPSC is a process of erasing somatic epigenomic memory and resetting its epigenome to a pluripotent state that is reminiscent of embryonic stem cells (ESCs). Reprogramming factors (Oct3/4, Klf4, Sox2, and c-Myc) can be delivered in multiple ways, which have been reviewed extensively elsewhere [112]. The optimal method for delivering the reprogramming factors is a virus-integration-free system (e.g., by Sendai virus) [113]. Fibroblasts from skin biopsies and blood cells are among the most commonly used source cells for iPSC reprogramming. Although the epigenomic memory of the source cells can be a concern in hiPSC, the fully reprogrammed iPSCs often have very similar epigenetic configurations to that of ESCs, and will have the full pluripotency to redifferentiate into all types of somatic cells.

It is important to determine the appropriate types of neuronal cells to derive from iPSC to model the SZ-relevant chromatin changes. iPSCs can be differentiated into multiple major types of neurons (dopaminergic, glutamatergic, or GABAergic) [114] that are relevant to SZ pathophysiology. iPSC can be efficiently differentiated into cortical excitatory neurons, mimicking a process of human cortical development [115,116]. The most difficult type of neurons to derive from iPSCs is the GABAergic interneuron, which may take a prolonged intrinsic timeline of up to 7 months [117]. By using different combinations of growth factors and small molecules in culture media, hiPSCs can be efficiently differentiated into midbrain dopaminergic [118], cortical glutamatergic [116], GABAergic inhibitory interneurons [117,119,120], and microglia [121]. As an alternative to using defined growth factors in culture, forced expression of exogenous TFs has also been employed to rapidly differentiate hiPSCs into functional neuronal lineages, such as the rapid differentiation of excitatory neurons via forced expression of NEUROD1 or NEUROG2 [122,123], or the GABAergic inhibitory interneurons via forced expression of ASCL1 and DLX2 [124]. This array of neuronal differentiation methods provides a rich resource to choose for modeling SZ-relevant chromatin changes, and for targeted epigenomic perturbation.

The hiPSC-derived neurons are a good alternative model to human postmortem brains for studying psychiatric disorders. First, compared to human brains and the emerging brain organoid model [125,126], iPSC-derived neurons are relatively homogeneous [110,127]. For instance, the forced exogenous expression of NGN2 gives rise to ~100% excitatory neurons in about 4 weeks [122,123]. Second, hiPSC-derived neuronal cells provide a model that enables studying molecular and cellular phenotypic changes, including the dynamic changes of chromatin states, in a temporal and cell type-specific manner that can recapitulate the early neurodevelopmental stages of SZ. This is important, because regulatory variants are often cell-type and developmental stage-specific [128–130]. Third, compared to postmortem brain tissues, that are well known to be confounded by tissue variability and environmental factors [66], hiPSC differentiation into neurons can be better controlled, thus the data maybe more reproducible. Lastly, compared to postmortem brains, hiPSC models are amenable to genetic modification or epigenomic perturbation, which is important for studying the functional impacts of disease-relevant chromatin dysregulation.

14.8.2 CRISPR-based approaches for genome/epigenome perturbation

hiPSC-derived neuronal cells provide a neurodevelopmental model for studying chromatin change relevant to SZ. For examining the effect of common disease risk variants on chromatin changes, directly comparing iPSC-neuron cultures of cases versus controls would require a technically and financially prohibitive number of iPSCs. This is because common variants often have small effects and there are substantial variation, in particular the variable genetic background between iPSC lines [131]. A prominent solution is the CRISPR/Cas nuclease-mediated genome editing system [132–137]. CRISPR/Cas editing enables the generation of isogenic iPSC neurons, differing only at a single SNP site or a specific DNA sequence of interest, representing a powerful design that can overcome possible confounding effects of variable genetic backgrounds when comparing differences between cells carrying risk versus nonrisk alleles [138–146].

The applications of CRISPR/Cas technology in the past few years have evolved rapidly (see review [147]). CRISPR/Cas can be used to modify a specific DNA sequence segment (e.g., insertion or deletion) or SNP site (i.e., precise single nucleotide editing), or epigenetically alter chromatin modification (e.g., histone methylation) without changing genomic DNA sequences. In these processes, the sequence-specific guide RNA (sgRNA) leads the Cas9 (for cutting DNA) or deactivated Cas9 (dCas9; without DNA cutting function) to the specific genomic region to be targeted. For CRISPR/Cas-mediated DNA sequence change, the classical approach gives rise to a double-strand break (DSB), and then achieves the sequence editing during DNA repair [132–137]. The most recent breakthrough technology, DNA base editor, can covert A:T to G:C with very high efficiency (~50%), without creating a DSB [148]. In this base editing approach, dCas9 is fused to a deaminase enzyme that can convert bases [148]. As the CRISPR/Cas9 and the base editing approaches change genomic DNA sequences, they are often referred to as genome editing, in contrast to epigenome editing, where no DNA sequences are altered.

For epigenetically altering chromatin modifications, the CRISPR-mediated epigenome editing is often referred to as "CRISPR interference" (CRISPRi) for transcriptional repression or "CRISPR activation" (CRISPRa) for transcriptional activation [147]. In CRISPRi, dCas9 is fused with a Kruppel-associated box (KRAB) effector domain to achieve transcriptional repression by spreading repressive histone modifications, such as H3K9me3. Other epigenome-modifying repressors can also be fused to dCas9 to achieve transcriptional repression, such as Lys-specific histone demethylase 1 (LSD1), HDACs, and DNA methyltransferases DNMT3A and MQ1 (see review [147]). For CRISPRa, dCas9 can be fused to a DNA demethylase (e.g., TET1), or to a histone acetyltransferase (e.g., p300), or to single activation effectors, such as VP64 [147] to achieve targeted transcriptional activation. These approaches enable epigenome perturbation of the chromatin regulatory architecture of genomic regions around a putative disease causal variant or risk gene.

14.8.3 CRISPR-based 2D and 3D chromatin perturbation relevant to SZ in hiPSC models

Using a CRISPR-based genome editing approach to perturb the regulatory sequences important for neuronal open chromatin (2D), or chromatin interaction (3D), in hiPSC models is conceptually and technically new in the SZ research field. We have recently studied genome-wide OCRs of hiPSC-derived glutamatergic neurons (assayed by ATAC-seq) [18]. At a leading SZ-risk locus, spanning *MIR137* [15], we observed hiPSC-derived neuronal OCRs absent in hiPSCs, or in fetal brain frontal cortex (DHS peak of Duke ENCODE) [18]. To functionally assess the relevance of noncoding sequences of OCRs in neuropsychiatric disorders, with the putative functional common SZ GWAS risk variant rs1198588 in this neuronal OCR, we employed CRIPSR/Cas9-mediated genome editing to correct both risk alleles of rs1198588 in a patient-derived iPSC line. Excitatory neurons, differentiated from the resulting isogenic hiPSC line that carries the nonrisk allele for rs1198588, significantly increased the chromatin accessibility of the noncoding *MIR137* promoter, which was correlated with increased *MIR137* expression, reduced dendritic complexity, and slowed neuronal maturation [18]. This is the first demonstration of the effect of a noncoding SZ GWAS risk variant in a neuronal OCR on adjacent gene expression through directly rescuing the risk allele in human neurons of an SZ patient, suggesting that neuronal OCRs and the noncoding disease variants within them may regulate key steps of neurodevelopment.

In another recent study, for the first time, Jiang and colleagues reported the epigenome perturbation of a 3D genome (Hi—C interaction matrices) relevant to SZ, in glutamatergic neurons derived from hiPSC [19]. They first showed that the 3D genome in human glutamatergic neurons derived from hiPSC is similar to that in NeuN$^+$ nuclei of the mouse cortex, and that such 3D chromatin structure in the brain was also conserved between mice and humans [80]. Specifically, at the *cPCDH* locus, the 3D chromatin interaction peak of *SETDB1*, overlaps with an SZ GWAS risk haplotype. To examine how these SZ-risk haplotype differences in *cPCDH* interaction frequencies translated into differential repressive potential, Jiang and colleagues introduced sgRNAs targeting the SZ-risk haplotype sequence into two stable NPC lines, one expressing a dCas9—KRAB fusion protein presumably tethering the KRAB-associated protein 1 (KAP1) repressor complex, and the other expressing dCas9—VP64 to dock the VP64 activator. They found that dCas9—KRAB perturbation at the SZ-risk haplotype/SETDB1 peak resulted in a robust multifold decrease in expression of *PCDHGB6*, but not *PCDHGA3*, while VP64 epigenomic editing increased expression of a subset of *cPCDH* genes [80]. These epigenome perturbation experiments suggest that SETDB1-interacting 3D chromatin has a repressive effect on protocadherin gene expression, which may be altered by the SZ-risk variants. To conclude, using hiPSC models in combination with CRISPR-based genome/epigenome perturbation may help with understanding the disease biology underlying abnormal chromatin signaling in SZ. It may also facilitate the development of novel therapeutic approaches that target chromatin signaling.

14.9 Therapeutic drugs that target chromatin structure and activity in SZ

Both typical antipsychotic drugs such as haloperidol, which primarily target the dopamine D2 receptor, and the atypical antipsychotic drugs such as clozapine, which target the D2 receptor and other neurotransmitter receptors, have been shown to be able to alter DNA methylation and/or histone modifications in peripheral blood, postmortem brains, or animal models (see review [39]). However, most of these studies are limited by the disease relevance of the chosen tissues/cells and experimental models, sample size, and assay robustness. For instance, the DNA hypomethylation observed in leukocytes of SZ patients treated with antipsychotics may be the consequence of chronic antipsychotic treatment that affects the transcription of genes related to methylation, it is not clear whether this methylation alteration is pathogenically related to SZ [149]. Similarly, olanzapine causes methylation changes in genes associated with DA neurotransmission, not only in the hippocampus and cerebellum, but also in the liver [149]. More convincing evidence for disease-relevant DNA methylation changes has been shown for some specific SZ candidate genes (e.g., *GAD67*, *RELN*) and brain-derived neurotrophic factor (*BDNF*) [149]. In prenatal restraint stress (PRS), with mice showing behavioral deficits reminiscent of behaviors of SZ patients, the behavioral deficits and the increased 5-methylcytosine (5 MC) and 5-hydroxymethylcytosine (5HMC) at *Gad67*, *Reln*, and *Bdnf* promoters in the frontal cortex were reversed by clozapine, but not by haloperidol [150]. Consistent with the methylation changes of these SZ candidate genes, clozapine, but not haloperidol, reduced the elevated levels of Dnmt1 and Tet1, as well as the elevated levels of DNMT1 binding to *Gad1*, *Reln*, and *Bdnf* promoters in PRS mice [150]. Despite the fact that a rodent model may not faithfully recapitulate the SZ-relevant human behavior [151], it would still be interesting to mechanistically examine whether this clozapine-induced DNA methylation affects other SZ-risk genes recently implicated by GWAS.

Antipsychotic drugs can also change histone methylation and acetylation [39]. Most studies have considered histone modification as a therapeutic target, specifically on HDAC inhibitors. The HDAC inhibitor valproate (VPA) has a long history of being used in the treatment of bipolar disorder, and has also been suggested for SZ treatment. VPA can potentiate the induction of DNA methylation at relevant promoters by atypical antipsychotic drug clozapine, in mouse models of SZ [149]. A mechanistic study, in both mouse and human frontal cortex, further showed that chronic atypical antipsychotics decreased histone acetylation at the promoter of metabotropic glutamate 2 receptor (*mGlu2*), which was accompanied by a serotonin 5-HT(2A) receptor-dependent increase of HDAC2 binding to the *mGlu2* promoter. This antipsychotic-induced repressive chromatin modification at the *mGlu2* promoter region downregulated *mGlu2* transcription [152]. HDAC inhibitors, such as VPA, can augment the therapeutic-like effects of atypical antipsychotics by preventing their effects on repressive histone modifications at the *mGlu2* promoter [152]. On the contrary, overexpression of HDAC2 in frontal cortex decreases *mGlu2* transcription and augments psychosis-like behavior [152]. In a recent study, deletion of Hdac2 in forebrain pyramidal neurons prevents the negative effects of antipsychotic treatment on synaptic remodeling and cognition [153]. These observations in rodent animals suggest that the use of HDAC2 inhibitors may be a promising therapeutic strategy for SZ.

Although intriguing in animal studies, using HDAC inhibitors for treating SZ [149,152] has yet to be clinically proven. The biggest concern is whether the MK-801-induced psychotic-like symptoms in these rodents are relevant to the pathogenic mechanisms of SZ [151]. Another concern is that the current HDAC inhibitors are not specific, likely acting on many targets in various tissues, which may cause substantial side effects. The nonspecific epigenetic effects may in part explain why commonly used antipsychotic drugs often cause strong side effects. Nonetheless, given the known link between abnormal chromatin signaling and SZ, and the reversible nature of chromatin modifications, identifying small molecules that can perturb the chromatin states of core SZ gene networks in a tissue-specific manner may aid in the development of novel therapeutic interventions.

14.10 Conclusion and perspectives

SZ is a complex disorder, involving both genetic and nongenetic factors. Both types of risk factors can influence the disease phenotype, through acting on chromatin signaling, and subsequently by perturbing the expression of interactive gene networks. Given the observed link between both 2D and 3D chromatin with SZ, it is conceivable that some of the core gene networks and/or some master regulators of SZ pathogenesis are related to chromatin function and signaling. Identifying such gene pathways and master regulators will not only help to understand the molecular mechanisms of this devastating disease, but also facilitate the development of novel and more effective drugs that target the dysregulated epigenome. Moreover, abnormal chromatin structural changes in the developing brain or peripheral tissues may be used as biomarkers for disease progression or drug efficacy for SZ.

Understanding the role of chromatin modifications and remodeling in SZ pathogenesis is an emerging field, facing a number of challenges, but also bringing enormous opportunities. Mechanistically, although the genetic risk factors strongly suggest the involvement of abnormal chromatin signaling in SZ, it remains elusive how chromatin remodeling affects disease phenotypes of SZ. Most knowledge on the pathogenic link between chromatin modifications and SZ is from studying peripheral blood and postmortem brains, both of which can easily be confounded by environmental factors. Mouse models have been useful for studying the link between chromatin signaling and SZ; however, given the different cortical organization between mice and human brains [154], the fidelity of rodent models remains an open question. Rapidly evolving new technologies and approaches, such as hiPSC models and CRISPR-based genome/epigenome editing, have started to make an inroad in understanding the disease biology related to chromatin modifications. However, it remains to be seen how hiPSC models, including the emerging hiPSC-based mini-brain [155], in combination with CRISPR technology, can help to close the gap between abnormality of chromatin signaling and SZ. Finally, even with a promising experimental model, devising more specific functional readouts, which can be used for high-throughput screening of drug candidates (small chemicals or RNAs) for brain disorders such as SZ, remains challenging. It is encouraging to see that more selective small molecules have been identified for targeting to histone acetylation in treating lineage-specific tumors [156]. Besides the traditional drug selection paradigms, single-cell technologies (e.g., transcriptomic profiles as a readout for pooled CRISPR screen such as Perturb-seq) or other functional readouts (e.g., single-cell Hi–C or ATAC-seq) [147] in single neuronal cells may help to unravel the pathogenic and therapeutic roles of chromatin modification and remodeling in SZ.

As an epigenetic mechanism, chromatin signaling plays an important role in modulating neurodevelopmental processes as well as adult neuronal activities and behaviors. Chromatin signaling may serve as a pivotal interface for mediating the effects of both genetic and nongenetic SZ factors, which may, in part, explain the high clinical heterogeneity and substantial pleiotropy of the disease. Finally, it is noteworthy that, while chromatin modifications and structure are affected by SZ genetic factors, chromatin structure also contributes to regional DNA mutation rates [157]. In this regard, given the prevalent somatic mutations (or mosaicism) in single human neurons during brain development and aging [158,159], it would be interesting to examine the effects of chromatin signaling on the accumulation of somatic mutations in human brain development, and its relevance to SZ and other neurodevelopmental disorders.

Acknowledgments

The work is supported by National Institutes of Health (NIH) grants MH102685, MH106575, MH116281 and DA041600.

References

[1] Owen MJ, Sawa A, Mortensen PB. Schizophrenia. Lancet 2016;388:86–97.

[2] Buckley PF, Miller BJ, Lehrer DS, et al. Psychiatric comorbidities and schizophrenia. Schizophr Bull 2009; 35:383–402.

[3] Gulsuner S, Walsh T, Watts AC, et al. Spatial and temporal mapping of de novo mutations in schizophrenia to a fetal prefrontal cortical network. Cell 2013;154:518–29.

[4] Purcell SM, Moran JL, Fromer M, et al. A polygenic burden of rare disruptive mutations in schizophrenia. Nature 2014;506:185–90.

[5] Fromer M, Pocklington AJ, Kavanagh DH, et al. De novo mutations in schizophrenia implicate synaptic networks. Nature 2014;506:179–84.

[6] Brennand KJ, Simone A, Tran N, et al. Modeling psychiatric disorders at the cellular and network levels. Mol Psychiatr 2012;17:1239–53.

[7] Penzes P, Cahill ME, Jones KA, et al. Dendritic spine pathology in neuropsychiatric disorders. Nat Neurosci 2011;14:285–93.
[8] Brennand KJ, Gage FH. Concise review: the promise of human induced pluripotent stem cell-based studies of schizophrenia. Stem Cell 2011;29:1915–22.
[9] Carlsson A, Lindqvist M. Effect of chlorpromazine or haloperidol on formation of 3methoxytyramine and normetanephrine in mouse brain. Acta Pharmacol Toxicol 1963;20:140–4.
[10] Purcell SM, Wray NR, Stone JL, et al. Common polygenic variation contributes to risk of schizophrenia and bipolar disorder. Nature 2009;460:748–52.
[11] Shi J, Levinson DF, Duan J, et al. Common variants on chromosome 6p22.1 are associated with schizophrenia. Nature 2009;460:753–7.
[12] Stefansson H, Ophoff RA, Steinberg S, et al. Common variants conferring risk of schizophrenia. Nature 2009;460:744–7.
[13] Genome-Wide Association Study (GWAS) Consortium. Genome-wide association study identifies five new schizophrenia loci. Nat Genet 2011;43:969–76.
[14] Ripke S, Schizophrenia Group of the Psychiatric Genomics Consortium. Psychiatric genomics consortium quadruples schizophrenia GWAS sample size to 35,000 cases and 47,000 controls. Abstract. American society of human genetics annual meeting, vol. 63; 2013.
[15] Schizophrenia Working Group of the Psychiatric Genomics Consortium. Biological insights from 108 schizophrenia-associated genetic loci. Nature 2014;511:421–7.
[16] Luger K, Mader AW, Richmond RK, et al. Crystal structure of the nucleosome core particle at 2.8 A resolution. Nature 1997;389:251–60.
[17] Felsenfeld G, Groudine M. Controlling the double helix. Nature 2003;421:448–53.
[18] Forrest MP, Zhang H, Moy W, et al. Open chromatin profiling in hiPSC-derived neurons prioritizes functional noncoding psychiatric risk variants and highlights neurodevelopmental loci. Cell Stem Cell 2017;21:305 e8–18 e8.
[19] Jiang Y, Loh YE, Rajarajan P, et al. The methyltransferase SETDB1 regulates a large neuron-specific topological chromatin domain. Nat Genet 2017;49:1239–50.
[20] King MC, Wilson AC. Evolution at two levels in humans and chimpanzees. Science 1975;188:107–16.
[21] Duan J, Sanders AR, Gejman PV. Genome-wide approaches to schizophrenia. Brain Res Bull 2010;83: 93–102.
[22] Thurman RE, Rynes E, Humbert R, et al. The accessible chromatin landscape of the human genome. Nature 2012;489:75–82.
[23] Bernstein BE, Stamatoyannopoulos JA, Costello JF, et al. The NIH Roadmap epigenomics mapping consortium. Nat Biotechnol 2010;28:1045–8.
[24] Dunham I, Kundaje A, Aldred SF, et al. An integrated encyclopedia of DNA elements in the human genome. Nature 2012;489:57–74.
[25] Ritchie GR, Dunham I, Zeggini E, et al. Functional annotation of noncoding sequence variants. Nat Methods 2014;11:294–6.
[26] Maurano MT, Humbert R, Rynes E, et al. Systematic localization of common disease-associated variation in regulatory DNA. Science 2012;337:1190–5.
[27] Degner JF, Pai AA, Pique-Regi R, et al. DNase I sensitivity QTLs are a major determinant of human expression variation. Nature 2012;482:390–4.
[28] Kohli RM, Zhang Y. TET enzymes, TDG and the dynamics of DNA demethylation. Nature 2013;502: 472–9.
[29] Ovenden ES, McGregor NW, Emsley RA, et al. DNA methylation and antipsychotic treatment mechanisms in schizophrenia: progress and future directions. Prog Neuro-Psychopharmacol Biol Psychiatry 2018;81: 38–49.

[30] Nohesara S, Ghadirivasfi M, Mostafavi S, et al. DNA hypomethylation of MB-COMT promoter in the DNA derived from saliva in schizophrenia and bipolar disorder. J Psychiatr Res 2011;45:1432–8.

[31] Noh JS, Sharma RP, Veldic M, et al. DNA methyltransferase 1 regulates reelin mRNA expression in mouse primary cortical cultures. Proc Natl Acad Sci U S A 2005;102:1749–54.

[32] Ruzicka WB, Subburaju S, Benes FM. Circuit- and diagnosis-specific DNA methylation changes at gamma-Aminobutyric acid-related genes in postmortem human Hippocampus in schizophrenia and bipolar disorder. JAMA Psychiatry 2015;72:541–51.

[33] Wockner LF, Noble EP, Lawford BR, et al. Genome-wide DNA methylation analysis of human brain tissue from schizophrenia patients. Transl Psychiatry 2014;4:e339.

[34] Aberg KA, McClay JL, Nerella S, et al. Methylome-wide association study of schizophrenia: identifying blood biomarker signatures of environmental insults. JAMA Psychiatry 2014;71:255–64.

[35] McClay JL, Shabalin AA, Dozmorov MG, et al. High density methylation QTL analysis in human blood via next-generation sequencing of the methylated genomic DNA fraction. Genome Biol 2015;16:291.

[36] Pidsley R, Viana J, Hannon E, et al. Methylomic profiling of human brain tissue supports a neurodevelopmental origin for schizophrenia. Genome Biol 2014;15:483.

[37] Jaffe AE, Gao Y, Deep-Soboslay A, et al. Mapping DNA methylation across development, genotype and schizophrenia in the human frontal cortex. Nat Neurosci 2016;19:40–7.

[38] Hannon E, Spiers H, Viana J, et al. Methylation QTLs in the developing brain and their enrichment in schizophrenia risk loci. Nat Neurosci 2016;19:48–54.

[39] Thomas EA. Histone posttranslational modifications in schizophrenia. Adv Exp Med Biol 2017;978:237–54.

[40] Ronan JL, Wu W, Crabtree GR. From neural development to cognition: unexpected roles for chromatin. Nat Rev Genet 2013;14:347–59.

[41] Pradeepa MM. Causal role of histone acetylations in enhancer function. Transcription 2017;8:40–7.

[42] Sharma RP, Rosen C, Kartan S, et al. Valproic acid and chromatin remodeling in schizophrenia and bipolar disorder: preliminary results from a clinical population. Schizophr Res 2006;88:227–31.

[43] Gavin DP, Kartan S, Chase K, et al. Reduced baseline acetylated histone 3 levels, and a blunted response to HDAC inhibition in lymphocyte cultures from schizophrenia subjects. Schizophr Res 2008;103:330–2.

[44] Tang B, Dean B, Thomas EA. Disease- and age-related changes in histone acetylation at gene promoters in psychiatric disorders. Transl Psychiatry 2011;1:e64.

[45] Gandal MJ, Haney JR, Parikshak NN, et al. Shared molecular neuropathology across major psychiatric disorders parallels polygenic overlap. Science 2018;359:693–7.

[46] Sun W, Poschmann J, Cruz-Herrera Del Rosario R, et al. Histone acetylome-wide association study of autism spectrum disorder. Cell 2016;167:1385–1397.e11.

[47] Forrest MP, Hill MJ, Kavanagh DH, et al. The psychiatric risk gene transcription factor 4 (TCF4) regulates neurodevelopmental pathways associated with schizophrenia, autism, and intellectual disability. Schizophr Bull 2017;44(5):1100–10.

[48] Schroeder FA, Gilbert TM, Feng N, et al. Expression of HDAC2 but not HDAC1 transcript is reduced in dorsolateral prefrontal cortex of patients with schizophrenia. ACS Chem Neurosci 2017;8:662–8.

[49] Bahari-Javan S, Varbanov H, Halder R, et al. HDAC1 links early life stress to schizophrenia-like phenotypes. Proc Natl Acad Sci U S A 2017;114:E4686–94.

[50] Vecera J, Bartova E, Krejci J, et al. HDAC1 and HDAC3 underlie dynamic H3K9 acetylation during embryonic neurogenesis and in schizophrenia-like animals. J Cell Physiol 2018;233:530–48.

[50a] Binda O. Epigenetics. On your histone mark, SET, methylate! 2013;8(5):457–63. https://doi.org/10.4161/epi.24451.

[51] Chase KA, Gavin DP, Guidotti A, et al. Histone methylation at H3K9: evidence for a restrictive epigenome in schizophrenia. Schizophr Res 2013;149:15–20.

[52] Gavin DP, Rosen C, Chase K, et al. Dimethylated lysine 9 of histone 3 is elevated in schizophrenia and exhibits a divergent response to histone deacetylase inhibitors in lymphocyte cultures. J Psychiatry Neurosci 2009;34:232–7.

[53] Akbarian S, Ruehl MG, Bliven E, et al. Chromatin alterations associated with down-regulated metabolic gene expression in the prefrontal cortex of subjects with schizophrenia. Arch Gen Psychiatr 2005;62: 829–40.

[54] Chase KA, Rosen C, Rubin LH, et al. Evidence of a sex-dependent restrictive epigenome in schizophrenia. J Psychiatr Res 2015;65:87–94.

[55] Kano S, Colantuoni C, Han F, et al. Genome-wide profiling of multiple histone methylations in olfactory cells: further implications for cellular susceptibility to oxidative stress in schizophrenia. Mol Psychiatr 2013;18:740–2.

[56] Huang HS, Matevossian A, Whittle C, et al. Prefrontal dysfunction in schizophrenia involves mixed-lineage leukemia 1-regulated histone methylation at GABAergic gene promoters. J Neurosci 2007;27:11254–62.

[57] Li Y, Kellis M. Joint Bayesian inference of risk variants and tissue-specific epigenomic enrichments across multiple complex human diseases. Nucleic Acids Res 2016;44:e144.

[58] Furey TS, Sethupathy P. Genetics. Genetics driving epigenetics. Science 2013;342:705–6.

[59] Stower H. Gene regulation: from genetic variation to phenotype via chromatin. Nat Rev Genet 2013;14:824.

[60] Kilpinen H, Waszak SM, Gschwind AR, et al. Coordinated effects of sequence variation on DNA binding, chromatin structure, and transcription. Science 2013;342:744–7.

[61] Kasowski M, Kyriazopoulou-Panagiotopoulou S, Grubert F, et al. Extensive variation in chromatin states across humans. Science 2013;342:750–2.

[62] Shi J, Marconett CN, Duan J, et al. Characterizing the genetic basis of methylome diversity in histologically normal human lung tissue. Nat Commun 2014;5:3365.

[63] Buenrostro JD, Giresi PG, Zaba LC, et al. Transposition of native chromatin for fast and sensitive epigenomic profiling of open chromatin, DNA-binding proteins and nucleosome position. Nat Methods 2013;10:1213–8.

[64] Fullard JF, Giambartolomei C, Hauberg ME, et al. Open chromatin profiling of human postmortem brain infers functional roles for non-coding schizophrenia loci. Hum Mol Genet 2017;26:1942–51.

[65] Handel AE, Gallone G, Cader MZ, et al. Most brain disease-associated and eQTL haplotypes are not located within transcription factor DNase-seq footprints in brain. Hum Mol Genet 2017;26(1):79–89.

[66] Lipska BK, Deep-Soboslay A, Weickert CS, et al. Critical factors in gene expression in postmortem human brain: focus on studies in schizophrenia. Biol Psychiatry 2006;60:650–8.

[67] Brennand K, Savas JN, Kim Y, et al. Phenotypic differences in hiPSC NPCs derived from patients with schizophrenia. Mol Psychiatry 2015 Mar;20(3):361–8.

[68] Su Y, Shin J, Zhong C, et al. Neuronal activity modifies the chromatin accessibility landscape in the adult brain. Nat Neurosci 2017;20:476–83.

[69] Dekker J, Marti-Renom MA, Mirny LA. Exploring the three-dimensional organization of genomes: interpreting chromatin interaction data. Nat Rev Genet 2013;14:390–403.

[70] Tang Z, Luo OJ, Li X, et al. CTCF-mediated human 3D genome architecture reveals chromatin topology for transcription. Cell 2015;163:1611–27.

[71] Rao SS, Huntley MH, Durand NC, et al. A 3D map of the human genome at kilobase resolution reveals principles of chromatin looping. Cell 2014;159:1665–80.

[72] Won H, de la Torre-Ubieta L, Stein JL, et al. Chromosome conformation elucidates regulatory relationships in developing human brain. Nature 2016;538:523–7.

[73] Szatkiewicz JP, O'Dushlaine C, Chen G, et al. Copy number variation in schizophrenia in Sweden. Mol Psychiatr 2014;19(7):762–73.

[74] Benes FM. Nicotinic receptors and functional regulation of GABA cell microcircuitry in bipolar disorder and schizophrenia. Handb Exp Pharmacol 2012:401–17. https://doi.org/10.1007/978-3-642-25758-2_13.
[75] Jones CK, Byun N, Bubser M. Muscarinic and nicotinic acetylcholine receptor agonists and allosteric modulators for the treatment of schizophrenia. Neuropsychopharmacology 2012;37:16–42.
[76] Cukier HN, Lee JM, Ma D, et al. The expanding role of MBD genes in autism: identification of a MECP2 duplication and novel alterations in MBD5, MBD6, and SETDB1. Autism Res 2012;5:385–97.
[77] Mendioroz M, Do C, Jiang X, et al. Trans effects of chromosome aneuploidies on DNA methylation patterns in human Down syndrome and mouse models. Genome Biol 2015;16:263.
[78] Garafola CS, Henn FA. A change in hippocampal protocadherin gamma expression in a learned helpless rat. Brain Res 2014;1593:55–64.
[79] Suderman M, McGowan PO, Sasaki A, et al. Conserved epigenetic sensitivity to early life experience in the rat and human hippocampus. Proc Natl Acad Sci U S A 2012;109(Suppl. 2):17266–72.
[80] Bharadwaj R, Peter CJ, Jiang Y, et al. Conserved higher-order chromatin regulates NMDA receptor gene expression and cognition. Neuron 2014;84:997–1008.
[81] Mitchell AC, Jiang Y, Peter C, et al. Transcriptional regulation of GAD1 GABA synthesis gene in the prefrontal cortex of subjects with schizophrenia. Schizophr Res 2015;167:28–34.
[82] Duan J, Shi J, Fiorentino A, et al. A rare functional noncoding variant at the GWAS-implicated MIR137/MIR2682 locus might confer risk to schizophrenia and bipolar disorder. Am J Hum Genet 2014;95:744–53.
[83] Network and Pathway Analysis Subgroup of Psychiatric Genomics Consortium. Psychiatric genome-wide association study analyses implicate neuronal, immune and histone pathways. Nat Neurosci 2015;18:199–209.
[84] Susser E, St Clair D, He L. Latent effects of prenatal malnutrition on adult health: the example of schizophrenia. Ann N Y Acad Sci 2008;1136:185–92.
[85] Torshizi AD, Armoskus C, Zhang S, et al. Reverse engineering of transcriptional networks uncovers candidate master regulators governing neuropathology of schizophrenia. bioRxiv 2017. https://doi.org/10.1101/133363.
[86] Fromer M, Roussos P, Sieberts SK, et al. Gene expression elucidates functional impact of polygenic risk for schizophrenia. Nat Neurosci 2016;19:1442–53.
[87] Lopez-Atalaya JP, Ito S, Valor LM, et al. Genomic targets, and histone acetylation and gene expression profiling of neural HDAC inhibition. Nucleic Acids Res 2013;41:8072–84.
[88] Parsa CF, Robert MP. Thromboembolism and congenital malformations: from Duane syndrome to thalidomide embryopathy. JAMA Ophthalmol 2013;131:439–47.
[89] Lang B, Alrahbeni TM, Clair DS, et al. HDAC9 is implicated in schizophrenia and expressed specifically in post-mitotic neurons but not in adult neural stem cells. Am J Stem Cells 2012;1:31–41.
[90] Kebir O, Chaumette B, Fatjo-Vilas M, et al. Family-based association study of common variants, rare mutation study and epistatic interaction detection in HDAC genes in schizophrenia. Schizophr Res 2014;160:97–103.
[91] Nyegaard M, Severinsen JE, Als TD, et al. Support of association between BRD1 and both schizophrenia and bipolar affective disorder. Am J Med Genet B Neuropsychiatr Genet 2010;153B:582–91.
[92] Severinsen JE, Bjarkam CR, Kiaer-Larsen S, et al. Evidence implicating BRD1 with brain development and susceptibility to both schizophrenia and bipolar affective disorder. Mol Psychiatr 2006;11:1126–38.
[93] Andreassen OA, Thompson WK, Dale AM. Boosting the power of schizophrenia genetics by leveraging new statistical tools. Schizophr Bull 2014;40:13–7.
[94] Fryland T, Christensen JH, Pallesen J, et al. Identification of the BRD1 interaction network and its impact on mental disorder risk. Genome Med 2016;8:53.
[95] McCarthy SE, Gillis J, Kramer M, et al. De novo mutations in schizophrenia implicate chromatin remodeling and support a genetic overlap with autism and intellectual disability. Mol Psychiatr 2014;19:652–8.

[96] Curtis D. Pathway analysis of whole exome sequence data provides further support for the involvement of histone modification in the aetiology of schizophrenia. Psychiatr Genet 2016;26:223–7.
[97] Singh T, Walters JTR, Johnstone M, et al. The contribution of rare variants to risk of schizophrenia in individuals with and without intellectual disability. Nat Genet 2017;49:1167–73.
[98] Cotney J, Muhle RA, Sanders SJ, et al. The autism-associated chromatin modifier CHD8 regulates other autism risk genes during human neurodevelopment. Nat Commun 2015;6:6404.
[99] Takata A, Xu B, Ionita-Laza I, et al. Loss-of-function variants in schizophrenia risk and SETD1A as a candidate susceptibility gene. Neuron 2014;82:773–80.
[100] Singh T, Kurki MI, Curtis D, et al. Rare loss-of-function variants in SETD1A are associated with schizophrenia and developmental disorders. Nat Neurosci 2016;19:571–7.
[101] Li Y, Jiao J. Histone chaperone HIRA regulates neural progenitor cell proliferation and neurogenesis via beta-catenin. J Cell Biol 2017;216:1975–92.
[102] Bassett AS, Scherer SW, Brzustowicz LM. Copy number variations in schizophrenia: critical review and new perspectives on concepts of genetics and disease. Am J Psychiatry 2010;167(8):899–914.
[103] Levinson DF, Duan J, Oh S, et al. Copy number variants in schizophrenia: confirmation of five previous findings and new evidence for 3q29 microdeletions and VIPR2 duplications. Am J Psychiatry 2011;168: 302–16.
[104] Marshall CR, Howrigan DP, Merico D, et al. Contribution of copy number variants to schizophrenia from a genome-wide study of 41,321 subjects. Nat Genet 2017;49:27–35.
[105] Franke M, Ibrahim DM, Andrey G, et al. Formation of new chromatin domains determines pathogenicity of genomic duplications. Nature 2016;538:265–9.
[106] Redin C, Brand H, Collins RL, et al. The genomic landscape of balanced cytogenetic abnormalities associated with human congenital anomalies. Nat Genet 2017;49:36–45.
[107] Zeitz MJ, Lerner PP, Ay F, et al. Implications of COMT long-range interactions on the phenotypic variability of 22q11.2 deletion syndrome. Nucleus 2013;4:487–93.
[108] Moy M, Duan J, Drigalenko E, et al. Transcriptome signature of schizophrenia-associated rare copy number variants (CNVs) in lymphoblastoid cell lines (LCLs). In: American society of human genetics meeting 2014 abstract; 2014.
[109] Panchision DM. Concise review: progress and challenges in using human stem cells for biological and therapeutics discovery: neuropsychiatric disorders. Stem Cell 2016;34:523–36.
[110] Wen Z, Christian KM, Song H, et al. Modeling psychiatric disorders with patient-derived iPSCs. Curr Opin Neurobiol 2016;36:118–27.
[111] Duan J. Path from schizophrenia genomics to biology: gene regulation and perturbation in neurons derived from induced pluripotent stem cells and genome editing. Neurosci Bull 2015;31:113–27.
[112] Brennand KJ, Landek-Salgado MA, Sawa A. Modeling heterogeneous patients with a clinical diagnosis of schizophrenia with induced pluripotent stem cells. Biol Psychiatry 2014;75:936–44.
[113] Seki T, Yuasa S, Fukuda K. Generation of induced pluripotent stem cells from a small amount of human peripheral blood using a combination of activated T cells and Sendai virus. Nat Protoc 2012;7:718–28.
[114] Wright R, Rethelyi JM, Gage FH. Enhancing induced pluripotent stem cell models of schizophrenia. JAMA Psychiatry 2014;71:334–5.
[115] Pasca SP, Portmann T, Voineagu I, et al. Using iPSC-derived neurons to uncover cellular phenotypes associated with Timothy syndrome. Nat Med 2011;17:1657–62.
[116] Shi Y, Kirwan P, Smith J, et al. Human cerebral cortex development from pluripotent stem cells to functional excitatory synapses. Nat Neurosci 2012;15:477–86. S1.
[117] Nicholas CR, Chen J, Tang Y, et al. Functional maturation of hPSC-derived forebrain interneurons requires an extended timeline and mimics human neural development. Cell Stem Cell 2013;12:573–86.

[118] Kriks S, Shim JW, Piao J, et al. Dopamine neurons derived from human ES cells efficiently engraft in animal models of Parkinson's disease. Nature 2011;480:547–51.

[119] Liu Y, Liu H, Sauvey C, et al. Directed differentiation of forebrain GABA interneurons from human pluripotent stem cells. Nat Protoc 2013;8:1670–9.

[120] Maroof AM, Keros S, Tyson JA, et al. Directed differentiation and functional maturation of cortical interneurons from human embryonic stem cells. Cell Stem Cell 2013;12:559–72.

[121] Muffat J, Li Y, Yuan B, et al. Efficient derivation of microglia-like cells from human pluripotent stem cells. Nat Med 2016;22:1358–67.

[122] Vierbuchen T, Ostermeier A, Pang ZP, et al. Direct conversion of fibroblasts to functional neurons by defined factors. Nature 2010;463:1035–41.

[123] Zhang Y, Pak C, Han Y, et al. Rapid single-step induction of functional neurons from human pluripotent stem cells. Neuron 2013;78:785–98.

[124] Yang N, Chanda S, Marro S, et al. Generation of pure GABAergic neurons by transcription factor programming. Nat Methods 2017;14:621–8.

[125] Pasca AM, Sloan SA, Clarke LE, et al. Functional cortical neurons and astrocytes from human pluripotent stem cells in 3D culture. Nat Methods 2015;12:671–8.

[126] Rigamonti A, Repetti GG, Sun C, et al. Large-scale production of mature neurons from human pluripotent stem cells in a three-dimensional suspension culture system. Stem Cell Reports 2016;6:993–1008.

[127] Brennand KJ, Simone A, Jou J, et al. Modelling schizophrenia using human induced pluripotent stem cells. Nature 2011;473:221–5.

[128] Nica AC, Ongen H, Irminger JC, et al. Cell-type, allelic, and genetic signatures in the human pancreatic beta cell transcriptome. Genome Res 2013;23:1554–62.

[129] Civelek M, Lusis AJ. Systems genetics approaches to understand complex traits. Nat Rev Genet 2014;15: 34–48.

[130] Paul DS, Soranzo N, Beck S. Functional interpretation of non-coding sequence variation: concepts and challenges. Bioessays 2014;36:191–9.

[131] Soldner F, Jaenisch R. Medicine. iPSC disease modeling. Science 2012;338:1155–6.

[132] Mali P, Yang L, Esvelt KM, et al. RNA-guided human genome engineering via Cas9. Science 2013;339: 823–6.

[133] Cong L, Ran FA, Cox D, et al. Multiplex genome engineering using CRISPR/Cas systems. Science 2013; 339:819–23.

[134] Wang H, Yang H, Shivalila CS, et al. One-step generation of mice carrying mutations in multiple genes by CRISPR/Cas-mediated genome engineering. Cell 2013;153:910–8.

[135] Fu Y, Foden JA, Khayter C, et al. High-frequency off-target mutagenesis induced by CRISPR-Cas nucleases in human cells. Nat Biotechnol 2013;31:822–6.

[136] Sander JD, Joung JK. CRISPR-Cas systems for editing, regulating and targeting genomes. Nat Biotechnol 2014;32:347–55.

[137] Shalem O, Sanjana NE, Hartenian E, et al. Genome-scale CRISPR-Cas9 knockout screening in human cells. Science 2014;343:84–7.

[138] Soldner F, Laganiere J, Cheng AW, et al. Generation of isogenic pluripotent stem cells differing exclusively at two early onset Parkinson point mutations. Cell 2011;146:318–31.

[139] Hockemeyer D, Wang H, Kiani S, et al. Genetic engineering of human pluripotent cells using TALE nucleases. Nat Biotechnol 2011;29:731–4.

[140] Sanjana NE, Cong L, Zhou Y, et al. A transcription activator-like effector toolbox for genome engineering. Nat Protoc 2012;7:171–92.

[141] Ding Q, Lee YK, Schaefer EA, et al. A TALEN genome-editing system for generating human stem cell-based disease models. Cell Stem Cell 2013;12(2):238–51.

[142] Reyon D, Tsai SQ, Khayter C, et al. FLASH assembly of TALENs for high-throughput genome editing. Nat Biotechnol 2012;30:460–5.
[143] Joung JK, Sander JD. TALENs: a widely applicable technology for targeted genome editing. Nat Rev Mol Cell Biol 2012;14:49–55.
[144] Bedell VM, Wang Y, Campbell JM, et al. In vivo genome editing using a high-efficiency TALEN system. Nature 2012;491:114–8.
[145] Schmid-Burgk JL, Schmidt T, Kaiser V, et al. A ligation-independent cloning technique for high-throughput assembly of transcription activator-like effector genes. Nat Biotechnol 2012;31:76–81.
[146] Choi SM, Kim Y, Shim JS, et al. Efficient drug screening and gene correction for treating liver disease using patient-specific stem cells. Hepatology 2013;57(6):2458–68.
[147] Montalbano A, Canver MC, Sanjana NE. High-throughput approaches to pinpoint function within the noncoding genome. Mol Cell 2017;68:44–59.
[148] Gaudelli NM, Komor AC, Rees HA, et al. Programmable base editing of A·T to G·C in genomic DNA without DNA cleavage. Nature 2017;551:464–71.
[149] Guidotti A, Grayson DR. DNA methylation and demethylation as targets for antipsychotic therapy. Dialogues Clin Neurosci 2014;16:419–29.
[150] Dong E, Tueting P, Matrisciano F, et al. Behavioral and molecular neuroepigenetic alterations in prenatally stressed mice: relevance for the study of chromatin remodeling properties of antipsychotic drugs. Transl Psychiatry 2016;6:e711.
[151] Hyman SE. Target practice: HDAC inhibitors for schizophrenia. Nat Neurosci 2012;15:1180–1.
[152] Kurita M, Holloway T, Garcia-Bea A, et al. HDAC2 regulates atypical antipsychotic responses through the modulation of mGlu2 promoter activity. Nat Neurosci 2012;15:1245–54.
[153] Ibi D, de la Fuente Revenga M, Kezunovic N, et al. Antipsychotic-induced Hdac2 transcription via NF-kappaB leads to synaptic and cognitive side effects. Nat Neurosci 2017;20:1247–59.
[154] Oh SW, Harris JA, Ng L, et al. A mesoscale connectome of the mouse brain. Nature 2014;508:207–14.
[155] Camp JG, Treutlein B. Human development: advances in mini-brain technology. Nature 2017;545:39–40.
[156] Lasko LM, Jakob CG, Edalji RP, et al. Discovery of a selective catalytic p300/CBP inhibitor that targets lineage-specific tumours. Nature 2017;550:128–32.
[157] Makova KD, Hardison RC. The effects of chromatin organization on variation in mutation rates in the genome. Nat Rev Genet 2015;16:213–23.
[158] Bae T, Tomasini L, Mariani J, et al. Different mutational rates and mechanisms in human cells at pregastrulation and neurogenesis. Science 2018;359(6375):550–5.
[159] Lodato MA, Rodin RE, Bohrson CL, et al. Aging and neurodegeneration are associated with increased mutations in single human neurons. Science 2018;359(6375):555–9.

CHAPTER

Gilles de la Tourette syndrome

15

Claudia Selvini[1], Stefano Cavanna[2], Andrea E. Cavanna[3,4,5]

Child Neuropsychiatry Unit, Department of Experimental Medicine, University of Insubria, Varese, Italy[1]; Department of Radiology, University of Turin, Turin, Italy[2]; Department of Neuropsychiatry, BSMHFT and University of Birmingham, Birmingham, United Kingdom[3]; School of Life and Health Sciences, Aston University, Birmingham, United Kingdom[4]; Sobell Department of Motor Neuroscience and Movement Disorders, UCL and Institute of Neurology, London, United Kingdom[5]

Chapter Outline

15.1 Introduction: Gilles de la Tourette syndrome and other tic disorders 332
- 15.1.1 Definition and diagnostic criteria of Gilles de la Tourette syndrome and other tic disorders 332
- 15.1.2 Epidemiology 333

15.2 Clinical presentation of tics 333
- 15.2.1 Shared characteristics of tics 333
- 15.2.2 Characteristics of motor tics 334
- 15.2.3 Characteristics of vocal/phonic tics 334
- 15.2.4 Characteristics of cognitive tics 334

15.3 Tic-related behavioral symptoms and health-related quality of life 334
- 15.3.1 Behavioral spectrum of Gilles de la Tourette syndrome 334
- 15.3.2 Obsessive—compulsive disorder 335
- 15.3.3 Attention-deficit and hyperactivity disorder 335
- 15.3.4 Health-related quality of life 335

15.4 Etiology and pathophysiology 335
- 15.4.1 Genetic factors 335
- 15.4.2 Environmental factors 336
- 15.4.3 Role of dopamine and cortico-striato-thalamo-cortical pathways 336
- 15.4.4 Possible role of chromatin regulation 337

15.5 Treatment strategies 339
- 15.5.1 Psychoeducation 339
- 15.5.2 Behavioral therapy 339
- 15.5.3 Pharmacotherapy 340
- 15.5.4 Other approaches 340

15.6 Conclusions: open questions and suggestions for future research 340

Acknowledgments 341

References 341

Chromatin Signaling and Neurological Disorders. https://doi.org/10.1016/B978-0-12-813796-3.00015-8

15.1 Introduction: Gilles de la Tourette syndrome and other tic disorders

15.1.1 Definition and diagnostic criteria of Gilles de la Tourette syndrome and other tic disorders

Gilles de la Tourette syndrome (GTS) is a neurodevelopmental condition characterized by the chronic presence of multiple motor tics and at least one vocal/phonic tic: sudden, rapid, recurrent, nonrhythmic movements, and vocalizations. The most frequently reported motor tics include eye blinking, facial grimacing, neck stretching, shoulder shrugging, abdominal contractions, and limb jerks; whereas, common vocal tics are grunting, sniffing, throat clearing, coughing, snorting, squeaking, and humming [1]. GTS was first described by George Gilles de la Tourette in 1885 [2] (Fig. 15.1) and currently belongs to the broader diagnostic category of tic disorders, which also includes persistent (chronic) motor or phonic tic disorder and provisional tic disorder, as well as the less commonly used diagnostic codes of other specified tic disorder and unspecified tic disorder [3].

From a neurological perspective, all tic disorders are hyperkinetic movement disorders characterized by the presence of motor and/or vocal/phonic tics, with different features [4]. Although in GTS both motor and vocal/phonic tics are present throughout life, in the persistent (chronic) motor or phonic tic disorder only motor or only vocal/phonic tics are present, respectively. In the provisional tic disorder, motor and/or vocal/phonic tics have been present for a short period of time (less than 1 year). Finally, other specified tic disorder and unspecified tic disorder are diagnostic codes used when the diagnostic criteria for other tic disorders are not met for specified reasons (e.g., atypical presentation and/or age at onset) or unspecified reasons (e.g., insufficient information) [3].

In patients diagnosed with GTS or other chronic tic disorders (persistent motor or phonic tic disorder), tics have been present for at least 1 year, typically with a fluctuating (waxing and waning) course. For young patients presenting with motor and/or phonic tics of less than 1 year since first tic

FIGURE 15.1

Georges Gilles de la Tourette (1857–1904). Attribution: By Georges_Gilles_de_la_Tourette.jpg: Unknown, the plate was photographed by an E. Pirou who might be Eugène Pirou (1841–1909). Derivative work: Mikhail Ryazanov [Public domain], via Wikimedia Commons. Copyright-free licensing: https://commons.wikimedia.org/wiki/File:Georges_Gilles_de_la_Tourette.png.

onset, the diagnosis of provisional tic disorder is usually considered. There is no specification in tic duration for other specified tic disorder and unspecified tic disorder [3].

Tic disorders typically begin in childhood, with an average age at onset between 4 and 6 years. Less commonly, tic onset occurs in adolescence, before the age of 18 years. New onset of tic symptoms in adulthood is rare and is typically associated with exposures to drugs (e.g., excessive cocaine use) or is the clinical expression of an underlying central nervous system insult (e.g., postviral encephalitis). Tic disorders are diagnosed on clinical grounds and require specialist assessments from experienced clinicians. In rare circumstances, the diagnosis of a tic disorder can require further tests to exclude other medical conditions (such as Huntington's disease) or substance abuse (such as stimulants) [1].

GTS is arguably the most complex and clinically relevant tic disorder, as it is accompanied by comorbid behavioral symptoms in about 90% of cases. The behavioral spectrum of GTS ranges from complex tic-like repetitive symptoms (i.e., self-injurious behaviors, non-obscene socially inappropriate behaviors, coprophenomena, echophenomena, and paliphenomena) to symptoms of attention-deficit and hyperactivity disorder (ADHD), obsessive–compulsive disorder (OCD), affective disorders, impulse control disorders, and personality disorders [5,6].

15.1.2 Epidemiology

The estimated prevalence figures for GTS in the general pediatric population range from 0.3% to 1%, as consistently shown by the findings of large epidemiological studies and meta-analyses [7–9]. Tics as isolated symptoms are estimated to be more common than GTS, potentially affecting around 5% of the general population at some point [10], although prevalence figures show a wide variability, ranging from 1% to 29% depending on the different study designs, methodologies, and sampled populations [11,12]. Males are more commonly affected than females, with a male:female ratio of 3:1 or 4:1 [1].

15.2 Clinical presentation of tics

15.2.1 Shared characteristics of tics

The complex and dynamic tic répertoire is characteristically experienced as an involuntary manifestation, although tics can be voluntarily suppressed for varying lengths of time, depending on the situation. Tics are often suggestible or inducible, for example, by exposure to other persons with tics or just by talking about tics. Importantly, in most patients, tics are preceded by distressing sensory experiences that are commonly referred to as "premonitory urges" [13]. Premonitory urges are unpleasant somatosensory sensations (tension, pressure, itch, stabbing pain, abdominal discomfort, heat, or cold), localized either within the individual muscles affected by the upcoming tics or somewhere else in the body or the head [14–17]. Tic expression provides transient relief from these unpleasant physical sensations, whereas tic suppression is characteristically accompanied by a sensation of mounting inner tension [18].

Tics usually present in bouts, with spontaneous fluctuations in both frequency and intensity; moreover, tics tend to decrease and often disappear during sleep [19]. The severity of tics depends in most cases on environmental factors. For instance, stressful events and excitement can exacerbate tics; whereas, relaxation and activities that require focused attention often decrease tic severity [20].

Tics can be classified according to type (i.e., motor, vocal/phonic), complexity (i.e., simple, complex), location, number, frequency, and duration (i.e., clonic, tonic) [21]. Of note, motor tics and vocal/phonic tics do not appear to reflect separate pathophysiological processes, and all tic symptoms are best viewed along a clinical and pathophysiological continuum. Tic disorders usually begin with simple motor tics, most commonly eye blinking [22], followed by other simple motor tics and more complex tics with a rostrocaudal distribution. In most patients with GTS phonic tics begin at a later stage, ordinarily in the early puberal period [23]. Improvement in tic severity by adulthood has been reported in a considerable proportion of cases across clinical studies [24].

15.2.2 Characteristics of motor tics

Motor tics result from contractions (often referred to as "twitches" by patients) affecting voluntary muscles [23]. Simple motor tics are of short duration (i.e., milliseconds) and are restricted to an individual muscle or a single muscle group. Complex motor tics tend to be of longer duration (i.e., seconds) and often have a compulsive nature, including a combination or pattern of simple motor tics, such as simultaneous head turning and shoulder shrugging. Through the involvement of multiple muscular segments, complex motor tics can result in whole body movements and/or abnormal gait. Sometimes complex motor tics can resemble intentional actions, as in the case of copropraxia (i.e., the involuntary production of obscene gestures) and echopraxia (i.e., the involuntary imitation of someone else's movement).

15.2.3 Characteristics of vocal/phonic tics

Vocal/phonic tics are defined as sounds generated by airflow through the vocal cords, mouth, or nose [23]. Simple vocal tics include vocalizations (e.g., "ah" and "ooh") and are more appropriately referred to as phonic tics when there is no actual involvement of the vocal cords (e.g., sniffing); vocal/phonic tics often result from the contraction of the diaphragm or the oropharynx muscles. Complex vocal/phonic tics are characterized by more elaborate sounds or words with an associated semantic content. Complex vocal/phonic tics include repeating one's own sounds or words (palilalia), repeating the last-heard word or phrase pronounced by someone else (echolalia), or uttering socially inappropriate words, including obscenities, ethnic, racial, or religious slurs (coprolalia) [25].

15.2.4 Characteristics of cognitive tics

Lastly, it has been proposed that some patients present with a further type of tics without physical correlates, sometimes referred to as "cognitive tics." Cognitive tics are described as repetitive thoughts that are not anxiety-driven, but are expressed in response to the urge to give in or act upon provocative auditory, visual, tactile, or inner stimuli [26,27]. Although their exact frequency is not known, cognitive tics encompass thought echophenomena, mental play, aimless counting, and repetitive thoughts with sexual or aggressive content that produce no fear. Of note, the boundary between certain cognitive tics and obsessions reported by patients with GTS can be difficult to establish [26,27].

15.3 Tic-related behavioral symptoms and health-related quality of life

15.3.1 Behavioral spectrum of Gilles de la Tourette syndrome

Although motor and vocal/phonic tics are the key features of GTS, most patients (about 90%) report the presence of a wide spectrum of associated behavioral symptoms. The most frequently reported behavioral problems range from complex tic-like symptoms (forced touching, self-injurious behaviors, nonobscene socially inappropriate behaviors, copro-, echo-, and paliphenomena) to comorbid OCD, ADHD, affective disorders, impulse control disorders, conduct disorder, oppositional-defiant disorders, and personality disorders. There appears to be also an association between GTS and sleep problems, and tics are reported more frequently in patients with autism spectrum disorders [5].

15.3.2 Obsessive–compulsive disorder

The reported percentage of patients with GTS who present with comorbid OCD varies from 11% to 80% [5,6] and tic-like OCD is believed to be intrinsically related to GTS. Obsessive–compulsive symptoms reported by patients with tics are often qualitatively different from obsessions and compulsions reported by patients diagnosed with primary OCD in the absence of tics [28]: compulsions are not usually driven by complex cognitions or anxiety and tend to be more frequently related to counting, symmetry, and "just right" behaviors, while obsessions are characterized by a higher prevalence of aggressive or inappropriate sexual thoughts and images [29].

15.3.3 Attention-deficit and hyperactivity disorder

The most frequent comorbid condition in children with GTS is ADHD (predominantly inattentive or combined subtype), which is diagnosed in up to 60% of young patients with GTS [30]. The diagnosis of comorbid ADHD in patients with GTS poses unique challenges, as the clinical presentations of these two neurodevelopmental disorders often overlap. The onset of ADHD symptoms in patients with GTS typically occurs around the age of 3–5 years, thus preceding the onset of tics. It is important to highlight that hyperactivity and impulsivity symptoms tend to improve by adulthood, whereas attention-deficit can persist throughout the lifespan [31].

15.3.4 Health-related quality of life

Both tic severity and co-occurring behavioral problems, particularly OCD and ADHD, have been shown to be potentially associated with significant impairment and personal distress in young patients with GTS, resulting in poorer health-related quality of life [32,33,33a]. Problems in daily life encompass the domains of motor function (e.g., frustration at not being in control of own movements and/or utterances), psychological well-being (e.g., lack of self-confidence), cognitive functions (e.g., difficulties with concentration), and obsessionality (e.g., repetitive thoughts). Moreover, comorbid behavioral problems including impulsivity [34] can lead to significant social difficulties, especially in childhood [35]. Similar findings have been reported from adult populations, with stronger correlations between health-related quality of life and tic severity (as well as presence of anxiety and depression) [36,37]. These observations are reflected in the multidimensional structure of the disease-specific quality of life scales used for young [38,39] and adult [40] patients with GTS.

15.4 Etiology and pathophysiology

15.4.1 Genetic factors

For a long time after its initial description, psychological models have been developed to explain the etiology and pathophysiology of GTS [41]. Modern etiological theories for GTS encompass genetic and environmental factors, such as infections and autoimmune dysfunction, prenatal and/or perinatal problems, psychosocial stressors, and androgen influences [6]. The concepts of genetic and etiological heterogeneity are in line with the findings of clinical studies, which suggest the existence of multiple phenotypes within the GTS spectrum: "pure" GTS (clinical presentations characterized by motor and vocal/phonic tics only), "full-blown" GTS (patients who report also complex tics and tic-related symptoms), GTS "plus" (patients diagnosed with comorbid psychiatric disorders) [5].

Convergent findings from twin, family and population studies have shown that genetic factors play an important role in the etiopathogenesis of GTS. Although initial studies suggested an autosomal dominant transmission model, polygenic and bilineal transmission models were subsequently postulated [42]. More recent studies have yielded strong evidence that GTS is a genetically heterogeneous disorder. Specifically, candidate gene studies have shown that several genes (i.e., *DRD2*, *DRD4*, *5-HT2C*, *SERT*) involved in the regulation of multiple neurochemical pathways (e.g., dopaminergic, serotonergic, and histaminergic circuitries) might be associated with the etiopathogenesis of GTS [43–49]. In recent years, new candidate genes (e.g., *SLITRK1*, *IMMP2L*, *CNTNAP2*, *NLGN4*) have been identified through linkage studies and structural genomic aberrations, in which very rare genetic variants with large effects were found in patients with GTS and families [50]. Interestingly, in addition to being involved in the above-mentioned neurochemical pathways, a few of these genes regulate neuronal activity through different pathways.

15.4.2 Environmental factors

Both GTS and associated comorbidities are characterized by lower heritability rates compared to other neuropsychiatric disorders, such as autism spectrum disorders, schizophrenia, and bipolar disorder, suggesting that environmental factors could play a role in pathogenesis of GTS and its comorbidities [51].

With regard to environmental factors, it has been hypothesized that GTS could be subsumed within a group of conditions called "pediatric autoimmune neuropsychiatric disorders associated with streptococcal infections" (PANDAS) [52] (Fig. 15.2). It has been shown that patients with GTS have a reduced frequency of regulatory T cells, which induce tolerance toward self-antigens. This mechanism could explain the predisposition to autoimmune responses in patients with GTS [53,54]. In addition to autoimmunity, an overall lowered immunity profile has been proposed as a possible etiological mechanism in patients with GTS. Specifically, some authors have demonstrated a form of dysgammaglobulinemia in patients with GTS showing low immunoglobulin A (IgA) levels [55]; whereas, the results of other studies have suggested that patients with GTS could have other forms of immune deficiency (decreased IgG3 and possibly also low IgM levels) [56]. The potential role of prenatal and perinatal events in the etiopathogenesis of GTS has been investigated in epidemiological studies and showed that complications during pregnancy, lower birth weight, maternal life stress during pregnancy, maternal smoking, severe nausea, and/or vomiting during the first trimester have all been found to be associated with GTS [16,17]. Finally, recent findings suggest that androgen exposure may also be an important factor in the etiopathogenesis of GTS and related disorders [57–59].

15.4.3 Role of dopamine and cortico-striato-thalamo-cortical pathways

The exact brain mechanisms associated with tic development and expression is not entirely known, although preliminary evidence from neurochemical and neuroimaging investigations suggests a

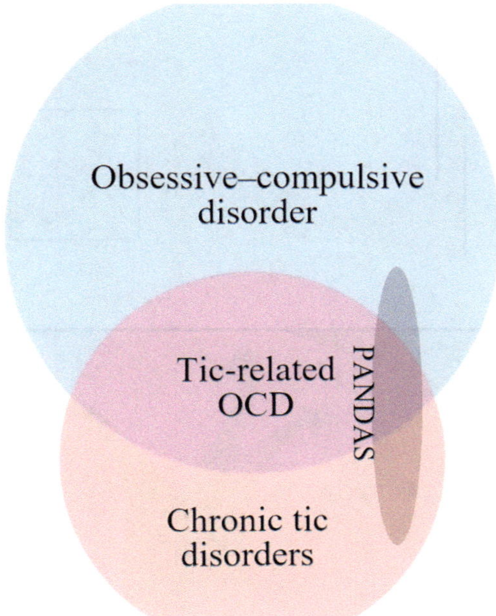

FIGURE 15.2

Proposed relationship between pediatric autoimmune neuropsychiatric disorders associated with streptococcal infections, tic disorders, and obsessive–compulsive disorder. Abbreviations: *OCD*, obsessive–compulsive disorder; *PANDAS*, pediatric autoimmune neuropsychiatric disorders associated with streptococcal infections. Attribution: By User: Eubulides (Own work) [GFDL (http://www.gnu.org/copyleft/fdl.html) or CC BY-SA 3.0 (https://creativecommons.org/licenses/by-sa/3.0)], via Wikimedia Commons. Copyright-free licensing: https://commons.wikimedia.org/wiki/File:Proposed_PANDAS.svg.

primary role for dysfunction of dopaminergic pathways within the basal ganglia and the cortico-striato-thalamo-cortical circuitries (Fig. 15.3). An alteration in striatal dopamine release in GTS may result in tic expression because of a focal excitatory abnormality within the striatum coupled with disinhibition of thalamo-frontal pathways [60,61]. The results of neuropathological studies have provided evidence for deficits in cerebral maturation, in particular at the level of striatal interneuron migration processes [62]. Finally, neuroimaging studies exploring the neural correlates of the premonitory urge to tic have revealed the involvement of both motor areas (e.g., supplementary motor area) and extramotor areas (e.g., insula, cingulate cortex) in tic generation [63].

15.4.4 Possible role of chromatin regulation

It has been established that there are no direct correlations between genotypes and phenotypes in neuropsychiatric disorders. The missing pieces of information between genomic variations and disease phenotypes include epigenetic mechanisms for the regulation of gene expression, such as structural modification of chromatin, posttranslational modification of histones (e.g., acetylation and methylation), chemical modification of DNA through methylation, or hydroxymethylation of cysteins, as well

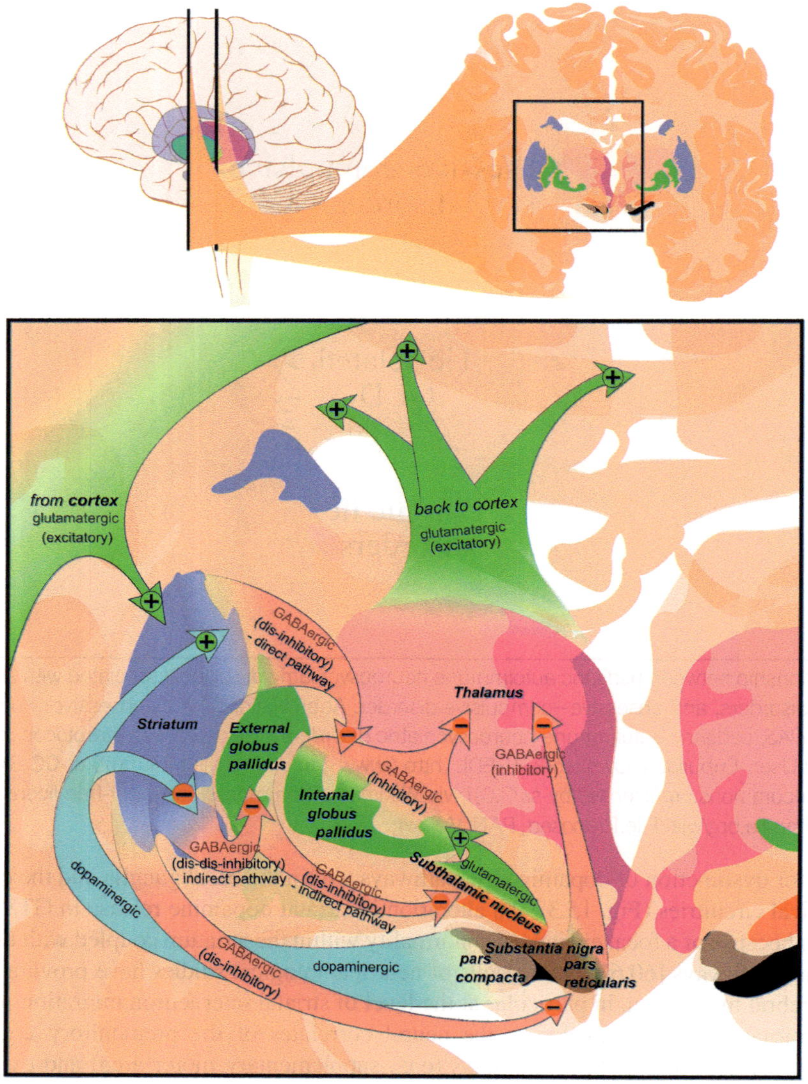

FIGURE 15.3

Role of dopaminergic neurotransmission within basal ganglia direct and indirect pathways. Positive and negative signs at the point of the arrows indicate, respectively, whether the pathway is excitatory or inhibitory in effect. Green arrows refer to excitatory glutamatergic pathways, red arrows refer to inhibitory GABAergic pathways, and turquoise arrows refer to dopaminergic pathways that are excitatory on the direct pathway and inhibitory on the indirect pathway. Abbreviation: *GABA*, gamma-aminobutyric acid. Attribution: By Mikael Häggström, based on images by Andrew Gillies/User:Anaru and Patrick J. Lynch [CC BY-SA 3.0 (https://creativecommons.org/licenses/by-sa/3.0)], via Wikimedia Commons. Copyright-free licensing: https://commons.wikimedia.org/wiki/File:Basal_ganglia_circuits.svg.

as expression of interfering noncoding RNAs [50]. Interestingly, the expression of highly specific subsets of fragile sites in association with GTS has been reported [64]. Moreover, Zilhão et al. [65] recently reported the results obtained in the first epigenome-wide association study of tic disorders. Phenotype and DNA methylation data were available in a large number of participants (1678 subjects; mean age = 41.5 years) in surveys at the Netherlands Twin Register and related biobank project. No probes reached genome-wide significance, with the strongest associated probe (cg15583738) being located in an intergenic region on chromosome 8. Interestingly, several of the top ranking probes were located in or nearby genes previously associated with neurological disorders (e.g., *GABBRI*, *BLM*, *ADAM10*), warranting further investigation of their possible role(s) in tic disorders. The top significantly enriched gene ontology terms among higher-ranking methylation sites included anatomical structure morphogenesis, developmental process, and cellular developmental process. By showing the possible role of DNA methylation in tic disorders, the results of this large study provided a first insight into the epigenetic mechanisms of these conditions. [66] suggested that epigenetic mechanisms (e.g., DNA methylation, histone modifications, and noncoding RNAs in the regulation of gene expression) may mediate the effect of environmental triggers on genetic background, thus leading to the onset of tic symptoms. However, these preliminary findings need to be extended to other neurobehavioral disorders to ascertain whether observed patterns can be considered representative of "chromatin endophenotypes" correlating with discrete sets of neurobehavioral symptoms. It is likely that a clearer picture of the dynamic control of neurobehavioral candidate coding regions may result from specific analyses that include chromatin levels of genomic information management [67].

15.5 Treatment strategies

15.5.1 Psychoeducation

The management of GTS is complex and requires regular input from an experienced clinician. It is important to highlight that a number of patients, in particular in childhood, do not require active treatment interventions for their tics beyond a comprehensive assessment and appropriate psychoeducation to the patient, family, and educators [11,12]. The targets of psychoeducation include improvement of tolerance for tic symptoms and stress reduction in patients with GTS. Psychoeducation should cover key information about the long- and short-term modulation of tics, the natural course of the condition, and possible coexisting behavioral problems.

15.5.2 Behavioral therapy

Active treatment interventions should be considered when tics interfere with daily life or recreational activities, when tics cause subjective discomfort (e.g., pain or injury), sustained social problems (e.g., social isolation or bullying) and emotional difficulties (e.g., reactive depressive symptoms), or functional interference (e.g., impairment of academic achievements) [68]. Treatment interventions encompass behavioral approaches, pharmacotherapy, and more invasive procedures such as deep brain stimulation for severe, refractory cases.

Over time, a number of behavioral approaches have been developed to assist patients with GTS in achieving better control over their tic symptoms [69]. Currently, the most commonly used behavioral strategies are habit reversal training and exposure and response prevention. Specifically, the most extensively investigated technique is habit reversal training, which is based on the subjective awareness of the premonitory urge for the individual tic that the patient decides to target. The ultimate goal is to enable the patient to habituate to the premonitory urge without tic expression, by developing appropriate competing responses [70–72,72a].

15.5.3 Pharmacotherapy

Currently, available pharmacological options can be viewed as symptomatic treatments that alleviate, but do not cure tic symptoms [73]. The high interindividual variability of symptoms, the temporal fluctuations of tics, and the potential interference from coexisting conditions are among the factors that make it difficult to precisely establish treatment effects for tic-suppressing agents. It is therefore not surprising that existing guidelines for the pharmacological management of tics are based on a rather limited evidence base [68,74,75]. Importantly, behavioral comorbidities, which can often cause more clinical impairment than the tics themselves, can be more responsive to treatment interventions, thereby leading to an improved health-related quality of life despite the persistence of tics.

In line with the evidence that the pathophysiology of GTS involves abnormalities at the level of central dopaminergic systems, the most commonly prescribed medications for tics have traditionally been dopamine antagonists (typical neuroleptics, such as haloperidol and pimozide, and atypical antipsychotics, such as risperidone) [6]. In clinical practice, over the last few decades, the use of atypical antipsychotics has been preferred, despite the relatively high frequent occurrence of adverse effects, including sedation and metabolic changes. More recently, aripiprazole, an antipsychotic agent with partial agonist properties at the level of dopamine receptors, has shown good potential for efficacy on tics with an acceptable tolerability profile [76]. A number of other agents acting on dopamine neurotransmission, ranging from substituted benzamides (e.g., sulpiride) to presynaptic dopamine depletors (e.g., tetrabenazine), have been shown to be potentially useful in the treatment of tics [77]. Alpha-2 adrenergic agonists (i.e., clonidine, guanfacine) are remarkably well tolerated and have an established role in the pharmacological management of GTS, especially in children with tics comorbid ADHD. Among antiepileptic drugs, the use of topiramate for the treatment of tics is justified by preliminary evidence that is particularly encouraging [78,79].

15.5.4 Other approaches

When standard treatment interventions fail to improve health-related quality of life because of poor efficacy and/or tolerability, the option of neuromodulation could be considered. Since the turn of the millennium, an increasing number of patients with severe, refractory forms of GTS have been referred to functional neurosurgery (deep brain stimulation of the thalamus or globus pallidus—pars interna) [80]. Further research is essential to better define both safety and efficacy parameters for the modulation of cortico-striato-thalamo-cortical pathways via this invasive procedure.

15.6 Conclusions: open questions and suggestions for future research

Over the last few decades, the scientific interest in GTS has increased considerably, with a number of key advances in our understanding of tic disorders, although it is undeniable that much is yet to be learnt. Indeed, several important aspects remain to be clarified, first of all the relationship between GTS and its comorbid behavioral conditions. For example, it is possible that tic disorders plus ADHD could represent a separate nosological entity, rather than a combination of two independent pathologies, however the pathophysiological mechanism that could explain this theory should be investigated more thoroughly. Overall, the pathogenetic aspects of GTS are largely unknown, although novel techniques and research methodologies are currently being employed to investigate the neural correlates of tic generation and tic expression. Furthermore, longitudinal neuroimaging studies are required to confirm preliminary findings suggesting alterations within specific brain regions and connectivity patterns in patients with chronic tic disorders. The elucidation of the genetic aspects of GTS has proven unexpectedly challenging. Although rare mutations have made important contributions to our understanding of GTS etiopathogenesis, future genetic studies are necessary to better define genotype–phenotype correlations within the GTS clinical spectrum. As both environmental and hormonal factors may be involved in GTS, epigenetic rather than genetic mechanisms may trigger tics: as in other neuropsychiatric conditions, it is likely that genetic mechanisms can "load the gun" and specific epigenetic mechanisms can "pull the trigger." Moreover, it would be useful to include functionally significant chromatin structural variation analyses when considering candidate genes for neurobehavioral disorders such as GTS. Despite the availability of multiple therapeutic strategies, the treatment of GTS is still suboptimal. The current cornerstone of tic management is pharmacotherapy, although the use of nonpharmacological approaches is increasing. Larger randomized controlled trials are essential to confirm the efficacy of both pharmacological and nonpharmacological therapeutic options. Finally, a particularly fruitful area for current and future research is the investigation of social cognition in GTS. A better understanding of the neural bases of the cognitive, emotional, and social aspects of this quintessentially neuropsychiatric condition is of paramount importance to clarify its pathophysiology and to improve patients' health-related quality of life.

Acknowledgments

AEC would like to acknowledge the ongoing support of Tourettes Action-UK and Tourette Association of America. This work is dedicated to Dr. S. Figatto.

References

[1] Cavanna AE, Seri S. Tourette's syndrome. Br Med J 2013;347:f4964.
[2] Gilles de la Tourette G. Etude sur une affection nerveuse caracterisee par la incoordination motrice accompagnee d'echolalie et de coprolalie. Arch Neurol 1885;9:19–42. 158-200.
[3] American Psychiatric Association. Diagnostic and statistical manual of mental disorders. 5th ed. American Psychiatric Association Press; 2013.
[4] Cavanna AE, Seri S. Chronic tic disorders: diagnosis, treatment and management. Clin Pharm 2016;8:2.
[5] Cavanna AE, Rickards H. The psychopathological spectrum of Gilles de la Tourette syndrome. Neurosci Biobehav Rev 2013;37:1008–15.
[6] Robertson MM. Tourette syndrome, associated conditions and the complexities of treatment. Brain 2000; 123:425–62.
[7] Centers for Disease Control and Prevention. Prevalence of diagnosed Tourette syndrome in persons aged 6-17 years in the United States. MMWR Morb Mortal Wkly Rep 2009;58:581–5.
[8] Scharf JM, Miller LL, Gauvin CA, Alabiso J, Mathews CA, Ben-Shlomo Y. Population prevalence of Tourette syndrome: a systematic review and meta-analysis. Mov Disord 2015;30:221–8.

[9] Stern JS, Burza S, Robertson MM. Gilles de la Tourette's syndrome and its impact in the UK. Postgrad Med 2005;81:12–9.

[10] Ong MT, Mordekar SR, Seal A. 15 minute consultation: tics and Tourette syndrome. Arch Dis Childhood Educ Pract Ed 2016;101:87–94.

[11] Robertson MM. The prevalence and epidemiology of Gilles de la Tourette syndrome. Part 1: the epidemiological and prevalence studies. J Psychosom Res 2008;65:461–72.

[12] Robertson MM. The prevalence and epidemiology of Gilles de la Tourette syndrome. Part 2: tentative explanations for differing prevalence figures in GTS, including the possible effects of psychopathology, aetiology, cultural differences, and differing phenotypes. J Psychosom Res 2008;65:473–86.

[13] Lang A. Patient perceptions of tics and other movement disorders. Neurology 1991;41:223–8.

[14] Crossley E, Cavanna AE. Sensory phenomena: clinical correlates and impact on quality of life in adult patients with Tourette syndrome. Psychiatr Res 2013;209:705–10.

[15] Crossley E, Seri S, Stern JS, Robertson MM, Cavanna AE. Premonitory urges for tics in adult patients with Tourette syndrome. Brain Dev 2014;36:45–50.

[16] Leckman JF. In search of the pathophysiology of Tourette syndrome. In: Bedard MA, Agid Y, Chouinard S, Fahn S, Korczyn AD, Lesperance P, editors. Mental and behavioral dysfunction in movement disorders. Totowa, New Jersey: Humana Press; 2003. p. 467–76.

[17] Leckman JF. Phenomenology of tics and natural history of tic disorders. Brain Dev 2003;25(Suppl. 1): S24–8.

[18] Eddy CM, Cavanna AE. Premonitory urges in adults with complicated and uncomplicated Tourette syndrome. Behav Modif 2014;38:264–75.

[19] Cohrs S, Rash T, Altmeyer S, Kinkelbur J, Kostanecka T, Rothenberger A, Rüther E, Hajak G. Decreased sleep quality and increased sleep related movements in patients with Tourette's syndrome. J Neurol Neurosurg Psychiatry 2001;70:192–7.

[20] Bloch MH, Leckman JF, Scott S, Stevenson J, Taylor J, Thapar A. Tic disorders. In: Rutter M, Bishop D, Pine DS, editors. Rutter's child and adolescent psychiatry. 5th ed. Blackwell; 2008. p. 719–36.

[21] van Woerkom TCAM, Cath DC, Coebergh JA. Tics and the Gilles de la Tourette syndrome. In: Wolters E, van Laar T, Berendse H, editors. Parkinsonism and related disorders. Amsterdam: VU University Press; 2007. p. 355–79.

[22] Martino D, Cavanna AE, Robertson MM, Orth M. Prevalence and phenomenology of eye tics in Gilles de la Tourette syndrome. J Neurol 2012;259:2137–40.

[23] Ganos C, Martino D. Tics and tourette syndrome. Neurol Clin 2015;33:115–36.

[24] Hassan N, Cavanna AE. The prognosis of Tourette syndrome: implications for clinical practice. Funct Neurol 2012;27:23–7.

[25] Eddy CM, Cavanna AE. 'It's a curse!': coprolalia in Tourette syndrome. Eur J Neurol 2013;20:1467–70.

[26] Cath DC, Hoogduin CAL, van de Wetering BJM, van Woerkom TCAM, Roos RAC, Rooymans HGM. Tourette Syndrome and obsessive-compulsive disorder: an analysis of associated phenomena. Adv Neurol 1992;58:33–41.

[27] Cath DC, van de Wetering BJM, van Woerkom TCAM, Hoogduin CAL, Roos RAC, Rooymans HGM. Mental play in Gilles de la Tourette's syndrome and obsessive–compulsive disorder. Br J Psychiatry 1992; 161:542–5.

[28] George MS, Trimble MR, Ring HA, Sallee FR, Robertson MM. Obsessions in obsessive compulsive disorder with and without Gilles de la Tourette syndrome. Am J Psychiatry 1993;150:93–7.

[29] Worbe Y, Mallet L, Golmard JL, Béhar C, Durif F, Jalenques I, Damier P, Derkinderen P, Pollak P, Anheim M, Broussolle E, Xie J, Mesnage V, Mondon K, Viallet F, Jedynak P, Ben Djebara M, Schüpbach M, Pelissolo A, Vidailhet M, Agid Y, Houeto JL, Hartmann A. Repetitive behaviours in patients with Gilles de la Tourette syndrome: tics, compulsions, or both? PLoS One 2010;5:e12959.

[30] Stewart SE, Illmann C, Geller DA, Leckman JF, King R, Pauls DL. A controlled family study of attention-deficit/hyperactivity disorder and Tourette's disorder. J Am Acad Child Adolesc Psychiatry 2006;45: 1354–62.

[31] Bloch MH, Craiglow BG, Landeros-Weisenberger A, Dombrowski PA, Panza KE, Peterson BS, Leckman JF. Predictors of early adult outcomes in pediatric-onset obsessive–compulsive disorder. Pediatrics 2009;124: 1085–93.

[32] Cavanna AE, David K, Bandera V, Termine C, Balottin U, Schrag A, Selai C. Health-related quality of life in Gilles de la Tourette syndrome: a decade of research. Behav Neurol 2013a;27:83–93.

[33] Eddy CM, Cavanna AE, Gulisano M, Calì P, Robertson MM, Rizzo R. The effects of comorbid obsessive-compulsive disorders and attention-deficit hyperactivity disorder on quality of life in Tourette syndrome. J Neuropsychiatry Clin Neurosci 2012;24:458–62.

[33a] Evans J, Seri S, Cavanna AE. The effects of Gilles de la Tourette syndrome and other chronic tic disorders on quality of life across the lifespan: a systematic review. Eur Child Adolesc Psychiatry 2016;25:939–48.

[34] Frank MC, Piedad J, Rickards H, Cavanna AE. The role of impulse control disorders in Tourette syndrome: an exploratory study. J Neurol Sci 2011;310:276–8.

[35] Robertson MM, Cavanna AE. Tourette syndrome: the facts. 2nd ed. Oxford: Oxford University Press; 2008.

[36] Elstner K, Selai CE, Trimble MR, Robertson MM. Quality of Life (QOL) of patients with Gilles de la Tourette's syndrome. Acta Psychiatr Scand 2001;103:52–9.

[37] Müller-Vahl K, Dodel I, Müller N, Münchau A, Reese JP, Balzer-Geldsetzer M, Dodel R, Oertel WH. Health-related quality of life in patients with Gilles de la Tourette's syndrome. Mov Disord 2010;25:309–14.

[38] Cavanna AE, Luoni C, Selvini C, Blangiardo R, Eddy CM, Silvestri PR, Calì PV, Seri S, Balottin U, Cardona F, Rizzo R, Termine C. The Gilles de la Tourette syndrome-quality of life scale for children and adolescents (C&A-GTS-QOL): development and validation of the Italian version. Behav Neurol 2013b;27: 95–103.

[39] Su MT, McFarlane F, Cavanna AE, Termine C, Murray I, Heidemeyer L, Heyman I, Murphy T. The English version of the Gilles de la Tourette syndrome-quality of life scale for children and adolescents (C&A-GTS-QOL). J Child Neurol 2017;32:76–83.

[40] Cavanna AE, Schrag A, Morley D, Orth M, Robertson MM, Joyce E, Critchley HD, Selai C. The Gilles de la Tourette syndrome quality of life scale (GTS-QOL): development and validation. Neurology 2008;71: 1410–6.

[41] Ferenczi S. Psychoanalytic observations on tic. Int J Psychoanal 1921;2:1–30.

[42] Ali F, Morrison KE, Cavanna AE. The complex genetics of Gilles de la Tourette syndrome: implications for clinical practice. Neuropsychiatry 2013;3:321–30.

[43] Barr CL, Wigg KG, Zovko E, Sandor P, Tsui LC. No evidence for a major gene effect of the dopamine D4 receptor gene in the susceptibility to Gilles de la Tourette syndrome in five Canadian families. Am J Med Genet 1996;67:301–5.

[44] Comings DE, Wu S, Chiu C, Ring RH, Gade R, Ahn C, MacMurray JP, Dietz G, Muhleman D. Polygenic inheritance of Tourette syndrome, stuttering, attention deficit hyperactivity, conduct, and oppositional defiant disorder: the additive and subtractive effect of the three dopaminergic genes - DRD2, D beta H, and DAT1. Am J Med Genet 1996;67:264–88.

[45] Gelernter J, Pauls DL, Leckman J, Kidd KK, Kurlan R. D2 dopamine receptor alleles do not influence severity of Tourette's syndrome: results from four large kindreds. Arch Neurol 1994;51:397–400.

[46] Lee CC, Chou IC, Tsai CH, Wang TR, Li TC, Tsai FJ. Dopamine receptor D2 gene polymorphisms are associated in Taiwanese children with Tourette syndrome. Pediatr Neurol 2005;33:272–6.

[47] Moya PR, Wendland JR, Rubenstein LM, Timpano KR, Heiman GA, Tischfield JA, King RA, Andrews AM, Ramamoorthy S, McMahon FJ, Murphy DL. Common and rare alleles of the serotonin transporter gene, SLC6A4, associated with Tourette disorder. Mov Disord 2013;28:1263–70.

[48] Tarnok Z, Ronai Z, Gervai J, Kereszturi E, Gadoros J, Sasvari-Szekely M, Nemoda Z. Dopaminergic candidate genes in Tourette syndrome: association between tic severity and 3′ UTR polymorphism of the dopamine transporter gene. Am J Med Genet B Neuropsychiatr Genet 2007;144:900−5.
[49] Yoon DY, Rippel CA, Kobets AJ, Morris CM, Lee JE, Williams PN, Bridges DD, Vandenbergh DJ, Shugart YY, Singer HS. Dopaminergic polymorphisms in Tourette syndrome: association with the DAT gene (SLC6A3). Am J Med Genet B Neuropsychiatr Genet 2007;144:605−10.
[50] Qi Y, Zheng Y, Li Z, Xiong L. Progress in genetic studies of Tourette's syndrome. Brain Sci 2017;7:134.
[51] Pinto R, Monzani B, Leckman JF, Rück C, Serlachius E, Lichtenstein P, Mataix-Cols D. Understanding the covariation of tics, attention-deficit/hyperactivity, and obsessive-compulsive symptoms: a population-based adult twin study. Am J Med Genet 2016;171:938−47.
[52] Madhusudan N, Cavanna AE. The role of immune dysfunction in the development of tics and susceptibility to infections in Tourette syndrome: a systematic review. Basal Ganglia 2013;3:77−84.
[53] Martino D, Dale RC, Gilbert DL, Giovannoni G, Leckman JF. Immunopathogenic mechanisms in Tourette syndrome: a critical review. Mov Disord 2009a;24:1267−79.
[54] Martino D, Defazio G, Giovannoni G. The PANDAS subgroup of tic disorders and childhood-onset obsessive-compulsive disorder. J Psychosom Res 2009b;67:547−57.
[55] Kawikova I, Grady BP, Tobiasova Z, Zhang Y, Vojdani A, Katsovich L, Richmand BJ, Park TW, Bothwell AL, Leckman JF. Children with Tourette's syndrome may suffer immunoglobulin A dysgammaglobulinemia: preliminary report. Biol Psychiatry 2010;67:679−83.
[56] Bos-Veneman NG, Olieman R, Tobiasova Z, Hoekstra PJ, Katsovich L, Bothwell AL, Leckman JF, Kawikova I. Altered immunoglobulin profiles in children with Tourette syndrome. Brain Behav Immun 2011;25:532−8.
[57] Alexander GM, Peterson BS. Testing the prenatal hormone hypothesis of tic-related disorders: gender identity and gender role behavior. Dev Psychopathol 2004;16:407−20.
[58] Peterson BS, Cohen DJ. The treatment of Tourette's syndrome: multimodal, developmental intervention. J Clin Psychiatr 1998;59:62−72.
[59] Peterson BS, Leckman JF, Scahill L, Naftolin F, Keefe D, Charest NJ, King RA, Hardin MT, Cohen DJ. Steroid hormones and Tourette's syndrome: early experience with antiandrogen therapy. J Clin Psychopharmacol 1994;14:131−5.
[60] Albin RL, Mink JW. Recent advances in Tourette syndrome research. Trends Neurosci 2006;29:175−82.
[61] Mink JW. Basal ganglia dysfunction in Tourette's syndrome: a new hypothesis. Pediatr Neurol 2001;25: 190−8.
[62] Felling RJ, Singer HS. Neurobiology of Tourette syndrome: current status and need for further investigation. J Neurosci 2011;31:12387−95.
[63] Cavanna AE, Black KJ, Hallett M, Voon V. Neurobiology of the premonitory urge in Tourette's syndrome: pathophysiology and treatment implications. J Neuropsychiatry Clin Neurosci 2017;29:95−104.
[64] Gericke GS, Simonic I, Cloete E, Buckle C, Becker PJ. Increased chromosomal breakage in Tourette syndrome predicts the possibility of variable multiple gene involvement in spectrum phenotypes: preliminary findings and hypothesis. Am J Med Genet 1995;60:444−7.
[65] Zilhão NR, Padmanabhuni SS, Pagliaroli L, Barta C, BIOS Consortium, Smit DJ, Cath D, Nivard MG, Baselmans BM, van Dongen J, Paschou P, Boomsma DI. Epigenome-wide association study of tic disorders. Twin Res Hum Genet 2015;18:699−709.
[66] Pagliaroli L, Vető B, Arányi T, Barta C. From genetics to epigenetics: new perspectives in Tourette syndrome research. Front Neurosci 2016;10:277.
[67] Gericke GS. Chromosomal fragility, structural rearrangements and mobile element activity may reflect dynamic epigenetic mechanisms of importance in neurobehavioural genetics. Med Hypotheses 2006;66: 276−85.

[68] Roessner V, Plessen KJ, Rothenberger A, Ludolph AG, Rizzo R, Skov L, Strand G, Stern JS, Termine C, Hoekstra PJ, the ESSTS Guidelines Group. European clinical guidelines for Tourette syndrome and other tic disorders. Part II: pharmacological treatment. Eur Child Adolesc Psychiatry 2011;20:173–96.
[69] Frank M, Cavanna AE. Behavioural treatments for Tourette syndrome: an evidence-based review. Behav Neurol 2013;27:105–17.
[70] Deckersbach T, Rauch S, Buhlmann U, Wilhelm S. Habit reversal versus supportive psychotherapy in Tourette's disorder: a randomized controlled trial and predictors of treatment response. Behav Res Ther 2006;44:1079–90.
[71] Himle MB, Woods DW, Piacentini JC, Walkup JT. Brief review of habit reversal training for Tourette syndrome. J Child Neurol 2006;21:719–25.
[72] Woods DW, Piacentini JC, Chang SW, Deckersbach T, Ginsburg GS, Peterson AL, Scahill LD, Walkup JT, Wilhelm S. Managing Tourette syndrome: a behavioral intervention for children and adults. New York: Oxford University Press; 2008.
[72a] Dutta N, Cavanna AE. The effectiveness of habit reversal therapy in the treatment of Tourette syndrome and other chronic tic disorders: a systematic review. Funct Neurol 2013;28:7–12.
[73] Gilbert D. Treatment of children and adolescents with tics and Tourette syndrome. J Child Neurol 2006;21: 690–700.
[74] Pringsheim T, Doja A, Gorman D, McKinlay D, Day L, Billinghurst L, Carroll A, Dion Y, Luscombe S, Steeves T, Sandor P. Canadian guidelines for the evidence-based treatment of tic disorders: pharmacotherapy. Can J Psychiatr 2012;57:133–43.
[75] Waldon K, Hill S, Termine C, Balottin U, Cavanna AE. Trials of pharmacological interventions for Tourette syndrome: a systematic review. Behav Neurol 2013;26:265–73.
[76] Gulisano M, Calì PV, Cavanna AE, Eddy C, Rickards H, Rizzo R. Cardiovascular safety of aripiprazole and pimozide in young patients with Tourette syndrome. Neurol Sci 2011;32:1213–7.
[77] Robertson MM, Stern JS. Gilles de la Tourette syndrome: symptomatic treatment based on evidence. Eur Child Adolesc Psychiatry 2000;9(Suppl. 1):S60–75.
[78] Cavanna AE, Nani A. Antiepileptic drugs and Tourette syndrome. Int Rev Neurobiol 2013;112:373–89.
[79] Jankovic J, Jimenez-Shahed J, Brown LW. A randomised, double-blind, placebo-controlled study of topiramate in the treatment of Tourette syndrome. J Neurol Neurosurg Psychiatry 2010;81:70–3.
[80] Eddy CM, Rickards HE, Cavanna AE. Treatment strategies for tics in Tourette syndrome. Ther Adv Neurol Disord 2011;4:25–45.